AF576054

UAV or Drones for Remote Sensing Applications in GPS/GNSS Enabled and GPS/GNSS Denied Environments

UAV or Drones for Remote Sensing Applications in GPS/GNSS Enabled and GPS/GNSS Denied Environments

Editors

Felipe Gonzalez Toro
Antonios Tsourdos

MDPI • Basel • Beijing • Wuhan • Barcelona • Belgrade • Manchester • Tokyo • Cluj • Tianjin

Editors

Felipe Gonzalez Toro
Airborne Systems Lab / QUT
Centre for Robotics
Queensland University
of Technology
Brisbane
Australia

Antonios Tsourdos
Autonomous and
Cyber-Physical Systems Centre
Centre for Autonomous and
Cyberphysical Systems
Cranfield
UK

Editorial Office
MDPI
St. Alban-Anlage 66
4052 Basel, Switzerland

This is a reprint of articles from the Special Issue published online in the open access journal *Sensors* (ISSN 1424-8220) (available at: www.mdpi.com/journal/sensors/special_issues/UAV_Drones_Remote_Sensing_Applications).

For citation purposes, cite each article independently as indicated on the article page online and as indicated below:

LastName, A.A.; LastName, B.B.; LastName, C.C. Article Title. *Journal Name* **Year**, *Volume Number*, Page Range.

ISBN 978-3-0365-1590-8 (Hbk)
ISBN 978-3-0365-1589-2 (PDF)

Contents

About the Editors

Felipe Gonzalez Toro

Dr Gonzalez is an Associate Professor at the School of Electrical Engineering and Robotics (EER) and a CI in QUT Centre for Robotics (QCR), Engineering Faculty with a passion for innovation in the fields of aerial robotics and automation. Gonzalez interest is in creating aerial robots, drones or UAVs that possess a high level of cognition using efficient on-board computer algorithms using advanced optimization and game theory approaches that assist us to understand and improve our physical and natural world. He is the co-author of several books in UAV based remote sensing and UAV based design and as of 2021 has published nearly 150 refereed papers. To date he has been awarded $40.1M in chief investigator / partner investigator grants ($6.5M total cash + in-kind contributions). This grant income represents a mixture of sole investigator funding, ARC DP, ARC LIEF, ARC Linkages , international, multidisciplinary collaborative grants and funding from industry.

Antonios Tsourdos

Professor Tsourdos obtained a MEng in Electronic, Control and Systems Engineering from the University of Sheffield (1995), an MSc in Systems Engineering from Cardiff University (1996) and a PhD in Nonlinear Robust Autopilot Design and Analysis from Cranfield University (1999).

He is the Head of the Centre for Autonomous and Cyber-Physical Systems in 2007, Professor of Autonomous Systems and Control in 2009, and Director of Research–Aerospace, Transport and Manufacturing in 2015.

Professor Tsourdos is Chair of the International Federation of Automatic Control (IFAC) Technical Committee on Aerospace Control, and a member of the American Institute of Aeronautics and Astronautics (AIAA) Unmanned Systems Integration and Outreach Committee; the Institute of Electrical and Electronics Engineers (IEEE) Control System Society Technical Committee on Aerospace Control (TCAC) and the IMechE Mechatronics, Informatics and Control Group board.

Preface to "UAV or Drones for Remote Sensing Applications in GPS/GNSS Enabled and GPS/GNSS Denied Environments"

Dear Colleagues,

The design of novel UAV systems and the use of UAV platforms integrated with robotic sensing and imaging techniques, as well as the development of processing workflows and the capacity of ultra-high temporal and spatial resolution data have enabled a rapid uptake of UAVs and drones across several industries and application domains.

The scope of this book provides a forum for high-quality peer-reviewed papers that broaden awareness and understanding of single and multiple UAV developments for remote sensing applications, and associated developments in sensor technology, data processing and communications, and UAV system design and sensing capabilities in GPS enabled and more broadly Global Navigation Satellite System (GNSS) enabled and GPS /GNSS denied environments.

Contributions include:

UAV-based photogrammetry, laser scanning, multispectral imaging, hyperspectral imaging, and thermal imaging;

UAV sensor applications, spatial ecology, pest detection, reef, forestry, volcanology, precision agriculture wildlife species tracking, search and rescue, target tracking, the monitoring of the atmosphere, chemical, biological, and natural disaster phenomena, fire prevention, flood prevention, volcanic monitoring, pollution monitoring, microclimates, and land use;

Wildlife and target detection and recognition from UAV imagery using deep learning and machine learning techniques;

UAV-based change detection;

Sense and avoid;

UAV navigation in cluttered environments;

Single and multiple UAVs or swarms in GPS /GNSS denied environments;

On-board decision making;

Real-time georeferencing for UAV-based imaging;

Radiometric and spectral calibration of UAV-based sensors;

UAV onboard data storage, transmission, and retrieval;

Collaborative strategies and architectures to control multiple UAVs and sensor networks for the purpose of remote sensing.

Felipe Gonzalez Toro, Antonios Tsourdos
Editors

Article

Scalable Distributed State Estimation in UTM Context

Marco Cicala [1,*], Egidio D'Amato [2], Immacolata Notaro [3] and Massimiliano Mattei [3]

[1] On-board Systems and ATM Department, Italian Aerospace Research Centre CIRA, 81043 Capua (CE), Italy
[2] Department of Science and Technology, University of Naples "Parthenope", 80143 Napoli, Italy; egidio.damato@uniparthenope.it
[3] Department of Engineering, University of Campania "L.Vanvitelli", 81031 Aversa (CE), Italy; immacolata.notaro@unicampania.it (I.N.); massimiliano.mattei@unicampania.it (M.M.)
* Correspondence: m.cicala@cira.it

Received: 30 March 2020; Accepted: 6 May 2020; Published: 8 May 2020

Abstract: This article proposes a novel approach to the Distributed State Estimation (DSE) problem for a set of co-operating UAVs equipped with heterogeneous on board sensors capable of exploiting certain characteristics typical of the UAS Traffic Management (UTM) context, such as high traffic density and the presence of limited range, Vehicle-to-Vehicle communication devices. The proposed algorithm is based on a scalable decentralized Kalman Filter derived from the Internodal Transformation Theory enhanced on the basis of the Consensus Theory. The general benefit of the proposed algorithm consists of, on the one hand, reducing the estimation problem to smaller local sub-problems, through a self-organization process of the local estimating nodes in response to the time varying communication topology; and on the other hand, of exploiting measures carried out nearby in order to improve the accuracy of the local estimates. In the UTM context, this enables each vehicle to estimate both its own position and velocity, as well as those of the neighboring vehicles, using both on board measurements and information transmitted by neighboring vehicles. A numerical simulation in a simplified UTM scenario is presented, in order to illustrate the salient aspects of the proposed algorithm.

Keywords: UAS traffic management; multiple UAV navigation; navigation in GPS/GNSS-denied environments; distributed state estimation; consensus theory

1. Introduction

Over the last few years, Small Unmanned Aircraft Systems (sUAS) have experienced a widespread diffusion both in military and civilian applications. Their diffusion is destined to grow even further, since they are capable of operating close to the ground and overcoming obstacles in all sorts of hazardous conditions forbidden to traditionally manned vehicles. sUAS offer new opportunities in different operational scenarios including public safety, search and rescue, disaster relief, infrastructure monitoring, precision farming and delivery of goods [1]. Most sUAV operations would take place in low-altitude, densely occupied airspace, over densely populated areas typical of urban scenarios.

The foreseen large-scale sUAV operations are not currently possible without drastically reducing safety levels of low-altitude airspace and a global need for new concepts and enabling technologies is clearly identified in the aeronautical community. These needs find influential formulations in NASA's UAS traffic management [2] (UTM) and the European Commission's U-space visions [3]. All the paradigms proposed so far assume a range of capabilities at an increasing level of autonomy, including Beyond Visual Line of Sight (BVLOS) operations, interactive planning, de-conflict operations with geo-fencing, collision and obstacle avoidance. A common element to these capabilities is the necessity to estimate the position and velocity of each vehicle.

sUAV navigation is typically based on the integration of low-cost Global Navigation Satellite System (GNSS) receivers and commercial-grade Micro-Electro-Mechanical Systems (MEMS)-based inertial and magnetic sensors [4,5]. In nominal conditions, these navigation systems can provide an accuracy of approximately 5–10 m [6], good enough to implement, many autonomous functionalities. In urban environments, this accuracy can be hindered by the presence of obstacles, or greater accuracy may be required in order to perform special operations.

This article deals with sUAV position and velocity estimation within the UTM context. If, on the one hand, the peculiar characteristics of a UTM scenario can be seen as a source of open issues to be faced with, if appropriately interpreted, on the other hand, it is possible to derive benefits compared to traditional estimation methods in terms of accuracy, availability and continuity. In fact, the high density of traffic, the presence of numerous and heterogeneous on-board sensors (in addition to GNSS and inertial and magnetic sensors, low-cost vision-based systems [7,8] and micro radars [9] are becoming increasingly widespread), the presence of vehicle-to-vehicle communication channels are all opportunities to be exploited.

The basic idea is simple and intuitive in nature. The presence of numerous on-board sensors provides a great deal of information on the state of each vehicle. The measurements of certain sensors, such as vision-based systems or radars, contain information not only regarding the vehicle hosting the sensors, but also related to other vehicles (e.g., relative distance). The exchange of information between neighboring vehicles can allow them to improve the estimates of their position and velocity. A typical condition that can benefit from this situation is the navigation of vehicles in an GNSS-denied zone, where position and velocity can be estimated thanks to relative position sensors with other vehicles flying nearby.

The idea is currently being widely investigated. In [10–12], a multiple vehicle configuration is proposed to improve navigation (and attitude estimation) performance of a chief vehicle exploiting differential GPS using information deriving from a formation of flying deputy vehicles. In [13–15], a GPS-denied condition is specifically addressed in similar multi-vehicle configurations. These works consider a fixed number of vehicles flying in formation. A peculiar feature of the UTM scenario is the non-preordained motion of the vehicles, free to fly in the airspace and the absence of hierarchical relationships between the vehicles, which must all have access to the same minimum navigation performance.

The objective of the article is to describe a novel methodology applicable to the navigation of a sUAS fleet, which exploits the typical features of the UTM system in order to improve performance with respect to traditional methods. Therefore, the fundamental assumptions characterizing the considered scenario include the absence of a Central Processing Unit (CPU) for information elaboration or distribution (decentralization), or of a vehicle that is hierarchically distinct from the others. Moreover, only locally relevant computation is required to take place in each local processing unit, allowing the number of nodes to grow arbitrarily, without exceeding local computational resources (scalability). Thus, the starting point is to translate the basic idea of multi-sensor multi-vehicle exploitation into formal terms, with an algorithm preserving optimality characteristic typical of many common State Estimators.

The Kalman Filter (KF) represents the cornerstone for optimal state estimation. In its classical implementation [16], it has an intrinsically-centralized structure, in which the CPU samples the measurements and performs the estimation process. Although possibly optimal, when applied to Large-Scale Multi-Sensor Systems, it does not provide a solution compliant with the previously discussed assumptions. The main Centralized Kalman Filter limitations are the high computational load overcharging the CPU when the size of the system increases and the high communication complexity when the spatial distribution of the system expands.

In order to guarantee the Kalman Filter adequate scalability and decentralization characteristics, many decentralized algorithms [17–24] have been proposed, based on multiple local Kalman Filters, one in each local processing unit. In order to take processing and communication limits into account, the local Kalman Filters must involve the computation and communication only of local quantities of dimension $n_l \ll n$, where n is the dimension of the global system.

Much of the existing studies [25–28] focus on large sensor network monitoring low-dimensional systems, and they mainly address the problem of how efficiently the available information is distributed. These solutions address scalability mainly looking to the dimension of the measurement signals, and not of the state of the system itself.

In other works [29,30], a reduced order Kalman Filters models have been proposed to specifically address the computation burden that arises from increasing the order of the global system. This research and other similar works address the issue of scalability in particular for fully connected or almost fully connected topologies. The algorithms based on this type of topology require long distance communication that reduces some of the benefits of decentralized architectures.

In order to address the problem over arbitrary communication networks, data fusion algorithms based on the Consensus Theory [31] are widely discussed in the literature. At the basis of consensus-based methodologies, there is the concept of covariance intersection [32], which represents a preliminary solution to the problem of merging the local estimates, so as to obtain a more accurate global estimate. All consensus-based methodologies can be interpreted as a generalization of the covariance intersection fusion rule. Consensus-based methodologies are iterative in nature. At the first iteration step, they conceptually correspond to the covariance intersection rule. When the number of iteration steps go to infinity, under certain conditions, they tend towards the centralized solution. This is a highly desirable characteristic.

A first form of Consensus-based algorithm for linear systems is the Consensus on Information (CI), discussed in [33]. The methodology derives from a decentralized estimation algorithm with stability properties guaranteed under collective observability and network-strong connectivity (thus ensuring a relaxation of the condition of full connectivity). Although these stability characteristics are guaranteed, even for only one single consensus step (in this case, the rule corresponds to the covariance intersection), the results obtained applying algorithms of this family do not tend towards a centralized solution for a number of consensus steps that tend towards infinity.

A different approach of Consensus-based estimation for linear systems is named Consensus on Measurements (CM). This method is discussed, among others, in [34,35], and differs from the CI for the quantities on which the consensus procedure is carried out. Unlike CI, Consensus on Measurements tends towards the equivalent centralized solution as the number of consensus steps goes to infinity, but does not guarantee stability unless the number of consensus step is sufficiently high.

In [36], a hybrid consensus approach is described, defined by the author as the Hybrid Consensus on Measurement and on Information (HCMCI), based both on CM and CI. The scope of the proposed approach is to combine their complementary benefits avoiding their main drawbacks. The HCMCI algorithm, which, among other things, extends the consensus-based solution to the non-linear case using an Extended Kalman Filter approach, appears to be a promising methodology for dealing with the problem of the distribution of information in systems of a more general topology.

Consensus-based methods address the issue of decentralization of the estimate by reducing the complexity of the communication system even in the case of systems that are not fully connected. However, these algorithms do not address the problem of scalability: in all the aforementioned methods, each local model has a cardinality equal to that of the global system. These solutions, therefore, do not address the problem of scalability as the size of the system grows.

Furthermore, a common element of consensus-based approaches is the stability criterion of the solution related to the system topology. Strong connectivity is a required condition. Without going into formalisms, this translates into the assumption that each vehicle is connected to the others at least through an indirect route, passing through the other vehicles. If the stability conditions, specifically relating to the system topology, continue to be verified in subsets, the proposed algorithms can be used locally. The topology of a system such as a fleet of sUAVs free to fly in space is highly variable over time. Therefore, it would be advantageous to have an appropriate clustering mechanism that consists of locally stable estimation systems formed of only a properly selected global state subset.

The methodology proposed in this article combines the results achieved for scalable decentralized systems obtained by the Internodal Transformation Theory methodology [29], with the advantages guaranteed by the use of consensus-based techniques [31]. The goal is to inherit the scalability properties from the former, the ability to distribute information within the strongly connected sub graphs from the latter, and to obtain, from the combined use of the two approaches, a self-clustering property, through which the local elements that form the global system self-aggregate, in order to form sub-systems in which the solution is stable.

2. Materials and Methods

This section introduces the fundamental notions for the definition of a decentralized estimation algorithm. Subsequently, it introduces the basic concepts of both the Internodal Transformation Theory and the Consensus Theory necessary for the proposed algorithm formulation. Finally, it defines the algorithm in its generic formulation identifying a possible application to the problem of estimating position and velocity for a sUAV fleet.

2.1. Problem Formulation

This article addresses Scalable Distributed State Estimation (SDSE) over a network consisting of nodes representing free-flying vehicles. Each vehicle has its own on-board sensors that can locally process sensor data and exchange data with other vehicles. The problem can be expressed as follows.

Let us consider a set of N flying vehicles. Communication topology between vehicles at any time k can be defined in terms of a directed graph $\mathcal{G} = \{\mathcal{V}, \mathcal{A}\}$, where $\mathcal{V} = \{\mathcal{V}_1, \ldots \mathcal{V}_N\}$ is the set of vehicles (nodes of the graph) and $\mathcal{A}$ is the set of pair describing a communication link from $\mathcal{V}_i$ to $\mathcal{V}_j$ (arcs): vehicle i can receive data from vehicle j if $\left(\mathcal{V}_i, \mathcal{V}_j\right) \in \mathcal{A}$. For each vehicle i, let be $\mathcal{A}^{\mathrm{i}} = \left\{\mathcal{V}_j \in \mathcal{V} : \left(\mathcal{V}_i, \mathcal{V}_j\right) \in \mathcal{A}\right\}$ the set of its neighbors.

Let $x \in \mathbb{R}^n$ be the global system state. State x is global in that it includes the information to describe the behavior of each vehicle.

Let us consider a non-linear dynamic model of the system in the state space form and discrete time domain:

$$x_k = f(x_{k-1}) + \omega_{k-1} \tag{1}$$

where f is the non-linear state transition function and $\omega_k \in \mathbb{R}^{\mathrm{n}}$ is the process noise. For each vehicle i, let us consider a set of non-linear measurements $z_k^i \in \mathbb{R}^{\mathrm{m}}$ given by:

$$z_k^i = h^i(x_k) + v_k^i \quad \mathcal{V}_i \in \mathcal{V} \tag{2}$$

where h^i is the local observation model and $v^i \in \mathbb{R}^{\mathrm{m}}$ is the measurement noise. Let us assume that the process and measurement noise ω and v^i are mutually uncorrelated zero-mean noise with covariance $\Omega_{k-1} = \mathbf{E}\left[\omega_{k-1}\omega_{k-1}{}^{\mathrm{T}}\right] > 0$ and $R_k^{\mathrm{i}} = \mathbf{E}\left[v_k^i v_k^{i\,\mathrm{T}}\right] > 0$.

The objective of a state estimation problem is to have, at each time k an estimation of the global system state $\hat{x}_k$ based on measurement $z_k = \left[z_k^{1\mathrm{T}}, \ldots, z_k^{N\mathrm{T}}\right]^{\mathrm{T}}$.

An SDSE problem introduces the concepts of Distribution and Scalability:

- In a Distributed State Estimation, each node has a locally-processed estimation of the global system state $\hat{x}_k^i$ based only on the local measurements z_k^i and date received form the adjacent vehicles j, such that $\mathcal{V}_j \in \mathcal{A}^i$
- In a Scalable State Estimation, a model distribution logic reduces the size of local estimation problems allowing each node to estimate only a subset of the global state.

2.2. The Basis of Decentralization

To address the problem, a fundamental result relating to the centralized problem can be briefly recalled. First, let us consider the overall system given the state dynamics (1) and the overall measurement model:

$$z_k = h(x_k) + v_k \tag{3}$$

where $v_k = \left[v_k^{1\mathrm{T}}, \ldots, v_k^{N\mathrm{T}}\right]^{\mathrm{T}}$.

The estimation of the state x at time k, given information up to and including time $(k - m)$ and the corresponding variance P are given by:

$$\hat{x}_{k|k-m} = \mathrm{E}[\ x_k \ \ |z_1, \ldots z_{k-m}\ ] \tag{4}$$

$$P_{k|k-m} = \mathrm{E}\left[\left(x_k - \hat{x}_{k|k-m}\right)^T \left(x_k - \hat{x}_{k|k-m}\right) |z_1, \ldots z_{k-m}\right] \tag{5}$$

being, in a recursive formulation, the most relevant cases those for $m = k-1$ and $m = k$.

The information filter, equivalent to the traditional Covariance Kalman Filter, provides a recursive estimation $\hat{x}_{k|k}$ for the state x at time k given the information $z_1, \ldots z_k$ up to time k. The information matrix Q and information state vector q can be defined [37] as the inverse of covariance matrix and as the product of the inverse of the covariance matrix and the state estimate, respectively:

$$Q_{k|k-m} = P_{k|k-m}^{-1} \tag{6}$$

$$q_{k|k-m} = Q_{k|k-m}\hat{x}_{k|k-m} \tag{7}$$

In terms of information space variables, the Extended Information Kalman Filter can be written in the following form [29], given without derivation:

prediction

$$\hat{x}_{k|k-1} = f\left(\hat{x}_{k-1|k-1}\right) \tag{8}$$

$$Q_{k|k-1} = \left[\frac{\partial f\left(\hat{x}_{k-1|k-1}\right)}{\partial x_k} \ Q_{k-1|k-1}^{-1} \ \frac{\partial f\left(\hat{x}_{k-1|k-1}\right)}{\partial x_k}^T + \Omega_{k-1}\right]^{-1} \tag{9}$$

$$q_{k|k-1} = Q_{k|k-1} \ \hat{x}_{k|k-1} \tag{10}$$

correction

$$q_{k|k} = q_{k|k-1} + i_k \tag{11}$$

$$Q_{k|k} = Q_{k|k-1} + I_k \tag{12}$$

where

$$I_k = \frac{\partial h\left(\hat{x}_{k|k-1}\right)}{\partial x_k}^T (R_k)^{-1} \frac{\partial h\left(\hat{x}_{k|k-1}\right)}{\partial x_k} \tag{13}$$

$$i_k = \frac{\partial h\left(\hat{x}_{k|k-1}\right)}{\partial x_k}^T (R_k)^{-1} \left[c_k + \frac{\partial h\left(\hat{x}_{k|k-1}\right)}{\partial x_k} \hat{x}_{k|k-1}\right] \tag{14}$$

$$c_k = z_k - h\left(\hat{x}_{k|k-1}\right) \tag{15}$$

The Global Information Filter applies to a fully centralized fusion architecture composed of a central processing unit directly connected to all sensing devices. One of the major advantages of the information filter formulation is its capability to be easily decentralized into a network of communicating nodes. In a decentralized fusion architecture, the system consists of different processing nodes able to perform the estimation of the global state on the basis of local observation and possible shared observations coming from other nodes. Let us first consider the case of a network of fully connected nodes N, which supposes that each vehicle is connected to all the other vehicles. Let us assume that each local node has a state model identical to the centralized model (1), i.e., each node performs a global state estimation $\hat{x}^i \in \mathbb{R}^n$. Local information matrix and information state vectors can be defined in each node i:

$$Q^i_{k|k-m} = \left(P^i_{k|k-m}\right)^{-1} \tag{16}$$

$$q^i_{k|k-m} = Q^i_{k|k-m}\hat{x}^i_{k|k-m} \tag{17}$$

The estimation algorithm for node i with information being communicated to it from other *(N-1)* nodes can be expressed in the following step analogous to the centralized case:

prediction

$$\hat{x}^i_{k|k-1} = f\left(\hat{x}^i_{k-1|k-1}\right) \tag{18}$$

$$Q^i_{k|k-1} = \left[\frac{\partial f\left(\hat{x}^i_{k-1|k-1}\right)}{\partial x_k}\left(Q^i_{k-1|k-1}\right)^{-1}\frac{\partial f\left(\hat{x}^i_{k-1|k-1}\right)}{\partial x_k}^T + \Omega^i_{k-1}\right]^{-1} \tag{19}$$

$$q^i_{k|k-1} = Q^i_{k|k-1}\,\hat{x}^i_{k|k-1} \tag{20}$$

correction

$$q^i_{k|k} = q^i_{k|k-1} + \sum_{j=1}^{N} i^j_k \tag{21}$$

$$Q^i_{k|k} = Q^i_{k|k-1} + \sum_{j=1}^{N} I^j_k \tag{22}$$

where

$$I^j_k = \frac{\partial h^j\left(\hat{x}^j_{k|k-1}\right)}{\partial x_k}^T \left(R^j_k\right)^{-1}\frac{\partial h^j\left(\hat{x}^j_{k|k-1}\right)}{\partial x_k} \tag{23}$$

$$i^j_k = \frac{\partial h^j\left(\hat{x}^j_{k|k-1}\right)}{\partial x_k}^T \left(R^j_k\right)^{-1}\left[c^j_k + \frac{\partial h^j\left(\hat{x}^j_{k|k-1}\right)}{\partial x_k}\hat{x}^j_{k|k-1}\right] \tag{24}$$

$$c^j_k = z^j_k - h^j\left(\hat{x}^j_{k|k-1}\right) \tag{25}$$

In the correction step expressed it can be assumed that each node begins with a common initial information state (e.g., $q^i_0 = 0\ Q^i_0 = 0\ \forall\ i$). The summations in Equations (21) and (22) are feasible because of the full connectivity of the system. Under these conditions, each local estimate is identical to the centralized system defined by the Equations (8)–(15).

The case of a fully connected system does not guarantee any real advantages with respect to the centralized case, both in terms of computational burden and of communication requirements. Nevertheless, it represents a starting point for the definition of any decentralized estimation algorithm based on a Kalman Filter.

2.3. Scalability

Let us first deal with the problem of scalability. In order to obtain the desired scalable solution, let us introduce the model distribution concepts as defined in [29]. Let us consider a local state for the node i related to the global state at time instant k by:

$$x_k^i = T_k^i x_k \tag{26}$$

where T_k^i is a linear nodal transformation matrix that select states or linear combinations of states from the global state vector. Using the Internodal Transformation Theory, it is possible to obtain a formulation in which each node performs the same estimation as the centralized formulation for a subset of the global state, while minimizing communication between nodes.

In order to derive a scalable solution, the inverse operation to model reduction is required. Generally speaking, T_k^i is rectangular and its ordinary inverse is not defined. Hence, the use of the generalized pseudo-inverse $\left(T_k^i\right)^+$ is required. The pseudo-inverse provides the solution to the problem of reconstructing the global state x_k starting from the local state x_k^i in node i, minimizing $\| x_k^i - T_k^i x_k \|$:

$$x_k = \left(T_k^i\right)^+ x_k^i \tag{27}$$

The geometrical interpretation of the reconstructed global state is a vector containing an unscaled relevant state and a zero in place of any irrelevant state to node i.

Let us introduce the concepts of information contribution at node i due to current observation from node j, defined as $i_k^{i|j}$, and the associated local information matrix $I_k^{i|j}$.

Error covariance at node i based on local observation from node j can be defined as:

$$P_{k|k}^{i|j} = \left(I_k^{i|j}\right)^+ \tag{28}$$

and the corresponding local estimate at node i based only on local observation from node j can be obtained from:

$$\hat{x}_{k|k}^{i|j} = P_{k|k}^{i|j} i_k^{i|j} \tag{29}$$

It is possible to rewrite [29] the distributed formulation of Equations (18)–(25) in an equivalent scalable form in which each node propagates only locally relevant states and exchanges only relevant information with any other node:

Prediction

$$\hat{x}_{k|k-1}^i = T_k^i f\left(\left(T_{k-1}^i\right)^+ \hat{x}_{k-1|k-1}^i\right) \tag{30}$$

$$\widetilde{Q}_{k-1|k-1}^i = T_k^i \left\{\left(T_{k-1}^i\right)^+ Q_{k-1|k-1}^i \left(T_{k-1}^i\right)\right\}\left(T_k^i\right)^+ \tag{31}$$

$$Q_{k|k-1}^i = \left[\frac{\partial f\left(\hat{x}_{k-1|k-1}^i\right)}{\partial x_k} \left(\widetilde{Q}_{k-1|k-1}^i\right)^{-1} \frac{\partial f\left(\hat{x}_{k-1|k-1}^i\right)^T}{\partial x_k} + \Omega_{k-1}^i\right]^{-1} \tag{32}$$

$$q_{k|k-1}^i = Q_{k|k-1}^i \hat{x}_{k|k-1}^i \tag{33}$$

Correction

$$q_{k|k}^i = q_{k|k-1}^i + \sum_{j=1}^N i_k^{i|j} \tag{34}$$

$$Q_{k|k}^i = Q_{k|k-1}^i + \sum_{j=1}^N I_k^{i|j} \tag{35}$$

In the correction phase (34), (35), it is assumed that each node i receives from the other nodes the local information $\left(\boldsymbol{i}_k^{i|j},\ \boldsymbol{I}_k^{i|j}\right)$. Each node is able to calculate the information contributions locally starting from local measures without relying on information communicated by the other nodes in a completely analogous way to what was done in Equations (23)–(25):

$$\boldsymbol{i}_k^{i|i} = \frac{\partial \boldsymbol{h}^i\left(\hat{\boldsymbol{x}}_{k|k-1}^i\right)}{\partial \boldsymbol{x}_k}^T \left(\boldsymbol{R}_k^i\right)^{-1} \left[\boldsymbol{c}_k^i + \frac{\partial \boldsymbol{h}^i\left(\hat{\boldsymbol{x}}_{k|k-1}^i\right)}{\partial \boldsymbol{x}_k} \hat{\boldsymbol{x}}_{k|k-1}^i\right] \tag{36}$$

$$\boldsymbol{I}_k^{i|i} = \frac{\partial \boldsymbol{h}\left(\hat{\boldsymbol{x}}_{k|k-1}^i\right)}{\partial \boldsymbol{x}_k}^T \left(\boldsymbol{R}_k^i\right)^{-1} \frac{\partial \boldsymbol{h}\left(\hat{\boldsymbol{x}}_{k|k-1}^i\right)}{\partial \boldsymbol{x}_k} \tag{37}$$

It is therefore necessary to look for transformations that locally carry out the following transformations in each node.

$$\boldsymbol{I}_k^{j|j} \rightarrow \boldsymbol{I}_k^{i|j} \qquad \forall j \neq i \tag{38}$$

$$\boldsymbol{i}_k^{j|j} \rightarrow \boldsymbol{i}_k^{i|j} \qquad \forall j \neq i \tag{39}$$

It is possible to show [29] that the Information Space Intermodal Transformation map can be schematized as in Figure 1, where:

$$\boldsymbol{V}_k^{ji} = \boldsymbol{T}_k^i \left(\boldsymbol{T}_k^j\right)^+ \tag{40}$$

$$\boldsymbol{T}_k^{ji} = \boldsymbol{I}_k^{i|j} \boldsymbol{V}_k^{ji} \left(\boldsymbol{I}_k^{j|i}\right)^+ \tag{41}$$

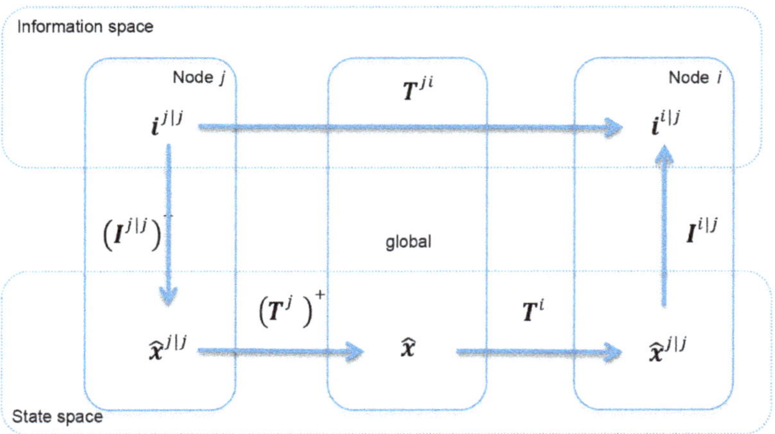

Figure 1. Information Space Intermodal Transformation map.

The information parameter in node $\mathcal{V}_i$ given only node $\mathcal{V}_j$ observation z_k^j can thus be been derived in each node starting from quantities, calculated locally as follows

$$\boldsymbol{I}_k^{i|j} = \left[\boldsymbol{T}_k^i \left[\boldsymbol{T}_k^{jT}\left(\boldsymbol{I}_k^{j|j}\right)\boldsymbol{T}_k^j\right]^+ \boldsymbol{T}_k^{iT}\right]^+ = \mathcal{F}_I^{j\rightarrow i}\left(\boldsymbol{I}_k^j\right) \tag{42}$$

$$\boldsymbol{i}_k^{i|j} = \boldsymbol{T}_k^{ji} \boldsymbol{i}_k^{j|j} = \mathcal{F}_i^{j\rightarrow i}\left(\boldsymbol{i}_k^{j|j}\right) \tag{43}$$

In order to carry out the transformations (42) and (43), each node j must therefore communicate the information to node i: $\left(\boldsymbol{T}_k^j, \boldsymbol{I}_k^{j|j}, \boldsymbol{i}_k^{j|j}\right)$.

The solution identified, as highlighted by Equations (34) and (35), formally still applies to a fully connected system. The process of minimizing communication between nodes is highly dependent on the choice of the matrices T_k^i of the model distributions. On the other hand, no hypothesis has been made so far about the criteria to be used to select these matrices. The effect of minimizing communications is evident by observing that $T_k^{ji} = T_k^{ij} = 0$ if two nodes do not share any common state. Therefore, the exchange of any information is not necessary between two nodes not sharing any common state. It is possible to extend this consideration to two or more sub-graphs that are individually strongly connected yet which are not connected to each other. By choosing a local state for each node that includes only the states of nodes belonging to its strongly connected subgraph, the need for communication between unconnected sub graphs is avoided. The selected algorithm then performs a sort of clustering of the estimation process selecting groups of nodes that require exchanging data (see Figure 2).

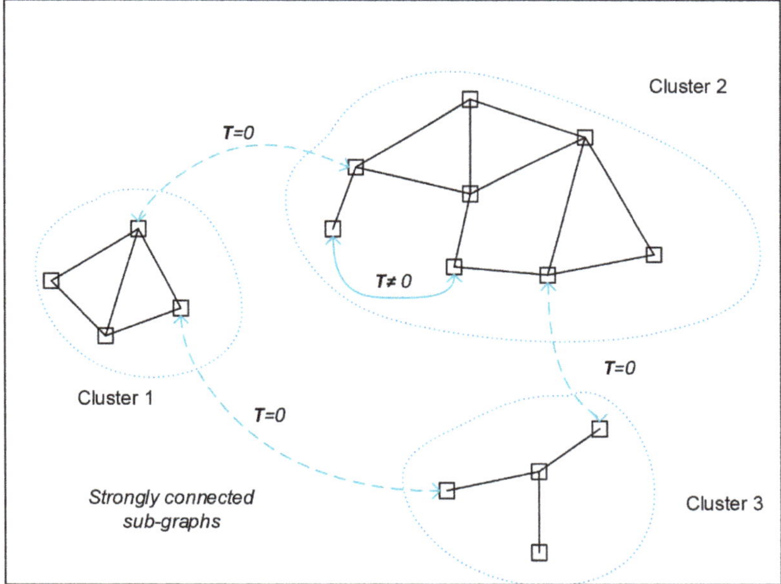

Figure 2. Internodal Transformation and graph connectivity.

2.4. Consensus Based Information Distribution

The algorithm discussed in the previous paragraph does not completely solve the problem of the information distribution: two nodes that share part of the local state must exchange information directly. In this way, the algorithm autonomously manages the formation and disintegration of connected sub-graphs, but a fully connectivity is required in each sub-graph. To overcome this problem, it is possible to use the Consensus Theory.

Let us consider a strongly connected subgraph $\widetilde{\mathcal{G}} = \{\widetilde{\mathcal{V}}, \widetilde{\mathcal{A}}\} \subseteq \mathcal{G}$ included in the global graph. The summation of generic terms χ_i distributed between its nodes:

$$X = \sum_{i \in \widetilde{\mathcal{A}}} \chi_i \tag{44}$$

can be obtained with a consensus averaging iterative process [38]:

$$\forall \mathcal{V}_i \in \widetilde{\mathcal{A}} \quad \chi_i^{(0)} = \chi_i$$

$$for\ \ell = 1, \ldots, L \qquad \chi_i^{(\ell+1)} = \sum_{j \in \widetilde{\mathcal{A}}} \mu_{ij} \chi_j^{(\ell)} \tag{45}$$

$$X = \widetilde{N} \chi_i^{(L)} \tag{46}$$

where $\widetilde{N}$ is the number of nodes of the strongly connected sub-graph.

With a proper choice of the μ_{ij}, in fact the solution converges to the average vector

$$\lim_{\ell \to \infty} \chi_i^{(\ell)} = \frac{1}{\widetilde{N}} \sum_{j \in \mathcal{A}_i} \chi_j \qquad \forall \mathcal{V}_i \in \widetilde{\mathcal{A}} \tag{47}$$

A possible choice of terms μ_{ij} is to select them as local-degree weights:

$$\mu_{ij} = \frac{1}{max\{d(i), d(j)\}}, \quad (\mathcal{V}_i, \mathcal{V}_j) \in \mathcal{A} \tag{48}$$

$$\mu_{ij} = 0, \quad (\mathcal{V}_i, \mathcal{V}_j) \notin \mathcal{A} \tag{49}$$

$$\mu_{ii} = 1 - \sum_{j \in \mathcal{A}_i} \mu_{ij} \tag{50}$$

where $d(i)$ is the degree of the node $\mathcal{V}_i$.

2.5. Algorithm Description

By applying to the Scalable Distributed algorithm defined by Equations (30)–(35), a consensus procedure to asymptotically obtain the summations in Equations (34) and (35), and adding a consensus procedure on a priori information pair to reproduce a Hybrid Consensus on Measurement and on Information (HCMCI) formulation [36], the following algorithm can be obtained:

$$\boldsymbol{UpdateT_k^i}$$

$$\boldsymbol{Local\ prediction}$$

$$\hat{\boldsymbol{x}}_{k|k-1}^i = \boldsymbol{T}_k^i \boldsymbol{f}\left(\left(\boldsymbol{T}_{k-1}^i\right)^+ \hat{\boldsymbol{x}}_{k-1|k-1}^i\right) \tag{51}$$

$$\widetilde{\boldsymbol{Q}}_{k-1|k-1}^i = \boldsymbol{T}_k^i \left\{\left(\boldsymbol{T}_{k-1}^i\right)^+ \boldsymbol{Q}_{k-1|k-1}^i \left(\boldsymbol{T}_{k-1}^i\right)\right\} \left(\boldsymbol{T}_k^i\right)^+ \tag{52}$$

$$\boldsymbol{Q}_{k|k-1}^i = \left[\frac{\partial \boldsymbol{f}\left(\hat{\boldsymbol{x}}_{k-1|k-1}^i\right)}{\partial \boldsymbol{x}_k} \left(\widetilde{\boldsymbol{Q}}_{k-1|k-1}^i\right)^{-1} \frac{\partial \boldsymbol{f}\left(\hat{\boldsymbol{x}}_{k-1|k-1}^i\right)}{\partial \boldsymbol{x}_k}^T + \boldsymbol{\Omega}_{k-1}^i\right]^{-1} \tag{53}$$

$$\boldsymbol{q}_{k|k-1}^i = \boldsymbol{Q}_{k|k-1}^i \hat{\boldsymbol{x}}_{k|k-1}^i \tag{54}$$

$$\boldsymbol{Consensus\ (oninformation)}\ \forall\ i$$

$$Initialization$$

$$\boldsymbol{Q}_{k|k-1}^{i\ (0)} = \boldsymbol{Q}_{k|k-1}^i\ \boldsymbol{q}_{k|k-1}^{i\ (0)} = \boldsymbol{q}_{k|k-1}^i \tag{55}$$

$$for\ \ \ell = 0, 1, \ldots .L$$

$$\forall\ \mathcal{V}_j \in \mathcal{A}^i \text{ receive } \boldsymbol{Q}_{k|k-1}^{j\ (\ell)}, \boldsymbol{q}_{k|k-1}^{j\ (\ell)}$$

$$\boldsymbol{Q}_{k|k-1}^{i\ (\ell+1)} = \sum_{j \in \mathcal{A}_i} \mu_{ij}\ \ \boldsymbol{T}_k^i \left\{\left(\boldsymbol{T}_k^j\right)^+ \boldsymbol{Q}_{k|k-1}^{j\ (\ell)} \left(\boldsymbol{T}_k^j\right)\right\} \left(\boldsymbol{T}_k^i\right)^+ \tag{56}$$

$$\boldsymbol{q}^{i}_{k|k-1}{}^{(\ell+1)} = \sum_{j\in\mathcal{A}_i} \mu_{ij}\ \ \boldsymbol{T}^{i}_{k}\left(\boldsymbol{T}^{j}_{k}\right)^{+}\boldsymbol{q}^{j}_{k|k-1}{}^{(\ell)} \tag{57}$$

$$\textbf{\textit{Localmeasure}}\ \forall\ i$$

$$Sample\boldsymbol{z}^{i}_{k}$$

$$\boldsymbol{c}^{i}_{k} = \boldsymbol{z}^{i}_{k} - \boldsymbol{h}^{i}\left(\hat{\boldsymbol{x}}^{i}_{k|k-1}\right) \tag{58}$$

$$\boldsymbol{I}^{i|i}_{k} = \frac{\partial \boldsymbol{h}^{i}\left(\hat{\boldsymbol{x}}^{i}_{k|k-1}\right)}{\partial \boldsymbol{x}_k}^{T}\left(\boldsymbol{R}^{i}_{k}\right)^{-1}\frac{\partial \boldsymbol{h}^{i}\left(\hat{\boldsymbol{x}}^{i}_{k|k-1}\right)}{\partial \boldsymbol{x}_k} \tag{59}$$

$$\boldsymbol{i}^{i|i}_{k} = \frac{\partial \boldsymbol{h}^{i}\left(\hat{\boldsymbol{x}}^{i}_{k|k-1}\right)}{\partial \boldsymbol{x}_k}^{T}\left(\boldsymbol{R}^{i}_{k}\right)^{-1}\left[\boldsymbol{c}^{i}_{k} + \frac{\partial \boldsymbol{h}^{i}\left(\hat{\boldsymbol{x}}^{i}_{k|k-1}\right)}{\partial \boldsymbol{x}_k}\hat{\boldsymbol{x}}^{i}_{k|k-1}\right] \tag{60}$$

$$\textbf{\textit{Consensus}}\ (\textbf{\textit{onmeasurement}})\ \forall\ i$$

$$Initialization$$

$$\boldsymbol{I}^{i\,(0)}_{k} = \boldsymbol{I}^{i}_{k}\ \ \boldsymbol{i}^{j\,(0)}_{k} = \boldsymbol{i}^{j}_{k} \tag{61}$$

$$for\ \ \ell = 0, 1, \ldots .L$$

$$\forall\ \mathcal{V}_j \in \mathcal{A}^{i}\ receive\ \boldsymbol{I}^{j|j\ (\ell)}_{k}, \boldsymbol{i}^{j|j\ (\ell)}_{k}$$

$$\boldsymbol{I}^{i|i\ (\ell+1)}_{k} = \sum_{j\in\mathcal{A}_i} \mu_{ij}\ \ \mathcal{F}^{j\to i}_{I}\left(\boldsymbol{I}^{j|j\ (\ell)}_{k}\right) \tag{62}$$

$$\boldsymbol{i}^{i|i\ (\ell+1)}_{k} = \sum_{j\in\mathcal{A}_i} \mu_{ij}\ \ \mathcal{F}^{j\to i}_{i}\left(\boldsymbol{i}^{j|j\ (\ell)}_{k}\right) \tag{63}$$

$$\textbf{\textit{Correction}}\ \forall\ i$$

$$\boldsymbol{Q}^{i}_{k|k} = \boldsymbol{Q}^{i}_{k|k-1} + N^{i}\boldsymbol{I}^{i|i\ (\mathrm{L})}_{k} \tag{64}$$

$$\boldsymbol{q}^{i}_{k|k} = \boldsymbol{q}^{i}_{k|k-1} + N^{i}\boldsymbol{i}^{i|i\ (\mathrm{L})}_{k} \tag{65}$$

$$\hat{\boldsymbol{x}}^{i}_{k|k} = \left(\boldsymbol{Q}^{i}_{k|k}\right)^{-1}\boldsymbol{q}^{i}_{k|k} \tag{66}$$

The logical building process of the Scalable DSE Algorithm is shown in Figure 3, together with the relationships with the algorithms from which it is derived.

The algorithm applies to a set of cooperating nodes (vehicle), in the sense that each node actively communicates information to neighboring nodes. The clustering property of the algorithm causes each node to estimate an automatically selected subset of the global state, based on the communication topology.

The mode selection update step can be approached in various ways. A simple way is based on a consensus procedure carried out on a local defined adjacency matrix initialized, in each node, as if the node were an isolated node.

$$a^{i}_{hh}{}^{(0)} = 1\quad if\ h = i\ ;\ otherwise\ \ a^{i}_{hk}{}^{(0)} = 0 \tag{67}$$

$$for\ \ell = 1, \ldots L\qquad \boldsymbol{A}^{i(\ell+1)} = \sum_{j\in\mathcal{A}_i} \mu_{ij}\boldsymbol{A}^{i(\ell)} \tag{68}$$

where $\boldsymbol{A}^{i}$ is a locally known adjacence matrix of elements a^{i}_{hk}. It is easy to verify that, for a proper number of consensus steps, the diagonal of each local adjacency matrix contains information on the nodes belonging to the same strongly connected sub graph (cluster).

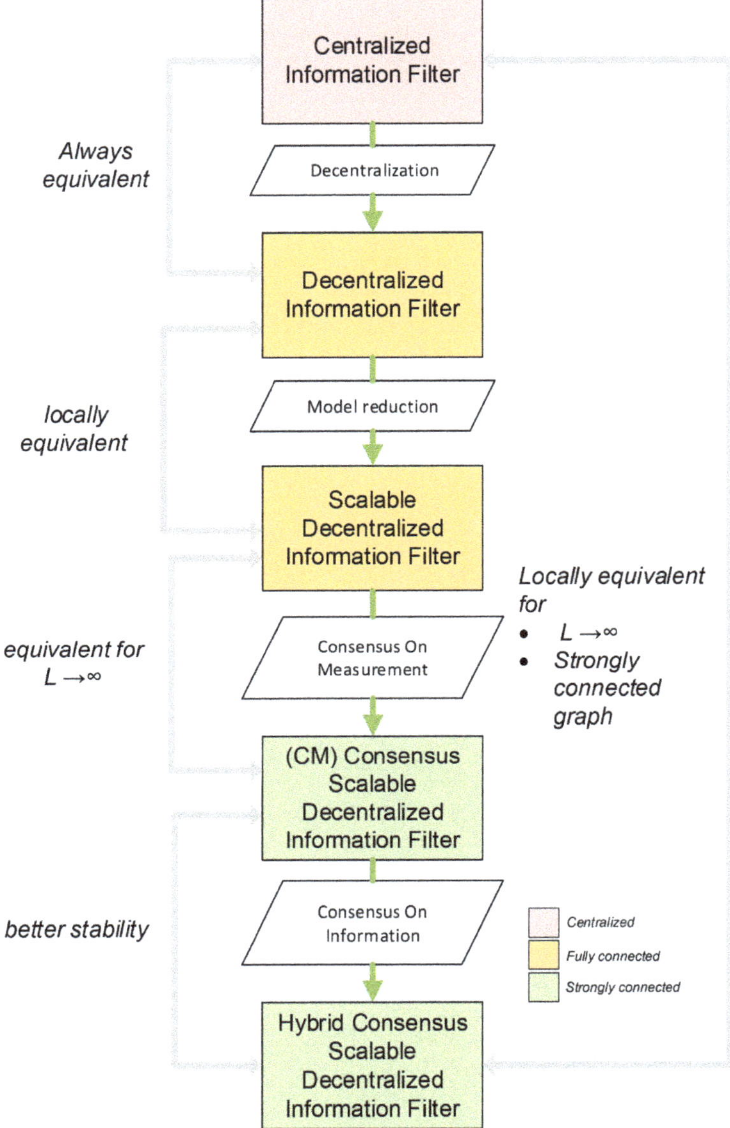

Figure 3. Scalable Distributed State Estimation (DSE) algorithm logical building process.

The equations in which the nodal transformation matrices explicitly appear, Equations (51), (52), (56), (57), (62), (63), perform the mode distribution operations. From an implementation perspective these equations, apparently onerous from a computational point of view, in the case of our interest, in which the matrices $\boldsymbol{T}_k^j$ make a simple selection of the states, are therefore reduced to rows and columns reordering and deletion operators. The topology of the system, and then the matrices $\boldsymbol{T}_k^j$ varies with time. When node is added to a cluster, Equations (51) and (52) include the initialization to null information pairs for the corresponding states. When a node is excluded from a cluster, Equations (51) and (52) correspond to the elimination of the corresponding rows and columns.

As far as the stability of the solution is concerned, the algorithm is equivalent to the algorithm described by Equations (30)–(35) for $L \to \infty$. On the other hand, the results achieved in the field of internodal transformations guarantee the equivalence of the local estimation and centralized state estimation limited to local retained states.

Stability can also be interpreted in a second way. During its evolution over time the system can be divided in clusters each strongly connected. Such clusters, as mentioned, varies over time through the disaggregation of a cluster or the fusion of two clusters. In each time interval in which a strongly connected cluster exists, the algorithm it is locally equivalent to an EFK algorithm, based on HCMCI consensus. Therefore, it is possible to consider locally the stability conditions applicable to that algorithm. These stability condition can be demonstrated in the linear time invariant form [36].

In the particular case that the following assumption are satisfied:

- The system is collectively observable;
- The network is strongly connected (i.e., any node is reachable from any other node through a directed path).

then, it is guaranteed that the estimation error is asymptotical bounded in mean square i.e., $\lim_{k\to\infty} \sup \mathbb{E}\left\{\| \hat{x}^i_{k|k} - x^i_k \|^2\right\} < \infty$. The HCMCI formulation ensures convergence to the centralized EKF in each strongly connected cluster.

2.6. sUAV Positon Estimation

Let us consider a set of N co-operative flying vehicles $\mathcal{V} = \{\mathcal{V}_1, \ldots \mathcal{V}_N\}$ free to move in a defined airspace of volume $\mathbb{V}$. A vehicle $\mathcal{V}_I$ can be located by means of its position $\boldsymbol{r}^i = \left(x^I, y^I, z^I\right)$ in an inertial reference system I. Its attitude is defined in terms of Euler's angles $\boldsymbol{\Theta}^i$ of a body reference frame B, with respect to inertial reference frame I.

In the proposed filter architecture, the chosen system dynamic model is a purely kinematic model not affected by any kind of model uncertainties. The state model equations for $\mathcal{V}_I$ assumes the following form:

$$\dot{\boldsymbol{r}}^i = \boldsymbol{v}^i \tag{69}$$

$$\dot{\boldsymbol{v}}^i = \boldsymbol{M}^i_{BI}\left(\boldsymbol{\Theta}^i\right)\boldsymbol{f}^i_B + \boldsymbol{G}^i_I \tag{70}$$

where vector $\boldsymbol{v}^i$ is the velocity vector in inertial reference system I; vector $\boldsymbol{f}^i_B$ is the specific force (force per unit mass) vector expressed in body frame B; $\boldsymbol{G}^i_I$ is the specific gravity force vector expressed in the inertial reference frame I and $\boldsymbol{M}^i_{BI}$ represents the rotation matrix from B to I.

We assume each that vehicle is equipped with a sensor suite including three type of sensors:

- Inertial Measurement Unit sensors (3-axes gyros, accelerometers and magnetometers);
- Absolute position and velocity (e.g., GPS);
- Relative position (e.g., visual based sensors, radar, radio frequency of time of flight).

Using these three types of sensors, each vehicle can locally generate the set of measurement included in one or more of the following equations:

$$\boldsymbol{y}^i_{acc} = \boldsymbol{f}^i_B \tag{71}$$

$$\boldsymbol{y}^i_{ahrs} = \boldsymbol{\Theta}^i \tag{72}$$

$$\boldsymbol{y}^i_{gnss} = \left[\boldsymbol{r}^i, \boldsymbol{v}^i\right] \tag{73}$$

$$\boldsymbol{y}^i_{rel} = \left\{\left(R^{ij}, \psi^{ij}, \Phi^{ij}\right),\ \mathcal{V}_j \in \mathcal{F}^i\right\} \tag{74}$$

More specifically, Equation (71) represent the specific force vector measured by accelerometers. Furthermore, it can be assumed that each vehicle has a local AHRS (Attitude and Heading reference system) capable of estimating attitude angles, expressed in Equation (72), by properly filtering the measurements coming from an Inertial Measurement Unit (IMU) sensors.

Equation (73) represents GNSS measures of inertial position and velocity. Finally, Equation (74) is representative of the relative position measurement. It is constituted by a triple including relative distance R, elevation Φ and azimuth ψ of all vehicles included into to Field of View $\mathcal{F}^i$ of the vehicle $\mathcal{V}_{\mathrm{i}}$. In this formulation, asynchronous measurements can occur both due to the different sampling times of the sensors, and for in and out Field of View transitions.

In order to complete the kinematic model, the following equations are added to the system: (69) and (70), to have an autonomous system capable to filter inertial sensors biases [39]:

$$\dot{f}_B^i = 0 \tag{75}$$

$$\dot{\mathbf{\Theta}}^i = 0 \tag{76}$$

Equations (75) and (76) are not restricting, due to the high update rates typical of the inertial measurements.

The state ξ^i of vehicle $\mathcal{V}_{\mathrm{i}}$ is described by:

$$\xi^i = \left[r^{iT}, v^{iT}, f_B^{iT}, \mathbf{\Theta}^{iT} \right]^T \tag{77}$$

The overall state ξ of system is

$$\xi = \left[\xi^{1T}, \xi^{2T}, \dots, \xi^{NT} \right]^T \tag{78}$$

Finally, it can be assumed that each vehicle is equipped with a bi-directional communication device whose range defines the connection graph, according to the rule that a vehicle $\mathcal{V}_j$ belong to neighbor $\mathcal{A}^i$ of vehicle $\mathcal{V}_i$ if its relative distance is less than the communication range.

3. Results

In order to highlight the most peculiar aspects of the proposed algorithm, which are those related to the decentralization, scalability and self-clustering of the local states in the nodes, the results are reported for a specific simplified, though sufficiently representative, scenario. The fundamental parameters, characteristic of the scenario, are summarized in Table 1. In order to facilitate the readability of the results, Equations (69), (70) are reduced to the two-dimensional case, i.e., two translational degrees of freedom and one angular degree of freedom. The measurements are asynchronous and obtained at different frequencies. The parameter Communication frequency in this context takes on the meaning of frequency of the consensus operation. The execution time of an entire consensus operation, consisting three Consensus steps ($L = 3$), is neglected.

Table 1. Scenario description.

Simulation Parameter	Value
Number of Vehicles	5
GNSS coverage	50%
GNSS frequency	1 Hz
Radar Range	500 m
Radar frequency	1 Hz
IMU frequency	10 Hz
Communication Range	600 m
Communication frequency	1 Hz

Figure 4 shows the position of the fleet, consisting of five vehicles, in four different time instants. The blue circles around each vehicle indicate the range of the relative position sensors; the red lines connecting two vehicles highlight the presence of an active communication link. The points in red indicate the absence of the GNSS signal, the position covered by the signal is reported in green.

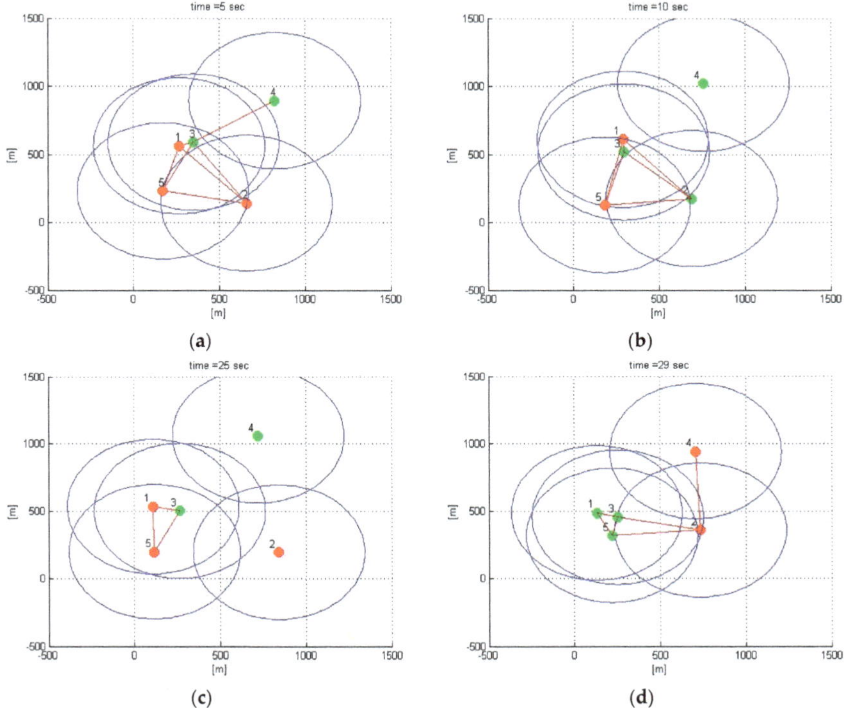

Figure 4. Fleet true position (**a**) t = 5 s (**b**) t = 10 s (**c**) t = 25 s (**d**) t = 30 s.

The coverage of the GNSS signal is distributed randomly with the only constraint of guaranteeing a ratio between GNSS denied areas and areas covered by the service equal to 50%. Even if it represents a simplified model, this distribution is representative of an urban canyon scenario.

The simulation shows a single cluster at time instant t = 5 s.; two clusters at time t = 10 s (Vehicle 4 remains separate from the others); three clusters at t = 25 s (Vehicles 4 and 2 separated from the others); once again, a unique strongly connected system at the time instant t = 30 s.

Table 2 summarizes, at the time instants shown in Figure 4, the states propagated from each vehicle. Within each cluster, collective observability is guaranteed by the presence of at least one vehicle with GNSS available [40].

Table 2. Model Distribution (estimated vehicle) for each vehicle.

	t = 5 s	**t = 10 s**	**t = 25 s**	**t = 30 s**
vehicle 1	1 2 **3 4**[1] 5	1 **2 3** 5	1 **3** 5	**1** 2 **3** 4 **5**
vehicle 2	1 2 **3 4** 5	1 **2 3** 5	2	**1** 2 **3** 4 **5**
vehicle 3	1 2 **3 4** 5	1 **2 3** 5	1 **3** 5	**1** 2 **3** 4 **5**
vehicle 4	1 2 **3 4** 5	1 **2 3** 5	**4**	**1** 2 **3** 4 **5**
vehicle 5	1 2 **3 4** 5	**4**	1 **3** 5	**1** 2 **3** 4 **5**

[1] in bold active GNSS.

Figures 5–7 show the estimates made locally by Vehicle 1. The EKF reference (black line) is obviously reported only in the case of self-estimation. It corresponds to the estimate made by the vehicle using only its own on-board sensors. In cross-estimation, whenever a vehicle is not in the same cluster of the estimating vehicle, the curve representing the estimate is interrupted and then resumed as soon as that vehicle returns to the cluster.

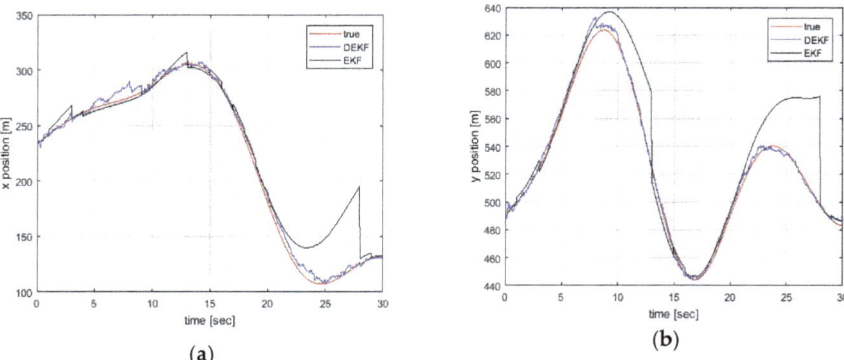

Figure 5. Position estimation performed by Vehicle 1 on its own position (**a**) x axis (**b**) y axis.

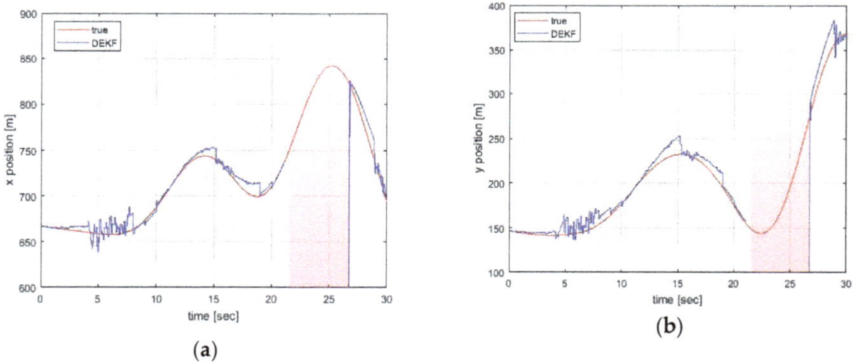

Figure 6. (**a**) Position estimation performed by Vehicle 1 on Vehicle 2 position (**a**) x axis (**b**) y axis.

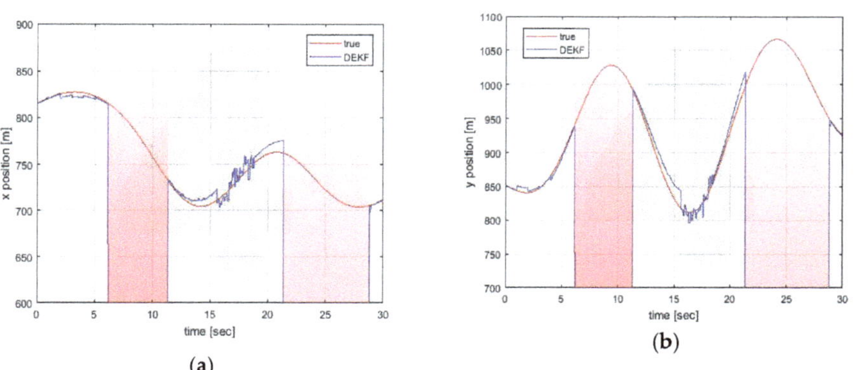

Figure 7. Position estimation performed by Vehicle 1 on Vehicle 4 position (**a**) x axis (**b**) y axis.

Three fundamental behaviors can be observed:

- The self-clustering mechanism is evident. Vehicle 2 is disconnected from the estimate around t = 20 s (Figure 6). Vehicle 4 shows two estimation black out windows (Figure 7) around t = 6 sec and around t = 21 s. In both cases, the estimate is correctly recovered when the vehicle rejoins the cluster of the estimating vehicle. When this happens, the transient is adequately managed in the average process of Equations (56) and (57), in which the contribution of each vehicle is weighed with its own covariance matrix.
- The self-estimation of Vehicle 1 is more accurate than the traditional centralized EKF estimate. Vehicle 1 periodically loses the GNSS signal. This leads to a widespread degradation of the estimate. It is particularly evident around t = 20 s, in which the GNSS denied persists for a longer interval. In the decentralized solution, the observability of the cluster 1-3-5 is guaranteed by the GNSS signal received by Vehicle 3 (see Figure 4c).
- The estimation process is generally good, with the exception of the intervals of time when some vehicles do not share the same cluster. However, this situation represents the case in which these vehicles cannot communicate with the node making the estimate. In this case, the vehicles cannot share their data to improve their estimates This condition is obviously not a limitation of the algorithm.

Similar conclusions can be drawn for the estimates made by the other vehicle. For illustrative purposes only, estimates for Vehicle 4 are reported in Figures 8–10.

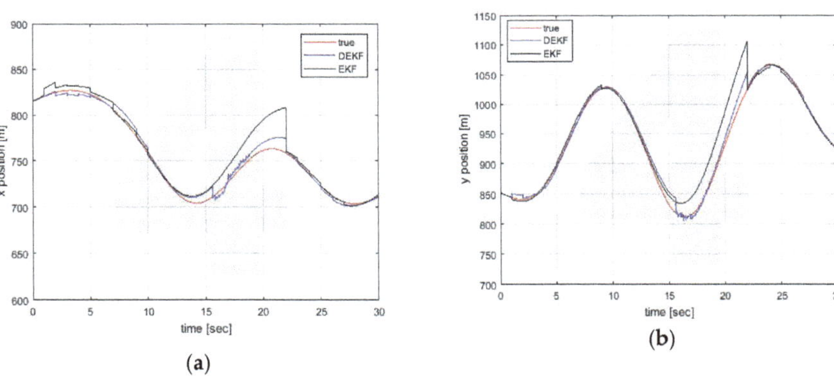

Figure 8. Position estimation performed by Vehicle 4 on its own position (**a**) x axis (**b**) y axis.

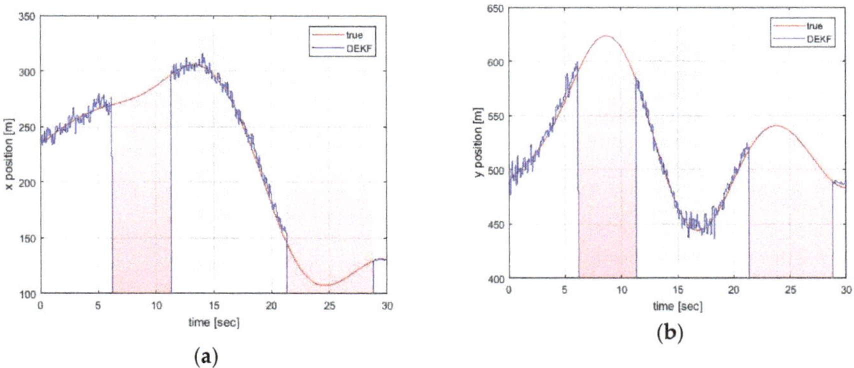

Figure 9. Position estimation performed by Vehicle 4 on Vehicle 1 position (**a**) x axis (**b**) y axis.

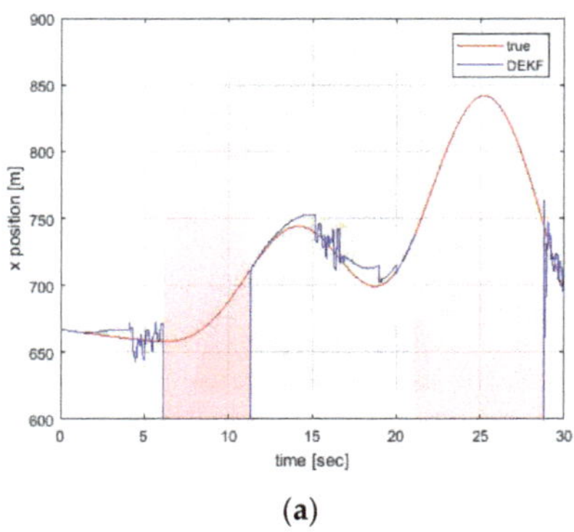

(a)

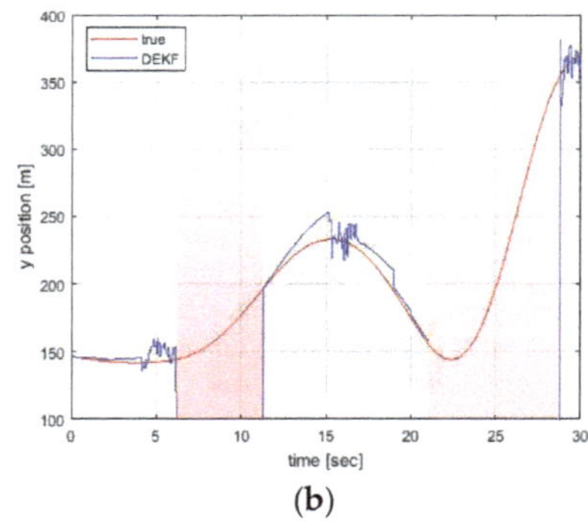

(b)

Figure 10. Position estimation performed by Vehicle 4 on Vehicle 2 position (**a**) x axis (**b**) y axis.

4. Conclusions

The article discussed a solution to the problem of Distributed State Estimation (DSE) in the framework of Kalman Filter-based algorithms. The proposed solution draws from the theoretical results derived from two different methodologies both related to the Kalman Filter theory: Internodal Transformation Theory and Consensus Theory.

From the former, the algorithm inherits the scalability property, that is the ability to decompose the global problem into different reduced order problems on a local level. From the latter, it inherits the ability to efficiently distribute information among local sensing and computational nodes.

A novel property, deriving from the fusion of the two methodologies, is the self-clustering property of the nodes which aggregate themselves in local estimation sub problems in response to the variation of the communication topology. The aggregation process is performed by each node through the information dynamically transmitted by its neighboring nodes, and is achieved by reaching an agreement resulting from a Consensus-based process.

The proposed algorithm makes it possible to obtain more accurate estimates than those obtainable individually from each node that uses only local measurements. Furthermore, the scalability property reduces the computational burden in each node by means of reducing the size of the local problems, while decentralization improves the communication efficiency, allowing each vehicle to exchange information only to the nearest vehicle.

The algorithm has a general validity, but assumes a specific meaning if applied to set of co-operating sUAVs equipped with heterogeneous on-board sensors and limited range communication devices. In this perspective, the algorithm is proposed as a formalization of an intuitive concept in which vehicles flying nearby other co-operative vehicles share its own on-board measurement to enable a better estimate of each vehicle.

A particularly significant condition is, for example, that in which a vehicle flying in a GNSS-denied zone can use the measurements of its position transmitted from another vehicle (i.e., from a micro radar or from a vision-based system) to correct the estimate regarding its own position. This scenario represents, among others, typical operational conditions of a fleet of sUAVs in a UTM context, in which registered vehicles can perform free flight operations in a potentially highly density airspace in urban environment. The application of the algorithm to a problem that involves the presence of non-collaborating elements, such as intruders, may be subject to future works.

The numerical examples shown have no ambitions of a complete numerical validation, but want to represent a clear example of the benefits that the proposed algorithm can guarantee.

Author Contributions: Conceptualization, methodology, investigation, formal analysis, creation of models, data curation and validation M.C., E.D., I.N. and M.M. Software development, writing, M.C. All authors have read and agreed to the published version of the manuscript.

Funding: This research was funded by REGIONE CAMPANIA, SCAVIR Project "Advanced Configurations Studies for an Innovative Regional Aircraft" CUP B43D18000210007.

Conflicts of Interest: The authors declare no conflict of interest.

References

1. Valavanis, K.; Vachtsevanos, G.J. *Handbook of Unmanned Aerial Vehicles*; Springer: Dordrecht, The Netherlands, 2015.
2. Kopardekar, P.; Rios, J.; Prevot, T.; Johnson, M.; Jung, J.; Robinson, J.E. Unmanned Aircraft System Traffic Management (UTM) Concept of Operations. In Proceedings of the 2016 16th AIAA Aviation Technology, Integration, and Operations Conference, Washington, DC, USA, 13–17 June 2016.
3. SESAR Joint Undertaking. Demonstrating RPAS integration in the European aviation system. *Tech. Rep.* **2016**, *1*, 1–24.
4. Farrell, J.A. *Aided Navigation: GPS with High Rate Sensors*; McGraw-Hill: New York, NY, USA, 2008.
5. Hasan, A.M.; Samsudin, K.; Ramli, A.R.; Azmir, R.S.; Ismaeel, S.A. A review of navigation systems (integration and algorithms). *Aust. J. Basic Appl. Sci.* **2009**, *3*, 943–959.
6. *GPS Standard Positioning Service Performance Standard*, 4th ed.; Office of the Secretary of Defense: New York, NY, USA, 2008.
7. Wu, A.; Johnson, E.; Kaess, M.; Dellaert, F.; Chowdhary, G. Autonomous flight in GPS-denied environments using monocular vision and inertial sensors. *J. Aerosp. Comput. Inf. Commun.* **2010**, *10*, 172–186.
8. Kaiser, M.K.; Gans, N.R.; Dixon, W.E. Vision-Based Estimation for Guidance, Navigation, and Control of an Aerial Vehicle. *IEEE Trans. Aerosp. Electron.* **2010**, *46*, 1064–1077. [CrossRef]
9. Scannapieco, A.; Renga, A.; Fasano, G.; Moccia, A. Ultralight radat for small and micro UAV navigation. *Int. Arch. Photogramm. Remote Sens. Spat. Inf. Sci.* **2017**, *42*, 333. [CrossRef]
10. Vetrella, A.R.; Fasano, G.; Renga, A.; Accardo, D. Cooperative UAV Navigation Based on Distributed Multi-Antenna GNSS, Vision, and MEMS Sensors. In Proceedings of the International Conference on Unmanned Aircraft Systems, Denver, CO, USA, 9–12 June 2015.
11. Vetrella, A.R.; Fasano, G.; Accardo, D. *Vision-Aided Cooperative Navigation for UAV Swarms*; AIAA Infotech@ Aerospace: San Diego, CA, USA, 2016.
12. Vetrella, A.R.; Fasano, G.; Accardo, D.; Moccia, A. Differential GNSS and Vision-Based Tracking to Improve Navigation Performance in Cooperative Multi-UAV Systems. *Sensors* **2016**, *16*, 2164. [CrossRef]
13. Yu, Y.; Peng, S.; Li, Q.; Dong, X.; Ren, Z. Cooperative Navigation Method Based on Adaptive CKF for UAVs in GPS Denied Areas. In Proceedings of the 2018 IEEE CSAA Guidance, Navigation and Control Conference (CGNCC), Xiamen, China, 10–12 August 2018; pp. 1–6.
14. Liu, X.; Xu, S. Multi-UAV Cooperative Navigation Algorithm Based on Federated Filtering Structure. In Proceedings of the 2018 IEEE CSAA Guidance, Navigation and Control Conference (CGNCC), Xiamen, China, 10–12 August 2018; pp. 1–5.
15. D'Amato, E.; Notaro, I.; Mattei, M.; Tartaglione, G. Attitude and Position Estimation for an UAV Swarm using Consensus Kalman Filtering. In Proceedings of the IEEE Metrology for Aerospace, Benevento, Italy, 4–5 June 2015.
16. Kalman, R. A new approach to linear filtering and prediction problems. *Trans. ASME J. Basic Eng.* **1960**, *82*, 35–45. [CrossRef]
17. Olfati-Saber, R.; Murray, R.M. Consensus problems in networks of agents with swtiching topology and time-delays. *IEEE Trans. Autom. Control* **2004**, *49*, 1520–1533. [CrossRef]
18. Olfati-Saber, R.; Murray, R.M. Distributed Kalman filtering for sensor networks. In Proceedings of the 46th IEEE Conference on Decision and Control, New Orleans, LA, USA, 12–14 December 2007; pp. 5492–5498.
19. Schizas, I.D.; Giannakis, G.B.; Roumeliotis, S.I.; Ribeiro, A. Consensus in ad hoc WSNs with noisy links—Part II: Distributed estimation and smoothing of random signals. *IEEE Trans. Signal Process.* **2008**, *56*, 1650–1666. [CrossRef]
20. Carli, R.; Chiuso, A.; Schenato, L.; Zampieri, S. Distributed Kalman filtering based on consensus strategies. *IEEE J. Sel. Areas Commun.* **2008**, *26*, 622–633. [CrossRef]
21. Calafiore, G.C.; Abrate, F. Distributed linear estimation over sensor networks. *Int. J. Control* **2009**, *82*, 868–882. [CrossRef]

22. Stankovic, S.S.; Stankovic, M.S.; Stipanovic, D.M. Consensus based overlapping decentralized estimation with missing observations and communication faults. *Automatica* **2009**, *45*, 1397–1406. [CrossRef]
23. Cattivelli, F.S.; Sayed, A.H. Diffusion strategies for distributed Kalman filtering and smoothing. *IEEE Trans. Autom. Control* **2010**, *55*, 2069–2084. [CrossRef]
24. Farina, M.; Ferrari-Trecate, G.; Scattolini, R. Distributed moving horizon estimation for linear constrained systems. *IEEE Trans. Autom. Control* **2010**, *55*, 2462–22475. [CrossRef]
25. Rao, B.; Durrant-Whyte, H. Fully decentralized algorithm for multisensor Kalman filtering. *IEE Proceedings-Control Theory Appl.* **1991**, *138*, 413–420. [CrossRef]
26. Saligrama, V.; Castanon, D. Reliable distributed estimation with intermittent communications. In Proceedings of the 45th IEEE Conference on Decision and Control, San Diego, CA, USA, 13-15 December 2006; pp. 6763–6768.
27. Chung, T.; Gupta, V.; Burdick, J.; Murray, R. On a decentralized active sensing strategy using mobile sensor platforms in a network. In Proceedings of the 43rd IEEE Conference on Decision and Control, Paradise Island, Bahamas, 14-17 December 2004; Volume 2, pp. 1914–1919.
28. Speranzon, A.; Fischione, C.; Johansson, K.H. Distributed and Collaborative Estimation over Wireless Sensor Networks. In Proceedings of the 45th IEEE Conference on Decision and Control, San Diego, CA, USA, 13–15 December 2006; pp. 1025–1030.
29. Mutambara, A.G.O. *Decentralized Estimation and Control for Multisensors Systems*; CRC Press LLC: Boca Raton, FL, USA, 1998; pp. 80–119.
30. Berg, T.; Durrant-Whyte, H. Model distribution in decentralized multi-sensor data fusion. In Proceedings of the American Control Conference, Boston, MA, USA, 2-28 June 1991; pp. 2292–2293.
31. Ren, W.; Beard, R. Overview of Consensus Algorithms in Cooperative Control. *Distrib. Consens. Multi-Veh. Coop. Control Springer* **2008**, 3–22. [CrossRef]
32. Julier, S.J.; Uhlmann, J.K. A non-divergent estimation algorithm in the presence of unknown correlations. In Proceedings of the 1997 American Control Conference, Albuquerque, NM, USA, 6 June 1997; Volume 4, pp. 2369–2373.
33. Battistelli, G.; Chisci, L.; Morrocchi, S.; Papi, F. An information theoretic approach to distributed state estimation. In Proceedings of the 18th IFAC World Congress, Milano, Italy, 28 August–2 September 2011; pp. 12477–12482.
34. Olfati-Saber, R.; Faxand, J.A.; Murray, R.M. Consensus and cooperation in networked multi-agent systems. *Proc. IEEE* **2007**, *95*, 49–54. [CrossRef]
35. Kamgarpour, M.; Tomlin, C. Convergence properties of a decentralized Kalman filter. In Proceedings of the 47th IEEE Conference on Decision and Control, Cancun, Mexico, 9–11 December 2008; pp. 3205–3210.
36. Battistelli, G.; Chisci, L.; Mugnai, G.; Farina, A. Consensus-based algorithms for distributed filtering. In Proceedings of the 51st IEEE Conference on Decision and Control, Maui, HI, USA, 10–13 December 2012; pp. 794–799.
37. Assimakis, N.; Adam, M.; Anargyros, D. Information Filter and Kalman Filter Comparison: Selection of the Faster Filter. *Int. J. Inf. Eng.* **2012**, *2*, 1–5.
38. Xiao, L.; Boyd, S. Fast Linear Iteration for Distributed Averaging. In Proceedings of the 42nd IEEE International Conference on Decision and Control (IEEE Cat. No.03CH37475), Maui, HI, USA, 9–12 December 2004.
39. Notaro, I.; Ariola, M.; D'Amato, E.; Mattei, M. HW VS SW Sensor Redundancy: Fault Detection and Isolation Observer based Approaches for Inertial Measurement Units. In Proceedings of the 29th Congress of International Council of Aeronautical Sciences, St. Petersburg, Russia, 9–12 September 2014.
40. Cicala, M.; D'Amato, E.; Notaro, I.; Mattei, M. Distributed UAV State Estimation in UTM context. In Proceedings of the 2019 6th International Conference on Control, Decision and Information Technologies (CoDIT), Paris, France, 23–26 April 2019; pp. 557–562.

Article

UAV Landing Based on the Optical Flow Videonavigation

Alexander Miller, Boris Miller *, Alexey Popov and Karen Stepanyan

Institute for Information Transmission Problems RAS, Bolshoy Karetny per. 19, Build.1, Moscow 127051, Russia; amiller@iitp.ru (A.M.); ap@iitp.ru (A.P.); KVStepanyan@iitp.ru (K.S.)

* Correspondence: bmiller@iitp.ru; Tel.: +7-495-6504781; Fax: +7-495-6500579

Received: 3 January 2019; Accepted: 10 March 2019; Published: 18 March 2019

Abstract: An automatic landing of an unmanned aerial vehicle (UAV) is a non-trivial task requiring a solution of a variety of technical and computational problems. The most important is the precise determination of altitude, especially at the final stage of approaching to the earth. With current altimeters, the magnitude of measurement errors at the final phase of the descent may be unacceptably high for constructing an algorithm for controlling the landing manoeuvre. Therefore, it is desirable to have an additional sensor, which makes possible to estimate the height above the surface of the runway. It is possible to estimate all linear and angular UAV velocities simultaneously with the help of so-called optical flow (OF), determined by the sequence of images recorded by an onboard camera, however in pixel scale. To transform them into the real metrical values it is necessary to know the current flight altitude and the camera angular position values. The critical feature of the OF is its susceptibility to the camera resolution and the shift rate of the observed scene. During the descent phase of flight, these parameters change at least one hundred times together with the altitude. Therefore, for reliable application of the OF one needs to coordinate the shooting parameters with the current altitude. However, in case of the altimeter fault presence, the altitude is also still to be estimated with the aid of the OF, so one needs to have another tool for the camera control. One of the possible and straightforward ways is the camera resolution change by pixels averaging in computer part which performed in coordination with theoretically estimated and measured OF velocity. The article presents results of such algorithms testing from real video sequences obtained in flights with different approaches to the runway with simultaneous recording of telemetry and video data.

Keywords: UAV; landing; optical flow; video navigation; Kalman filter

---

1. Introduction

Today the development of optoelectronic devices and data transmission systems allows to use them in remote control systems of unmanned aerial vehicles (UAV). In case of remote control which is carried out by an operator the characteristics of the optical devices and the transmitted image must correspond to the capabilities of human vision. However, in case of the UAV autonomous flight, the optical systems and the onboard computer must work together solving the problems of identifying the observed objects with their coordinates. It means that requirements to characteristics of optoelectronic devices are different from remote flight control case. The UAV control system which includes an optoelectronic system (OES) and an onboard computer determines the movement of the onboard video camera and identifies the objects observed in the field of view of the camera [1]. There are several approaches to the usage of OES [2]. The first one is to detect and to track the movement of specific local areas (reference points) on the image by analogy with human vision [3,4]. With this approach, it is easy to transform the recorded images into metric values for the control system. Some examples of this approach to UAV navigation are as follows:

- the tracking of singular points and development of the algorithm based on determining of their angular coordinates and establishing of correspondence of their images in preliminary uploaded template map based on RANSAC methodology described in [5];
- conjugation of the rectilinear objects segments such as walls of buildings and roads are in [6];
- fitting of characteristic curvilinear elements [7];
- matching of textured and coloured areas [8,9];
- matching of epipolar lines, such as runways, at landing [10].

One possible reference to this approach is *sparse OF*. An example of this approach to the UAV landing in case of the smoke occlusions basing on the natural terrain landmarks is given in [11]. The second approach is based on non-metric analysis and is built by analogy with a vision of insects or birds [12]. Nowadays an approach that uses information containing in the vector field of the image motion (in other words in the optical flow (OF)) is developing. OF is well known to engineers developing cameras for shooting from moving carriers where unreduced OF leads to the resolution degradation and needs reduction either optomechanically [13,14] or electronically [15]. At the same time in the UAV application area, the OF contains the information about the carrier's velocities and thereby may serve as an additional sensor for the UAV navigation.

OF is the projection of the camera's motion onto the focal plane. OF generates the nonuniform field of image shifts velocities which is known as *dense optical flow*. In case of tracking the displacement of some reference points on the underlying surface, it is common to use the *sparse optical flow* term. The methodology of the OF computation had been developed long ago, mainly for estimation of quality for various optomechanical image shift compensation systems performance [16]. Nowadays there are various examples of the OF usage in UAV applications such as:

- in landing at the unknown hazardous environment with the choice of the landing place [17];
- in experimental landing with the aid of special landing pads [18];
- vision based and mapping for landing with the aid of model predictive control [19];
- in landing manoeuvring of the UAV [20,21];
- in tracking tasks of linear objects such as communication lines and pipelines on the terrain [22];
- even in usual manoeuvring [23];
- slope estimation for autonomous landing [24];
- in distance estimation with application to the UAV landing with the aid of the mono camera [25].

Our research team is working on video navigation as an additional navigation tool for UAV of aeroplane type. We consider OF as an additional tool for estimation of the UAV velocity in the absence of specific regions on the earth surface which can serve as beacons for estimation of the UAV position. The specific feature of the OF is that it gives information about velocities only but not on position. Therefore, the bias of the position estimates which inevitably exists at the beginning of the flight path can only increase without additional intermediate corrections. Correct filtering algorithm can reduce this bias but cannot eradicate it. It means that it is essential to evaluate the specific noises related to the OF estimation, so in our experiments, we perform flights where qualified pilot performs the series of approaches to the runway, during which one can carefully record the video sequences and corresponding telemetry data. These data serve as a source of information about specific OF noises. Filtering method fuses the data at UAV altitude estimation during landing.

In the existing literature, the OF usage at landing generally relates to the copter type UAV [11]. Here it is possible to coordinate velocity of descent with the observable video sequence. There are a series of works demonstrating successful approaches to the copter control based on the divergence observation. Here the divergence of the OF field serves as a measure of the approach velocity to the earth surface. Almost all articles were presenting the successful application of the OF relate to micro air vehicles (MAV), where OF used with supplemental range meter. For example, in [26] authors consider a Vertical-Take-OFF-and-Landing UAV (VTOL UAV) with an Inertial Measurement Unit

(IMU) equipped with suitable filtering algorithms providing reliable estimates of the UAV pose and rotational velocities. Moreover, this UAV equipped with an optoelectronic camera serving as an additional passive sensor whose output is enriching in-flight information. These authors also assume that the UAV control system has an image processing capability to identify a target and to compute OF over the full image basing on well-known Horn and Schunck approach [27].

In [28] a nonlinear controller is presented for a VTOL UAV that exploits a measurement of the average OF to enable hovering and landing onto a moving platform such as the deck of a seagoing vessel.

In a detailed review of copter type UAV's [11] video control presents a variety of approaches to the usage of natural landmarks and the OF approaches. The principal differences between UAVs of a copter type and even MAV and standard sized UAVs lay in their sizes and velocities. When small-sized UAV is approaching the obstacles with low velocity, the latter can be controlled by the OF signal, since the divergence of the OF field serves as a measure of approaching speed [29–32]. On the contrary, the standard sized UAVs are landing onto the runway and approaching to the earth by standard glissade during which the altitude changes from hundreds of meters to zero and velocity reduces from dozens of meters per second to meters per second and finally to zero. Simultaneously the rate of the image motion changes on the same orders which lead to the resolution degradation, so the detection of the natural landmarks like in [11] becomes impossible. An application of the OF techniques which determine the image velocity via analysis of the evolution of local image areas needs the coordination of the resolution level of OES with the current speed. The measure of such coordination is obtainable by comparison of measured and theoretically calculated OF velocities.

Even if there are communications related to successful usage of the OF with the aid of two vertically displaced cameras [33,34], more careful analysis shows that with real images and flights the bias between the estimated and real values of the OF achieves unacceptable values [35].

That is why the approach developed for MAV do not apply to the aeroplane landing where the descent velocity cannot be arbitrary controlled. Moreover, the OES resolution when approaching to the earth changes about a hundred times from the start of a glissade. By this reason, the standard OF evaluation does not work accurately if the OES resolution does not reduce in coordination with the UAV altitude. A changing of the camera focal length is somewhat tricky since it needs an additional controlled optomechanical system. Meanwhile, the resolution can be changed by averaging pixels in coordination with estimated OF velocity obtained from the video sequence and its comparison with the value calculated theoretically from exact OF formulas based on current altitude estimation. The principal aim of this article is to demonstrate this approach via real video sequences obtained in real flights. There are two ways to use video navigation in autonomous UAV flights. The first one is a detection of the terrain objects with known coordinates, a definition of their aspect angles and finally, a determining the current UAV coordinates. We demonstrated this approach in [5]. However, if such objects are absent in the field of view, one can determine the current velocity and coordinates by filtering, and the OF is a good way for that. So the combination of direct measurements and OF provides the continuous tracking of the UAV trajectory. This way is convenient in so-called cruise part of the flight, where the altitude and velocity remain almost constant. The landing is the different issue, since the altitude changes on few orders as well as the scale of the picture, therefore, small details become apparent and affect the accuracy of the determining of the OF [36]. That is why at landing phase one needs to adapt the camera resolution and the frame rate following the current altitude. Creation of such cameras is a separate problem, though one can change the resolution by averaging the image signal. In this article, we show how to create such averaging algorithm, which together with filtering based estimation guarantees the reliable evaluation of the altitude at the descent from the altitude of 300 m to 5 m.

The structure of the article is as follows. In Section 2, we present the theory of the OF computation. In Section 3, we give the algorithm of the UAV motion estimation with the aid of the OF and Kalman filtering. Section 4 gives the results of the OF obtained during the real UAV flight at the landing phase and presents the approach to the resolution adaptation from current altitude estimation. In Section 5 we give the algorithms of the averaging scale switching based on the comparison of exact OF values (theoretically calculated) and their estimates with the aid of Lucas-Kanade (L–K) algorithm [37]. It shows that the difference between estimated and calculated OF values may serve as a sensor for the control of averaging. The results of numerical tests showing the reliable tracking of the altitude up to the 5m provided. Section 6 is conclusions where we discuss the results.

2. OF Computation: Theory

Since the end of the 70s, many mentions of the OF formulas appeared in the literature, see for example [38]. However, these formulas usually relate to the case of fixed camera position and cannot take into account possible inclination of the line of sight which occurs either during the plane turn (azimuth and roll angles) or during the landing-descent (pitch angle). Meanwhile, the case of a camera with a stabilised line of sight is much more complicated. As an example, the copter needs some inclination angle in order to create propulsion in the flight direction. Hence the pitch angle correction is necessary as it presented in the example in [39]. In recent years navigation by computation of the camera path and the distance to obstacles with the aid of field of image motion velocities (i.e., OF) became highly demanded particularly in the area of relatively small and even micro UAV. Video sequences captured by onboard camera give the possibility of the onboard OF calculation with the aid of relatively simple algorithms like Lucas-Kanade [37] and even more sophisticated ones using higher order terrain illuminance approximations [40–42].

Theory of the OF computation in general flight conditions is in [43–45]. On that basis, the specialised software IMODEL (version 8.2.o, proprietary) developed and various scenarios of the OF usage in the UAV navigation presented. The IMODEL permits to analyse both the *direct* and *inverse* problems of the OF computation. The *direct* problem is the calculation of the image motion at any point of the field of view for the general orientation of the line of sight. The *inverse* one is the estimation of the OF field for given modelled moving landscape registered by the camera virtually. Moreover, in the implementation stage, the calculus precision problem arises. On one side the higher level filtering algorithm described in [45] and the exact parameters for it affect the resulting precision of the estimated parameters of UAV position. On the other side, the precision achieved is the consequence of the chosen low-level method to determine OF.

An example of the software application is the evaluation of the altitude estimation algorithm with the aid of two vertically displaced cameras [33,35] which shows the presence of bias in the altitude estimation.

A general approach to the OF computation is based on the pinhole camera geometrical model assuming the flight over a flat surface [39,43,44] (see Figures 1 and 2).

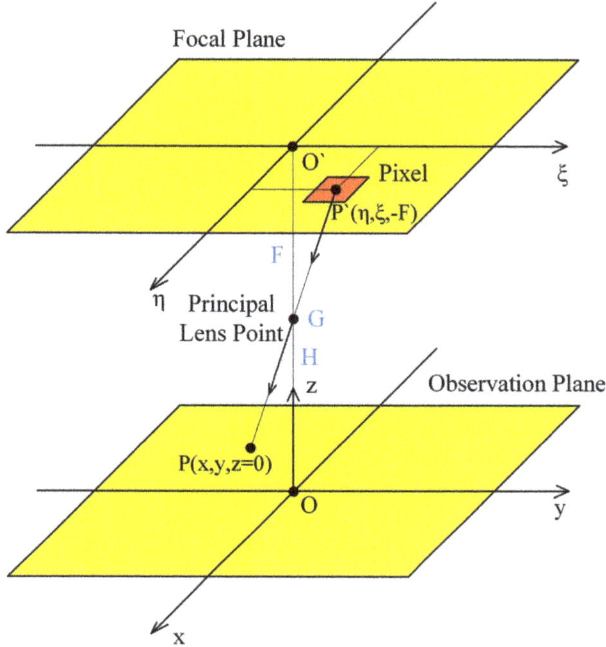

Figure 1. The picture shows the projection of the image plane onto the earth surface. The orientation of the line of sight is changing by rotation at point G.

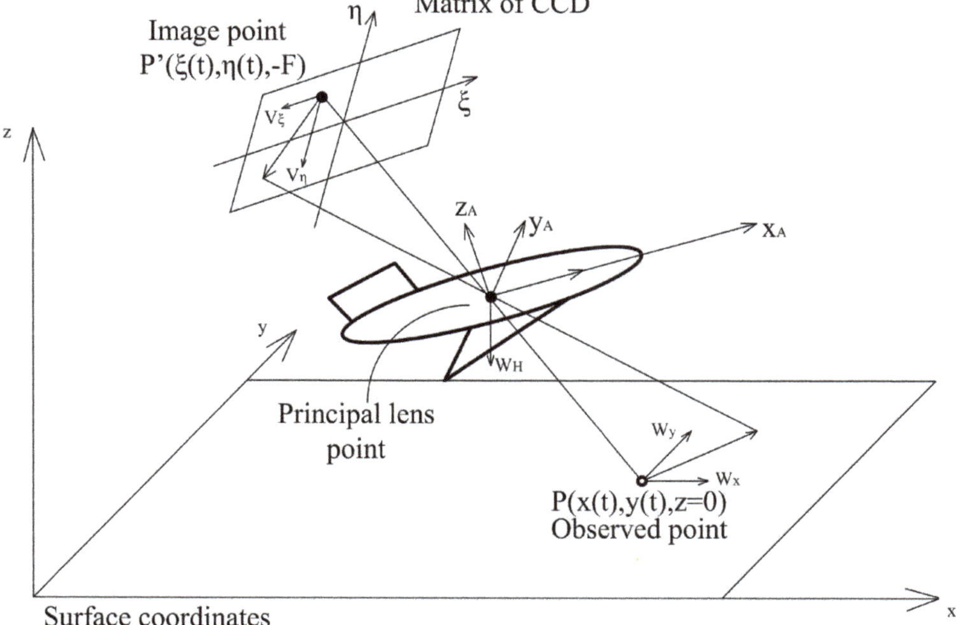

Figure 2. Image motion velocities. **CCD** is the charge coupled devices matrix in the focal plane.

We use the camera coordinate system $(\xi, \eta, -F)$, where (ξ, η) are the pixel coordinates, ξ corresponds to the direction of flight, η is perpendicular to ξ, and F is the lens focal length, and the third coordinate directs to the principal camera point. As it follows from the general theory [16,45] the OF velocities (V_ξ, V_η) equal to

$$\begin{pmatrix} V_\xi \\ V_\eta \end{pmatrix} = \begin{pmatrix} \dfrac{\partial \xi}{\partial t} \\ \dfrac{\partial \eta}{\partial t} \end{pmatrix} \Bigg|_{x=x(\xi,\eta,\Lambda(t));y=y(\xi,\eta,\Lambda(t))}, \tag{1}$$

where dependence on $\Lambda(t)$ includes the current attitude of the vehicle and orientation of the line of sight. For in-flight computations, one needs to use more extended formula including $W_x, W_y, W_H, \omega_p, \omega_r, \omega_y$ which is the following matrix equation:

$$\begin{pmatrix} V_\xi \\ V_\eta \end{pmatrix} = D_1(\xi, \eta, \Lambda(t)) \begin{pmatrix} W_x \\ W_y \\ W_H \end{pmatrix} + D_2(\xi, \eta, \Lambda(t)) \begin{pmatrix} \omega_p \\ \omega_r \\ \omega_y \end{pmatrix}, \tag{2}$$

where $(W_x, W_y)^T$ are the velocities components of the UAV in the plane parallel to the earth surface, $(\omega_p, \omega_r, \omega_y)^T$ are the rotation velocities relative to the line of sight, and W_H is the vertical velocity component of the UAV. Exact formulas may found in [39]. By using the estimations of the left-hand side (LHS) of (2) either via Lucas-Kanade (L–K) [37] algorithm or any other, one gets the additional sensor for the UAV velocities estimation. The principal difficulties are the implicit dependence on $\Lambda(t)$ including the line of sight orientation, given by angles (p, r, y) conditionally called the *pitch, roll, yaw*, and the current altitude H.

Therefore, the observation algorithm requires estimation of all mentioned above parameters by Kalman filtering via observation of their derivatives from (2).

The system (2) contains observable variables in the LHS and a set of six variables to be estimated in the right-hand side (RHS). For a sufficiently large field of view, one can evaluate velocities in a vast amount of pixels, while the values are the same for any point. It gives the possibility to solve the issue for example via least square method since the observation model is linear in velocities, and it is possible to estimate unobservable variables via Kalman filtering [45].

Remark 1. *In our approach, we use OF just as an additional sensor to estimate the altitude only. The reason is that in these experiments we use the data obtained with the aid of a series of landing approaches made by a qualified pilot of the light aeroplane. Other parameters such as velocity and orientation angles can be obtained from inertial navigation system (INS) with sufficient accuracy, so we use their nominal values with small corrections made with the aid of Kalman filter (see below* (5), (6)*). The current value of the altitude is necessary for estimation via OF. In the real flight of the UAV all possible navigation sensors must be used in the fusion with the OF. Here we just test the ability of the OF solely, therefore the real approaches to the runway have been performed by a pilot of a light aeroplane with recording of the video images synchronized with telemetry data from INS and satellite navigation system (SNS).*

3. Estimation of the UAV Motion by the OF and the Kalman Filtering

The model of UAV motion described below. It includes the UAV dynamic model and generic measurements model which based on the OF estimation of the UAV attitude velocities.

3.1. The UAV Linear Velocity Estimation

The UAV velocity vector $\mathbf{V} = col(V_x, V_y, V_z)$ by coordinates x, y, z :

$$\mathbf{V}(t_{k+1}) = \mathbf{V}(t_k) + \mathbf{a}(t_k)\Delta t + \mathbf{W}(t_k), \tag{3}$$

where t_k is the current time, $t_k = t_0 + k\Delta t$,

$\mathbf{a}(t_k) = col(a_x, a_y, a_z)$ — the vector of accelerations coming from INS (inertial navigation system),

$\mathbf{W}(t_k)$ — is the vector of the current perturbations in UAV motion.

We assume that the components of the perturbation vector are white noises with variances $(\sigma_x^2, \sigma_y^2, \sigma_z^2)$.

Velocity measurements using OF have the following general form:

$$\mathbf{m}_V(t_k) = \mathbf{V}(t_k) + \mathbf{W}_V(t_k), \tag{4}$$

where $\mathbf{W}_V(t_k)$ — are uncorrelated white noises with variances $(\sigma_{V_x}^2, \sigma_{V_y}^2, \sigma_{V_z}^2)$.

Consider relations (3) and (4) for the velocity along x axis:

$$V_x(t_{k+1}) = V_x(t_k) + a_x(t_k)\Delta t + W_x(t_k),$$

$$m_{V_x}(t_k) = V_x(t_k) + W_{V_x}(t_k).$$

Velocity along x axis estimation on the $k+1$ step:

$$\hat{V}_x(t_{k+1}) = K_x(t_{k+1}) m_{V_x}(t_{k+1}) + (1 - K_x(t_{k+1}))\tilde{V}_x(t_{k+1}),$$

$$\tilde{V}_x(t_{k+1}) = \hat{V}_x(t_k) + a_x(t_k)\Delta t.$$

Kalman filter gives the estimation of $\hat{V}_x$:

$$\begin{aligned} \hat{V}_x(t_{k+1}) &= K_x(t_{k+1}) m_{V_x}(t_{k+1}) + (1 - K_x(t_{k+1}))(\hat{V}_x(t_k) + a_x(t_k)\Delta t), \\ K_x(t_{k+1}) &= \frac{\hat{p}^{V_xV_x}(t_k) + \sigma_x^2}{\hat{p}^{V_xV_x}(t_k) + \sigma_x^2 + \sigma_{V_x}^2}, \\ \hat{p}^{V_xV_x}(t_{k+1}) &= \frac{\sigma_{V_x}^2(\hat{p}^{V_xV_x}(t_k) + \sigma_x^2)}{\hat{p}^{V_xV_x}(t_k) + \sigma_x^2 + \sigma_{V_x}^2}. \end{aligned} \tag{5}$$

The formulae for $\hat{V}_y$ and $\hat{V}_z$ are analogous.

Remark 2. *Kalman filter coefficients have been derived from empirically registered standard deviations of noise processes. They assumed to be constant during the whole duration of landing. Only the scale rate changes, but after application of the least squares method for estimation of velocities in the RHS of (1) for the large number of observations the remaining noise in observation of the LHS of (1) could be assumed constant for various averaging scales.*

3.2. The UAV Angles and Angular Velocities Estimation

UAV angular position estimation is given by three angles $\theta(t_k), \varphi(t_k), \gamma(t_k)$ that are (pitch, roll and yaw, respectively), angular velocities $\omega_p(t_k), \omega_r(t_k), \omega_y(t_k)$ and angular accelerations $a_p(t_k), a_r(t_k), a_y(t_k)$.

Pitch angle and pitch angular velocity dynamics described by the following relations:

$$\theta(t_{k+1}) = \theta(t_k) + \omega_p(t_k)\Delta t + a_p(t_k)\frac{\Delta t^2}{2},$$

$$\omega_p(t_{k+1}) = \omega_p(t_k) + a_p(t_k)\Delta t + W_p(t_k).$$

where $W_p(t_k)$ — is the white noise with variance σ_p^2.

The pitch angular velocity measurement using the OF has the following form:

$$m_p(t_k) = \omega_p(t_k) + W_{\omega_p}(t_k),$$

where $W_{\omega_p}(t_k)$ — is the noise in the angular velocity measurements using OF, which is the white noise with variance $\sigma_{\omega_p}^2$.

Similarly to the linear velocity estimation we get the pitch angle $\theta(t_k)$ and pitch angular velocity $\omega_p(t_k)$ estimations:

$$\begin{aligned}
\hat{\theta}(t_{k+1}) &= \hat{\theta}(t_k) + \hat{\omega}_p(t_k)\Delta t + a_p(t_k)\frac{\Delta t^2}{2} \\
\hat{\omega}_p(t_{k+1}) &= K_p(t_{k+1})m_p(t_{k+1}) + (1 - K_p(t_{k+1}))(\hat{\omega}_p(t_k) + a_p(t_k)\Delta t), \\
K_p(t_{k+1}) &= \frac{\hat{p}^{\omega_p\omega_p}(t_k) + \sigma_p^2}{\hat{p}^{\omega_p\omega_p}(t_k) + \sigma_p^2 + \sigma_{\omega_p}^2}, \\
\hat{p}^{\omega_p\omega_p}(t_{k+1}) &= \frac{\sigma_{\omega_p}^2(\hat{p}^{\omega_p\omega_p}(t_k) + \sigma_p^2)}{\hat{p}^{\omega_p\omega_p}(t_k) + \sigma_p^2 + \sigma_{\omega_p}^2}.
\end{aligned} \tag{6}$$

The formulae for $\hat{\varphi}, \hat{\gamma}$ and $\hat{\omega}_r, \hat{\omega}_y$ are analogous.

3.3. Joint Estimation of the UAV Attitude

Jointly (5) and (6) give the estimation $\hat{\mathbf{V}}$ of the attitude parameters vector, namely:

$$\mathbf{V} = col(V_x, V_y, V_z, \omega_p, \omega_r, \omega_y).$$

This vector measured via OF for each pixel in each frame, which gives according to (2) the overdetermined system of linear equation for **V** entries.

Since all noises are uncorrelated the covariance matrix P for **V** is diagonal with entries (5), (6).

Integration of velocities **V** gives the estimation of the UAV position which used for estimation of matrices D_1, D_2 in (2).

3.4. Discussion of the Algorithm

Many uncertainties remain in the algorithm described above. They may be evaluated only in the real flight with real data. The principal uncertainty is the noise level in formulas (2) and its dependence on the altitude of flight. Experiments show that when approaching to the earth, the tiny details become very important and corrupt the OF estimation. In order to suppress these details, one needs to change resolution but to keep velocity estimation to be correct. In the real flight by averaging the pixel samples one can change resolution, but it is necessary to do it in coordination with the current altitude. Our experiments show that it is possible to change the resolution by comparison of the L–K OF estimation and theoretically calculated OF velocity through the current estimation of the altitude. That is a heuristic approach, and only experiments with real data may show either is it appropriate or not. The difference between two ways of the OF calculation is that L–K does not use the data from INS and even the estimation of the altitude obtained on previous steps with the aid of Kalman filtering. On the other hand, the OF calculation from theoretical model uses the current UAV attitude estimation. In the next section, we demonstrate an approach to the development of such an algorithm to control the scaling.

4. OF Estimation in the Real Flight

4.1. Examples of the OF Estimation in Approaching to the Earth

In our experiments, we used standard HERO 5 camera (GoPro Inc., San Mateo, CA, USA) and did not change its parameters during the flight. Since the aim of our research is the analysis of possible applications of video navigation in UAV, the camera is at the nose part of the light aeroplane. This aeroplane performed the series of approaches to the runway with accurate recording of the images and telemetric data. During the descent, the resolution of the camera sharpens, and it captures more and more tiny details which prevent good estimation of the image movement. Thereby, the estimation of the OF, which behaves more or less regularly at relatively high altitude, becomes absolutely chaotic when it approaches the earth surface. It is possible to observe this effect in the sequence of frames where the image motion velocities calculation done L–K estimates (see Figure 3), where the image is in almost natural scale with averaging 2 × 2. It is necessary to coordinate the camera resolution with the current altitude of flight to avoid this effect. Another effect is the increasing image velocity and therefore the additional blurring of the image, which needs the coordinated increasing of the frame rate. Of course, all such features usually do not present in standard cameras and their usage in navigation demands the developments of special observation tools. However, some issues may solve by onboard image processing. It means the artificial change of the resolution by averaging. The effect of the averaging is in Figures 3 and 4.

Figure 3. Optical flow estimated at landing with averaging 2 × 2 at the altitudes 195 m (**right**), 50 m (**centre**), and 20 m (**left**).

Figure 4. Optical flow estimated at landing with averaging 8 × 8 at the altitudes 195 m (**right**), 50 m (**centre**), and 20 m (**left**).

One can see the estimation of the OF becomes more and more regular with increasing the averaging scale, especially at low altitudes though it is impossible to use the high averaging scale for all altitudes of flight. At low altitudes, the blurring due to the image shift really needs the increasing of the frame rate. At high altitudes, where the averaging leads to the decreasing of the resolution, one cannot get a good quality of the OF estimation.

4.2. Switching of Scaling by Means of the Altitude Estimation

Another difficulty with the use of the OF is the change of scale and thereby the image shift rate caused by the change of UAV velocity and the flight altitude. Just as an example one can see the estimate of the OF based on Lucas-Kanade [37] algorithm obtained during the real flight at landing from 300 m to the runway (see Figures 5 and 6). As a result, the estimate of altitude, based on the OF estimated by the image of natural scale works more or less satisfactory up to the height 150 m and gives absolutely wrong values after (see Figure 7). Therefore, one needs to manipulate the zoom of the camera (change of focal length) in order to coordinate it with the characteristics of the OF. Meanwhile, one can use another approach, and to change the OES resolution by pixels averaging. The averaging of pixels made the OF calculation somewhat regular even at low altitudes (see Figure 8) with averaging by 16×16 at the altitude $\approx$30 m. One can compare it with Figure 6 It provides the reliable work of the algorithm up to the height of approximately 5–10 m (see Figures 9 and 10).

Figure 5. Optical flow estimated by Lucas-Kanade algorithm at the beginning of glissade, height $\approx$ 200 m, level of averaging 2×2. One can observe a rather regular nature of the optical flow which permits to estimate the flight parameters of the unmanned aerial vehicle with more or less high accuracy.

Figure 6. Optical flow estimated by Lucas-Kanade algorithm at the end of glissade, height $\approx$ 30 m, level of averaging 2×2. One can observe the very chaotic nature of the optical flow which prevents the estimation of the flight parameters.

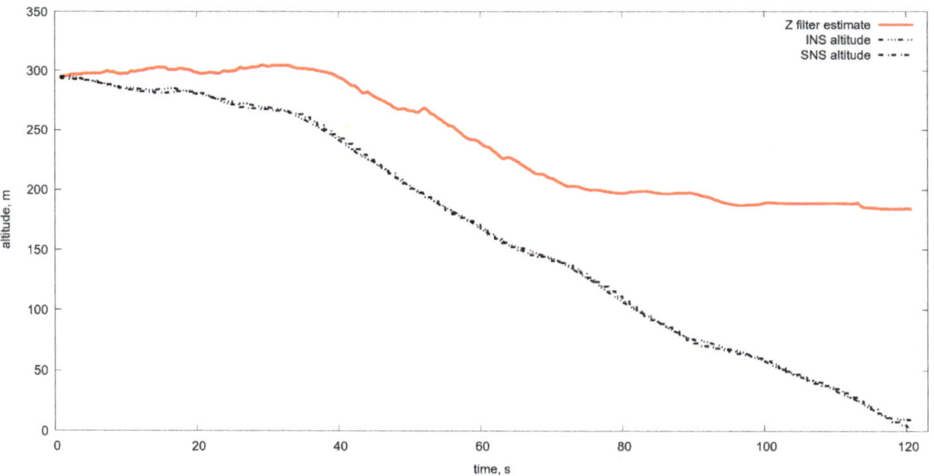

Figure 7. Estimated altitude with the aid of images registered by optoelectronic system without averaging in comparison with the real one given by inertial navigation system with satellite measurements. One can see that estimation on the basis of the Lucas-Kanade algorithm without averaging does not work in the entire range of the altitudes less than 300 m.

In recent work [46], we realised the idea of the scaling control using the current estimation of the altitude. Generally, it is difficult to evaluate the effect of this approach since the estimate of altitude without the use of other sensors is based just on the OF estimation which depends on the altitude itself.

Figure 8. Optical flow estimated by Lucas-Kanade algorithm at the end of glissade, height $\approx$ 30 m, level of averaging 16×16. By comparing with Figure 6 one can observe much more regular behaviour of the optical flow which gives a better estimation of the flight parameters.

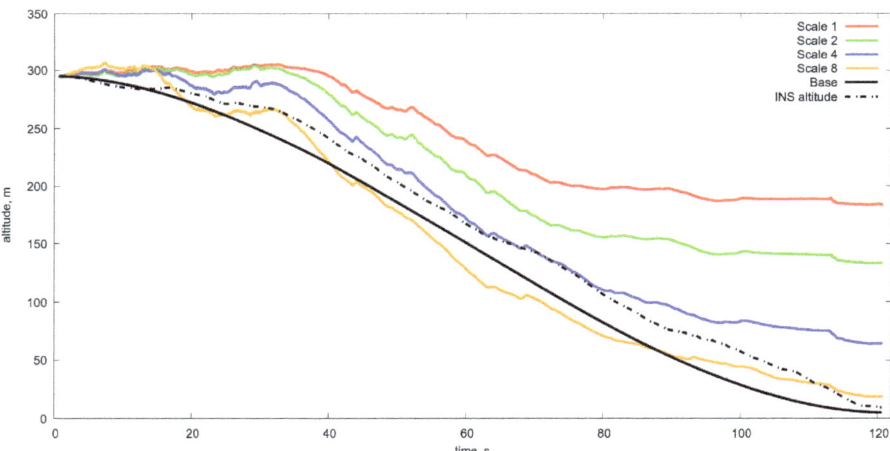

Figure 9. Estimated altitude in comparison with the real one obtained with a various level of resolution with averaging from 1×1 to 8×8. One can see that only the averaging 8×8 gives the acceptable accuracy of the altitude estimation in the range from 300 m to 50 m. Scale 1, 2, 4, 8 estimated altitude obtained by optical flow and Kalman filtering with different level of averaging, namely from 1 to 8. Base is the program motion. INS is the altitude from the inertial navigation system.

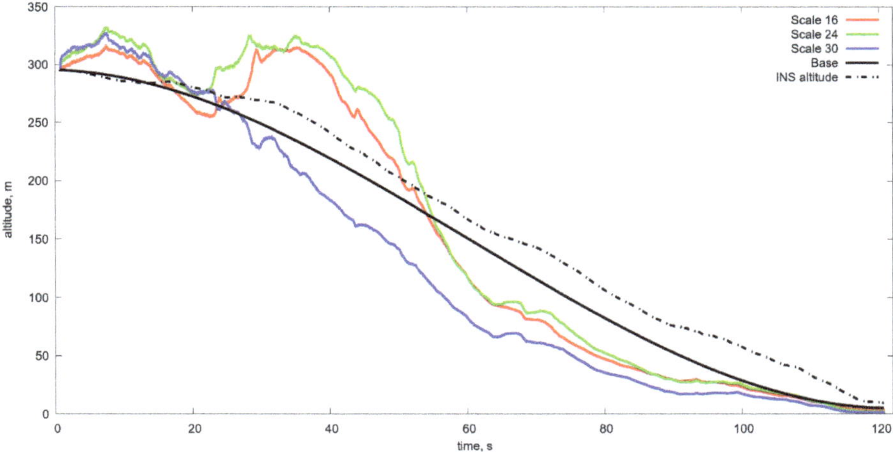

Figure 10. Estimated altitude in comparison with the real one obtained with a various level of resolution with averaging from 16 × 16 to 30 × 30. One can see that the averaging greater than 16 × 16 gives the acceptable accuracy of the altitude estimation in the range from 50 m to 5 m. Scale 16, 24, 30 estimated altitude obtained by optical flow and Kalman filtering with different level of averaging, namely from 16 to 30. Base is the program motion. INS is the altitude from the inertial navigation system.

The series of experiments based on the real data shows that it is possible to adapt the level of the averaging in order to expand the range of altitude estimation by the OF.

4.3. Test of the Algorithm Based on the Scale Switching via Current Altitude Estimation

As follows from the above consideration one needs to change the resolution of OES in coordination with the height of flight, and of course, the observed image is to be taken into account. Generally it is a nontrivial problem which needs consideration shortly, however now one can suggest the empirical algorithm for the averaging scale changing, based on the data which we have at hands. This empirical algorithm works using current height estimation as described in Table 1 [46].

Table 1. Change of the averaging level as a function of the height estimate.

Height	Scale	Height of Switch
300–50 m	4 × 4	150 m
150–80 m	8 × 8	80 m
80–50 m	16 × 16	50 m
50–5 m	24 × 24	5 m

One can see that the switching of scale improves the altitude estimation and permits to extend the algorithm operating range (see Figure 11).

Therefore, the possible solution of the OF usage is the changing of the averaging level during the descent. However, what kind of measurements could serve as a sensor for such switching of averaging? In the previous work [46], we examined the switching utilising the altitude estimation. However, at low altitudes all measurements and estimates become unreliable, so we tried to compare the OF velocities computed by L–K algorithm with the OF velocities calculated with the exact formulas and current estimation of the altitude. The parameters of the scaling change algorithm are in Table 1 [46]. The corresponding result of the altitude estimation is in Figure 11. One can observe satisfactory work of the OF in the altitude estimation. However, it is not a clean experiment since the estimation is done using the knowledge of the current altitude which is necessary to transform the OF data into real velocities.

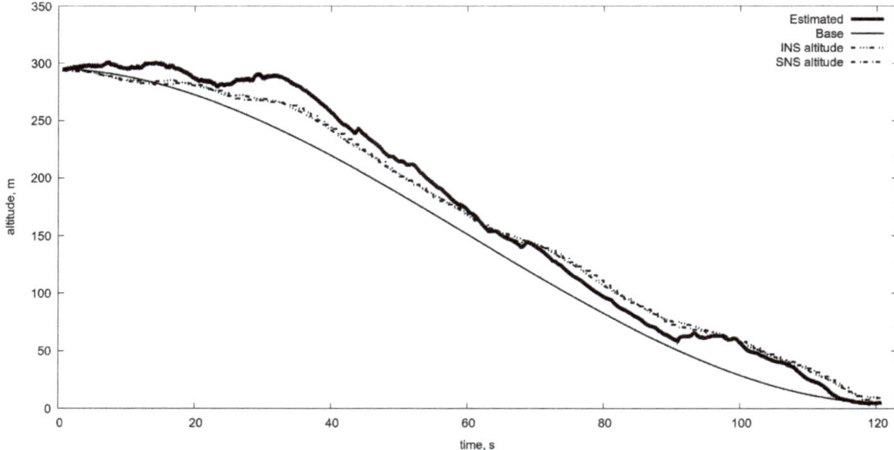

Figure 11. Estimated altitude in comparison with the real one obtained with the aid of averaging algorithm parameters described in Table 1. Estimated is the altitude obtained by optical flow and Kalman filtering. Base is the program motion. INS is the altitude from the inertial navigation system. SNS is the altitude from the satellite navigation system.

5. Scale Switching by the Comparison of Calculated and Estimated OF

5.1. Comparison of Calculated and Estimated OF

Here we use another sensor for the scaling switching that is the difference between the estimated and the calculated OF velocities.

The results of experiments related to the comparison of exact and L–K estimated velocities are in Figures 12 and 13. One can observe that the OF becomes useless when close to the earth, though the coordinated increasing of scaling level enlarges the range of reliable measurements.

Remark 3. *In all these pictures and below velocity is measured in $F * 10^{-3}$units/s, where F is the lens focal length in meters.*

In Figures 12 and 13, the exact OF value calculated using (1) with known values of the current flight parameters obtained from INS and corrected with the aid of Kalman filtering. The L–K parameters estimated on from the current video sequence registered by the onboard camera.

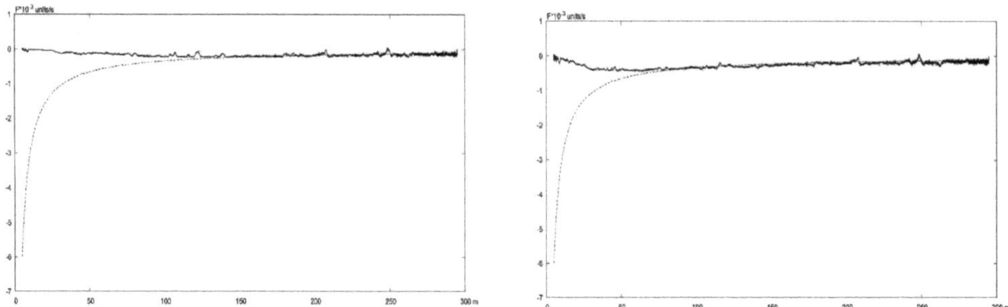

Figure 12. Optical flow velocities via Lucas-Kanade (bold) in comparison with exact values (dots) calculated for different level of scaling: **left**– 4 × 4, **right**–8 × 8.

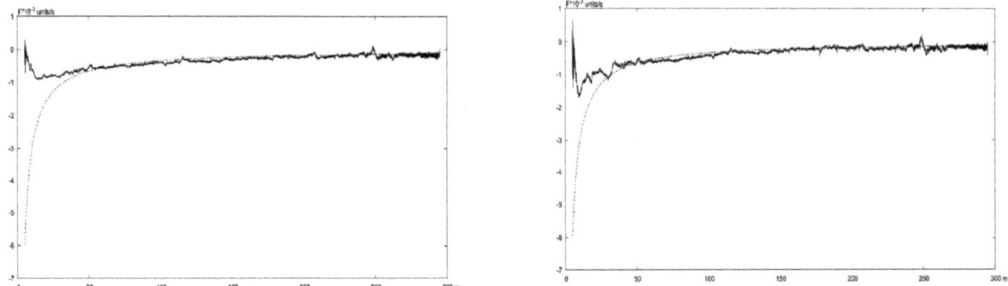

Figure 13. Optical flow velocities via Lucas-Kanade (bold) in comparison with exact values (dots) calculated for different level of scaling: **left**–16 × 16, **right**–30 × 30.

5.2. OF Estimation as a Sensor for Scaling Switch

As specified in the Introduction, we used the series of video sequences captured during the series of descents, where we estimated the OF velocities with the aid of L–K algorithm and compared it with theoretical values corresponding to the current aeroplane altitude. It is evident that the difference increases in approaching to the earth surface, meanwhile the OF permits to evaluate velocity at 25–30 m of height, though the difference becomes unacceptable below. It means that the noise in the velocities estimation via OF exposed to considerable perturbations which corrected by filtering utilising the dynamical model. The usage of Kalman filter based on the UAV dynamical model, observations of the OF, and the control accelerations permit to estimate the altitude in the range from 30–5 m. The picture in Figure 14 shows how to coordinate the scaling with the chosen threshold, which is equal to 0.2 unit/s, starting from averaging 4 × 4 at the beginning of glissade and increasing up to 30 ×30 when the aeroplane is close to the earth, approximately at the height of 5 m. Red vertical lines show the switch of scaling based on the difference of estimated and calculated OF velocities. Figure 15 shows the altitude estimate based only on the OF and filtering. Of course, it is just an experiment demonstrating the ability of the OF in such a complicated situation. In reality, it must combine with other altimeters, but if they fail to work accurately due to some reasons, the OF could serve as a reserved one.

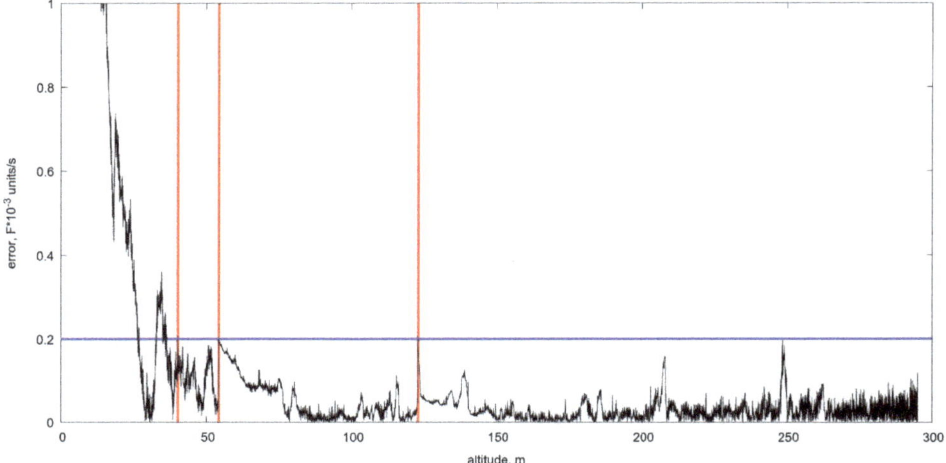

Figure 14. The difference (error) of the optical flow velocities via Lucas-Kanade and exact values, calculated at different altitudes from 300 m to 0 m. Vertical red lines show moments of switching of the averaging levels, that are: 4×4, 8×8, 16×16, and 30×30 from right to left.

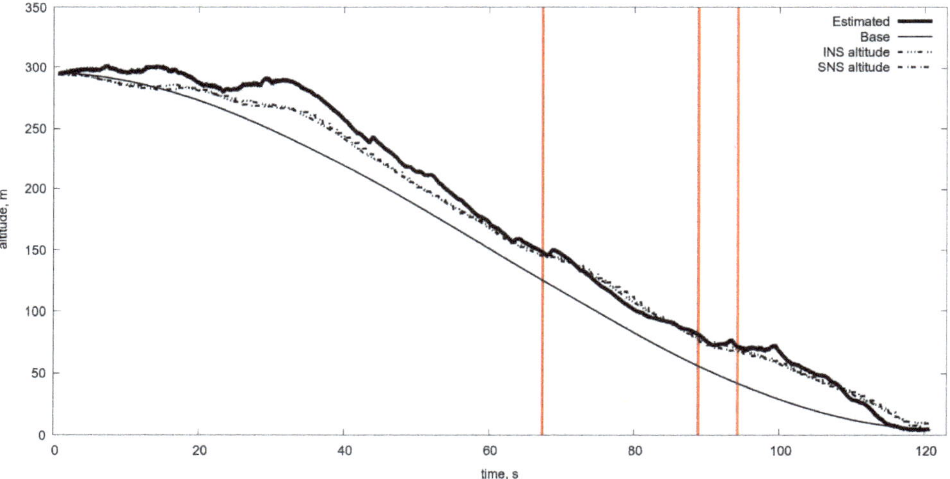

Figure 15. The altitude estimation via optical flow with switching averaging scale. The estimation of the altitude is made on the basis of standard Kalman filtering by fusion of data from the control system and the optical flow measurements. Of course, at low altitudes, less than 25–30 m, the OF measurements are corrupted by very high noise, but the usage of dynamical model and data from control system permit estimate reliably the altitude up to 5 m. Estimated is the altitude obtained by OF and Kalman filtering. Base is the program motion. INS is the altitude from the inertial navigation system. SNS is the altitude from the satellite navigation system.

Finally, the comparison of the altitude descent curves can be carried out. The program trajectory is the reference to the other ones. The results of the altitude tracking error are in Table 2. The rightmost column represents overall statistics for the new scale switching algorithm. The scale switch occurs when OF velocity error raises above level 0.2 in Figure 14. Conducted trial shows that the altitude estimate through the video sequence is quantitatively comparable with two other ones.

Table 2. Sample statistics of the altitude tracking error to the program trajectory.

	INS	SNS	Video
Mean, m	17.40	17.43	23.32
Median, m	19.39	19.75	23.85
Minimum, m	−3.36	−4.41	−1.32
Maximum, m	30.04	31.78	46.87
Standard deviation	9.33	10.02	11.60
Standard error	0.12	0.13	0.15
Final error, m	4.42	−1.21	−0.15

Remark 4. *In these experiments, the estimation with the aid of L–K algorithm done by standard software and needs the frames from the video sequence and the current scale size in pixels for calculation of the OF velocities. The OF exact value was calculated with the aid of* (2) *with values of flight parameters obtained from INS corrected by Kalman filter* (5),(6). *One can observe that the difference of exact and estimated OF velocities gives threshold points which are different in comparison with Table* 1.

6. Conclusions

In summary, the article presents the investigation of the OF usage as a sensor at the UAV landing. In general, it needs the adaptation of shooting parameters to the current altitude and velocity of flight that leads to the necessity of changing the camera characteristics such as virtual resolution and the frame rate. However, one can resolve the problem by using image processing such as change the resolution by controllable scaling. By using the difference of the OF velocity estimated by L–K algorithm and calculated via exact formulas and filtered by Kalman estimate as a sensor of the scale switching, it becomes possible to achieve the reliable altitude estimation up to 5 m. It shows how to provide the data fusion of the OF and filtering in the complicated problem of the UAV landing.

Author Contributions: Conceptualization, A.M., B.M.; methodology, K.S.; software, K.S., A.P.; validation A.P.

Funding: This research was partially funded by Russain Academy of Sciences and Foundation for Basic Research.

Acknowledgments: The authors would like to thank Russian Foundation for Basic Research for partial support of the research by grants 16-31-60049, and 16-08-01076.

Conflicts of Interest: The authors declare no conflict of interest.

Abbreviations

The following abbreviations are used in this manuscript:

UAV	Unmanned Aerial Vehicle
INS	Inertial navigation system
SNS	Satellite navigation system
OF	Optical flow
LHS	Left hand side
RHS	Right hand side
L–K	Lucas–Kanade (algorithm)
OES	Optoelectronic system

References

1. Aggarwal, J.; Nandhakumar, N. On the computation of motion from sequences of images—A review. *Proc. IEEE* **1988**, *76*, 917–935. doi:10.1109/5.5965. [CrossRef]
2. Konovalenko, I.; Kuznetsova, E.; Miller, A.; Miller, B.; Popov, A.; Shepelev, D.; Stepanyan, K. New Approaches to the Integration of Navigation Systems for Autonomous Unmanned Vehicles (UAV). *Sensors* **2018**, *18*, 3010. doi:10.3390/s18093010. [CrossRef] [PubMed]

3. Lowe, D.G. Object recognition from local scale-invariant features. In Proceedings of the Seventh IEEE International Conference on Computer Vision, Kerkyra, Greece, 20–27 September 1999; Volume 2, pp. 1150–1157. doi:10.1109/ICCV.1999.790410. [CrossRef]
4. Lowe, D.G. Distinctive Image Features from Scale-Invariant Keypoints. *Int. J. Comput. Vis.* **2004**, *60*, 91–110. doi:10.1023/B:VISI.0000029664.99615.94. [CrossRef]
5. Karpenko, S.; Konovalenko, I.; Miller, A.; Miller, B.; Nikolaev, D. UAV Control on the Basis of 3D Landmark Bearing-Only Observations. *Sensors* **2015**, *15*, 29802–29820. doi:10.3390/s151229768. [CrossRef]
6. Kunina, I.; Terekhin, A.; Khanipov, T.; Kuznetsova, E.; Nikolaev, D. Aerial image geolocalization by matching its line structure with route map. In Proceedings of the Ninth International Conference on Machine Vision (ICMV 2016), Nice, France, 17 March 2017; Volume 10341. doi:10.1117/12.2268612.
7. Savchik, A.V.; Sablina, V.A. Finding the correspondence between closed curves under projective distortions. *Sens. Syst.* **2018**, *32*, 60–66. doi:10.7868/S0235009218010092. (In Russian) [CrossRef]
8. Kunina, I.; Teplyakov, L.; Gladkov, A.; Khanipov, T.; Nikolaev, D. Aerial images visual localization on a vector map using color-texture segmentation. In Proceedings of the Tenth International Conference on Machine Vision (ICMV 2017), Vienna, Austria, 13–15 November 2017; Volume 10696. doi:10.1117/12.2310138.
9. Teplyakov, L.; Kunina, I.A.; Gladkov, A. Visual localisation of aerial images on vector map using colour-texture segmentation. *Sens. Syst.* **2018**, *32*, 26–34. doi:10.7868/S0235009218010055. (In Russian) [CrossRef]
10. Ovchinkin, A.; Ershov, E. The algorithm of epipole position estimation under pure camera translation. *Sens. Syst.* **2018**, *32*, 42–49. doi:10.7868/S0235009218010079. (In Russian) [CrossRef]
11. Cesetti, A.; Frontoni, E.; Mancini, A.; Zingaretti, P.; Longhi, S. A Vision-Based Guidance System for UAV Navigation and Safe Landing using Natural Landmarks. *J. Intell. Robot. Syst.* **2010**, *57*, 233. [CrossRef]
12. Sebesta, K.; Baillieul, J. Animal-inspired agile flight using optical flow sensing. In Proceedings of the 2012 IEEE 51st Conference on Decision and Control (CDC), Maui, HI, USA, 10–13 December 2012; pp. 3727–3734. doi:10.1109/CDC.2012.6426163. [CrossRef]
13. Miller, B.M.; Fedchenko, G.I.; Morskova, M.N. Computation of the image motion shift at panoramic photography. *Izvestia Vuzov. Geod. Aerophotogr.* **1984**, *4*, 81–89. (In Russian)
14. Miller, B.M.; Fedchenko, G.I. Effect of the attitude errors on image motion shift at photography from moving aircraft. *Izvestia Vuzov. Geod. Aerophotogr.* **1984**, *5*, 75–80. (In Russian)
15. Miller, B.; Rubinovich, E. Image motion compensation at charge-coupled device photographing in delay-integration mode. *Autom. Remote Control* **2007**, *68*, 564–571. doi:10.1134/S0005117907030162. [CrossRef]
16. Kistlerov, V.; Kitsul, P.; Miller, B. Computer-aided design of the optical devices control systems based on the language of algebraic computations FLAC. *Math. Comput. Simul.* **1991**, *33*, 303–307. doi:10.1016/0378-4754(91)90109-G. [CrossRef]
17. Johnson, A.; Montgomery, J.; Matthies, L. Vision Guided Landing of an Autonomous Helicopter in Hazardous Terrain. In Proceedings of the 2005 IEEE International Conference on Robotics and Automation, Barcelona, Spain, 18–22 April 2005; pp. 3966–3971. doi:10.1109/ROBOT.2005.1570727. [CrossRef]
18. Merz, T.; Duranti, S.; Conte, G. Autonomous Landing of an Unmanned Helicopter based on Vision and Inertial Sensing. In *Experimental Robotics IX;* Ang, M.H., Khatib, O., Eds.; Springer: Berlin/Heidelberg, Germany, 2006; pp. 343–352, ISBN 978-3-540-28816-9, doi:10.1007/11552246_33.
19. Templeton, T.; Shim, D.H.; Geyer, C.; Sastry, S.S. Autonomous Vision-based Landing and Terrain Mapping Using an MPC-controlled Unmanned Rotorcraft. In Proceedings of the 2007 IEEE International Conference on Robotics and Automation, Roma, Italy, 10–14 April 2007; pp. 1349–1356. doi:10.1109/ROBOT.2007.363172. [CrossRef]
20. Serra, P.; Le Bras, F.; Hamel, T.; Silvestre, C.; Cunha, R. Nonlinear IBVS controller for the flare maneuver of fixed-wing aircraft using optical flow. In Proceedings of the 49th IEEE Conference on Decision and Control, Atlanta, GA, USA, 15–17 December 2010; pp. 1656–1661. doi:10.1109/CDC.2010.5717829. [CrossRef]
21. McCarthy, C.; Barnes, N. A Unified Strategy for Landing and Docking Using Spherical Flow Divergence. *IEEE Trans. Pattern Anal. Mach. Intell.* **2012**, *34*, 1024–1031. doi:10.1109/TPAMI.2012.27. [CrossRef] [PubMed]
22. Serra, P.; Cunha, R.; Silvestre, C.; Hamel, T. Visual servo aircraft control for tracking parallel curves. In Proceedings of the 2012 IEEE 51st Conference on Decision and Control (CDC), Maui, HI, USA, 10–13 December 2012; pp. 1148–1153. doi:10.1109/CDC.2012.6426240. [CrossRef]

23. Liau, Y.S.; Zhang, Q.; Li, Y.; Ge, S.S. Non-metric navigation for mobile robot using optical flow. In Proceedings of the 2012 IEEE/RSJ International Conference on Intelligent Robots and Systems (IROS), Vilamoura, Portugal, 7–12 October 2012; pp. 4953–4958. doi:10.1109/IROS.2012.6386221. [CrossRef]
24. de Croon, G.; Ho, H.; Wagter, C.; van Kampen, E.; Remes, B.; Chu, Q. Optic-Flow Based Slope Estimation for Autonomous Landing. *Int. J. Micro Air Veh.* **2013**, *5*, 287–297. doi:10.1260/1756-8293.5.4.287. [CrossRef]
25. De Croon, G.C.H.E. Monocular distance estimation with optical flow maneuvers and efference copies: a stability-based strategy. *Bioinspir. Biomim.* **2016**, *11*, 016004. doi:10.1088/1748-3190/11/1/016004. [CrossRef]
26. Rosa, L.; Hamel, T.; Mahony, R.; Samson, C. Optical-Flow Based Strategies for Landing VTOL UAVs in Cluttered Environments. *IFAC Proc. Vol.* **2014**, *47*, 3176–3183. doi:10.3182/20140824-6-ZA-1003.01616. [CrossRef]
27. Horn, B.K.P.; Schunck, B.G. Determining Optical Flow. *Artif. Intell.* **1981**, *17*, 185–203.[CrossRef]
28. Herissé, B.; Hamel, T.; Mahony, R.; Russotto, F. Landing a VTOL Unmanned Aerial Vehicle on a Moving Platform Using Optical Flow. *IEEE Trans. Robot.* **2012**, *28*, 77–89. doi:10.1109/TRO.2011.2163435. [CrossRef]
29. Pijnacker Hordijk, B.J.; Scheper, K.Y.W.; de Croon, G.C.H.E. Vertical landing for micro air vehicles using event-based optical flow. *J. Field Robot.* **2018**, *35*, 69–90. doi:10.1002/rob.21764. [CrossRef]
30. Ho, H.W.; de Croon, G.C.H.E.; van Kampen, E.; Chu, Q.P.; Mulder, M. Adaptive Control Strategy for Constant Optical Flow Divergence Landing. *arXiv* **2016**, arXiv:1609.06767.
31. Ho, H.W.; de Croon, G.C.H.E.; van Kampen, E.; Chu, Q.P.; Mulder, M. Adaptive Gain Control Strategy for Constant Optical Flow Divergence Landing. *IEEE Trans. Robot.* **2018**, *34*, 508–516. [CrossRef]
32. Fantoni, I.; Sanahuja, G. Optic-Flow Based Control and Navigation of Mini Aerial Vehicles. *J. Aerosp. Lab* **2014**. doi:10.12762/2014.AL08-03. [CrossRef]
33. Chahl, J.; Rosser, k.; Mizutani, A. Vertically displaced optical flow sensors to control the landing of a UAV. *Proc. SPIE* **2011**, *7975*, 797518. doi:10.1117/12.880715. [CrossRef]
34. Rosser, K.; Fatiaki, A.; Ellis, A.; Mizutani, A.; Chahl, J. Micro-autopilot for research and development. In *16th Australian International Aerospace Congress (AIAC16)*; Engineers Australia: Melbourne, Australia, 2015; pp. 173–183.
35. Popov, A.; Miller, A.; Stepanyan, K.; Miller, B. Modelling of the unmanned aerial vehicle navigation on the basis of two height-shifted onboard cameras. *Sens. Syst.* **2018**, *32*, 19–25. doi:10.7868/S0235009218010043. (In Russian) [CrossRef]
36. Miller, A.; Miller, B. Stochastic control of light UAV at landing with the aid of bearing-only observations. In Proceedings of the Eighth International Conference on Machine Vision (ICMV 2015), Barcelona, Spain, 19–21 November 2015; Volume 9875. doi:10.1117/12.2228544.
37. Lucas, B.D.; Kanade, T. An Iterative Image Registration Technique with an Application to Stereo Vision. In *Proceedings of the 7th International Joint Conference on Artificial Intelligence*; IJCAI: Vancouver, BC, Canada, 1981; Volume 2, pp. 674–679.
38. Chao, H.; Gu, Y.; Gross, J.; Guo, G.; Fravolini, M.L.; Napolitano, M.R. A comparative study of optical flow and traditional sensors in UAV navigation. In Proceedings of the 2013 American Control Conference, Washington, DC, USA, 17–19 June 2013; pp. 3858–3863. doi:10.1109/ACC.2013.6580428. [CrossRef]
39. Popov, A.; Miller, A.; Miller, B.; Stepanyan, K. Optical flow and inertial navigation system fusion in the UAV navigation. In Proceedings of the Unmanned/Unattended Sensors and Sensor Networks XII, Edinburgh, UK, 26–29 September 2016; Volume 9986. doi:10.1117/12.2241204.
40. Farnebäck, G. Fast and accurate motion estimation using orientation tensors and parametric motion models. In Proceedings of the 15th International Conference on Pattern Recognition, IPCR-2000, Barcelona, Spain, 3–8 September 2000; Volume 1, pp. 135–139. doi:10.1109/ICPR.2000.905291. [CrossRef]
41. Farnebäck, G. Orientation estimation based on weighted projection onto quadratic polynomials. In Proceedings of the 5th International Fall Workshop, Vision, Modeling, and Visualization, Saarbrücken, Germany, 22–24 November 2000; Max-Planck-Institut für Informatik: Saarbrücken, Germany, 2000; pp. 89–96.
42. Farnebäck, G. Two-Frame Motion Estimation Based on Polynomial Expansion. In Proceedings of the 13th Scandinavian Conference on Image Analysis, SCIA 2003, Halmstad, Sweden, 29 June–2 July 2003; pp. 363–370. doi:10.1007/3-540-45103-X_50. [CrossRef]
43. Popov, A.; Miller, A.; Miller, B.; Stepanyan, K. Estimation of velocities via optical flow. In Proceedings of the 2016 International Conference on Robotics and Machine Vision, Moscow, Russia, 14–16 September 2016; Volume 10253. doi:10.1117/12.2266365.

44. Popov, A.; Miller, B.; Miller, A.; Stepanyan, K. Optical Flow as a Navigation Means for UAVs with Opto-electronic Cameras. In Proceedings of the 56th Israel Annual Conference on Aerospace Sciences, Tel-Aviv and Haifa, Israel, 9–10 March 2016.
45. Miller, B.M.; Stepanyan, K.V.; Popov, A.K.; Miller, A.B. UAV navigation based on videosequences captured by the onboard video camera. *Autom. Remote Control* **2017**, *78*, 2211–2221. doi:10.1134/S0005117917120098. [CrossRef]
46. Miller, A.; Miller, B.; Popov, A.; Stepanyan, K. Optical Flow as a navigation means for UAV. In Proceedings of the 2018 Australian New Zealand Control Conference (ANZCC), Melbourne, Australia, 7–8 December 2018; pp. 302–307. doi:10.1109/ANZCC.2018.8606563. [CrossRef]

Article

UAV Flight and Landing Guidance System for Emergency Situations †

Joon Yeop Lee ‡, Albert Y. Chung ‡, Hooyeop Shim ‡, Changhwan Joe ‡, Seongjoon Park and Hwangnam Kim *

School of Electrical Engineering, Korea University, Seoul 136-713, Korea; charon7@korea.ac.kr (J.Y.L.); aychung@korea.ac.kr (A.Y.C.); hooyp@korea.ac.kr (H.S.); chjoe01@korea.ac.kr (C.J.); psj900918@korea.ac.kr (S.P.);

* Correspondence: hnkim@korea.ac.kr; Tel.: +82-2-3290-4821

† This mansucript is extension version of the conference paper: Chung, A.Y.; Lee, J.Y.; Kim, H. Autonomous mission completion system for disconnected delivery drones in urban area. In Proceedings of the 2017 IEEE 7th Conference on Robotics and Biomimetics (ROBIO), Macau, China, 5–8 December 2017.

‡ These authors contributed equally to this work.

Received: 30 August 2019; Accepted: 14 October 2019; Published: 15 October 2019

Abstract: Unmanned aerial vehicles (UAVs) with high mobility can perform various roles such as delivering goods, collecting information, recording videos and more. However, there are many elements in the city that disturb the flight of the UAVs, such as various obstacles and urban canyons which can cause a multi-path effect of GPS signals, which degrades the accuracy of GPS-based localization. In order to empower the safety of the UAVs flying in urban areas, UAVs should be guided to a safe area even in a GPS-denied or network-disconnected environment. Also, UAVs must be able to avoid obstacles while landing in an urban area. For this purpose, we present the UAV detour system for operating UAV in an urban area. The UAV detour system includes a highly reliable laser guidance system to guide the UAVs to a point where they can land, and optical flow magnitude map to avoid obstacles for a safe landing.

Keywords: UAV; laser guidance; emergency landing; particle filter; optical flow

---

1. Introduction

The unmanned aerial vehicle (UAV) has a wide operating radius and can be equipped with diverse sensing and actuation devices [1]. Through these advantages, UAVs can perform various roles and missions. Recently, a variety of studies have been carried out to apply UAV to real life, such as collecting vehicle traffic information [2,3], surveillance systems for cities [4–6], constructing network infrastructure [7,8], and delivering products [9]. In particular, many companies are investing in the field of delivery systems using UAV. Amazon Prime Air patented a UAV delivery system [10]; DHL tested its parcel copter in Germany; and various delivery companies such as UPS, USPS, Swiss Post, SF Express and Ali Baba started researches related to UAV delivery system. These delivery systems often deliver small items ordered by people, so UAVs often operate in densely populated cities. This is possible because UAV can move freely in three dimensions, so it can deliver goods to destinations in high-rise buildings or apartments.

However, the city has many elements that interfere with the UAVs' flight. Buildings make it difficult to obtain a line of control (LOC) for UAV control and degrade the accuracy of GPS-based positioning. There are studies that the use of UAV is difficult due to the low stability of GPS in urban areas [11,12]. Without GPS, the autonomous waypoint flight of the UAV is not possible. Also, if wind drift occurs, UAV will not be able to calibrate its position. UAV can hardly plan its flight for missions when GPS service is degraded. In addition, RF interference and multi-path effects can interfere with the flight of UAV in the urban area [13]. In this case, the signal for manipulating the UAV is disturbed, so that the operator cannot maneuver the UAV remotely. UAVs that are thus disturbed during their flight are more likely to cause accidents in urban areas. Therefore, we need a flight and landing guidance system for public safety against UAVs that lose connection and position (and even malfunctioned UAVs).

For these purposes, we propose the UAV detour system that satisfies the following conditions.

- UAV should be allowed to continue its mission in situations where GPS is not available or the network is disconnected;
- The system that guides the UAVs' flight should be able to overcome multi-path fading and interference that may occur in urban areas;
- When the UAV is landing, the UAV must be able to land safely while avoiding obstacles.

The UAV detour system that satisfies the aforementioned conditions has two subsystems. The first is the flight guidance system, and the second is the safe landing system. The flight guidance system can guide UAV to desired landing points using laser devices that are free of multi-path effects and other interference common in radio waves. When a UAV with flight guidance system detects a laser, the UAV moves in the direction of the incoming laser. Also, the flight guidance system is based on a particle filter. The other system, safe landing system provides obstacle avoidance for UAV. The safe landing system was developed based on our previous work [14]. In addition, it has been improved for the use of optical flow magnitude maps, which make it more stable on a low-power computing board in UAV than in previous work. When the UAV is flying or landing, the UAV can extract the optical flow from the images taken by the mounted camera. The safe landing system analyzes the optical flow magnitude map to identify obstacles and maneuver the UAV away from obstacles. Overall, the safe landing system allows UAV to safely land without collisions even when GPS or network assistance is unstable. The UAV detour system was installed on a real UAV and tested in an actual environment to see if it could steer UAV and land it safely in urban area [15–17].

The rest of the paper is organized as follows. Section 2 reviews related work and background theory. Detailed explanation of the UAV detour system is described in Section 3. Section 4 describes the detailed implementation of the UAV detour system, and Section 5 presents the experiments and evaluation of the proposed system. Section 6 describes the limitations of UAV detour system and plans to improve it. Finally, Section 7 concludes the paper.

2. Preliminaries

2.1. Related Work

There have been a number of previous studies related to the unmanned aerial vehicle (UAV) detour system. Descriptions of related techniques are described below.

2.1.1. Laser Guidance Systems

Lasers have the advantage of being able to distinguish easily from other light sources. Focusing on the advantages of lasers, there have been studies to use lasers in the guidance of UAV.

Vadim et al. proposed a system for delivering GPS-based flight information to the UAVs via a laser [18]. The information transmitted through the laser includes both flight path and landing location information. However, the authors did not implement the system on real UAVs. The system proposed by Shaqura et al. aims to detect the laser point by applying image processing on images taken by the camera mounted on the UAV. The laser guides UAV to landing point [19]. However, since these works provide flight information based on the UAVs' GPS, it is difficult to expect successful operation in an environment where the GPS signal is degraded due to the environment.

In an emergency situation, there was an attempt to guide a UAV based on a laser without relying on GPS. The system proposed by Jang et al. used a laser to hold the current position of UAVs in case of deteriorated GPS signal [20]. This paper assumes that the UAVs are flying in an electric wave-shaded region, preventing the drift of the UAVs and holding the position through a number of photo-resistors. The system presented in this paper does not guide over long-distance, but shows that flight stability can be improved by using laser even in the environment without GPS.

Most of the proposed laser guidance systems were GPS dependent or had a short guidance range. However, our proposed flight guidance system was designed to take advantage of the long-range of lasers, assuming flight in a GPS-denied environment. The proposed system can reliably recognize the laser and determine the flight direction through the particle filter, which has been modified to be suitable for the real UAV flight. Particle filter had proven to be useful in state estimation problems such as simultaneous localization and mapping (SLAM) of robot research area [21]. In addition, Hightower et al. implemented a multi-sensor localization system based on particle filter and presented performance comparison showing that it is practical to run a particle filter on handheld devices [22]. Research on particle filters has continued over the last few decades and has been applied to address non-Gaussian distributions in various fields.

2.1.2. Obstacle Avoidance Studies

In order for a UAV to fly or land in an urban area, the UAVs must be able to avoid obstacles. There have been studies to implement obstacle avoidance based on optical flow. Lorenzo et al. proposed an optical flow-based landing system [23]. However, the system proved its performance only by simulation. Souhila et al. applied obstacle avoidance based on optical flow to robots [24]. The algorithm proposed in this study determines that there is an obstacle if an optical flow value imbalance is detected while the robot is moving. Similarly, Yoo et al. applied an algorithm to the UAV navigation system to avoid obstacles based on the imbalance of optical flow values [25]. However, these studies had limited movement because they could only maneuver the robot or UAV in the left and right directions. Our proposed safe landing system can avoid the obstacles in all directions based on optical flow, and can effectively cope with the complex environment and various obstacles in the urban area. Miller et al. estimated reliable altitude using the difference between the optical flow velocity and calculated via exact formulas [26]. Estimation of altitude is tested from video sequences obtained in flights, but altitude is not calculated during the actual flight. Herisse et al. presented a nonlinear controller for the vertical landing of a VTOL UAV using the measurement of average optical flow with the IMU data [27]. Since VTOL is assumed to be equipped with a camera and IMU, there is a difference between our research in the case of using only cameras.

In addition, there have been studies to avoid obstacles with various techniques. Mori et al. showed a technique for determining obstacles, using the SURF [28]. In the paper, the authors proposed a system that determines obstacles when a large image difference is detected by comparing each image frame by frame. The author-proposed system used the phenomenon in which objects nearer to the camera become larger as the camera moves towards the objects. Hrabar et al. used stereo vision to identify obstacles [29]. The authors created a 3D map of the surrounding terrain using the images shown around them. This assisted UAVs to avoid obstacles. Ferrick et al. used LIDAR [30]. LIDAR utilizes laser to estimate the distance to nearby objects in real-time and detect the approaching object. In addition, as shown in Kendoul's survey paper [31], there are various techniques that allow UAVs to avoid obstacles through autopilot in the event of GPS or communication failures.

2.1.3. Autonomous UAV Landing Systems

There have also been studies that use image processing for autonomous landing. In order for the UAV to recognize the landing point, several studies have proposed to use landing markers at landing points.

Bi et al. proposed a system for calculating the relative position of a UAV with a marker and landing a UAV toward the marker's position [32]. Lange et al. proposed a system that controls the speed of the UAVs by estimating the relative altitude and position of the marker [33]. Venugopalan et al. placed a landing marker on the autonomous marine vehicle to land the UAVs on it [34].

Using the marker makes it easy for the UAVs to identify the landing point and allows more precise landing. In order to land the UAV in an area where the markers are not ready, some of the studies have used optical flow to identify the landing area. Cesetti et al. proposed a system for identifying safe landing sites using optical flow [35]. In this paper, the depth map of the ground is drawn using optical flow and feature matching. Then, Cesetti et al. analyzed the flatness of the depth map to determine the safe landing area. Similar to [35], Edenbak et al. proposed a system for identifying safe landing points [36]. This system identifies the structure of the ground via optical flow.

Our proposed UAV detour system does not require any special markers to land, and automatically determine the flat ground that can be landed. In addition, the proposed system includes the flight guidance system that can guide UAVs to landing points even over long distances. The proposed system was modified to be suitable for real UAV and proved its performance through actual experiments.

2.2. Background

This section briefly introduces the particle filter, the core theory of our proposed flight guidance system, and the optical flow method used in the safe landing systems.

2.2.1. Particle Filter Theory

The sequential Monte Carlo method, also known as particle filtering or bootstrap filtering, is a technique for implementing a recursive Bayesian filter by Monte Carlo simulations. The key idea is to represent the required posterior density function by a set of random samples with associated weights and to compute estimates based on these samples and weights. As the number of samples becomes very large, this Monte Carlo characterization becomes an equivalent representation to the usual functional description of the posterior pdf [37].

2.2.2. Optical Flow Method

Optical flow is defined as the pattern of apparent motion of objects in a visual scene caused by the relative motion between an observer and a scene. In addition, Lucas–Kanade algorithm is one of the simple optical flow techniques which can provide an estimate of the movement of interesting features in successive images of a scene. Lucas–Kanade algorithm bases on two assumptions that two images between frames are separated by a small time increment and movement of objects is not significant. Moreover that the spatially adjacent pixels tend to belong in same object and have identical movements, constant brightness [38].

3. Methods

3.1. System Overview

As shown in Figure 1, the UAV detour system has two major subsystems. One of the subsystems is the flight guidance system, that uses the laser to maneuver the UAV in the desired direction. The flight guidance system detects the laser with a light sensor mounted on the UAV. However, the laser that is detected by the light sensor is not linear. To reliably detect the laser and determine the direction the UAV will fly, the flight guidance system uses particle filter theory to estimate the direction based on the direction the light is coming from.

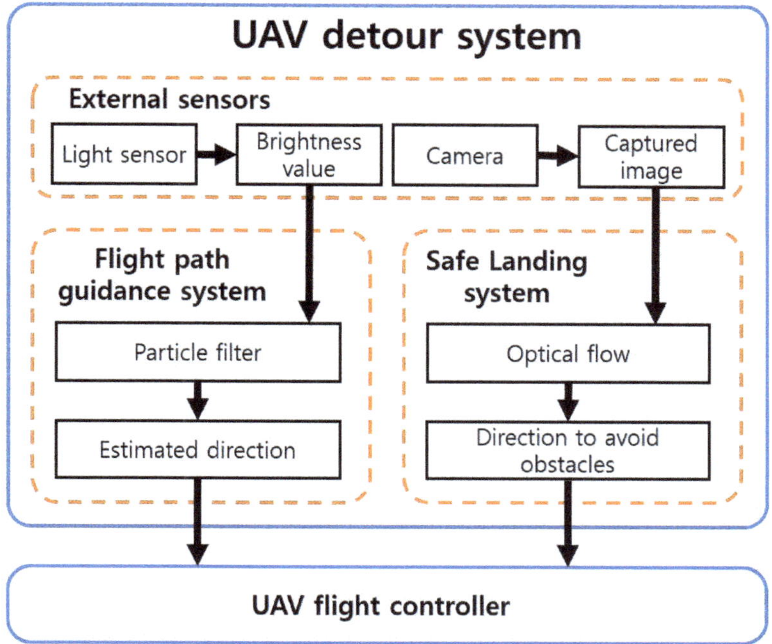

Figure 1. Overview of the unmanned aerial vehicle (UAV) detour system.

The other subsystem is the safe landing system that can identify obstacles on the landing area with an optical flow magnitude map. The safe landing system obtains images through a camera mounted on the UAV. By analyzing the image, optical flow information can be obtained. The safe landing system analyzes the magnitude of the obtained optical flow to determine the obstacles below the UAV. When the UAV determines that there are obstacles at the landing area, the UAV lands with avoiding obstacles. After operations of subsystems, the maneuver commands of the UAV generated from the two subsystems are transferred to the UAV controller so that the UAV detour system can be applied to the movement of the actual UAV.

3.2. Flight Guidance System for UAV

In order to increase the accuracy of flight direction through laser detection, a modified particle filter was used in the flight guidance system. A variety of adjustments have been applied to the modified particle filter, considering that it operates on a UAV.

3.2.1. Particle Filter Based Flight Guidance

In the flight guidance system, when the measured brightness value exceeds the threshold at multiple sensors, the sensor with the brightest light (laser) is identified and finally controls the next flight direction of the UAV. If the UAV detects the incoming laser, UAV computes the bearing angle of the laser and moves at a given constant speed to the detected direction. However, moving the UAV and detecting the direction of the laser during moving cannot be linearly described, we used particle filter to improve the accuracy of laser identification. The particle filter recursively estimates the sequence of system states (approximated direction to the source of the light) from the sensor measurements. Filtering via sequential importance sampling (SIS) consists of recursive propagation of importance weights $w_{\mathrm{k}}^{\mathrm{i}}$ and support points $l_{\mathrm{k}}^{\mathrm{i}}$ as each measurement is received sequentially [37].

Figure 2 shows a brief overview of the particle-applied flight guidance system. The blue circles represent the weight of each particle. As shown in Figure 2a, the particles are uniformly distributed in all directions. Then, as shown in Figure 2b, the weight of the particle in the direction in which the laser is detected is getting larger. As can be seen in Figure 2c, more particles are placed where there is a larger weight value. At this time, the flight guidance system prepare for situations in which the direction of the light changes by placing a small number of particles on the sensors in different directions. Finally, the UAV can fly to the direction of the most particles.

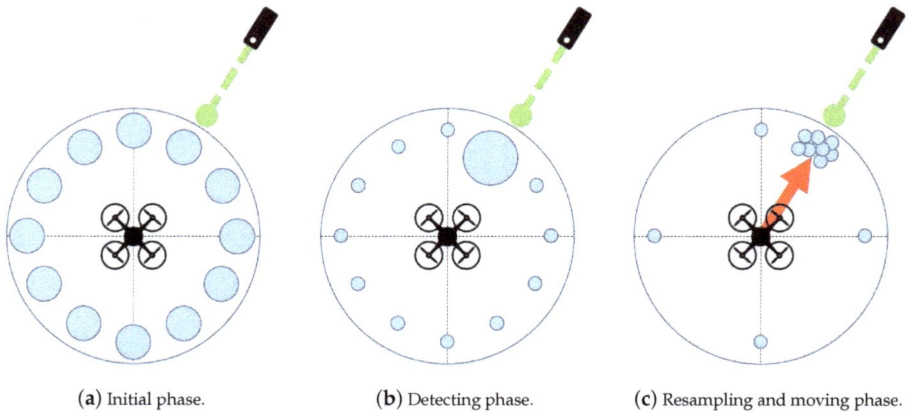

(**a**) Initial phase. (**b**) Detecting phase. (**c**) Resampling and moving phase.

Figure 2. Examples of particle-applied flight guidance algorithm.

To guide the UAV exactly in the intended direction, they must maintain information about the direction along which the laser will guide. So, the target direction can be described by the state vector l_k as follows:

$$l_{\mathrm{k}} = [x_{\mathrm{k}}, y_{\mathrm{k}}, \theta_{\mathrm{k}}]^{\mathrm{T}},$$

where k is the discrete-time, T denote transpose, x_k and y_k denote Cartesian coordinates at $time = k$ which can be calculated from the bearing angle θ_{k} pointing to the laser light source.

In order to successfully guide the UAV using laser light, the operator must continually irradiate the laser in one direction unless an exceptional situation occurs. It can be expected that the previous state values will be changed slightly. Thus, the state transition equation for the flight guidance system can be written as

$$l_{(\mathrm{k}+1)} = F_{\mathrm{k}} l_{\mathrm{k}} + v_{\mathrm{k}},$$

where v_{k} is the white Gaussian process noise and F_{k} is the state transition matrix can be written as

$$F_k = \begin{bmatrix} \frac{cos(\pi/2-(\theta_{\mathrm{k}}+v_{\mathrm{k}}))}{x_{\mathrm{k}}} & 0 & 0 \\ 0 & \frac{sin(\pi/2-(\theta_{\mathrm{k}}+v_{\mathrm{k}}))}{y_{\mathrm{k}}} & 0 \\ 0 & 0 & \frac{\theta_{\mathrm{k}}+v_{\mathrm{k}}}{\theta_{\mathrm{k}}} \end{bmatrix}.$$

Based on this matrix, we update the current state's bearing angle θ_{k} by adding the process noise v_{k} to the current value and then calculate the next state's coordinates $x_{(\mathrm{k}+1)}$ and $y_{(\mathrm{k}+1)}$. On the one hand, the measurement equation can also be defined as

$$z_{\mathrm{k}} = h(l_{\mathrm{k}}) + e_{\mathrm{k}},$$

where z_{k} is the measurement and e_{k} is the white Gaussian measurement noise. The nonlinear measurement function $h(l_{\mathrm{k}})$ can be denoted as

$$h(l_{\mathrm{k}}) = \theta_{\mathrm{k}} = tan^{-1}\frac{y_{\mathrm{k}}}{x_{\mathrm{k}}},$$

when sensors detect the light incidence, they convert the direction of the light into polar coordinates in two dimension. Therefore, we define $h(l_k)$ as a part of measurement equations with formulas changing Cartesian coordinates into polar coordinates and assume that the radius is always 1 when transforming the coordinate system. Note that the flight guidance system was developed to maneuver UAV in two dimensions but it can be extended to maneuver UAV in three dimensions for UAV flight on building rooftops or in complex terrain.

3.2.2. Resampling Method of Particle Filter

There are numerous versions of the resampling method in the field of a particle filter. Thus, it is important to choose an efficient method because each resampling method has different complexity depending on the operational algorithm. Considering the simplicity of implementation and the efficiency of the algorithm, systematic resampling was applied to the flight guidance system [37]. The most serious problem is that some particles with large weights are inevitably selected during the resampling and sample impoverishment occurs as the diversity of the sample decreases. As mentioned earlier, the flight guidance system must accurately estimate the direction that the UAV should travel. Therefore, under the assumption that the incoming direction of light is constant, it may be helpful to have less sample diversity. However, if the direction of the light guiding the UAV is suddenly changed, the exact direction cannot be estimated due to the effect of the sample impoverishment which becomes too severe. To solve sample impoverishment, the flight guidance system maintains a minimum level of sample diversity and combines the sample dispersion process with the resampling method. The sample dispersion allows a given portion of the total number of samples to be redistributed evenly over the entire state space. In the resampling algorithm, to determine how severe the weight degeneracy is before conducting the resampling process, it estimates an effective sample size $\hat{N}_{\text{eff}}$ as follows

$$\hat{N}_{\text{eff}} = \frac{1}{\sum_{i=1}^{N_s} (w_k^i)^2},$$

where w_k^i is the normalized weight, N_s is the number of particles. After a certain number of recursion steps, all but one particle have negligible normalized weights, which is called weight degeneracy. As the degeneracy phenomenon becomes more severe, the $\hat{N}_{\text{eff}}$ value approaches zero. Therefore, if $\hat{N}_{\text{eff}}$ is smaller than the predefined threshold N_T (e.g., $N_T = N_s/2$), the resampling process is executed to mitigate the degeneracy phenomenon. Here, lowering the N_T value solves the weight degeneracy problem, but the resampling process is frequently performed, resulting in system performance problems. On the contrary, if the value of N_T is higher than $N_s/2$, the opposite situation occurs and it can be regarded as a trade-off relationship. In addition, after setting all weight values to $1/N_s$ during the resampling process, the sample dispersion is executed to avoid the problem of sample impoverishment. In the flight guidance system, only 10% of the total number of samples are uniformly dispersed across the state space by selecting the bearing angle θ_k of each state vector within the range of direction values in which light can be irradiated.

3.2.3. Delay Reduction Analysis through Sample Dispersion Modeling

The sample dispersion method was applied to the particle filter of the flight guidance system so that the UAV can respond to incoming laser in various directions. Figure 3a shows simulation result of particle filter with sample dispersion. Assuming that the incident angle of the laser light varies suddenly about 180 degrees, the particle filter using the sample dispersion accurately estimated the value close to the measurement almost twice as fast as the particle filter without the sample dispersion. Figure 3b shows root mean square error (RMSE) performance of our simulation result. A large error values are shown due to a sudden change of the incident angle, but a particle filter with sample dispersion minimizes error about twice as fast as the general particle filter.

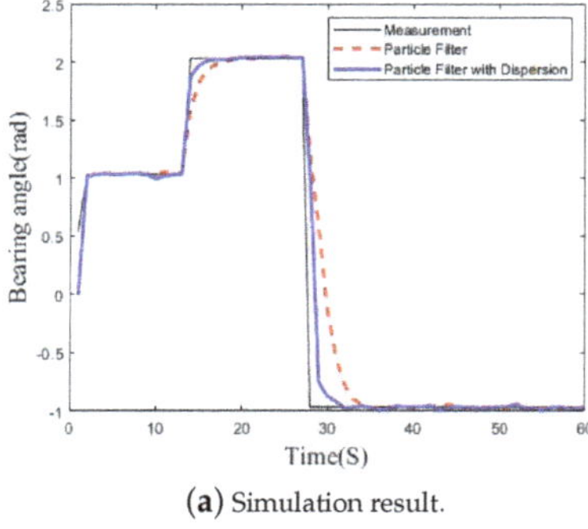

(a) Simulation result.

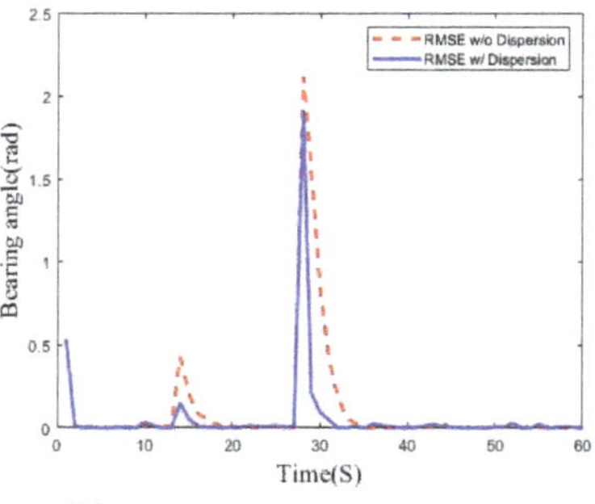

(b) Comparison of RMSE performance.

Figure 3. Performance of particle filter with sample dispersion.

3.2.4. Optimal Number of Particles through Modeling

Determining the statistically efficient number of samples is very important in the particle filter, as it is possible to estimate the expected value more accurately as the number of particles increases, but at the same time the computational complexity also increases. So we simulated the number of particles suitable for calculating the direction of the UAV flight and the results are shown in Table 1. Through this simulation, we can obtain the error between the constant measurements and the estimations from the particle filters with a different number of particles. When the number of particles is 500 or greater, the error with the measurements is less than 1 degree.

Table 1. Comparison of root mean square error (RMSE) performance depending on the number of particles.

Number of Particles	100	500	2000	10,000
RMSE (rad)	0.047	0.011	0.004	0.002
RMSE (degree)	2.712	0.609	0.204	0.129

In the flight guidance system, the direction of flight can be accurately calculated even with a small number of particles. In addition, considering that UAVs run the proposed system based on single-board computer with a low processing capacity using Lithium-ion batteries, the number of particles suitable for the flight guidance system is 500.

3.3. *Safe Landing System for UAV*

In order to allow the UAV to make a safe landing, the safe landing system was able to identify obstacles based on the optical flow and avoid the obstacles.

3.3.1. Optical Flow Based Obstacle Avoidance

When a UAV with a downward facing camera descends, the image sequences captured from the camera produce optical flow that spreads out from a point called focus of expansion (FOE). By locating FOE and analyzing the patterns of the optical flow, optical flow magnitude module calculates optical flow magnitude to estimate the heading of the UAV and the structure of the environment beneath the UAV. Figure 4 illustrates an optical flow map processed from a descending UAV with downward facing camera on a flat surface. Figure 5 illustrates the operation of calculating optical flow. The magnitude of the optical flow can be formulated with the following equation:

$$|\overrightarrow{OF}_{x,y}| \approx \alpha|\overrightarrow{v}|\tan\theta. \tag{1}$$

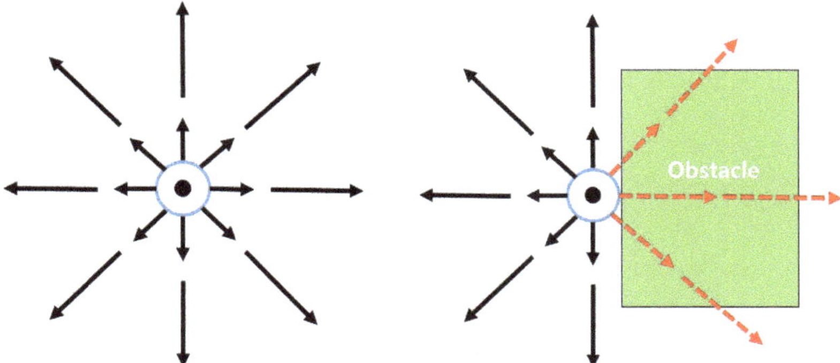

(**a**) Example of an optical flow observed in descending UAV.

(**b**) Example of an optical flow magnitude in field with obstacles.

Figure 4. Examples of optical flow magnitude map.

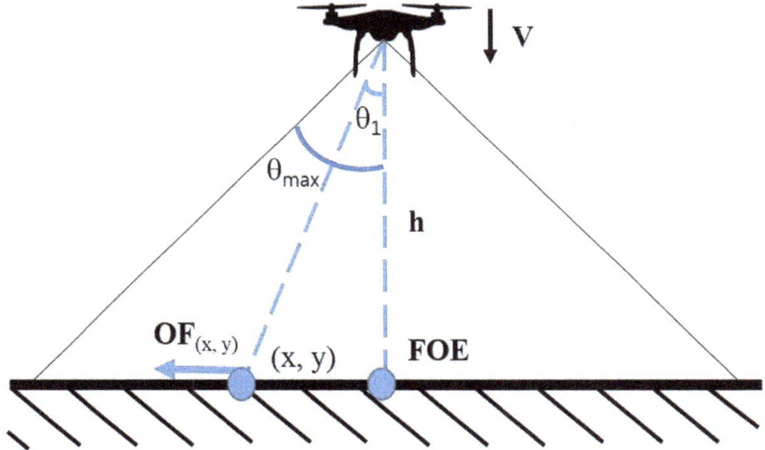

Figure 5. Optical flow of a descending UAV.

In Equation (1), $\overrightarrow{OF}_{x,y}$ is the optical flow value at (x, y), $\overrightarrow{v}$ is the descending speed of the UAV, θ is the angle that between the camera to FOE and camera to point (x, y), and α is the scale factor of the camera. Since θ is greater for the points farther away from FOE, the magnitude of the optical flow for points farther away from FOE is greater than for the points closer to FOE. Furthermore, points with the same distance from FOE will have the same optical flow magnitude.

By analyzing the optical flow magnitude map, the optical flow magnitude module estimates the structure of the ground. If the surface beneath is flat, the optical flow magnitude observed around the FOE is balanced. For surfaces where obstacle exists, the magnitude of optical flow shows deformation at the location where obstacle exist as shown in Figure 4b. Locations where the altitude is higher than the rest, the angular acceleration of θ becomes greater than other locations that lie on the same distance from FOE in the image. The increased angular acceleration of θ results in greater optical flow magnitude and the optical flow magnitude map shows the unbalanced magnitude of optical flows around FOE.

3.3.2. Optical Flow Modeling

The safe landing system system identifies obstacles through the optical flow. However, if the magnitude of optical flow is small, it is difficult for the UAV to detect the obstacle. Because the UAV fly at high altitudes, the variation in optical flow is very small. Therefore, we have confirmed through modeling that the magnitude of optical flow measured in the UAV is large enough to detect obstacles prior to experiments using real UAVs.

To model the optical flow, we assume that the UAV is equipped with a camera with a $2 \times \theta_{max}$ viewing angle, as shown in Figure 5. When we look at the point at the camera to calculate the optical flow at the angle of θ_1, the coordinates $OF_{(x,y)}$ shown can be expressed by the following equation:

$$OF_{(x,y)} = h \times \tan(\theta_1). \tag{2}$$

The magnitude of optical flow($\overrightarrow{OF}_{x,y}$) can be obtained by calculating the point at which the $OF_{(x,y)}$ point is observed at the image pointing to the camera after the UAV descend to the d altitude, and the equation is as follows:

$$\begin{aligned} \overrightarrow{OF}_{x,y} &= \frac{h \times \tan(\theta_1)}{(h-d) \times \tan(\theta_{max})} - \frac{h \times \tan(\theta_1)}{h \times \tan(\theta_{max})} \\ &= \frac{d \times \tan(\theta_1)}{(h-d) \times \tan(\theta_{max})}. \end{aligned} \tag{3}$$

If an obstacle with a height of g exists, the magnitude of the optical flow measured at the UAV increases by the height of the obstacle. For example, if there is an obstacle to the right of the camera image, the magnitude of optical flow difference is calculated as follows:

$$\Sigma\|\overrightarrow{W_L}\| - \Sigma\|\overrightarrow{W_R}\| = \frac{d \times \tan(\theta_1)}{(h-d) \times \tan(\theta_{max})} - \frac{d \times \tan(\theta_1)}{(h-g-d) \times \tan(\theta_{max})}. \tag{4}$$

$\Sigma\|\overrightarrow{W}\|$ indicates the magnitude of optical flow in either the left or right region of the camera image. In order to verify the magnitude of optical flow, we set up a simulated flight environment similar to the real UAV experiment. When the UAV equipped with a camera with an angle of view of 90° (θ_{max} = 45°) is flying, the angle for determining the optical flow(θ_1) is set to 30°. When there is an obstacle with a height of 1 m (g = 1), the UAV measures the optical flow while lowering the altitude by 1 m (d = 1). Figure 6 shows how the magnitude of the optical flow($\overrightarrow{OF}_{x,y}$) is measured when the UAV is at an altitude of 10 m to 20 m.

As a result of the modeling, when the UAV is flying at a height of 10 m, the magnitude of optical flow is 1.50 times higher than that of the flat surface when the obstacle is present. Also, when the height of the UAV was 20 m, the magnitude of optical flow showed a difference of 1.18 times. These results show that the obstacle can be identified through the magnitude of optical flow at the height at which the UAV is normally flying.

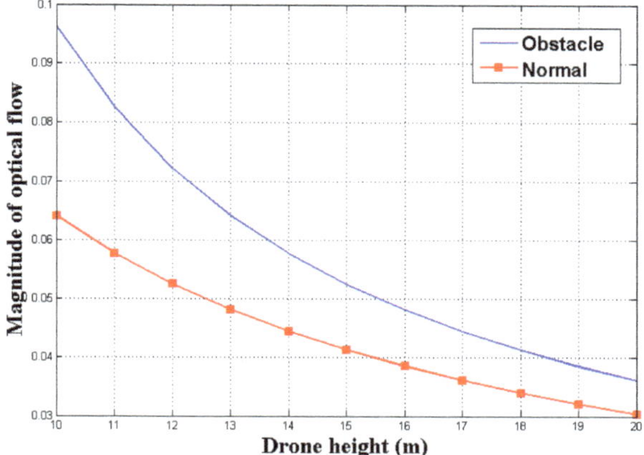

Figure 6. Magnitude of optical flow modeling.

4. System Implementation

This section details the algorithms of subsystems and implementation techniques applied to the actual UAV. As shown in Figure 7, UAV detour system consists of two major subsystems.

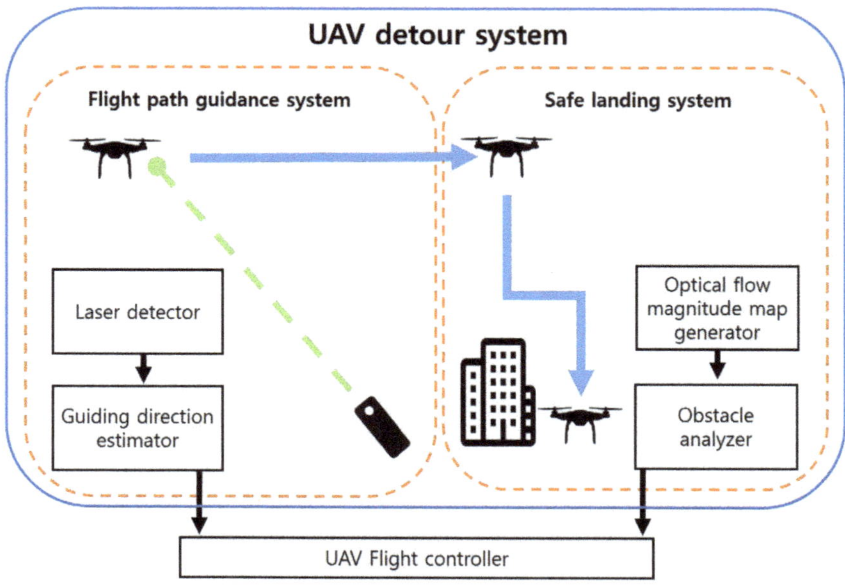

Figure 7. Overall systems and modules.

4.1. Flight Guidance System

4.1.1. Laser Detector

The laser detector identifies the incoming direction of the laser through light sensors. The laser detector consists of 12 light sensors arranged in a circular shape, which identifies the incoming direction of the laser. By assigning 30 degrees to each sensor, 12 sensors can cover 360 degrees in all directions. If the higher brightness is measured above a certain threshold than the initial brightness, the light sensor determines that the laser light has been received. In addition, when laser light is detected, the middle value of the range assigned to each sensor is returned.

Furthermore, as the distance increases, the area where light enters becomes larger, so that adjacent sensors can detect light at the same time. Also, the sensors can be affected by the momentary reflection or scattering of other light. In this case, the angular range of each sensor is added up and then the middle value is returned. Then, the laser detector transmits the bearing angle of incoming laser to the guiding direction estimator that returns the direction where the UAV is guided.

4.1.2. Guiding Direction Estimator

The guiding direction estimator operates the particle filtering based on the measurement value received from the laser detector to approximate the direction in which the UAV will be guided. The full algorithm of guiding direction estimator is presented in Algorithm 1.

Algorithm 1 Guiding direction estimator.

Initialization
discrete-time $k = 0$
for $i = 1, i{+}{+}, i == N_s$ (N_s = number of particles) **do**
  Initialize_state_vector l_0^i
end for
while measurment $z_k = true$ **do**
Weight update
  for $i = 1, i{+}{+}, i == N_s$ **do**
    $l_k^i \sim p(l_k | l_{k-1}^i)$
    $\tilde{w}_k^i \sim \tilde{w}_{k-1}^i p(z_k | l_k^i)$
  end for
Normalize
  for $i = 1, i{+}{+}, i == N_s$ **do**
    $w_k^i = \tilde{w}_k^i / \sum_{i=1}^{N_s} \tilde{w}_k^i$
  end for
Resampling based on effective sample size
  $\hat{N}_{eff} = 1 / \sum_{i=1}^{N_s} (w_k^i)^2$
  if then $\hat{N}_{eff} < N_T$
    Resampling (l_k^i)
    Sample dispersion $\theta_k^i = \theta_k^i + u[-\frac{\pi}{2}, \frac{\pi}{2}]$
  end if
Flight direction estimation
  $S_k = \sum_{i=1}^{N_s} w_k^i l_k^i$
  Operate_flight_controller $\leftarrow$ **Get_direction**(S_n)
  $k = k + 1$
end while

The Algorithm 1, guiding direction estimator, works in the following steps. First, in the initialization phase, the system initializes the bearing angle θ_k to have different values throughout the entire state space to evenly distribute each particle in all directions. Then, the system checks to see if it has received the measurement from the laser detector because it will start the estimation process with a particle filter after the laser light is detected. Second, the flight guidance system draws the samples from the transitional prior $p(l_k | l_{k-1}^i)$, because we have chosen the proposal distribution $q(l_k | l_{k-1}^i, z_k)$ as the transitional prior. Third, the selection of the proposal distribution can simplify the weight update equation and update the weight using the likelihood $p(z_k | l_k)$. Fourth, in the normalization step, the weight of each sample $(\tilde{w}_k^i)$ is divided by the total sum to make the sum of the normalized weights $\sum w_k^i$ to be 1. Fifth, the system calculates the effective sample size $\hat{N}_{eff}$ and determines that the weight degeneracy problem becomes severe when $\hat{N}_{\text{eff}}$ is less than the threshold N_{T}. In this case, the system performs the resampling process and applies the sample dispersion method described in Section 3.2.2 to quickly respond to drastic changes in measurements. Finally, the guiding direction estimator obtains an estimated bearing angle (S_{k}) and transmits the direction calculated from the estimated bearing angle to the flight controller. Then, the flight controller maneuvers the UAV based on the direction.

4.2. Safe Landing System

This subsection describes the optical flow that is the basis of our obstacle avoidance and describes the two modules installed in the safe landing system, the optical flow magnitude map generator and obstacle analyzer.

4.2.1. Optical Flow Magnitude Map Generator

Optical flow magnitude map generator calculates optical flows between frames from the images obtained by the camera. In this process, we considered the situation that the optical flows should be computed on low-power computing boards mounted on UAV. As computing the optical flow for all pixels and drawing a magnitude map [39] has a large load to run on the low-power computing board, it is inappropriate for real-time operation. Therefore, the Lucas–Kanade algorithm [38] could be considered, which sets up a pixel window for each pixel in one frame and finds a match to this window in the next frame for specific pixels extracted with some standards. However, the Lucas–Kanade algorithm has the challenge that it cannot calculate large movement.

Therefore, the optical flow magnitude map generator used the iterative Lucas–Kanade method with pyramids [40], which can supplement this disadvantage. The specific pixels used for the iterative Lucas–Kanade determined by goodFeaturesToTrack function on openCV [41], which detects the strong corner on the image which is easy to trace its movement. Thus, using the benefit of calculating only for certain pixels, not for the entire pixels, the load for computing board can be reduced and does not cause performance degradation on operating.

4.2.2. Obstacle Analyzer

An obstacle analyzer determines the existence of obstacles, and two criteria can be considered. One is the magnitude of optical flow and the second is the feature point, both are from optical flow magnitude map generator. In Section 3.3.2, we modeled magnitude of optical flow. It shows that if an obstacle exists, it has a larger magnitude than the normal, and the closer it is, the larger it becomes. Also, the feature points extracted form goodFeaturesToTrack function are extracted mainly on the obstacles, because obstacles not only have a visual difference in color or pattern against landing point, but also the difference in height against the landing point. In addition, the image in which the obstacle exists creates more feature points than the flat image. Therefore, the greater the number of feature points and the larger the optical flow magnitude, the higher the probability that the obstacle actually exists. In our system, we used the *metric* to multiply the optical flow by the feature point and use it to identify obstacles. Using this *metric*, the location of the obstacle can be determined depends on where it is located in the image obtained through the camera. The image is divided into $\text{m} \times \text{m}$ arrays of segments, creating a total of m^2 segments per image. The value of m can be freely selected according to the experimental situation, such as 3, 5, and 7. In evaluation, m is set to 3 and the image is divided into 3×3, nine segments. As the obstacles that UAVs face in urban canyons would be large in size (e.g., trees and buildings), setting m to 3 is considered to be sufficient to identify the location of the obstacle and avoid it. The value of the *metric* is derived significantly from the segmented screen where the obstacle is located, and the UAV recognizes that the location obstacle exists, and then flies to the opposite direction.

Algorithm 2 presents the algorithm of obstacle analyzer. When the frames come in through the camera, the obstacle analyzer uses the extracted feature points and calculated magnitude of optical flows from the optical flow magnitude map generator. First, the obstacle analyzer divides the location of feature points in several directions according to the coordinates of the feature points. The directions can be multiple directions, and in Algorithm 2, the directions are set to eight directions. Second, the obstacle analyzer adds the magnitude of optical flow at the feature point to the direction, and repeats this process on every point. By this method, the optical flow for a particular direction becomes proportional to the magnitude and the number of feature points, and if the value is greater than the empirical static threshold value O_T, it is determined that an obstacle exists in a particular direction.

Algorithm 2 Obstacle analyzer.

```
while optical_flow_exists == true do
    for i = 0, i++, i < number_of_directions do
        if location_of_feature_point(x,y) == DIRECTION(i) then
            optical_flow_DIRECTION(i) += magnitude_optical_flow(x,y)
        end if
        if optical_flow_DIRECTION(i) > O_T & Variance > V_T then
            exist_of_obstacle_DIRECTION(i) = true
        end if
    end for
end while
```

During the process, the magnitude of optical flow in the segmented screen can be measured evenly large when the camera is facing the ground without obstacle after avoiding it. The first reason for this case is because the strong corners that affect the goodFeaturesToTrack function are even on every obstacle-free ground, and the second is because of the tendency that the magnitude of optical flow measured at each segment screen can be similar on the flat ground. For these reasons, if only the magnitude of optical flow is used, unintentional situations where the obstacle analyzer misunderstands a flat ground as an obstacle can occur. In order to prevent this case, the obstacle analyzer additionally utilized another metric, the variance of optical flow magnitude values measured at each segment screen. When analyzing the optical flow map in an obstacle environment, the difference between the optical flow magnitude values measured on the segment screen with obstacles and the segment screen without obstacles is huge. Therefore, in the obstacle environment, the variance of optical flow magnitude values increases. Thus, the obstacle analyzer only performs detection of obstacle in situations that variance of optical flow magnitude is greater than the empirical threshold value V_T. Both O_T and V_T should be adapted to the actual environments, and automatically calibrating the threshold value is left for future work, as mentioned in Section 6.

5. Experiments and Demonstrations

5.1. Experimental Setup

The implementation of the proposed system is based on our previous work [42,43]. For flight guidance system, twelve Cadmium Sulfide (CDS) light sensors were attached to the UAV in different directions. Also, a camera for the safe landing system was mounted on the UAV. Figure 8 shows a prototype of the UAV for proposed system. We adopted DJI's F550 ARF KIT for the frame and HardKernel's ODROID XU4 for the processing unit. The processing unit is connected to the light sensors, a camera, and a communication interface. In experiments, the camera was ODROID USB-CAM 720P, which has 720 p resolution, 30 fps with color scale, and θ_{max} was measured experimentally at about 21°.

Figure 8. Prototype implementation of the system.

5.2. Flight Guidance System Demonstration

Prior to the guidance experiment, we confirmed that the light sensor can detect the laser in various environments. We measured the brightness of the light under a sunny, cloudy, night, and indoor with fluorescent light while the light sensor and the lasers were 15 m and 30 m away. For this experiment, we used a commercial laser. Table 2 shows the average of the measured brightness values. This experiment shows that laser can be distinguished under any environments. Even in the sunny, the brightest environment of all environments, the laser detector was able to identify the laser. Through this experiment, we were able to determine the threshold value to discriminate the laser. We also set the number of particles to 500, as mentioned in Section 3.2.4, and set the threshold (N_T) for the sample dispersion to 250, which is half the number of particles.

Table 2. Measurement of light intensity in various environments.

	Distance (15 m)		Distance (30 m)	
	Without Laser (lx)	**Laser Projected (lx)**	**Without Laser (lx)**	**Laser Projected (lx)**
Sunny	10,820.45	30,306.2	11781.6	20,264.17
Cloudy	7031.69	31,509.4	7250.12	15,792
Night	2.31	32,870.2	2.82	14,827.4
Indoor (flourescent light)	202.59	41,238.2	133.4	24,834

We used the UAV shown in Figure 8 to confirm that the flight guidance system is working properly. As shown in Figure 9a, the laser was aimed at the UAV in the autonomous flight. In addition, as shown in Figure 9b, we confirmed that the UAV was flying in the guided direction. The demonstration video can be seen on the following link [15].

(a) Detect laser from light sensor attached to UAV.

(b) The UAV fly in the direction guided by the laser.

Figure 9. Laser guidance demonstration.

In an additional experiment, we measured the time lag from the moment the laser was emitted toward the UAV to estimate the coordinates to move along that direction. In particular, the time lag includes resampling and direction estimation as well as updating the state and weight of each particle. We also calculated the RMSE of the difference between the first estimated flight direction and the constant measurements right after the initialization phase, and Table 3 shows the results depending on the number of particles. In Table 3, the average time lag of the currently implemented system with 500 particles was 91.7 milliseconds. The impact of this result is expected to be negligible when the UAV with a loss of GPS signal is hovering in place. Furthermore, as the number of particles decreased, the time lag was reduced, while the accuracy of the flight direction was significantly compromised. On the contrary, as the number of particles increased, the accuracy was improved, but the time was delayed too much. As mentioned in Section 3.2.4, it is very important to select and implement the optimal number of particles for each system.

Table 3. Comparison of performance depending on the number of particles in real experiment.

Number of Particles	100	500	2000	10,000
Average time lag (ms)	27.2	91.7	397.1	1642.4
RMSE (rad)	0.0568	0.0239	0.0118	0.0047

5.3. Safe Landing System Demonstration

The safe landing system proved its performance through UAV landing experiments. In this experiment, when the UAV was lowering its altitude for landing, an optical flow magnitude map was generated from the image taken by the downward-facing camera. At this time, we confirmed whether the safe landing system can detect obstacles based on the optical flow magnitude map and whether the UAV can move in the direction of avoiding obstacles.

Figure 10 represents the optical flow magnitude map of the image viewed by the camera facing downward of the UAV. In Figure 10a, the green dot represents the feature point, and the green line represents the optical flow calculated at the feature point. As the goodFeaturesToTrack function detects the strong corner and tends to detect from obstacles that are visually different from the floor and higher in height, Figure 10 shows the feature points mainly presented on the obstacle. As shown in Figure 10a, the left side of the tree is the highest obstacle, and the UAV obtains the highest optical flow magnitude values from the tree on the left side. As shown in Figure 10b, the OpenCV on the UAV recognized that the nearest obstacle was on the left. In this experiment, the safe landing system successfully maneuvered the UAV to avoid obstacles. Also, the frame rate was 31.1 FPS which is about five times faster than the previous work [14], 5.9 FPS which calculates the optical flow for all pixels.

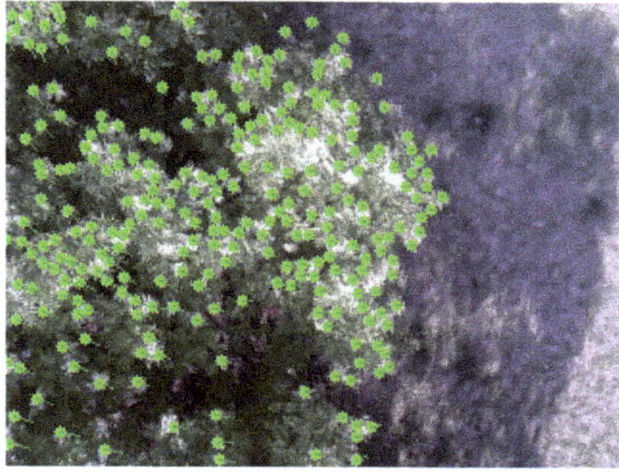

(a) Optical flow of a surface with an obstacle.

(b) Operation of obstacle avoidance.

Figure 10. Optical flow magnitude map of obstacle avoidance.

The variance measured during the experiment also shows that the UAV was able to recognize obstacles successfully. Figure 11 shows the variance of the optical flow magnitude in each segmented screen during the experiment. During obstacle avoidance, obstacles are detected in the left segmented screen, resulting in high optical flow magnitude in the left segmented screen. Therefore, the variance value of segmented screens was high until 10 s when the UAV was avoiding the obstacles. After the UAV completely avoided the obstacle, the sharp increase in variance at 10 s is shown in Figure 11. This is a temporary phenomenon that occurs while creating a new feature point on the ground because there are no more obstacles. After creating the feature points of the ground, the variance of the optical flow magnitude was measured low because the optical flow magnitude is evenly measured on each segmented screen. The full demonstration can be seen in the following link [16].

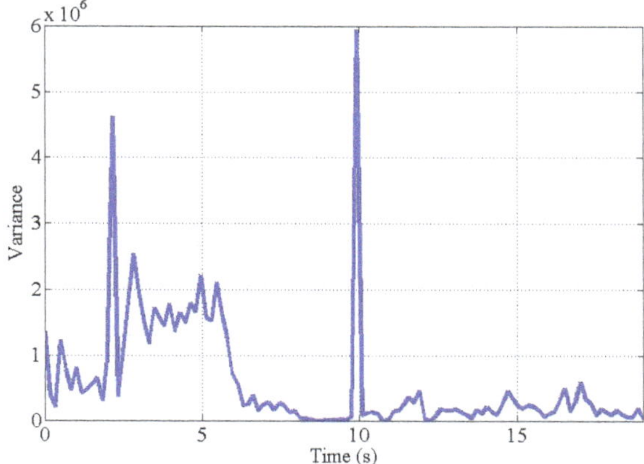

Figure 11. Variance of optical flow magnitude.

5.4. *UAV Detour System Demonstration*

Finally, we demonstrated the UAV detour system that consists of the flight guidance system and safe landing system. In this demonstration, the operator used a laser to guide the UAV to the landing point where the operator stood. The UAV's flight guidance system identified the laser, guided the UAV to the landing point, and then proceeded to land. Since the operator was standing at the landing point of the UAV, the safe landing system recognized the operator as an obstacle, then automatically avoided the operator and landed safely at that point. The demonstration video of the UAV detour system can be seen on the following link [17].

6. Future Work

The flight guidance system can detect laser, but cannot identify malicious lasers that are intended to interfere with UAV movement. To solve this problem, as future work, we will develop a paring system so that only certified lasers can take control of the UAV. We plan to improve the system through a bandpass filter so that the sensor can identify lasers with a specific wavelength. If the data bit is transmitted through a laser, an encryption technique can be applied to the laser. This improvement will allow the UAV to identify the laser containing the certified data bit and move in that direction. Also, in an environment where a line of sight (LoS) is not secured, it is difficult to guide UAV with a laser. To cope with this environment, we are developing a system that guides UAV with extra media that can be used even if LoS is not secured (e.g., ultrasound). In addition, The flight guidance system requires the operator to operate the laser. To solve this inconvenience, we will develop an improved landing point system that identifies UAVs through image processing and automatically aims the laser. Overall, we will improve the flight guidance system to suit the delivery system in an urban area.

The safe landing system will be extended to automatically avoid obstacles that UAVs can encounter during the entire process of takeoff, flying, and landing to perform their mission in an urban area. In addition, the values we set as the threshold (e.g., O_T, V_T) should be set in response to various circumstances. We are setting it as a future goal to make automatic calibration through machine learning.

7. Conclusions

UAVs can perform various missions. Some of UAVs performing missions are capturing video or collecting information over an extensive area, and some UAVs perform missions in urban areas such as delivery UAVs. However, in urban areas, buildings weaken the GPS signal, and there are many obstacles that disturb the UAVs' flight. Therefore, UAV flying in urban area requires additional systems to fly in the absence of GPS or to avoid obstacles. This paper proposes the UAV detour system considering UAVs performing missions in urban areas. The UAV detour system allows the UAV to fly and land in situations where GPS or networks are disconnected. The flight guidance system, which is one of the subsystems of the UAV detour system, maneuvers the UAV by using a laser that is not disturbed by various radio waves or signal interference. Another subsystem, the safe landing system, identifies obstacles based on optical flow, allowing the UAV to avoid obstacles when landing. Finally, the proposed subsystems were tested on a prototype UAV. The performance of subsystems were verified by successfully performing flight guidance and obstacle avoidance landing.

Author Contributions: conceptualization, A.Y.C. and H.K.; methodology, A.Y.C. and H.S.; software, H.S., C.J. and S.P.; validation, J.Y.L. and H.K.; formal analysis, A.Y.C., J.Y.L. and H.S.; investigation, A.Y.C. and H.S.; resources, H.K.; data curation, C.J. and H.S.; writing—original draft preparation, A.Y.C. and J.Y.L.; writing—review and editing, J.Y.L. and H.K.; visualization, J.Y.L. and C.J.; supervision, H.K.; project administration, H.K.; funding acquisition, H.K.

Funding: This research was supported by Unmanned Vehicles Advanced, the Unmanned Vehicle Advanced Research Center (UVARC) funded by the Ministry of Science, ICT and Future Planning, Republic of Korea (NRF-2016M1B3A1A01937599), and in part by "Human Resources program in Energy Technology" of the Korea Institute of Energy Technology Evaluation and Planning(KETEP) granted financial resource from the Ministry of Trade, Industry and Energy, Republic of Korea (No. 20174030201820).

Conflicts of Interest: The authors declare that they have no conflicts of interest.

References

1. Park, S.; Lee, J.Y.; Um, I.; Joe, C.; Kim, H.T.; Kim, H. RC Function Virtualization-You Can Remote Control Drone Squadrons. In Proceedings of the 17th Annual International Conference on Mobile Systems, Applications, and Services, Seoul, Korea, 17–21 June 2019; ACM: New York, NY, USA, 2019; pp. 598–599.
2. Hart, W.S.; Gharaibeh, N.G. Use of micro unmanned aerial vehicles in roadside condition surveys. In Proceedings of the Transportation and Development Institute Congress 2011: Integrated Transportation and Development for a Better Tomorrow, Chicago, IL, USA, 13–16 March 2011; pp. 80–92.
3. Cheng, P.; Zhou, G.; Zheng, Z. Detecting and counting vehicles from small low-cost UAV images. In Proceedings of the ASPRS 2009 Annual Conference, Baltimore, MD, USA, 9–13 March 2009; Volume 3, pp. 9–13.
4. Jensen, O.B. Drone city—Power, design and aerial mobility in the age of "smart cities". *Geogr. Helv.* **2016**, *71*, 67–75. [CrossRef]
5. Jung, J.; Yoo, S.; La, W.; Lee, D.; Bae, M.; Kim, H. Avss: Airborne video surveillance system. *Sensors* **2018**, *18*, 1939. [CrossRef] [PubMed]
6. Bae, M.; Yoo, S.; Jung, J.; Park, S.; Kim, K.; Lee, J.; Kim, H. Devising Mobile Sensing and Actuation Infrastructure with Drones. *Sensors* **2018**, *18*, 624. [CrossRef] [PubMed]
7. Chung, A.Y.; Jung, J.; Kim, K.; Lee, H.K.; Lee, J.; Lee, S.K.; Yoo, S.; Kim, H. Poster: Swarming drones can connect you to the network. In Proceedings of the 13th Annual International Conference on Mobile Systems, Applications, and Services, MobiSys 2015, Florence, Italy, 18–22 May 2015; Association for Computing Machinery, Inc.: New York, NY, USA, 2015; p. 477.
8. Park, S.; Kim, K.; Kim, H.; Kim, H. Formation control algorithm of multi-UAV-based network infrastructure. *Appl. Sci.* **2018**, *8*, 1740. [CrossRef]
9. Farris, E.; William, F.M.I. System and Method for Controlling Drone Delivery or Pick Up during a Delivery or Pick Up Phase of Drone Operation. U.S. Patent 14/814,501, 4 February 2016.
10. Kimchi, G.; Buchmueller, D.; Green, S.A.; Beckman, B.C.; Isaacs, S.; Navot, A.; Hensel, F.; Bar-Zeev, A.; Rault, S.S.J.M. Unmanned Aerial Vehicle Delivery System. U.S. Patent 14/502,707, 30 Seotember 2014.
11. Paek, J.; Kim, J.; Govindan, R. Energy-efficient Rate-adaptive GPS-based Positioning for Smartphones. In Proceedings of the 8th International Conference on Mobile Systems, Applications, and Services, San Francisco, CA, USA, 15–18 June 2010; MobiSys '10; ACM: New York, NY, USA, 2010; pp. 299–314, doi:10.1145/1814433.1814463. [CrossRef]
12. QUARTZ Amazon Drones won't Replace the Mailman or FedEx Woman any Time soon. Available online: http://qz.com/152596 (accessed on 28 September 2019).
13. Nguyen, P.; Ravindranatha, M.; Nguyen, A.; Han, R.; Vu, T. Investigating cost-effective rf-based detection of drones. In Proceedings of the 2nd Workshop on Micro Aerial Vehicle Networks, Systems, and Applications for Civilian Use, Singapore, 26 June 2016; ACM: New York, NY, USA, 2016; pp. 17–22.
14. Chung, A.Y.; Lee, J.Y.; Kim, H. Autonomous mission completion system for disconnected delivery drones in urban area. In Proceedings of the 2017 IEEE International Conference on Robotics and Biomimetics (ROBIO), Macau, China, 5–8 December 2017; pp. 56–61.
15. Lee, J.Y.; Shim, H.; Park, S.; Kim, H. Flight Path Guidance System. Available online: https://youtu.be/vjX0nKODgqU (accessed on 2 October 2019).
16. Lee, J.Y.; Joe, C.; Park, S.; Kim, H. Safe Landing System. Available online: https://youtu.be/VSHTZG1XVLs (accessed on 28 September 2019).
17. Lee, J.Y.; Shim, H.; Joe, C.; Park, S.; Kim, H. UAV Detour System. Available online: https://youtu.be/IQn9M1OHXCQ (accessed on 2 October 2019).
18. Stary, V.; Krivanek, V.; Stefek, A. Optical detection methods for laser guided unmanned devices. *J. Commun. Netw.* **2018**, *20*, 464–472. [CrossRef]
19. Shaqura, M.; Alzuhair, K.; Abdellatif, F.; Shamma, J.S. Human Supervised Multirotor UAV System Design for Inspection Applications. In Proceedings of the 2018 IEEE International Symposium on Safety, Security, and Rescue Robotics (SSRR), Philadelphia, PA, USA, 6–8 August 2018; pp. 1–6.
20. Jang, W.; Miwa, M.; Shim, J.; Young, M. Location Holding System of Quad Rotor Unmanned Aerial Vehicle (UAV) using Laser Guide Beam. *Int. J. Appl. Eng. Res.* **2017**, *12*, 12955–12960.

21. Fox, D.; Burgard, W.; Kruppa, H.; Thrun, S. A probabilistic approach to collaborative multi-robot localization. *Auton. Robot.* **2000**, *8*, 325–344. [CrossRef]
22. Hightower, J.; Borriello, G. Particle filters for location estimation in ubiquitous computing: A case study. In Proceedings of the International conference on ubiquitous computing, Nottingham, UK, 7–10 September 2004; Springer: Berlin/Heidelberg, Germany, 2004; pp. 88–106.
23. Rosa, L.; Hamel, T.; Mahony, R.; Samson, C. Optical-flow based strategies for landing vtol uavs in cluttered environments. *IFAC Proc. Vol.* **2014**, *47*, 3176–3183. [CrossRef]
24. Souhila, K.; Karim, A. Optical flow based robot obstacle avoidance. *Int. J. Adv. Robot. Syst.* **2007**, *4*, 13–16. [CrossRef]
25. Yoo, D.W.; Won, D.Y.; Tahk, M.J. Optical flow based collision avoidance of multi-rotor uavs in urban environments. *Int. J. Aeronaut. Space Sci.* **2011**, *12*, 252–259. [CrossRef]
26. Miller, A.; Miller, B.; Popov, A.; Stepanyan, K. UAV Landing Based on the Optical Flow Videonavigation. *Sensors* **2019**, *19*, 1351. [CrossRef] [PubMed]
27. Herissé, B.; Hamel, T.; Mahony, R.; Russotto, F.X. Landing a VTOL unmanned aerial vehicle on a moving platform using optical flow. *IEEE Trans. Robot.* **2011**, *28*, 77–89. [CrossRef]
28. Mori, T.; Scherer, S. First results in detecting and avoiding frontal obstacles from a monocular camera for micro unmanned aerial vehicles. In Proceedings of the 2013 IEEE International Conference on Robotics and Automation, Karlsruhe, Germany, 6–10 May 2013; pp. 1750–1757.
29. Hrabar, S. 3D path planning and stereo-based obstacle avoidance for rotorcraft UAVs. In Proceedings of the 2008 IEEE/RSJ International Conference on Intelligent Robots and Systems, Nice, France, 22–26 September 2008; pp. 807–814.
30. Ferrick, A.; Fish, J.; Venator, E.; Lee, G.S. UAV obstacle avoidance using image processing techniques. In Proceedings of the 2012 IEEE International Conference on Technologies for Practical Robot Applications (TePRA), Woburn, MA, USA, 23–24 April 2012; pp. 73–78.
31. Kendoul, F. Survey of advances in guidance, navigation, and control of unmanned rotorcraft systems. *J. Field Robot.* **2012**, *29*, 315–378. [CrossRef]
32. Bi, Y.; Duan, H. Implementation of autonomous visual tracking and landing for a low-cost quadrotor. *Opt.-Int. J. Light Electron Opt.* **2013**, *124*, 3296–3300. [CrossRef]
33. Lange, S.; Sünderhauf, N.; Protzel, P. A vision based onboard approach for landing and position control of an autonomous multirotor UAV in GPS-denied environments. In Proceedings of the 2009 International Conference on Advanced Robotics, Munich, Germany, 22–26 June 2009; pp. 1–6.
34. Venugopalan, T.; Taher, T.; Barbastathis, G. Autonomous landing of an Unmanned Aerial Vehicle on an autonomous marine vehicle. In Proceedings of the 2012 Oceans, Hampton Roads, VA, USA, 14–19 October 2012; pp. 1–9.
35. Cesetti, A.; Frontoni, E.; Mancini, A.; Zingaretti, P.; Longhi, S. A vision-based guidance system for UAV navigation and safe landing using natural landmarks. In Proceedings of the Selected papers from the 2nd International Symposium on UAVs, Reno, NV, USA, 8–10 June 2009; Springer: Berlin/Heidelberg, Germany, 2010; pp. 233–257.
36. Eendebak, P.; van Eekeren, A.; den Hollander, R. Landing spot selection for UAV emergency landing. In Proceedings of the SPIE Defense, Security, and Sensing, Baltimore, MD, USA, 29 April–3 May 2013; International Society for Optics and Photonics: Bellingham, WA, USA, 2013; p. 874105.
37. Ristic, B.; Arulampalam, S.; Gordon, N. Beyond the Kalman filter. *IEEE Aerosp. Electron. Syst. Mag.* **2004**, *19*, 37–38.
38. Lucas, B.D.; Kanade, T. An iterative image registration technique with an application to stereo vision. In Proceedings of the 7th International Joint Conference (IJCAI) 1981, Vancouver, BC, Canada, 24–28 August 1981.
39. Farnebäck, G. Two-frame motion estimation based on polynomial expansion. In Proceedings of the Scandinavian Conference on Image Analysis, Halmstad, Sweden, 29 June–2 July 2003; Springer: Berlin/Heidelberg, Germany, 2003; pp. 363–370.
40. Bouguet, J.Y. Pyramidal implementation of the affine lucas kanade feature tracker description of the algorithm. *Intel Corp.* **2001**, *5*, 4.

41. Bradski, G. The opencv library. *Dr Dobb's J. Software Tools* **2000**, *25*, 120–125.
42. Yoo, S.; Kim, K.; Jung, J.; Chung, A.Y.; Lee, J.; Lee, S.K.; Lee, H.K.; Kim, H. A Multi-Drone Platform for Empowering Drones' Teamwork. Available online: http://youtu.be/lFaWsEmiQvw (accessed on 28 September 2019).
43. Chung, A.Y.; Jung, J.; Kim, K.; Lee, H.K.; Lee, J.; Lee, S.K.; Yoo, S.; Kim, H. Swarming Drones Can Connect You to the Network. Available online: https://youtu.be/zqRQ9W-76oM (accessed on 28 September 2019).

Article

A Framework for Multiple Ground Target Finding and Inspection Using a Multirotor UAS

Ajmal Hinas [1,*], Roshan Ragel [2], Jonathan Roberts [1] and Felipe Gonzalez [1]

1 Robotics and Autonomous Systems, Queensland University of Technology (QUT), Brisbane City QLD 4000, Australia; Jonathan.roberts@qut.edu.au (J.R.); felipe.gonzalez@qut.edu.au (F.G.)

2 Department of Computer Engineering, University of Peradeniya (UOP), Peradeniya 20400, Sri Lanka; roshanr@pdn.ac.lk

* Correspondence: ajmal.hinas@hdr.qut.edu.au or ajmalhinas@gmail.com; Tel.: +61-0470534175

† This paper is an expanded version of "Multiple Ground Target Finding and Action Using UAVs" published in Proceedings of 2019 IEEE Aerospace Conference, Big Sky, MT, USA, 2–9 March 2019, and "A Framework for Vision-Based Multiple Target Finding and Action Using Multirotor UAVs" published in Proceedings of the 2018 International Conference on Unmanned Aircraft Systems (ICUAS 2018), Dallas, TX, USA, 12–15 June 2018.

Received: 26 November 2019; Accepted: 31 December 2019; Published: 3 January 2020

Abstract: Small unmanned aerial systems (UASs) now have advanced waypoint-based navigation capabilities, which enable them to collect surveillance, wildlife ecology and air quality data in new ways. The ability to remotely sense and find a set of targets and descend and hover close to each target for an action is desirable in many applications, including inspection, search and rescue and spot spraying in agriculture. This paper proposes a robust framework for vision-based ground target finding and action using the high-level decision-making approach of Observe, Orient, Decide and Act (OODA). The proposed framework was implemented as a modular software system using the robotic operating system (ROS). The framework can be effectively deployed in different applications where single or multiple target detection and action is needed. The accuracy and precision of camera-based target position estimation from a low-cost UAS is not adequate for the task due to errors and uncertainties in low-cost sensors, sensor drift and target detection errors. External disturbances such as wind also pose further challenges. The implemented framework was tested using two different test cases. Overall, the results show that the proposed framework is robust to localization and target detection errors and able to perform the task.

Keywords: unmanned aerial vehicle; unmanned aerial system; vision-based navigation; search and rescue; vision and action; OODA; inspection; target detection; remote sensing

1. Introduction

Small, unmanned aerial systems (UASs) now have advanced waypoint-based navigation capabilities, which enable them to collect surveillance [1], wildlife ecology [2] and air quality data [3] in new ways. One capability that is not available in most off-the-shelf UASs is the ability for a higher level of onboard decision-making for vision and action.

The ability to remotely sense and find a set of targets, and descend and hover close to each target is desirable in many applications, including spot spraying in agriculture (such as pesticide spraying for brown planthopper (BPH) [4] and spear thistle weed [5]), inspection and search and rescue operations in a larger area with limited human resources, or in difficult terrain, such as searching for survivors or an aviation black box.

Figure 1 is an illustrative example of a search scenario for survivors. The task can be divided into multiple steps. Initially, the UAS searches for ground targets at search height h_s (1). If target/s are found, the UAS picks a target, based on a criterion such as first detected target and changes its original path and moves toward the target and descends (2). The UAS then hovers above the target closely (h_h = 0.5–2.5) to inspect the target (3). After inspection, the UAS climbs slightly (<2 m) and reaches height h_c and moves laterally to the next target (4). After reaching the next target, the UAS descends (5) and inspects that target (3). The UAS repeats steps 3, 4 and 5 with each target.

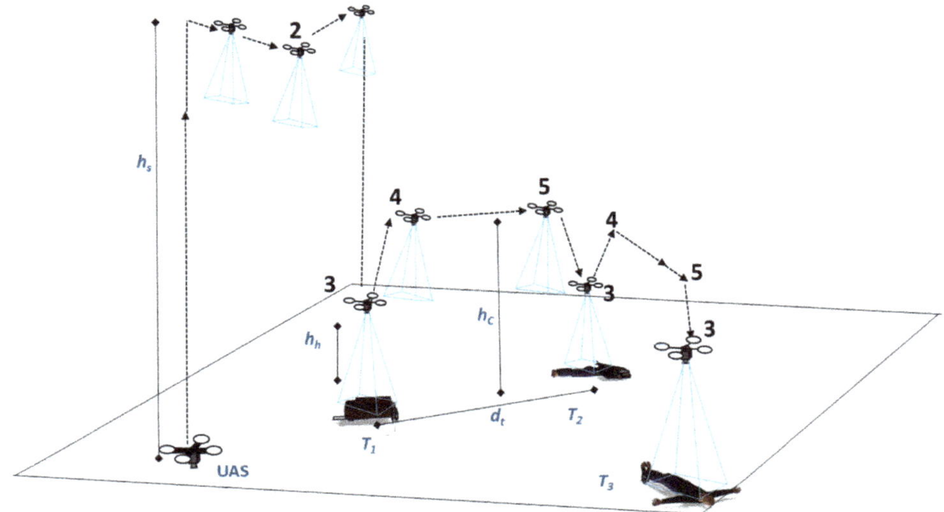

Figure 1. Illustrative search for survivors scenario with multiple objects scattered on the ground. The numbers 1–5 indicate the steps performed in the mission. Steps 3–5 are repeated for each target.

Very accurate target detection is needed to perform this task. However, target detection algorithms have a certain degree of uncertainty in real outdoor environments. Typical single frequency GPS receivers have an accuracy of 5–6 m [6,7]. Another challenge is that a greater localization accuracy needed to act on targets. In this paper, our aim is to develop and demonstrate a framework for autonomous vision-based ground target finding and action using a multirotor UAS that is robust to the (i) errors and uncertainties in the localization of the UAS; (ii) errors and uncertainties in localization of the target; (iii) target detection errors and; (iv) the effects of external disturbances such as wind.

The framework uses the high-level decision-making approach of observe, orient, decide and act (OODA) [8–10] and a series of techniques to perform the necessary navigation and control and target tracking. The proposed framework was implemented as a modular software system and demonstrated in two different test cases.

The contributions of the paper are summarized as follows: (1) An onboard framework for finding and acting on multiple ground targets which is robust to errors in localization of the UAS, localization of the targets and target detection, and external disturbances; (2) A computationally efficient voting-based scheme to avoid false detections in vision-assisted navigation or target tracking using UASs.

2. Related Work

The development of the multiple target detection and action framework is closely related to previous work on vision-based landing, target detection, target tracking, infrastructure inspection and OODA.

The vision-based landing task of UASs is a well-explored area of research. In this task, the UAS guides and descends towards a detected landing pad using vision. Mostly, a single, known target is used to detect the landing area such as a landing pad, and the accuracy of a few meters from the landing pad is usually acceptable. Specialized patterns, such as "H", are common to improve detection [4–6] or control the UAS [11]. Techniques such as Image-Based Visual Servoing (IBVS) and the relative distance-based approaches [7] are common. However, these techniques have limitations for multiple targets when the targets have unknown shapes and sizes or are similar.

Zarudzki et al. [12] explored multiple target tracking using an IBVS scheme and different colors for the targets to separate them. However, the IBVS schemes are limited when all or a subset of the targets are visually similar.

Many researchers have explored the tracking of a moving target [8–11]. For example, Greatwood et al. [13] used the relative location of an unmanned ground vehicle (UGV) to track it. However, relative location-based approaches are limited for multiple similar targets. In another study, Vanegas et al. [14] demonstrated finding a UGV and tracking it in GPS-denied and cluttered environments using a Partially Observable Markov Decision Process (POMDP) formulation and known obstacle position to correct the UAS pose. However, such knowledge is not available in our work. Tracking of a user-selected object was also demonstrated by Cheng et al. [14]. The authors used a UAS fixed with a gimbaled camera and a Kernelized Correlation Filter (KCF)-based visual tracking scheme to track the selected object. All of these works are limited to a single known target.

Vision-based or assisted navigation of UASs is also a well-studied area of research [15–17]. Stefas et al. [18] studied the navigation of a UAS inside an apple orchard. Missions following water channels [19] and roads [20] have also been explored. Other inspection tasks using UASs are power line inspection [21,22] and railway track inspection [23]. In these studies, a common approach is to follow the continuous image feature lines present in the observed structures. However, this research investigates the problem of finding and acting on multiple discrete targets.

In other works for railway semaphore inspection [24] and pole inspection [25], visual control techniques have been explored to fly the UAS around a pole-like structure. However, finding and approaching the target is not considered in these studies.

Vision-based ground target finding and action systems have been studied for a single target [5,26,27]. Multiple ground target finding and action capability have also been demonstrated [28,29]. However, those works did not consider the uncertainties in target detection.

Most previous studies used a control system approach to control the UAS. Another possible approach is a cognitive model-based approach of OODA [30,31]. For example, Karim and Heinze [32] implemented and compared two distinct autonomous UAS controllers. They found OODA is more intuitive and extensible than the control system approach and allows asynchronous processing of information between stages. The work of Karim and Heinze was limited to a simple mission of choosing an alternative waypoint. Priorities in developing a framework include design intuitiveness, extendability, and scalabilities. In our framework, the OODA approach is used to control the UAS.

UASs have been explored for search and rescue (SAR) missions in other related work [33–35]. Hoai and Phuong [36] studied anomaly color detection algorithms on UAS images for SAR work. Niedzielski et al. [37] conducted SAR field experiments to guide ground searchers in finding a lost person. In these studies, the focus was limited to identifying the targets from UAS images, and the computation was primarily performed at an offboard computation infrastructure. Some researchers have also documented that the communication delay between the UAS and the offboard computation infrastructure is a severe limitation. However, our focus is to develop an autonomous system that can perform onboard detection in real time and closely approach the detected target to perform a close inspection for more informed decision-making.

The assessment of autonomy level is an important aspect to compare and evaluate the performance of the system presented in this paper. Several models have been proposed to assess the autonomy level of the systems, including autonomy level evaluation [38] formulated by the National Aeronautics and Space Administration (NASA), the Autonomous Control Levels (ACL) method [39] proposed by the U.S. military, and the Autonomy Levels for Unmanned Systems (ALFUS) framework [40] by the National Institute of Standards and Technology (NIST)—each model has its drawbacks [41].

The autonomy level evaluation formulated by NASA is used in this research as it is simple and easy to use for comparison of unmanned systems platforms, regardless of their operating environment.

In summary, the task of vision-based multiple ground target finding and action involves solving problems in detection, tracking, navigation, and action on a set of discrete ground targets where visual uniqueness of the targets cannot always be guaranteed, and targets may also move out of the camera's field of view (FOV).

3. Framework

The main contribution of this paper is a navigation framework. This framework was developed and implemented as a modular software system. The current implementation of the framework has six modules: main module, image capture module, a target detection module, mapping module, autopilot driver module and external sensor module. Figure 2 describes the software system architecture.

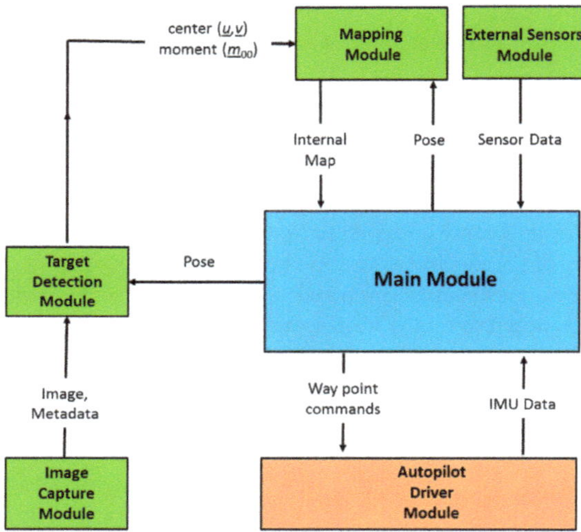

Figure 2. Software system architecture.

The image capture module (Section 3.1), target detection module (Section 3.2), mapping module (Section 3.3), and the main module (Section 3.5) were developed in C++, and Mavros [42] is used as the autopilot driver module. It is available as a ROS [43] package. The main module performs the orient and decide functions of the OODA loop. The following subsections describe each module.

3.1. Image Capture Module

The image capture module captures images from the attached camera sensor and sends the captured images and image metadata, such as timestamp and resolution, to the target detection module.

3.2. Target Detection Module

The target detection module (Figure 3) detects targets and extracts interesting features of the target, such as the center and size, and passes the information to the mapping module. In a ground target finding and action mission, the UAS flies through different sets of altitudes depending on which step (Figure 1) the UAS performs. A static target detection algorithm may not perform well for all altitudes. Therefore, the target detection module receives the UAS pose information from the main module and adjusts the detection method according to the altitude of the UAS for more accurate detection.

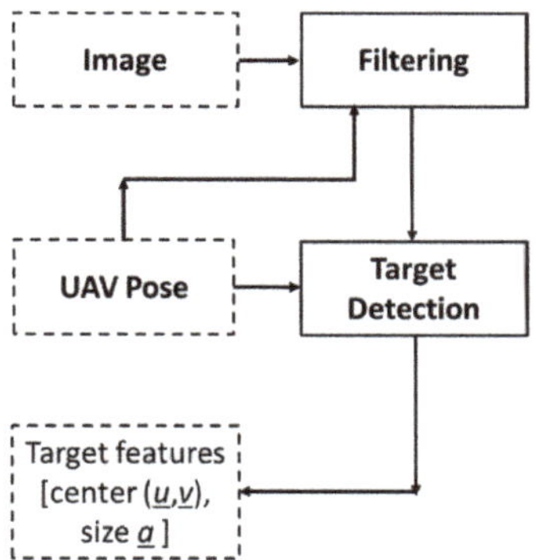

Figure 3. Target detection module.

The target detection module performs two different functions: filtering and target detection. The filtering function rejects unsuitable images from the processing pipeline and helps to reduce some processing cycles of the onboard computer. The target detection error increases with the rotation rate of the UAS. Therefore, the criterion given in Equation (1) is used to discard the images.

$$\exists n \in \mathbb{Z}^+, \quad |\phi_n - \phi_{n-1}| > \alpha \vee |\theta_n - \theta_{n-1}| > \alpha \vee |\psi_n - \psi_{n-1}| > \alpha \tag{1}$$

Here,ϕ_n, θ_n and ψ_n are the roll, pitch and yaw angles of the UAS in the nth frame, and α is the threshold value determined experimentally. UAS images might also be susceptible to motion blur depending on the camera and UAS speed. Therefore, the filtering function can be implemented to include automatic blur detection algorithms described in [44] to filter those images, which is beneficial when more complex target detection algorithms are used. However, our current implementation did not use these methods, considering the processing overhead.

3.3. Mapping Module

The mapping module creates an internal map of all detected targets. The map helps to track the targets when they go outside the FOV. The map is also used to compensate for the effects of localization errors due to GPS noise and drift. Figure 4 shows an image (Figure 4a) and the corresponding internal map (Figure 4b) in a graphical illustration. In Figure 4a, the fourth target T_4 is outside the FOV.

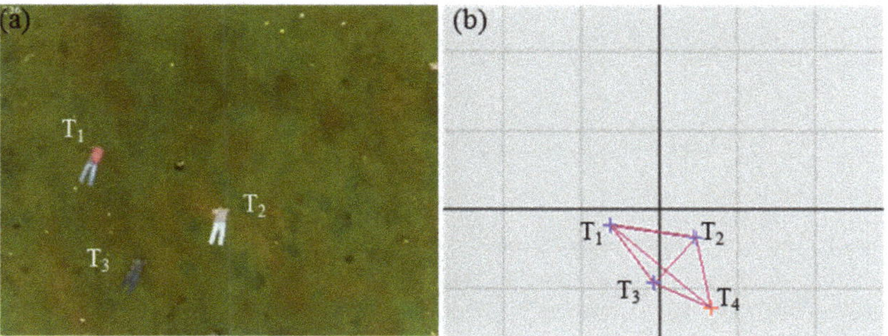

Figure 4. Image with targets (**a**) and corresponding internal map in graphical illustration (**b**).

The mapping module creates an internal map through the following steps:

3.3.1. Estimate Target Position

The estimate target position step estimates the 3D position of the target in the inertial frame from the outputs of the target detection module. Our previous paper [27] describes the detailed steps of the method.

3.3.2. Track Targets

The track targets step matches the target between each image frame. The estimated position of each target may vary throughout the mission due to inaccuracies and sensor drift. However, the variation between adjacent frames is small; therefore, targets are matched using the Euclidean distance between the estimated target positions in adjacent frames.

Target $T_{i,n}$ and target $T_{j,n-1}$ are the same target if:

$$\exists n, i, j \in \mathbb{Z}^+, \sqrt{(\hat{x}_{i,n} - \hat{x}_{j,n-1})^2 + (\hat{y}_{i,n} - \hat{y}_{j,n-1})^2} < d \quad (2)$$

where $T_{i,n}$ is the ith detected target from the nth frame, $(\hat{x}_{i,n}, \hat{y}_{i,n})$ is the position estimation of the ith detected target from the nth image frame, and d is the gating distance.

The filtering stage keeps d within a reasonable limit.

3.3.3. Build Internal Map of Adjacent Targets

When the UAS descends towards a selected target for an action, other targets go out of the camera FOV. However, the target selected for the action can be visible most of the time. An internal map of the targets is built using weighted graph representation to use this property. This map also tracks the locations of the targets already registered, but is undetected in specific frames (false negatives).

The target $T_{i,m}$ and the target $T_{j,n}$ are adjacent if:

$$\exists\, m, n, i, j \in \mathbb{Z}^+, m = n \wedge i \neq j \quad (3)$$

Figure 5 shows an example of an internal map using four targets. Nodes T_1–T_4 represent the targets, and the weights d_1–d_6 represent the distance between them.

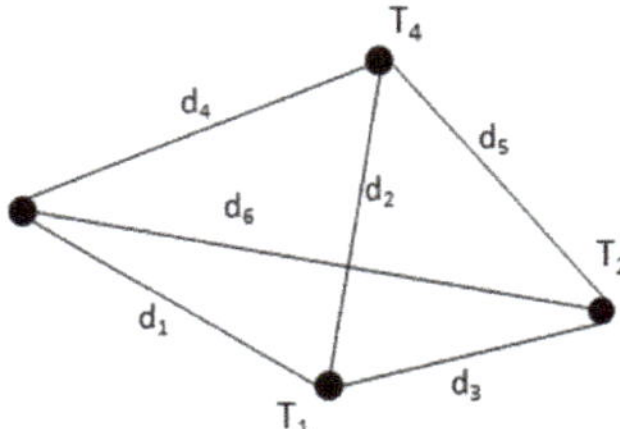

Figure 5. An example of the internal map using four targets.

3.3.4. Remove Duplicates

Due to the errors in estimating the target's position, duplicate nodes (targets) are introduced in the internal mapping step. This step removes these duplicate nodes and their vertices from the map. The node T_i and its vertices are removed if:

$$\exists n, i, j \in \mathbb{Z}^+, \sqrt{(\hat{x}_{i,m} - \hat{x}_{j,n})^2 + (\hat{y}_{i,m} - \hat{y}_{j,n})^2} < d \quad (4)$$

3.3.5. Update Position

Due to detection errors in the target detection module, an already registered target may not be detected in adjacent frames, or the target may go outside the camera FOV when the UAS is flying at a lower altitude. In this case, the positions of the previously registered but not visible (not detected/out of FOV) targets are updated in this step by considering one of the visible targets as a base target. The updated location of the other target is given by:

$$\exists\, n, i, j, k \in \mathbb{Z}^+,\ i \neq k$$

$$\hat{x}_{i,n} = \hat{x}_{k,n} + D\hat{x}_{k,j} \tag{5}$$

$$\hat{y}_{i,n} = \hat{y}_{k,n} + D\hat{y}_{k,j} \tag{6}$$

Here the kth target is a base target.$D\hat{x}_{k,j}$, $D\hat{y}_{k,j}$ are the distance between the kth and ith targets along the x and y axis, respectively. Here the distances are estimated using the jth frame.

3.3.6. Remove False Targets

Very accurate target detection is needed for reliable navigation of the UAS because false positives and false negatives may change the UAS's path and make the entire operation inefficient. All target detection algorithms have a certain degree of uncertainty unless the target is very simple. The degree of uncertainty may increase with the distance between the target and the UAS. In an outdoor environment, other factors such as fog, time of day, and lighting changes may also increase this uncertainty.

It is found that false positives are usually not persistent across all images. A weighted voting scheme is used to reject false positives. In this scheme, each target T_i holds votes V_i. If the target T_i is detected in a frame, it casts w_d number of votes:

$$\exists n, i, j \in \mathbb{Z}^+,\ V_i \leftarrow V_i + w_d \tag{7}$$

where w_d can be a simple constant value or modeled as a complex function of other parameters such as distance, weather, time of day and illumination conditions, which determine the certainty of detection. For example, detection at a lower altitude may be assigned more votes than detection from a higher altitude. The optimal function of w_d is specific to each particular application.

If the target T_i is not detected in a frame but its location is inside the FOV, it casts $-w_n$ number of votes (Equation (8)). The value of w_n can also be determined similar to w_d:

$$\exists\, i, w_n \in \mathbb{Z}^+,\ V_i \leftarrow V_i - w_n \tag{8}$$

The target T_i is removed if:

$$\exists\, i \in \mathbb{Z}^+,\ V_i < G \tag{9}$$

The value of the G is experimentally determined.

3.4. External Sensors Module

The external sensor module allows the connection of sensors such as sonar directly to the onboard computer for additional state estimation information. In our implementation, only an ultrasonic module is used as an external sensor to compensate for altitude measurement errors induced by strong gusts in barometric height measurement used in typical low cost autopilots.

3.5. Main Module

A finite state machine (FSM) model is used to implement the main module of the system. Figure 6 shows the finite state machine. The state of the system is controlled by following an OODA loop. Here, a state represents the current belief of the system. According to the observations received from other modules, the system orients and transitions to the next suitable state.

Figure 6. The finite state machine (FSM) of the main module.

If a target or targets are observed in the search state, the system is updated to the *move to target* state. In the *re-estimate target position* state, the UAS moves laterally towards the target and updates the target position. In the *descend* state, the UAS descends by a predefined amount of height. In the *adjust* state, the UAS aligns its x, y position above the target by the proportional controller defined by Equations (10) and (11). This controller also assists the UAS to maintain its hovering position against external disturbances such as wind.

$$x_n = x_c + (U_o - u)K/R_u \tag{10}$$

$$y_n = y_c + (V_o - v)K/R_v \tag{11}$$

In these equations, (x_n, y_n) is the output position of the controller, (x_c, y_c) is the current UAS position and (U_o, V_o) is the optical center of the camera sensor in pixels. (u, v) is the center of the target in pixels, K is proportional gain, and R_u and R_v are the horizontal and vertical resolutions of the image.

In the *climb* state, the UAS increases its height by a small amount (<2 m) and transitions into the *confirm target* state. The *confirm target* state confirms the availability of the target. If the availability of the target is not confirmed, it will be removed from the internal map, considering it as a false positive. Actions such as inspection or spraying are performed in the *action* state. After completing the action, target T_i is selected as the next target if:

$$\exists\, i, j, l \in \mathbb{Z}^+, (D_i = \min_{j=\{1\ldots l\}} D_j) \wedge V_i > H \tag{12}$$

where D_i is the distance to the target T_i and V_i is the number of votes held by the target T_i. H is the minimum number of votes that qualify the target as a valid target for a visit. If there is no target to satisfy the criterion (12), the UAS lands at a predefined position.

3.6. Autopilot Driver Module

The autopilot driver module acts as an intermediary between the autopilot firmware and the onboard computer system. This module provides data such as autopilot state, inbuilt sensor data and battery state to the main module. Commands from the main module are also translated into autopilot specific control commands.

4. Experiments

4.1. Hardware System

A quadrotor UAS was developed to conduct outdoor field experiments using a Pixhawk 2 open source autopilot and a DJI F450 frame. The Pixhawk consists of accelerometers, gyroscopes, barometer sensors and other devices on a single board. Figure 7 shows the hardware system used in the experiments. The hardware system architecture shows the components integrated into the UAS and their interconnection, with images of important components. The Raspberry Pi3 onboard computer, camera and sonar are attached under the UAS.

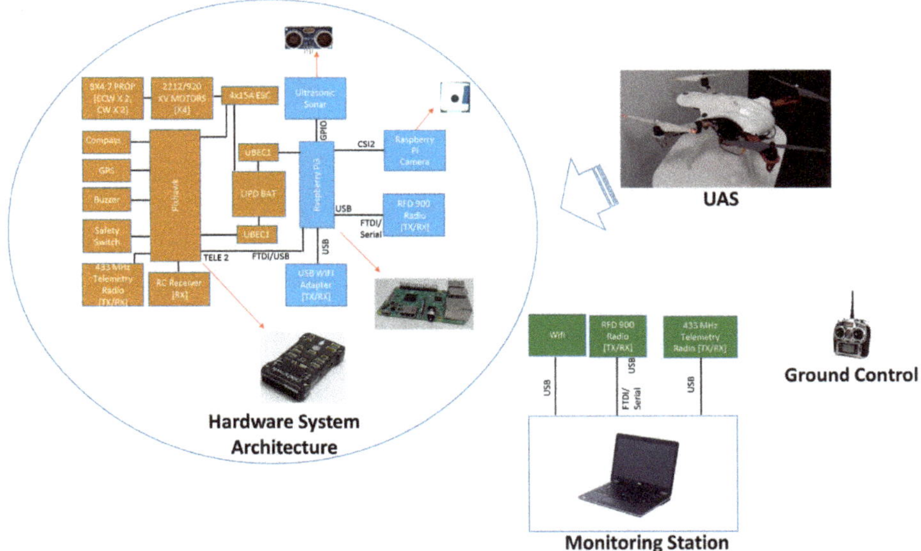

Figure 7. Hardware system used in the experiments.

The implemented framework was fully executed in the onboard Raspberry Pi3 computer. A ground control RC transmitter was used only to trigger the system.

4.2. Field Experiments

As mentioned already, our focus is to develop and demonstrate a framework for vision-based multiple ground target finding and action using a multirotor UAS that is robust to the (i) errors and uncertainties in localization of the UAS; (ii) errors and uncertainties in localization of the target; (iii) target detection errors; and (iv) effects of external disturbances such as wind. In this paper, we have validated our framework using two different test cases: (1) Finding and inspection of various type of objects scattered on the ground in search and rescue after a plane crash. (2) Finding and inspection of multiple red color ground objects.

Figure 8 shows the scenario used for both test cases. The UAS has to take off to the search height h_s (e.g., 40 m) and fly from the waypoint A to the waypoint B and search for targets. If any target/s are detected, the UAS picks a target and changes its flight path, descends and hovers above each target autonomously, as shown in Figure 1, steps 3, 4 and 5. Six targets were scattered randomly. The search boundary was 50 m × 50 m. The test was repeated multiple times, and the exact positions of the targets were randomly changed for different tests. However, a minimum separation distance of 4 m was kept between the targets, and the visibility of all targets in the FOV of the camera at search height h_s (40 m) was maintained. Arbitrary shapes in different colors indicates the targets in Figure 8.

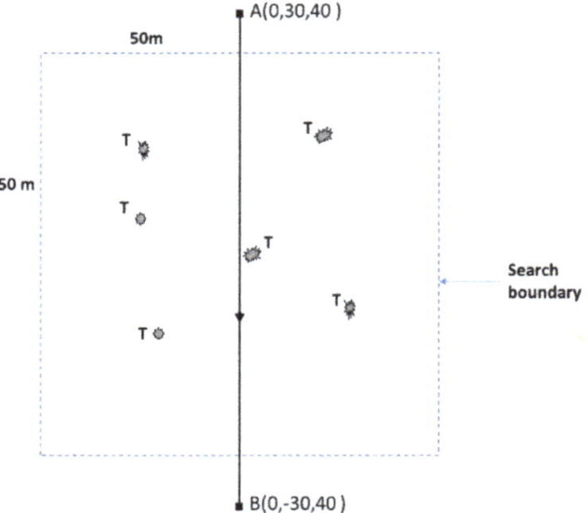

Figure 8. Test case scenario. The dotted square (50 m × 50 m) indicates the search boundary. T indicates scattered targets within the search boundary. Arbitrary shapes of the targets are used to indicate that the targets are not defined by their shape. Waypoints are indicated by (•).

The experiments presented here used several parameters related to the framework and the hardware system used in the experiments. Table 1 lists these parameter values.

Table 1. Parameters used in the experiments.

Paramete	Value
Search height (h_s)	40 m
Threshold rotation rate (α)	0.8 deg/s
Gating distance (d)	2 m
Camera frame rate	10 Hz
Camera resolution	640 × 480
Proportional gain (K)	2
Votes for a detection (w_d)	1
Votes for a non-detection (w_n)	−1
Removal cut-off (G)	−2
Valid target cut-off (H)	5

Parameters camera resolution and frame rate describe the camera used. Other parameters, such as threshold rotation rate, gating distance, proportional gain, votes for a detection, votes for a non-detection, removal cut-off and valid target cut-off are related to the framework and determined using simulation and trial-and-error experiments.

These framework-related parameters have impact on the performance of the framework. For example, the gating distance d was set to 2 m in our experiments. Setting a larger value may label two different targets as the same target while setting a smaller value may label the same target in adjacent frames as different targets. The valid target cut-off H was set as 5. Therefore, if a target holds less than five votes, it will not be considered as a possible candidate for a visit or inspection. Hence, a target that appeared in less than five images will be ignored by the UAS. Reducing this value may increase the chance of false targets to become a valid target. Similar effects are possible with the value for removal cut-off G. These values can be adjusted depending on the accuracy of the target detection method used. For an example, a lower value of valid target cut-off H is adequate for more accurate target detection algorithm.

4.2.1. Test Case 1: Finding and Inspection of Objects Scattered on the Ground in Search and Rescue after a Plane Crash

In this test case, a scenario was formulated to represent objects scattered in a search and rescue mission after a plane crash. The unmanned aerial system (UAS) must autonomously find these targets and conduct a close inspection at a low altitude. This capability is very useful in search and rescue operations in a larger area with limited human resources or in difficult terrains, such as searching for survivors or a plane's black box. Figure 9a shows objects scattered at the experimental site from the ground view. Figure 9b shows the test site and targets from a UAS image. Images from Malaysia Airlines flight MH17 crash site [45–47] were used as references. Multiple objects such as clothing to represent people, and suitcases and backpacks were used as targets. In this test case, the targets have different shapes, colors and sizes, and the possibility of false detection is relatively high because of the generalized target detection algorithm.

Figure 9. (**a**) Objects scattered at the experimental site in the test scenario; (**b**) Top view of the experimental site, and target objects from a UAS image.

In order to detect various types of objects present in a vegetated area, the target detection algorithm illustrated in Figure 10 was implemented in the target detection function of the target detection module (see Figure 3).

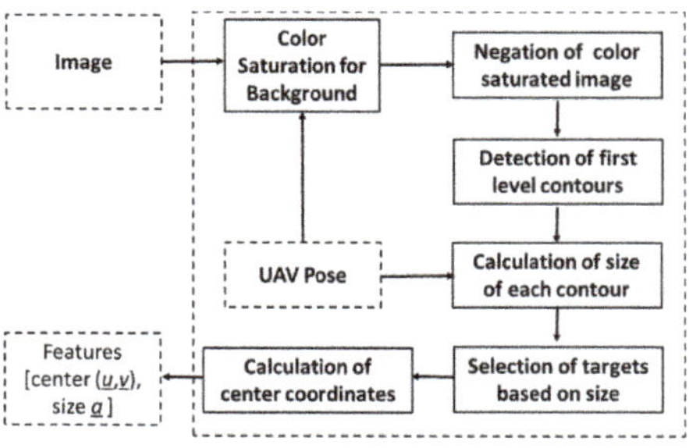

Figure 10. Implementation of the target detection function in the search and rescue test case.

In the first step, the image is saturated for the background color (e.g., green). In the second step, the resulting binary image is negated. In the third step, first level contours are detected. Next, the approximate size a (area on the ground) of each contour is calculated using Equation (13).

$$a = m_{00}\frac{z^2}{f^2} \tag{13}$$

Here m_{00} is the 0th moment, z is the altitude of the UAS, and f is the focal length of the camera. In a binary image, 0th moment m_{00} is defined by:

$$m_{00} = \sum_{u=0}^{u=w}\sum_{v=0}^{v=h} J(u,v) \tag{14}$$

where w and h are the width and height of the blob, respectively. $J(u,\ v)$ is the value of the pixel at $(u,\ v)$.

In the next step, the targets were selected based on size. Very small and large targets were omitted. The size of targets must be configured according to the specific application. The primary step of the detection algorithm is color saturation for the background, such as green vegetation. A simple detection algorithm was selected in order to show the capability of the framework in the presence of a high number of false positives.

Test case 1 was repeated five times. Table 2 summarizes the results. The UAS visited all six targets in three of the five flight tests. One target was missed in two tests because the number of votes acquired by the target was less than the minimum number of votes H to qualify to visit the target. An analysis of experimental data also showed that in three tests, the system (main module) did not activate the action state and considered one of the targets as a false detection. A detailed investigation revealed that this was due to detection failure at low altitude (2–4 m). A possible reason might be the auto adjustment settings of the Raspberry Pi camera used.

Table 2. Summary of results from flight tests.

Test No.	Number of Targets	Number of Targets Visited	Number of Targets Not Inspected (Action Failure)
1	6	6	1
2	6	5	0
3	6	5	0
4	6	6	1
5	6	6	1

Figure 11 shows the selected images from test 5.

Figure 12 shows the results from test 5. In Figure 12a, the UAS searches for the targets while flying from waypoint A to B. At point C, the first target is recognized, and the finite state machine state changes to *move to target* state. The UAS moves towards the first target designated T_1 (step 2, Figure 1) and performs an inspection (hovering above the target for 3 s). Next, the UAS climbs to the cruising height h_c (4 m) and moves to the next nearest adjacent target T_5. The process repeats until the UAS has visited all six identified targets.

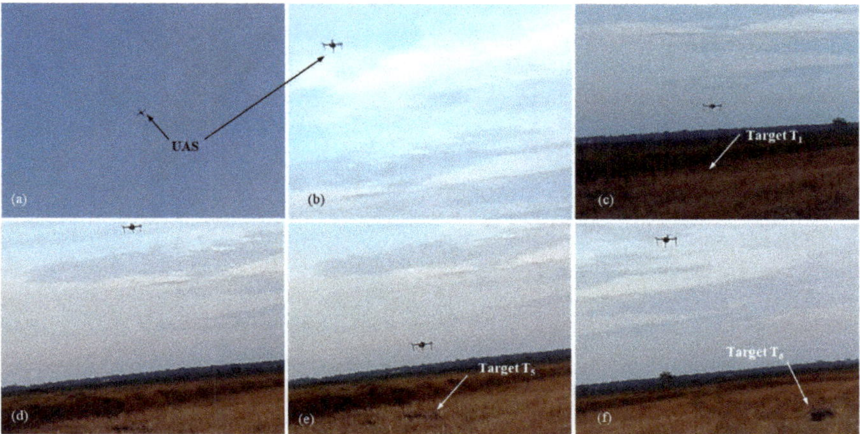

Figure 11. Pictures from test case 1-test 5: (**a**) UAS searches ground target; (**b**) UAS descends towards target T_1; (**c**) UAS inspects target T_1; (**d**) UAS moves towards target T_5; (**e**) UAS inspects target T_5; (**f**) UAS moves towards target T_6.

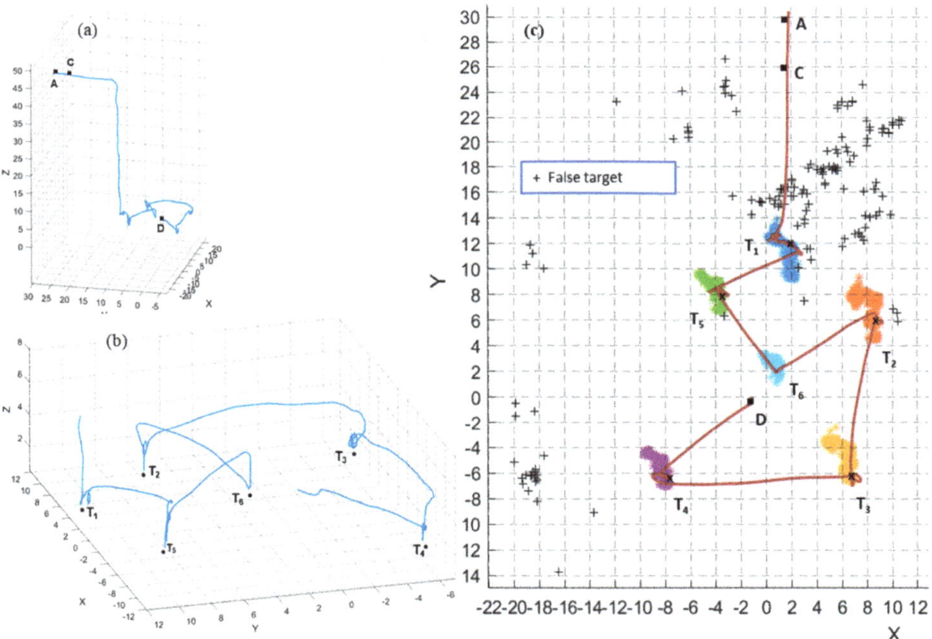

Figure 12. Results from flight test 5: (**a**,**b**) 3D view of the flight trajectory in full and zoomed view of the visiting targets and action part; (**c**) Top view of the flight trajectory and scatter plot of estimated target positions in the full mission of flight test 5. (**+**) indicates false targets. (**x**) indicates the estimated position of the target when the UAS was performing the inspection (hovering above the target). Waypoint B is omitted to increase the clarity of the plots.

However, in this test, the action state is not activated for target T_6. The target detection module failed to detect target T_6 in the confirm state of the finite state machine. T_6 was considered a false target, and the UAS moved to the next target T_2. This shows the autonomous re-planning ability of the system without getting stuck with target T_6.

Figure 12c illustrates the top view of the flight trajectory and the scatter plot of the estimated target positions throughout the full mission. The plot shows several false targets. Moreover, ideally, the position of true targets T_1–T_6 must appear as points. However, they appear like a cloud due to the variation in the estimates because of the localization errors. The highest variation calculated as the furthest distance between two estimates is 5 m in T_2. In total, 117 different possible targets were detected throughout the mission, but only six of them—the actual targets—were visited by the UAS. Therefore, the developed framework can overcome localization errors and target detection errors in target finding and action tasks.

4.2.2. Test Case 2: Finding and Inspection of Multiple Red Color Ground Objects

In this test case, the UAS must fly according to the same scenario described in Figure 8 and find multiple red color ground objects and approach them and conduct a close inspection at a low altitude. Six similar red color circles with 0.2 m radius were used as targets. For this test case, only the target detection function of the target detection module was modified. Instead of the detection algorithm described in Figure 10, a simple red color detection algorithm was implemented. Similar targets were used in order to demonstrate that the framework's capability, even in the absence of additional information such as unique color and shape for tracking and control. Figure 13 shows the test site and targets from a UAS image. Figure 14 shows the selected images from one of the tests. Test case repeated five times as before. UAS has visited and inspected all six targets in all five tests. A link to the video of the test is given at the end of the paper. Due to the similarity to the test case 1 results, a detailed discussion of the results is avoided.

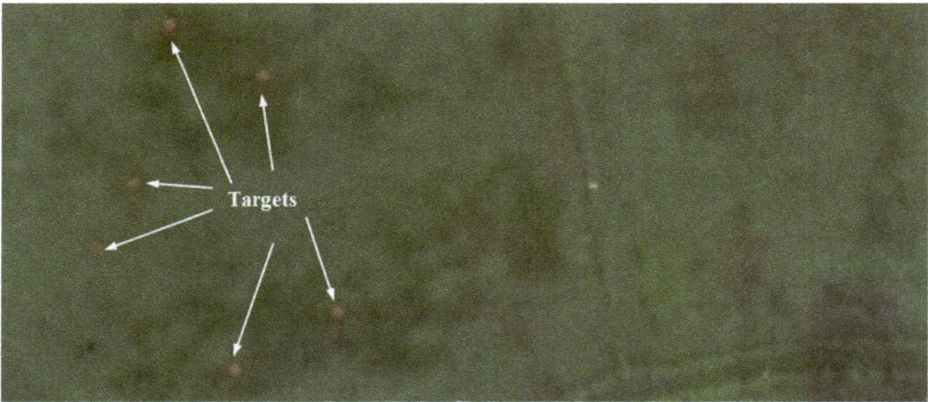

Figure 13. Top view of the experimental site and target objects from a UAS image.

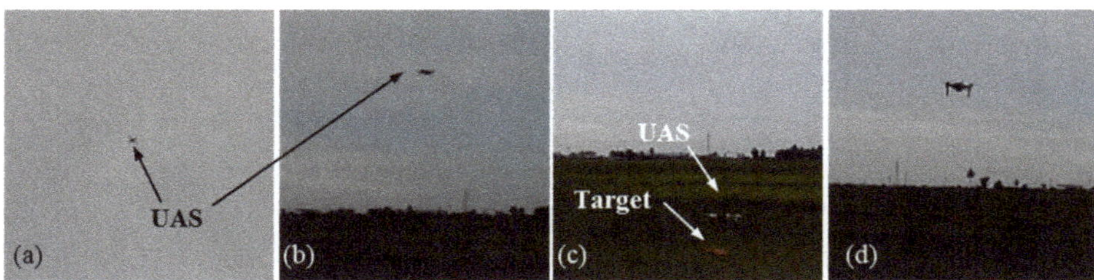

Figure 14. Pictures from a test in test case 2: (**a**) UAS searches ground targets; (**b**) UAS descends towards targets; (**c**) UAS inspects a target; (**d**) UAS moves towards another target for inspection.

The test case 2 was conducted under heavy rain cloud. However, all the tests were successful is implicates the robustness of the framework to such conditions as well.

5. Discussion

To our knowledge, no previous work has attempted multiple ground target finding and action using UAS. Related work on vision-based landing provides some insights. However, the techniques used in those studies are mostly limited to a single target and are only demonstrated at low altitudes (<5 m) where the effects of detection and localization errors have minimal impact. One possible technique for multiple target finding and action is to estimate the targets' positions from the image and guide the UAS towards each target. However, this technique requires a more precise and accurate localization of the target as well as the UAS and a 100% accurate target detection. Given the dynamic nature of the task and the outdoor environment, a 100% accurate target detection is nearly impossible [34,48]. For an example, we observed a large number of false detections when sudden lighting changes occurred due to the movement of rainy clouds. Moreover, image-based target positioning had errors due to factors such as lens distortion, imaging sensor resolution, GPS accuracy, barometric altimeter resolution and target height [49]. GPS sensors used in typical, low-cost UASs are also subject to drift [27]. This study has developed a framework to address these challenges with less computational requirements and validated the framework using two different test cases with different implementations of the target detection module. The navigation framework proposed in this paper can address these challenges independent from detection algorithm thus a reasonable accuracy from detection algorithms is adequate to perform the task.

When we consider one potential application of the framework such as search and rescue, in earlier works [33–37], UAS was limited in its use as only as a monitoring device. The framework proposed in this research further extends the capabilities of UASs to more active participation, such as conducting a detailed closer inspection or delivering emergency kits. In other studies on railway semaphore inspection [23] and pole inspection [24], the UAS was manually positioned at a workable distance, and researchers did not address the part of finding and approaching the target for inspection. The framework proposed in this research may be used in conjunction with those techniques to increase autonomy. This framework has opened the whole new avenue of research where UASs can be used for close inspection of ground targets for more informed decision-making or can be used to perform an action on the target.

The system presented in this paper uses a higher level decision-making approach OODA for control and has no operator involvement from the mission start to end, and can automatically re-plan the task in the absence of identified targets or if new targets appear. This ability meets level 4 autonomy, according to the autonomy level evaluation method of NASA [38]. However, the framework's lack of the capability to avoid obstacles means it cannot perceive the environment widely as required by level 4 autonomy. This requirement needs to be addressed in the future to achieve a fully compatible level 4 autonomy.

The proposed framework has limitations. The voting-based scheme employed to remove false detections assumes that false detections are not always persistent. However, this might not be true for all target detection methods. Techniques used for target tracking assume errors and drift in sensors are gradual over time, which might fail when abrupt changes occur. The framework also assumes the targets are located in reasonably flat terrain. Although this assumption may be appropriate for agricultural land, further research is needed to address targets located in hilly or sloping terrain.

6. Conclusions

This paper proposed a framework for autonomous, multiple ground target-finding and action using a low-cost, multirotor unmanned aerial system. The framework can be effectively used in a variety of UAS applications with suitable detection algorithms in fully and supervised autonomous missions such as inspection, search and rescue and spot spraying in agriculture. Uncertainties and errors introduced by the onboard sensors and external disturbances such as wind pose significant challenges in real outdoor environments. A framework was proposed using the high-level decision-making approach of OODA, a series of techniques, and a modular software system to address these challenges.

Two different test cases were formulated to search and inspect multiple ground objects scattered on the ground. Each test case was repeated multiple times to show that the framework can overcome the errors and uncertainties and perform the task successfully. The detection algorithm is simple compared to the state-of-the-art algorithm used in precision agriculture or search and rescue work. There are two reasons for the simple algorithm: one is to demonstrate the capability of the framework with a high number of false detections, and the other one is the limited computation capacity of the Raspberry Pi used in our UAS.

Ongoing work focuses on integrating a sprayer system to a suitable UAS platform, identifying a test site, and obtaining approvals to test the system in paddy fields for brown planthopper control application.

A video of the flight tests can be found at https://www.youtube.com/watch?v=QVmHo_uUkfU.

Author Contributions: A.H. developed the software and contributed to methodology, investigation, data curation, formal analysis, validation and writing (original draft). R.R. contributed to supervision and writing (review and editing). J.R. contributed to supervision. F.G. conceptualized the overarching idea and contributed to funding acquisition, project administration, supervision and writing (review and editing). All authors have read and agreed to the published version of the manuscript.

Funding: This research was partially fuded by University Grant Commission Sri Lanka grant number ucg/dric/qut2015/seusl/01 and the APC was funded by Felipe Gonzalez.

Conflicts of Interest: The authors declare no conflicts of interest.

References

1. Kontitsis, M.; Valavanis, K.P.; Tsourveloudis, N. A Uav Vision System for Airborne Surveillance. In Proceedings of the IEEE International Conference on Robotics and Automation (ICRA '04), New Orleans, LA, USA, 26 April–1 May 2004; Volume 17, pp. 77–83.
2. Gonzalez, L.F.; Montes, G.A.; Puig, E.; Johnson, S.; Mengersen, K.; Gaston, K.J. Unmanned aerial vehicles (uavs) and artificial intelligence revolutionizing wildlife monitoring and conservation. *Sensors* **2016**, *16*, 97. [CrossRef] [PubMed]
3. Villa, T.F.; Salimi, F.; Morton, K.; Morawska, L.; Gonzalez, F. Development and validation of a uav based system for air pollution measurements. *Sensors* **2016**, *16*, 2202. [CrossRef] [PubMed]
4. Bandara, P.T.; Jayasundara, M.U.P.; Kasthuriarachchi, R.; Gunapala, R. *Brown plant hopper (bph) on rice In Pest Management Decision Guide: Green and Yellow List*; CAB International: Oxfordshire, England, 2013.
5. Alsalam, B.H.Y.; Morton, K.; Campbell, D.; Gonzalez, F. Autonomous Uav with Vision Based on-Board Decision Making for Remote Sensing and Precision Agriculture. In Proceedings of the 2017 IEEE Aerospace Conference, Big Sky, MT, USA, 4–11 March 2017; pp. 1–12.
6. Gps Accuracy. Available online: https://www.gps.gov/systems/gps/performance/accuracy/ (accessed on 6 June 2019).
7. Lieu, J. Australia Is Spending $200 Million to Make Its GPS More Accurate. Available online: https://mashable.com/2018/05/09/gps-australia-investment-accuracy/#SsV9QnDCksqI (accessed on 22 September 2018).
8. Mindtools.com. Ooda Loops Understanding the Decision Cycle. Available online: https://www.mindtools.com/pages/article/newTED_78.htm (accessed on 10 May 2016).
9. Thoms, J. Autonomy: Beyond ooda? In Proceedings of the 2005 The IEE Forum on Autonomous Systems (Ref. No. 2005/11271), Stevenage, UK, 28 Novenber 2005; p. 6.
10. Plehn, M.T. Control Warfare: Inside the Ooda Loop. Master's Thesis, Air University, Maxwell Airforce Base, AL, USA, 2000.
11. Yang, S.; Scherer, S.A.; Zell, A. An onboard monocular vision system for autonomous takeoff, hovering and landing of a micro aerial vehicle. *J. Intell. Robot. Syst.* **2013**, *69*, 499–515. [CrossRef]
12. Zarudzki, M.; Shin, H.S.; Lee, C.H. An image based visual servoing approach for multi-target tracking using an quad-tilt rotor uav. In Proceedings of the 2017 International Conference on Unmanned Aircraft Systems (ICUAS), Miami, FL, USA, 13–16 June 2017; pp. 781–790.

13. Greatwood, C.; Bose, L.; Richardson, T.; Mayol-Cuevas, W.; Chen, J.; Carey, S.J.; Dudek, P. Tracking control of a uav with a parallel visual processor. In Proceedings of the 2017 IEEE/RSJ International Conference on Intelligent Robots and Systems (IROS), Vancouver, BC, Canada, 24–28 September 2017; pp. 4248–4254.
14. Vanegas, F.; Bratanov, D.; Powell, K.; Weiss, J.; Gonzalez, F. A novel methodology for improving plant pest surveillance in vineyards and crops using uav-based hyperspectral and spatial data. *Sensors* **2018**, *18*, 260. [CrossRef] [PubMed]
15. Perez-Grau, F.J.; Ragel, R.; Caballero, F.; Viguria, A.; Ollero, A. An architecture for robust uav navigation in gps-denied areas. *J. Field Robot.* **2018**, *35*, 121–145. [CrossRef]
16. Zhang, J.; Wu, Y.; Liu, W.; Chen, X. Novel approach to position and orientation estimation in vision-based uav navigation. *IEEE Trans. Aerosp. Electron. Syst.* **2010**, *46*, 687–700. [CrossRef]
17. Zhang, J.; Liu, W.; Wu, Y. Novel technique for vision-based uav navigation. *IEEE Trans. Aerosp. Electronic Syst.* **2011**, *47*, 2731–2741. [CrossRef]
18. Stefas, N.; Bayram, H.; Isler, V. Vision-based uav navigation in orchards. *IFAC Pap.* **2016**, *49*, 10–15. [CrossRef]
19. Brandao, A.S.; Martins, F.N.; Soneguetti, H.B. A vision-based line following strategy for an autonomous uav. In Proceedings of the 2015 12th International Conference on Informatics in Control, Automation and Robotics (ICINCO), Colmar, France, 21–23 July 2015; SciTePress: Colmar, Alsace, France; pp. 314–319.
20. Ramirez, A.; Espinoza, E.S.; Carrillo, L.R.G.; Mondie, S.; Lozano, R. Stability analysis of a vision-based uav controller for autonomous road following missions. In Proceedings of the 2013 International Conference on Unmanned Aircraft Systems (ICUAS), Piscataway, NJ, USA, 28–31 May 2013; IEEE: Piscataway, NJ, USA; pp. 1135–1143.
21. Araar, O.; Aouf, N. Visual servoing of a quadrotor uav for autonomous power lines inspection. In Proceedings of the 22nd Mediterranean Conference on Control and Automation, Palermo, Italy, 16–19 June 2014; pp. 1418–1424.
22. Mills, S.J.; Ford, J.J.; Mejías, L. Vision based control for fixed wing uavs inspecting locally linear infrastructure using skid-to-turn maneuvers. *J. Intell. Robot. Syst.* **2011**, *61*, 29–42. [CrossRef]
23. Máthé, K.; Buşoniu, L. Vision and control for uavs: A survey of general methods and of inexpensive platforms for infrastructure inspection. *Sensors* **2015**, *15*, 14887–14916. [CrossRef] [PubMed]
24. Máthé, K.; Buşoniu, L.; Barabás, L.; Iuga, C.I.; Miclea, L.; Braband, J. In Proceedings of the Vision-based control of a quadrotor for an object inspection scenario. In Proceedings of the 2016 International Conference on Unmanned Aircraft Systems (ICUAS), Arlington, VA, USA, 7–10 June 2016; pp. 849–857.
25. Sa, I.; Hrabar, S.; Corke, P. Inspection of pole-like structures using a visual-inertial aided vtol platform with shared autonomy. *Sensors* **2015**, *15*, 22003–22048. [CrossRef] [PubMed]
26. Choi, H.; Geeves, M.; Alsalam, B.; Gonzalez, F. Open source computer-vision based guidance system for uavs on-board decision making. In Proceedings of the 2016 IEEE Aerospace Conference, Big Sky, MT, USA, 5–12 March 2016; pp. 1–5.
27. Hinas, A.; Roberts, J.; Gonzalez, F. Vision-based target finding and inspection of a ground target using a multirotor uav system. *Sensors* **2017**, *17*, 2929. [CrossRef] [PubMed]
28. Hinas, A.; Ragel, R.; Roberts, J.; Gonzalez, F. Multiple ground target finding and action using uavs. In Proceedings of the IEEE Aerospace Conference 2019, Big Sky, MT, USA, 2–9 March 2019.
29. Hinas, A.; Ragel, R.; Roberts, J.; Gonzalez, F. A framework for vision-based multiple target finding and action using multirotor uavs. In Proceedings of the 2018 International Conference on Unmanned Aircraft Systems (ICUAS), Dallas, TX, USA, 12–15 June 2018; pp. 1320–1327.
30. Theunissen, E.; Tadema, J.; Goossens, A.A.H.E. Exploring network enabled concepts for u(c)av payload driven navigation. In Proceedings of the 2009 IEEE/AIAA 28th Digital Avionics Systems Conference, Orlando, FL, USA, 23–29 October 2009; pp. 3–15.
31. Sari, S.C.; Prihatmanto, A.S. Decision system for robosoccer agent based on ooda loop. In Proceedings of the 2012 International Conference on System Engineering and Technology (ICSET), Bandung, Indonesia, 11–12 September 2012; pp. 1–7.
32. Karim, S.; Heinze, C. Experiences with the design and implementation of an agent-based autonomous uav controller. In Proceedings of the 4rd International Joint Conference on Autonomous Agents and Multiagent Systems (AAMAS 2005), Utrecht, The Netherlands, 25–29 July 2005; Association for Computing Machinery: Utrecht, The Netherlands; pp. 59–66.

33. Doherty, P.; Rudol, P. A uav search and rescue scenario with human body detection and geolocalization. In Proceedings of the 20th Australian Joint Conference on Artificial Intelligence, Gold Coast, Australia, 2–6 December 2007; pp. 1–13.
34. Rudol, P.; Doherty, P. Human body detection and geolocalization for uav search and rescue missions using color and thermal imagery. In Proceedings of the 2008 IEEE Aerospace Conference, Big Sky, MT, USA, 1–8 March 2008; IEEE: Piscataway, NJ, USA, 2008; p. 8.
35. Sun, J.; Li, B.; Jiang, Y.; Wen, C.-y. A camera-based target detection and positioning uav system for search and rescue (sar) purposes. *Sensors* **2016**, *16*, 1778. [CrossRef] [PubMed]
36. Hoai, D.K.; Phuong, N.V. Anomaly color detection on uav images for search and rescue works. In Proceedings of the 2017 9th International Conference on Knowledge and Systems Engineering (KSE), Hue, Vietnam, 19–21 October 2017; pp. 287–291.
37. Niedzielski, T.; Jurecka, M.; Miziński, B.; Remisz, J.; Ślopek, J.; Spallek, W.; Witek-Kasprzak, M.; Kasprzak, Ł.; Świerczyńska-Chlaściak, M. A real-time field experiment on search and rescue operations assisted by unmanned aerial vehicles. *J. Field Robot.* **2018**, *35*, 906–920. [CrossRef]
38. Young, L.; Yetter, J.; Guynn, M. System analysis applied to autonomy: Application to high-altitude long-endurance remotely operated aircraft. In *Infotech@aerospace*; American Institute of Aeronautics and Astronautics: Arlington, VA, USA, 2005.
39. *Unmanned Aerial Vehicles Roadmap 2000–2025*; Office Of The Secretary Of Defense: Washington, DC, USA, 2001.
40. Huang, H.-M. Autonomy levels for unmanned systems (alfus) framework: Safety and application issues. In Proceedings of the 2007 Workshop on Performance Metrics for Intelligent Systems, Gaithersburg, MD, USA, 28–30 August 2007; pp. 48–53.
41. Durst, P.J.; Gray, W.; Nikitenko, A.; Caetano, J.; Trentini, M.; King, R. A framework for predicting the mission-specific performance of autonomous unmanned systems. In Proceedings of the 2014 IEEE/RSJ International Conference on Intelligent Robots and Systems, Chicago, IL, USA, 14–18 September 2014; pp. 1962–1969.
42. Ros.org. Mavros. Available online: http://wiki.ros.org/mavros (accessed on 2 June 2017).
43. Ros.org. Ros. Available online: http://wiki.ros.org (accessed on 6 June 2017).
44. Sieberth, T.; Wackrow, R.; Chandler, J.H. Automatic detection of blurred images in uav image sets. *ISPRS J. Photogramm. Remote Sens.* **2016**, *122*, 1–16. [CrossRef]
45. Oliphant, R. Mh17 Crash Site. The Long Grass Hides the Carnage. Many Other Photos Cannot in Conscience Be Published. Available online: https://twitter.com/RolandOliphant/status/999596340949680128 (accessed on 28 June 2019).
46. Valmary, S. Shelling Adds to Mh17 Nightmare for East Ukraine Village. Available online: https://news.yahoo.com/shelling-adds-mh17-nightmare-east-ukraine-village-193407244.html (accessed on 28 June 2019).
47. Mh17 Ukraine Plane Crash: Additional Details Revealed. Available online: https://southfront.org/wp-content/uploads/2016/11/CIMG0997.jpg (accessed on 28 June 2019).
48. Pena, J.M.; Torres-Sanchez, J.; Serrano-Perez, A.; de Castro, A.I.; Lopez-Granados, F. Quantifying efficacy and limits of unmanned aerial vehicle (uav) technology for weed seedling detection as affected by sensor resolution. *Sensors* **2015**, *15*, 5609–5626. [CrossRef] [PubMed]
49. Nisi, M.; Menichetti, F.; Muhammad, B.; Prasad, R.; Cianca, E.; Mennella, A.; Gagliarde, G.; Marenchino, D. Egnss high accuracy system improving photovoltaic plant maintenance using rpas integrated with low-cost rtk receiver. In Proceedings of the Global Wireless Summit, Aarhus, Denmark, 27–30 November 2016.

Article

Airborne Visual Detection and Tracking of Cooperative UAVs Exploiting Deep Learning

Roberto Opromolla *, Giuseppe Inchingolo and Giancarmine Fasano

Department of Industrial Engineering, University of Naples Federico II, Piazzale Tecchio 80, 80125 Naples, Italy; giu.inch@gmail.com (G.I.); giancarmine.fasano@unina.it (G.F.)

* Correspondence: roberto.opromolla@unina.it; Tel.: +39-081-768-3365

Received: 3 September 2019; Accepted: 5 October 2019; Published: 7 October 2019

Abstract: The performance achievable by using Unmanned Aerial Vehicles (UAVs) for a large variety of civil and military applications, as well as the extent of applicable mission scenarios, can significantly benefit from the exploitation of formations of vehicles able to fly in a coordinated manner (swarms). In this respect, visual cameras represent a key instrument to enable coordination by giving each UAV the capability to visually monitor the other members of the formation. Hence, a related technological challenge is the development of robust solutions to detect and track cooperative targets through a sequence of frames. In this framework, this paper proposes an innovative approach to carry out this task based on deep learning. Specifically, the You Only Look Once (YOLO) object detection system is integrated within an original processing architecture in which the machine-vision algorithms are aided by navigation hints available thanks to the cooperative nature of the formation. An experimental flight test campaign, involving formations of two multirotor UAVs, is conducted to collect a database of images suitable to assess the performance of the proposed approach. Results demonstrate high-level accuracy, and robustness against challenging conditions in terms of illumination, background and target-range variability.

Keywords: unmanned aerial vehicles; UAV swarms; visual detection; visual tracking; machine vision; deep learning; YOLO

1. Introduction

Nowadays, Unmanned Aerial Vehicles (UAVs) represent a reliable and affordable tool suitable for many applications, such as intelligence, surveillance and reconnaissance (ISR) [1], aerial photography [2], infrastructure monitoring [3], search and rescue [4], and precision agriculture [5]. Indeed, the exponential growth in the use of these vehicles has been driven, in recent years, by the miniaturization and cost reduction of electronic components (e.g., microprocessors, sensors, batteries and wireless communication units) which have supported the development of small and micro UAVs [6]. However, while these systems have demonstrated notable advantages being cheap, flexible and easy to deploy, they still have limitations in practical scenarios due to their lower payload capacity, shorter endurance and operational range. These aspects have led to the concept of UAV swarms, i.e., formations of UAVs able to fly in a coordinated manner to provide increased mission performance, reliability, robustness and scalability while ensuring a reduced mission cost [7,8]. For instance, the coordinated flight of multiple UAVs can allow faster area coverage [9,10], improve navigation performance under nominal GNSS coverage [11,12], or enable safe flight in GNSS-challenging areas [13,14].

Clearly, some key challenges must be addressed to unleash the full potential of UAV swarms, thus achieving performance improvement, with respect to missions carried out by standalone UAVs. These challenges may include, but are not limited to, cooperative guidance/control, communication, navigation and situational awareness. In view of guidance and control issues, UAV swarms can act as either centralized or decentralized networks depending on whether the main processing and decision-making functions are concentrated on a single node of the network or distributed among all the nodes [15,16]. Communication issues are also important, and different solutions may be exploited to ensure adequate UAV-UAV datalink, such as Wi-Fi, Ultra/Very High Frequency, LTE, Satcom or based on airborne transponder [17,18], where the choice of the most convenient strategy depends on factors like application scenario, cost budgets and types of UAV involved (e.g., small vs. large UAVs). Finally, regarding navigation and situational awareness, collaborative and relative sensing issues may arise. Collaborative sensing is the problem of exploiting onboard sensors and processing resources in a cooperative way to obtain specific goals, like object detection and tracking, surveillance and data collection [19]. Instead, relative sensing is aimed at getting information about the relative and absolute state of multiple UAVs, and it can have critical importance to enable safe control of the formation or to improve navigation performance when the formation operates in complex environments [20,21].

This paper focuses its attention on relative sensing aspects. Indeed, in some mission scenarios, it may be extremely important that each UAV is able to detect and track the other members of the swarm obtaining information such as relative distance and bearing (Line-of-Sight, LOS). Considering the cooperative nature of the formation, this task can be carried out using various solutions based on acoustic sensors, Radio-Frequency ranging, LIDAR and visual cameras [22–25]. In this respect, despite their limited applicability in absence of adequate illumination, visual cameras are extremely versatile sensors, being low-cost, lightweight and able to provide highly-accurate LOS estimates. Clearly, this latter aspect is achieved if image processing algorithms, able to determine the target position in the image plane with pixel-level accuracy, are developed. This goal involves complex technical challenges. Some of the most important are listed below:

- *Sun illumination*. Indeed, wide portions of the field-of-view (FOV) may be saturated if the Sun direction is quasi-frontal with respect to the camera. Also, the object appearance (e.g., color and intensity) can change as a function of the Sun direction.
- *Local background*. Detection and tracking algorithms may suffer from sudden variations of the local background, e.g., caused by continuous passages of the target above/below the horizon.
- *Target scale*. Depending on the mission scenario, the target distance can vary from tens to hundreds of meters. Consequently, the image processing algorithms must be able to deal with significant variations of the size of FOV region occupied by the target.

In this framework, artificial intelligence has recently produced a breakthrough in the computer vision field, especially due to the advancements of deep learning techniques [26,27]. Hence, this paper aims to analyze the potential of such approaches for airborne visual detection and tracking of cooperative UAVs. Specifically, an original image processing architecture is developed, which exploits the You Only Look Once (YOLO) object detection system [28] as main processing block. A peculiar characteristic of this method is that machine-vision algorithms are aided by navigation data (e.g., own-ship attitude and target distance) some of which are available thanks to the cooperative nature of the formation (e.g., exchanged through a communication link). Validation and performance assessment of the proposed algorithmic architecture are carried out using a database of images collected by means of an experimental campaign carried out using formations of two multirotor UAVs, which are indicated in the following as tracker (the one with the camera on board) and target.

The remainder of the paper is organized as follows. Section 2 reviews the state of the art regarding techniques for airborne visual detection and tracking of flying vehicles, and it highlights the contribution of this work. Section 3 describes in detail the proposed image processing architecture. Section 4 presents the experimental setups and the flight-test data collected for training the deep neural network, and for testing of the overall detection and tracking architecture. Section 5 details the achieved results based on ad-hoc defined performance metrics. Finally, Section 6 gives the conclusions, while also discussing lessons learnt and providing indications about future work.

2. Related Work

In the applications of interest to this work, the tracker and the target are both flying vehicles. Thus, extremely fast variations of the target appearance may occur in the image plane. This aspect must be considered in the design of image processing algorithms for airborne detection and tracking of a moving target. An overview of state-of-the-art approaches from the open literature is provided in the following. The main classification criterion is related to the fact that the target UAV can be either cooperative or non-cooperative.

The issues of visual detection and tracking of non-cooperative target UAVs are mainly treated in the frame of Sense and Avoid (SAA) [29–33] and counter-drone applications [34,35]. In SAA scenarios, the main requirement is to maximize the declaration range (i.e., the range at which a firm track of the target is generated) to satisfy safety margins regarding the time needed to perform an avoidance maneuver [29]. Consequently, most research efforts are addressed to very long-range scenarios, where a target occupies only very few pixels in the image plane. The detection task may be carried out exploiting different approaches. Image processing architectures based on the morphological filtering operator are mainly proposed [30,31]. However, intruders may also be detected by computing the difference between two subsequent images (corrected thanks to an homography estimation based on corner features extracted under the horizon) [32]. More recently, the deep learning concept has also been applied to this task by training a Convolutional Neural Network (CNN), and combining this detector with a Bottom-Hat morphological filter [33]. A known issue for standalone optical detection systems is the average number of false detections. This problem is typically addressed by filtering architectures (e.g., Extended Kalman Filters or Particle Filters) designed to generate target tracks from multiple subsequent detections. Recently, the Deep Learning concept has also been applied for counter-drone applications, i.e., to detect intruder, non-cooperative UAVs using one or multiple visual cameras fixed on ground [34,35].

With regards to visual detection and tracking of cooperative UAVs, the interval of relative distances can vary significantly during operation. Hence, the target can either occupy a few pixels or a wide portion of the image plane. While this is certainly a challenge if the range variations occur fast enough, on the other side it gives the possibility to exploits a wider range of image processing approaches, which can be mainly classified into direct- and feature-based methods. Direct-based methods exploit pixel-wise estimation of the local gradient of the image intensity to treat the detection process as the problem to estimate the transformation (registration) which aligns two frames in the best way, i.e., based on a purposely defined error metric function [36]. Instead, feature-based methods exploit the extraction of salient features and the use of proper descriptors to recognize the portion of the image containing the target [37]. Techniques adopting the machine-learning or deep-learning concepts can also be put under this category since, the trained network can be interpreted as a detector and classifier of features [38]. With specific attention to approaches exploited to detect UAVs with airborne cameras the following works can be mentioned. A trained cascade classifier designed for detection and distance estimation of a small quadrotor UAV can be found in [37]. Experimental tests, both outdoor and indoor, show satisfactory performance and demonstrate real-time operation but the method is designed to deal only with relatively-short distances (i.e., below 25 m). The vision-based architecture presented in [39] is designed to detect and track multiple target UAVs using a camera on board a custom, delta-wing, small UAV. First, the background motion is estimated using a perspective transformation model. Then, corner features are extracted from the background-less image and moving objects are detected using optical flow. Though the method is designed to be robust against false alarms, it suffers missed detections for targets below the horizon. A motion-compensation-based approach for airborne detection of UAVs is presented in [40]. Specifically, the main idea is that the detection process should exploit both appearance and motion cues. This is done by combining deep learning (based on a CNN) with an original motion compensation method, both applied over a sequence of frames (namely a spatio-temporal cube). This approach shows better performance than methods based only on either machine learning or motion compensation, but it is better tailored to detect flying vehicles at long range (e.g., for collision avoidance purposes). An object detection and tracking approach which integrates a salient object detector within the Kalman filtering framework is proposed in [41]. Though, it shows better tracking performance than state-of-the-art methods, the authors highlight limitations in dealing with detection of flying targets when placed against a cluttered background. Finally, a recent deep learning based approach, proposed for visual detection and tracking of a drogue object at short range, in the frame of UAV aerial refueling, can be found in [42]. Though some of the abovementioned methods exploit the knowledge of the target, they do not take full advantage of the cooperative nature of the formation. This can be done installing on board artificial markers, such as active Light Emission Diodes (LEDs) [43] and passive retro reflectors [44] or exploiting color-based information if the target has a highly-distinguishable color signature [45,46]. Some limits can be highlighted for these solutions. Specifically, the use of active LEDs involves an increase in the power and weight requirements for the target UAV; passive markers are visible only at relatively-short distances; color-based methods can be negatively affected by illumination conditions which may cause a variation in the color signature. To cope with these issues, machine-vision algorithms could be aided using absolute and relative navigation data, as proposed in [47,48], where the image processing function is based on the combination of template matching and morphological filtering concepts.

The visual detection and tracking approach presented in this work recalls this idea, but navigation data available on board the tracker UAV and cooperative navigation information received from the target UAV are used to aid an original image processing architecture which relies on the deep-learning concept. Specifically, the use of a deep-learning approach is motivated by the need to improve performance (i.e., minimize missed detections without increasing false alarms) when the background is extremely cluttered (e.g., below the horizon) with respect to other machine vision approaches. The innovative points are related to both detection and tracking and are conceived to improve LOS estimation accuracy and ensure adequate robustness against the challenges caused by Sun illumination, local background and target scale. The former aspect, which is extremely important in cooperative applications, is entrusted to an original technique to refine the location of the bounding box which identifies the estimated target position with respect to the one detected by a state-of-the-art DL-based object detector (i.e., YOLO in this work). Regarding the robustness aspect, the use of navigation data not only enables the proposed approach to deal with an extremely wide interval of target-tracker distances, but, as in [47,48], it also allows reducing the occurrence of false alarms and the computational effort thanks to the possibility to focus the target search process over a limited portion of the image plane both during detection and tracking.

3. Detection and Tracking Architecture

The proposed architecture is composed of two processing blocks, namely a detector and a tracker, whose goal is to estimate the position of the target UAV on the focal plane of the camera installed on board of the tracker UAV. In the following, the region of the image plane occupied by the target is defined as a rectangular bounding box (BB), from which the unit vector representing the target LOS in the Camera Reference Frame (CRF) can be derived as the direction corresponding to the geometric center of the BB. A state diagram describing this architecture is shown in Figure 1.

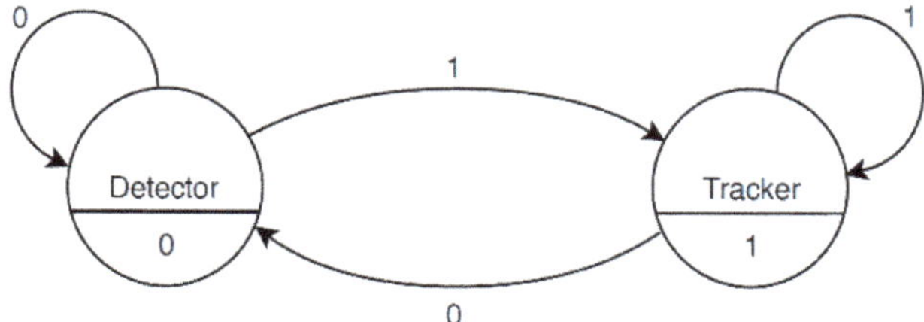

Figure 1. State diagram of the proposed architecture. Two cases are possible for the state of the system, namely target detected (1) or not detected (0). This relatively simple architecture is justified by the cooperative nature of the assumed multi-UAV system.

The state of the system can be 0 or 1 depending on whether the target position on the image plane at the previous frame is available or not. In the former case, the target identification process is entrusted to the detector which cannot use visual information from the previous frame. As soon as a detection is confirmed, the state becomes 1 and the identification process is entrusted to the tracker, which exploits the knowledge of the target position on the image plane at the previous frame. Clearly, if the tracker fails at detecting the target (missed detection) or if the target is outside the camera FOV, the state of the system returns to 0 and the identification process must be re-initialized by the detector. In the following, the algorithms implemented within the detector and tracking blocks are described in detail. Both these algorithms take advantage of navigation data (e.g., own-ship attitude and relative positioning) to obtain a reliable prediction of the target position on the image plane at the current frame.

3.1. DL-Based Detector

In case of first frame acquisition, or if no information about the target position in the image plane is available from the previous frame, the target search is carried out by the detector block, which exploits the algorithmic strategy whose main steps, inputs and output are highlighted in Figure 2.

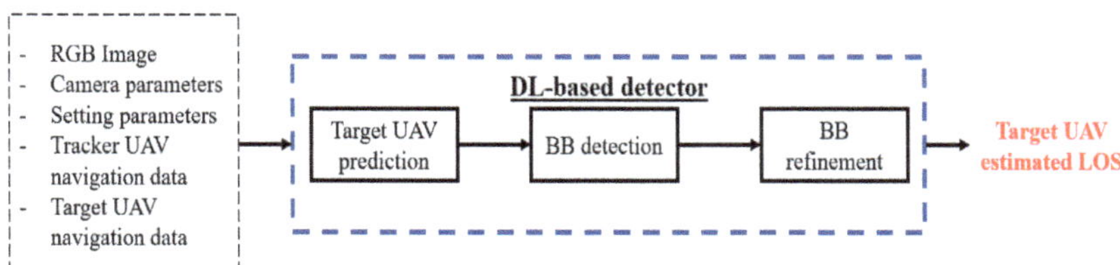

Figure 2. Scheme summarizing the algorithmic strategy characterizing the proposed DL-based detector. The input parameters are listed within a black (dashed) rectangular box. The processing blocks are enclosed within black rectangular boxes. The final output is highlighted in red.

First, the *Target UAV prediction* block is used to generate a tentative solution for the target UAV projection on the image plane. Then, the main processing block (*BB detection*) uses this prediction to identify a limited portion of the image plane within which the search for the BB containing the target is carried out exploiting the YOLO object detection system [28]. The BB produced in output by the *BB detection* block is finally used by the *BB refinement* block to obtain a more accurate estimate of target UAV LOS. Detailed information about each of these functions is provided in the following sub-sections.

3.1.1. Target UAV Prediction

This processing step takes advantage of absolute navigation data of the tracker UAV, as well as of target-tracker relative navigation data to predict the projection of the target UAV in the image plane. The relative navigation data are available thanks to the cooperative nature of the proposed visual architecture. Indeed, if a reliable communication link is used to transmit absolute navigation data from the target to the tracker, their relative position vector in the Earth Centered Earth Fixed (ECEF) reference frame ($\underline{\rho}_e$) and, consequently, their relative distance (ρ) can be estimated. Specifically, this relative positioning information can be obtained with different strategies.

- Computing the difference between the two position vectors in ECEF ($\underline{P}_{e,tracker}$ and $\underline{P}_{e,target}$ for the tracker and target, respectively), obtained using either the GNSS position fix or, more in general, the positioning solution available to the onboard autopilot.
- Exploiting Differential GNSS (DGNSS) and Carrier-Phase DGNSS (CDGNSS) algorithms.

The choice of the method, which clearly has an impact on the accuracy level of the relative position estimate, depends on the quality of local GNSS coverage for the two vehicles as well as on the type of exchanged data (since differential and carrier-phase differential algorithms require pseudo-range and carrier-phase measurements). Regarding the results presented in this paper, the former approach is exploited to obtain the target-tracker range and relative position vector.

Independently of the adopted solution for relative positioning, the estimated value of $\underline{\rho}_e$ allows obtaining a prediction of the target-tracker relative position vector in CRF ($\underline{\rho}_c$) using Equation (1):

$$\underline{\rho}_c = \underline{\underline{R}}_{bc}\left(\underline{\underline{R}}_{nb}\underline{\underline{R}}_{en}\underline{\rho}_e + \underline{t}_{cb}\right) \tag{1}$$

where $\underline{\underline{R}}_{en}$ is the rotation matrix from ECEF to the local North-East-Down reference frame (NED), $\underline{\underline{R}}_{nb}$ is the rotation matrix from NED to the Body Reference Frame (BRF) of the tracker UAV, while $\underline{\underline{R}}_{bc}$ and $\underline{t}_{cb}$ are the rotation matrix from BRF to CRF and the camera-to-body relative position vector expressed in BRF, respectively. $\underline{\underline{R}}_{en}$ and $\underline{\underline{R}}_{nb}$ can be obtained using the onboard navigation data of the tracker UAV. Specifically, the former depends on the position in ECEF of the tracker UAV, while the latter is an attitude rotation matrix which is also typically parametrized by a 321 sequence of Euler angles (i.e., heading, pitch and roll). Finally, $\underline{\underline{R}}_{bc}$ and $\underline{t}_{bc}$ depend on the camera mounting geometry, and they can be computed by carrying out an ad-hoc extrinsic calibration procedure either off-line [49] or on-line [50]. After computing $\underline{\rho}_c$, the horizontal (u_{pr}) and vertical (v_{pr}) pixel coordinates of the target projection on the image plane can be derived using the intrinsic calibration parameters, i.e., focal length, principal point, skew coefficient and distortion (radial and tangential) coefficients, typically computed carrying out an off-line calibration procedure [51].

3.1.2. Bounding Box Detection

Recently, the problem of using Deep Learning algorithms to address computer vision applications has been deeply investigated [52]. In this framework, several object detection systems based on Deep Learning have been developed, such as the Single Shot MultiBox Detector (SSD) [53], Faster R-CNN [54], RefineDet [55], RFBNet [56] and YOLO [28,57,58]. A detailed overview of object detection systems using Deep Learning including comparisons in terms of accuracy and efficiency can be found in [59,60].

Within the proposed architecture, the target detection task is entrusted to a state-of-the-art object detector, i.e., the YOLO v2 neural network [57]. This DL-based network is obtained using the MATLAB Deep Learning Toolbox [61] (MathWorks, Natick, Massachusetts, USA) starting from a pre-trained CNN, i.e., ResNet-50 [62]. Given a dataset of RGB images taken with the target UAV at different distances relative to the tracker, the training dataset is obtained by randomly cropping each of these images around the target position. Specifically, the cropped image region used for training is a square of 150-by-150 pixels.

By taking advantage of the predicted target projection obtained at the previous processing step, it is possible to restrain the portion of the image plane to which the trained detector must be applied. Clearly, the shape and the size of the search area should be set considering the uncertainty characterizing each term contributing to the application of Equation (1), i.e., the positions of both the two UAVs, and the attitude of the tracker. Focusing on the uncertainty in the attitude parameters, it is important to highlight that the heading, pitch and roll estimates do not have the same accuracy level. Specifically, due to the possibility to exploit gravity direction, considering performance of low-cost IMUs on board of small UAVs, the uncertainties in the roll and pitch estimates provided by the onboard autopilot are typically bounded at the degree-level [63], while the error in the heading estimate can be significantly larger (up to several degrees) [64]. For this reason, the target search can be limited to a horizontal stripe in the image plane. Of course, the size of the stripe, in general, must account for UAV attitude dynamics, camera frame rate and synchronization uncertainties. In this framework, recent results from our previous work show that if the size of the stripe (v_{max}) is at least 100 pixels, it is possible to minimize the possibility to accidently remove the target from the analyzed search area [48]. Considering that the YOLO v2 network requires an input RGB image at least as large as the training data, v_{max} is here set to 150 pixels. So, the defined horizontal stripe can be subdivided into squared regions (search windows) having the same size as the training datasets (150 × 150 pixels), as shown in Figure 3. The reference point used to define each search window is the upper-left corner. The horizontal separation (d_u) between two consecutive search windows, i.e., the distance between two consecutive reference points, is a setting parameter for the user. Clearly, the maximum allowable value of d_u is 150 pixels, since it avoids losing FOV coverage. This value allows minimizing the number of search windows (N_w), and consequently, the algorithm's computational effort. However, it can be convenient to slightly reduce d_u, thus having a partial overlap between consecutive search windows. Although this solution causes a slight increase in the computational effort (due to the increase in N_w), it also limits the risk that the target is cut by the border of two consecutive search windows as shown, for instance, in Figure 4a. If this occurs, the YOLO v2 detector cannot find "good-quality" BBs, i.e., characterized by large overlap with the target actual position, thus compromising the probability of correct detection. So, a value lower than 150 pixels can be set to d_u to ensure that at least one search window fully contains the target UAV, as shown, for instance, in Figure 4b where d_u is 100 pixels.

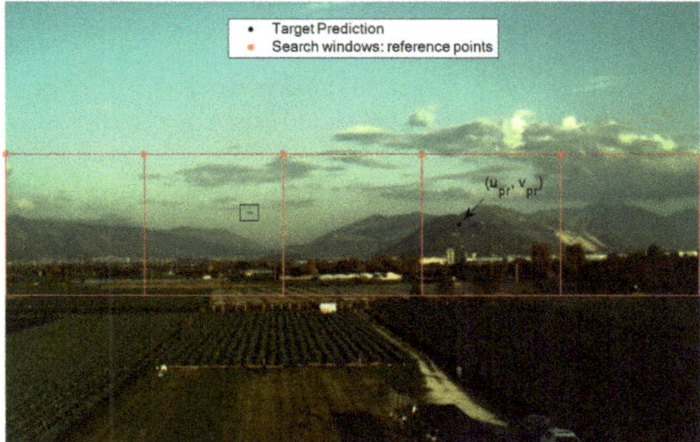

Figure 3. DL-based detector: example of search windows definition for a 752 × 480-pixels RGB image. d_u is set to 150 pixels (N_w = 5). The target UAV position is highlighted by a black box.

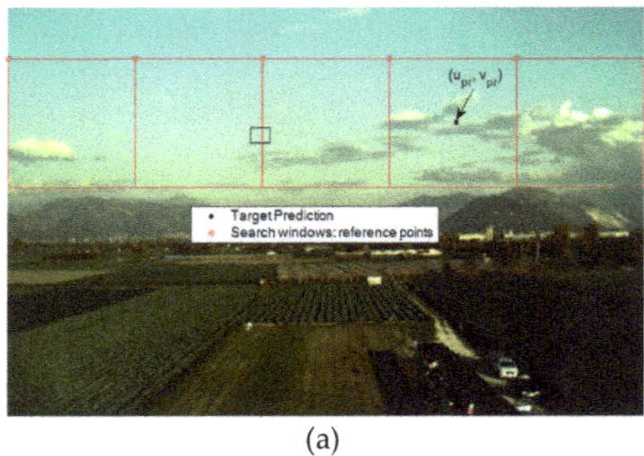

(a)

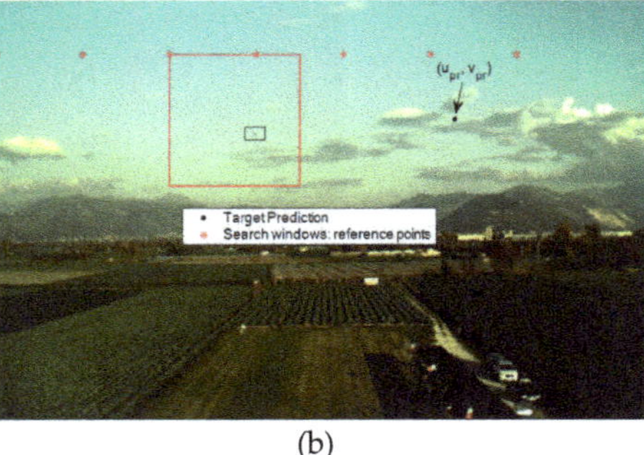

(b)

Figure 4. DL-based detector: example of search windows definition for a 752 × 480-pixels RGB image. The target UAV position is highlighted by a black box. (**a**) d_u is set to 150 pixels (N_w = 5). The target UAV projection on the image plane is cut by the border between the second and third search window. (**b**) d_u is set to 100 pixels (N_w = 7). Only the third search window, which fully contains the target UAV, is highlighted for the sake of clarity.

Once the target search area is defined, the YOLO v2 detector is applied to each search window independently. This operation produces a large set of BBs representing candidate image regions potentially containing the target according to a specific level of confidence (S), i.e., the score measuring the accuracy of the BB. Specifically, the score encodes both the probability that the class of the target object appears in the box and how well the predicted box fits the object [28]. At this point, if the maximum value of S (S_{max}) is higher than a detection threshold (τ_{det}), to be assigned by the user, the detection process is deemed successful (i.e., the state of the system becomes 1 according to the diagram in Figure 1) and the corresponding BB is sent to the *BB refinement* block.

3.1.3. Bounding Box Refinement

The reasoning behind the introduction of this processing block is that, despite the BB selected at the previous stage may indeed contain the target, there may be a horizontal and/or vertical offset between the center of this BB (from which the target LOS can be estimated knowing the intrinsic calibration parameters of the camera) and the actual center of the target. Such situations tend to degrade the angular accuracy in the estimated target LOS which is extremely important in several tasks involving cooperative UAVs, e.g., to enable highly-accurate cooperative navigation solutions [11].

This phenomenon can also be interpreted as a relatively-poor overlap between the estimated BB and a reference BB (i.e. the one representing the ground truth). The quality of this overlap can be measured using the Intersection over Union (*IOU*) as performance metric. An example is shown in Figure 5, where the *IOU* between the detected and reference BB is 0.538.

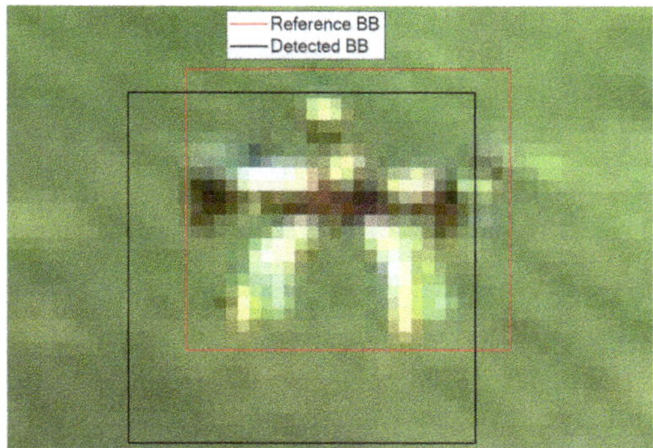

Figure 5. Example of best YOLO detection. *IoU* = 0.538. The reference BB, obtained using a supervised approach, is approximately centered at the geometric center of the target. The detected BB is the output of the DL-based detector.

In this case, the estimated position of the target UAV (i.e., the center of the detected BB) is characterized by an offset of about 5 pixels, in both the horizontal and vertical directions. This does not allow taking full advantage of the camera angular resolution, since the errors in the azimuth and elevation angles (see Section 5.1 for details on the estimation of these angles) corresponding to the target LOS in CRF, i.e., 0.21° and 0.26°, respectively are quite larger than the camera IFOV (0.05°). Hence, an approach based on standard machine vision tools is applied to update the BB, whose main steps are highlighted in Figure 6.

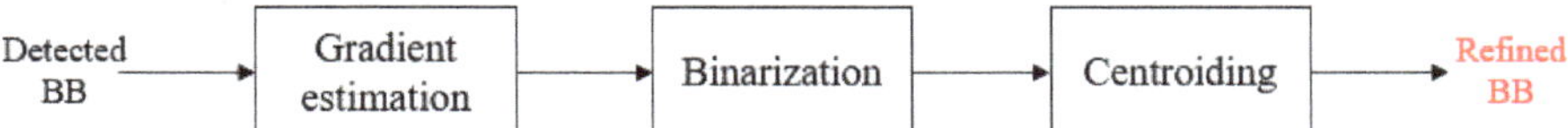

Figure 6. Main steps of the image processing approach to refine the detected BB. An image crop is obtained from the detected bounding box. The gradient operator is applied within this image portion and the gradient image is then binarized. Finally, the centroid of the set of pixels highlighted in the binarized image is computed and a refined BB is centered around this point.

The input can be either the detected BB or an enlarged image portion, i.e., centered at the BB center but with height and width increased by a factor c (e.g., between 1.5 and 2). This latter option is useful to avoid the risk that the target is not fully contained in the analyzed area due to the limited overlap of the BB.

An example of application of this procedure is provided by Figure 7. First, the local intensity gradient is computed using the Sobel operator. The resulting gradient image, in which the target is highlighted with respect to the background, is then binarized. Finally, the centroid of the resulting binary image is computed. By centering the refined BB at this point, as shown in Figure 8, it is possible to increase the *IoU* up to 0.747, and the angular errors in the azimuth and elevation angles obtained from the target position on the image plane become $6{\cdot}10^{-4\circ}$ and 0.05°, respectively. Considering that the reference BB, i.e., the ground truth (which is obtained manually by a human operator), has an inherent sub-pixel level uncertainty, the very low value obtained for the azimuth error ($6{\cdot}10^{-4\circ}$) implies that the refinement process can achieve sub-pixel accuracy (in the horizontal direction only in this case).

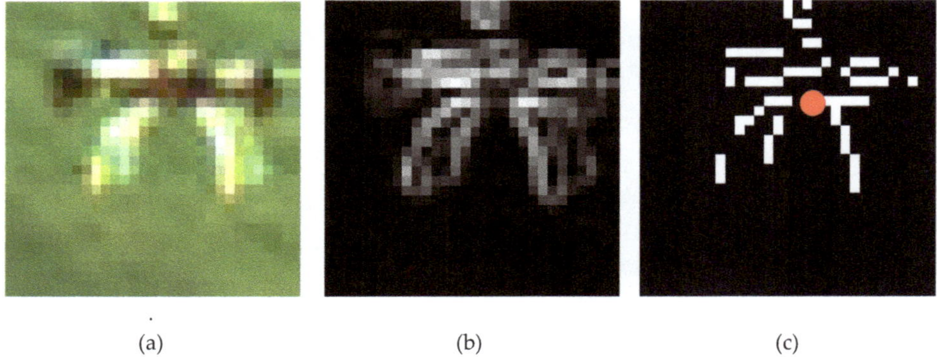

Figure 7. Example of application of the *BB refinement* block. In this case, the factor *c* is 1. (**a**) Detected BB. (**b**) Result of gradient estimation. (**c**) Result of binarization and centroid calculation (highlighted by a red dot).

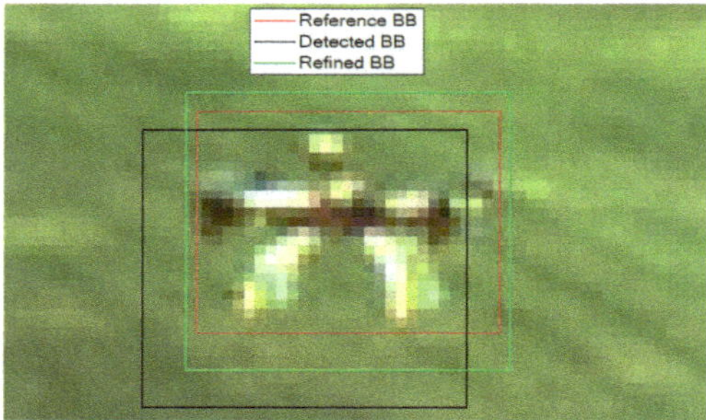

Figure 8. Result of the *BB refinement* algorithm. The *IoU* of the refined BB is 0.747.

3.2. DL-Based Tracker

If a detection is confirmed (according to the thresholding criterion presented in Section 3.1.2), the process of searching for the target projection on the image plane of a new frame, is entrusted to a DL-based tracker function. The proposed algorithmic strategy is composed of the same processing blocks presented and described for the DL-based detector. However, consistent differences are relevant to the *Target UAV prediction* and the *BB detection* blocks, mainly due to the possibility to exploit the prior information about the target projection on the image plane available at the previous frame. Specifically, if $\underline{\rho}_{c,los(k\text{-}1)}$ is the unit vector representing the LOS of the target UAV in CRF estimated at the k-1th frame, the predicted LOS at the current (kth) frame ($\underline{\rho}_{c,los(k)}$) can be computed as follows:

$$\underline{\rho}_{c,los(\mathrm{k})} = \underline{\underline{R}}_{bc}\underline{\underline{R}}_{nb}^{\ k}\left(\underline{\underline{R}}_{bn}^{\ k-1}\underline{\underline{R}}_{cb}\underline{\rho}_{c,los(\mathrm{k}-1)}\right) \tag{2}$$

where $\underline{\underline{R}}_{nb}^{\ k-1}$ and $\underline{\underline{R}}_{nb}^{\ k}$ are the attitude rotation matrixes for the tracker UAV computed by its onboard navigation system. This relation produces highly accurate predictions of the LOS of the target UAV at the current frame provided that the target-tracker relative dynamics is smooth while the frame rate is adequately high, so that the variation of the relative unit vector in NED between two successive frames is negligible. In general, when dealing with particularly aggressive relative dynamics, a more accurate prediction can be obtained by using the velocities of the two UAVs estimated by the corresponding onboard navigation systems.

The result of Equation (2) can then be used to estimate u_{pr} and v_{pr} using the intrinsic calibration parameters of the camera. Hence, the YOLO v2 object detector can be applied to a single search window centered at the predicted target projection on the image plane. The horizontal and vertical sizes of this search window, i.e., $d_{u,tr}$ and $d_{v,tr}$, respectively, can be set by the user considering that their minimum value is 150 pixels (as the minimum size for the input of the YOLO v2 detector is defined by the size of the training dataset). The selection of larger values for $d_{u,tr}$ and $d_{v,tr}$ can be convenient when dealing with fast relative dynamics relatively to the frame rate and, especially, if the target UAV is flying at close range with respect to the tracker (thus occupying large portions of the FOV).

Clearly, this solution allows significantly reducing the number of computations with respect to the DL-based detector. Some examples of search area definition during the tracking process for flight tests characterized by different target, camera (detector and optics) and frame rate are shown in Figure 9. These examples are relative to extreme cases, i.e., the target-tracker relative dynamics is particularly aggressive relatively to the frame rate, thus causing a larger prediction error. The frames depicted in Figure 9a,b are acquired using a 752 × 480-pixel detector at 1 fps. Due to the small size of the target (far range scenario), $d_{u,tr}$ and $d_{v,tr}$ are set to 150 pixels. Instead, the frames depicted in Figure 9c,d are acquired using a 1600 × 1200-pixel detector at 7 fps. Due to the large size of the target (close range scenario), $d_{u,tr}$ and $d_{v,tr}$ are set to 300 pixels. Finally, it is possible to notice that if the prediction is too close to the borders of the image plane, the search window cannot be centered at that position, as in Figure 9a, but rather it must be adjusted to the size of the detector.

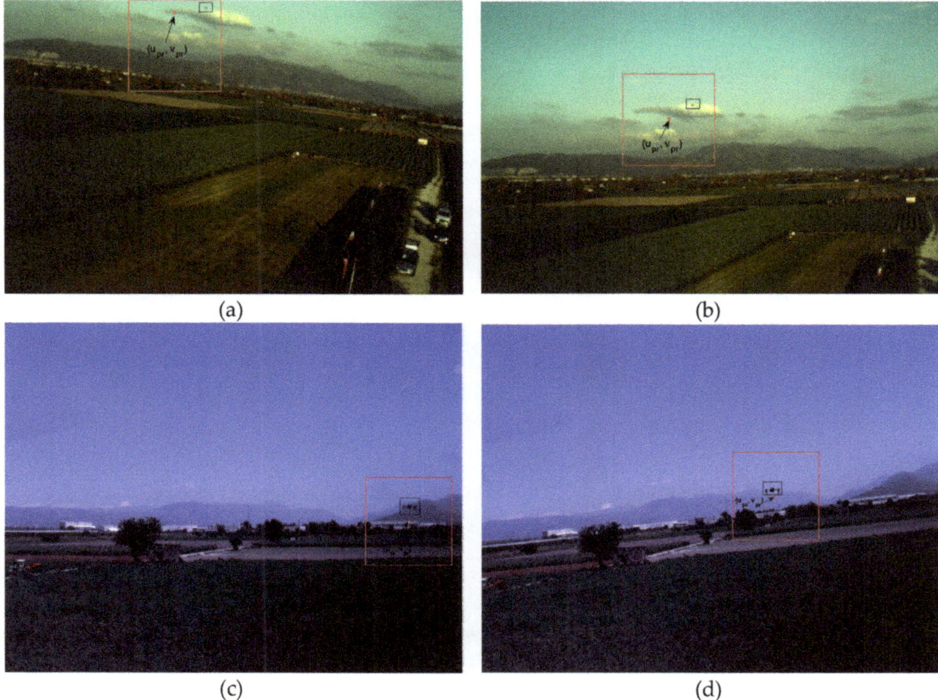

Figure 9. Examples of prediction of the target UAV projection on the image plane (highlighted by a red dot) carried out by the DL-based tracker. The search area is drawn as a red square. The target UAV is enclosed in a black box. (**a**,**b**) Far range scenario (target range ≈ 116 m); $d_{u,tr}$ = $d_{v,tr}$ = 150 pixels; prediction error ≈ 45 pixels. (**c**,**d**) Close range scenario (target range ≈ 20 m); $d_{u,tr}$ = $d_{v,tr}$ = 300 pixels; prediction error ≈ 20 pixels.

Once the YOLO v2 object detector is applied to the defined search window, as for the DL-based detector, the BB characterized by the maximum score is selected. The confirmation of this detection is again carried out applying a thresholding criterion but a different threshold (τ_{tr}) with respect to the detection phase can be used. Clearly, also in this case, the confirmed BB is sent to the *BB refinement* block before estimating the target LOS and proceeding to the next frame.

4. Flight Test Campaign

A campaign of flight tests involving formations of two UAVs has been recently conducted to build a database of images suitable for training the deep neural network composing the YOLO v2 object detector exploited within the proposed visual-based architecture, as well as for assessing its performance for detection and tracking of cooperative UAVs. The database can be divided in two parts (A and B) obtained using different rotorcraft UAVs equipped with different cameras.

For Database A, the role of the tracker UAV is played by the Pelican^{TM} quadcopter depicted in Figure 10a. This UAV, produced by Ascending Technologies (http://www.asctec.de), is equipped with an autopilot characterized by a 1000 Hz update rate in the inertial guidance system, an onboard computer (AscTec Mastermind^{TM}) featuring an Intel i7 processor, a GNSS receiver and a low-cost IMU unit based on MEMS sensors. This vehicle is customized by the installation of an additional GNSS receiver (LEA-6T produced by Ublox^{TM}) and antenna to enable the acquisition of raw GNSS measurements (e.g., pseudoranges) not accessible from the AscTec autopilot, while a miniaturized CMOS camera (mvBlueFOX-MLC-200wc produced by Matrix Vision^{TM}) with 752-by-480 pixel resolution is mounted on the top of the Pelican structure in frontal looking configuration for detection and tracking purposes. The camera is equipped with a 6.5-mm focal length optics, with a resulting IFOV of about 0.05° Both these auxiliary sensors are connected to the AscTec Mastermind via a USB link. The role of the target is played by a X8+ octocopter produced by 3D Robotics^{TM} (https://3dr.com/) depicted in Figure 10b. As in the case of the Pelican, the X8+ is customized by the installation of an auxiliary GNSS system, which is connected via USB to an embedded CPU (i.e., the Odroid XU4^{TM}) for data acquisition and storage.

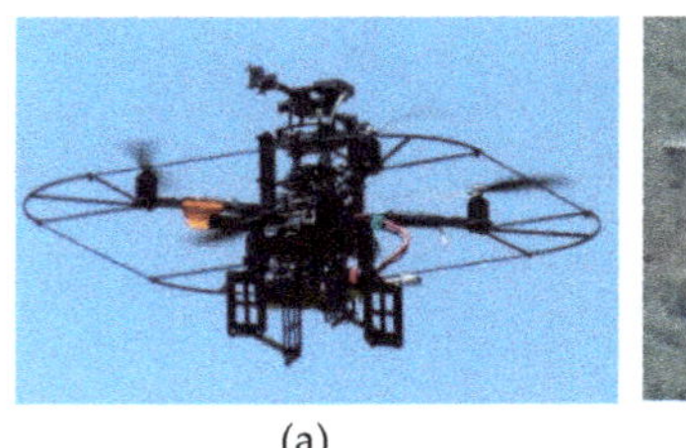
(a)

(b)

(c)

Figure 10. UAVs exploited for the flight test campaign. (**a**) Tracker UAV for database A: customized Pelican by Ascending Technologies. (**b**) Target UAV for database A: customized X8+ by 3D Robotics. (**c**) Target and tracker UAV for database B: customized M100 by DJI.

For Database B, the roles of tracker and target are played by two customized versions of the DJI M100^{TM} quadrotor depicted in Figure 10c. They are both equipped with an onboard computer (Intel NUC^{TM} with i7 CPU) and an additional GNSS single frequency receiver (uBlox LEA-M8T^{TM}) with raw measurements capabilities and an auxiliary GNSS antenna. The tracker vehicle is also equipped with an off-the-shelf CMOS camera for vision-based detection and tracking purposes (Flea FL3-U3-20E4C-C produced by FLRI^{TM}) characterized by 1600-by-1200 resolution in pixels and maximum frame rate of 59 fps. The camera is equipped with an 8-mm focal length optics, with a resulting IFOV of about 0.03°. Again, the auxiliary visual and GNSS sensors are connected to the Intel NUC via USB.

This flight test campaign is conducted using the approach of data acquisition for off-line processing. Specifically, an ad-hoc developed acquisition software is used to save all the sensor data with an accurate time-tag based on the CPU clock. This time-tag is associated with GNSS measurements, which include the GPS time, with very small latency. This solution has a key role to enable the use of a cooperative approach since it allows accurately synchronizing all data acquired on each flying vehicle.

Clearly, with regards to a real-time implementation of the proposed approach, a V2V communication link shall be foreseen to enable the exchange of navigation information required to estimate the target-tracker relative range and line of sight in ECEF or NED. However, the amount of this information (e.g., GNSS data) is limited to a few Kbits per second. Hence, this constraint can be easily met by standard V2V communication systems [65].

4.1. Database A

This database is composed of images collected during three Flight Tests (FT), that are indicated as FT1-A, FT2-A and FT3-A in the following discussion. The adopted testing strategy is aimed at assessing the capability of the proposed vision-based architecture to detect and track a cooperative vehicle flying at relatively-far range, i.e., when it occupies only a limited portion of the FOV. Indeed, the target-tracker relative distance which is computed as the difference between the synchronized GNSS position fixes provided by the receivers on board of the two vehicles, varies from a minimum of 53 m (the target occupies a rectangular region of about 15 × 10 pixels) to a maximum of 153 m (the target occupies very few pixels). Moreover, during each FT, a series of 360° turns is commanded to the tracker UAV to verify whether the proposed architecture can autonomously recognize when the target falls outside the FOV and to redetect it once the rotation is ended. Due to this testing strategy, the total number of collected images (N) can be divided in frames either containing (N_{IN}) or not containing the target (N_{OUT}). The acquisition software executed on board of the tracker UAV is designed so that the image acquisition is triggered by the reception of a GNSS package. This solution allows getting corresponding visual and positioning data with a negligible latency and an update rate of 1 Hz. Due to this choice, the capability of the camera to acquire and save images at higher frame rates (e.g., around 20–30 Hz) is not exploited. However, this solution allows testing the proposed vision-based architecture against more complex conditions for tracking since the target position in the FOV experiences a much faster dynamic. Main information about the FTs are synthetized in Table 1, where this latter phenomenon is quantitatively measured by the mean and standard deviation of the target displacement between consecutive frames (Δu and Δv for the horizontal and vertical direction, respectively), which is of the order of tens of pixels.

Table 1. Statistics of the FTs for database A.

FT	N	N_{IN}	N_{OUT}	ρ (m)		$\Delta u/\Delta v$ (pixel)	
				Mean	Std	Mean	Std
1	398	376	22	84.2	17.1	38/24	59/26
2	361	319	42	105.9	19.5	30/26	50/26
3	381	323	58	120.2	19.5	33/21	42/21

During all these flights, the local background (with respect to the position of the tracker in the FOV) is characterized by a significant variability in terms of color and clutter level due to the presence of mountains and vegetation.

Even when the target is above the horizon, the background conditions can be challenging when the target UAV is hindered by clouds or if most of the FOV is saturated due to sudden variation of the Sun illumination conditions. A small set of images with a zoom on the target is collected in Figure 11 to provide examples of the above-mentioned challenges to the effectiveness of the proposed approach.

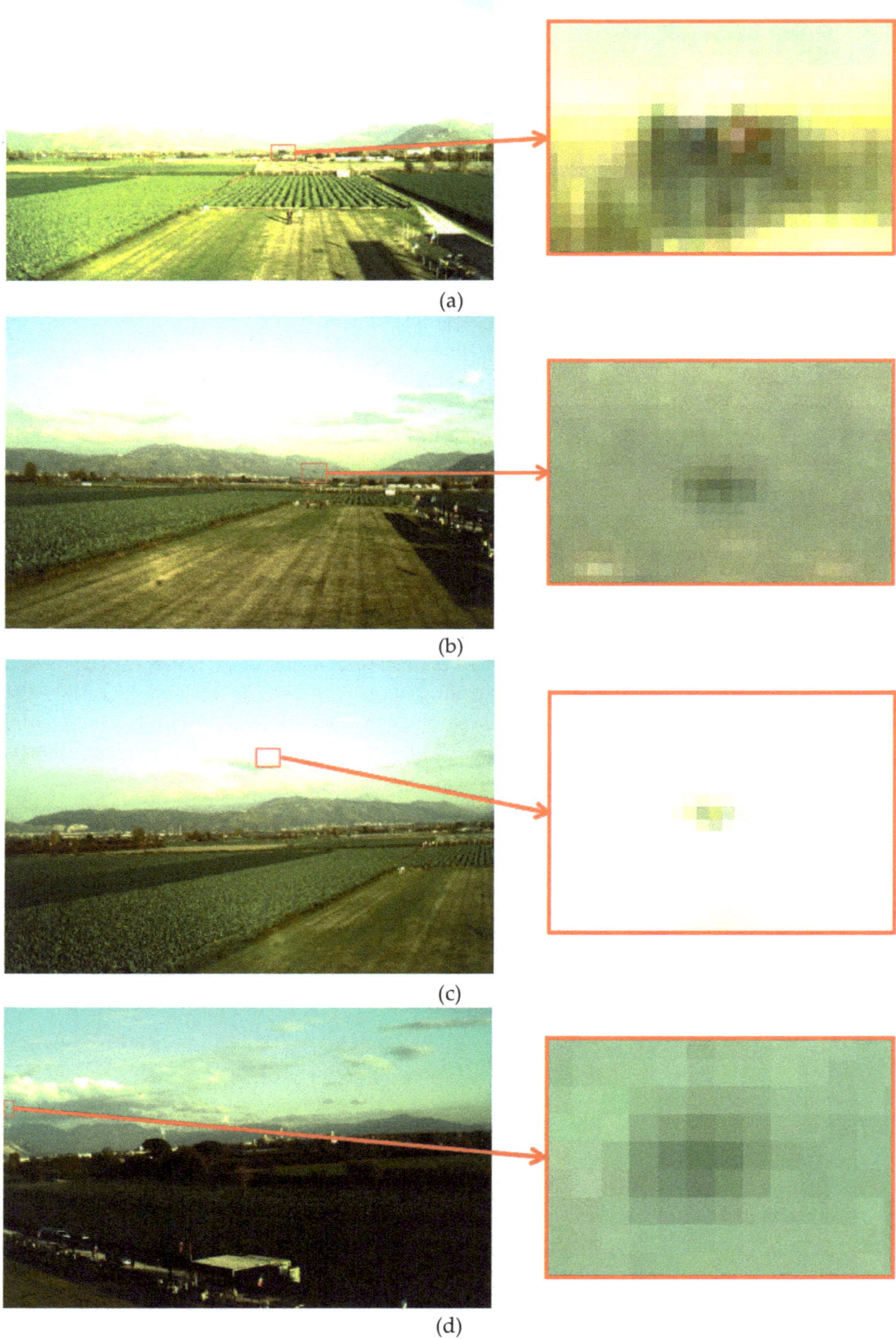

Figure 11. Example of images from FT3-A. The target (i.e., the X8+ octocopter) occupies a few pixels as highlighted by the zoom on the right side of each figure. (**a**,**b**). Target below the horizon. (**c**,**d**) Target above the horizon hindered by clouds.

4.2. Database B

This database is composed of images collected during two FTs, that will be indicated as FT1-B and FT2-B in the following. The former FT allows assessing the performance of the proposed vision-based architecture in a wider interval of ranges, i.e., from a minimum of 25 m to a maximum of 300 m. The capability to image the target UAV at longer distance with respect to the Database A is due to the better resolution of the camera on board of the tracker UAV (the IFOV is 0.03° instead of 0.05°). Instead, FT2-B is a coordinated flight at close range, i.e., the two UAVs move in the same direction while keeping the target-tracker distance in a short range of variation (from 17 m and 23 m).

A different data acquisition software is executed on board of the tracker UAV for these FTs, which allows collecting images with an update rate of 7 Hz. It is developed in Robot Operating System (ROS) and it includes a custom ROS node coded in C++ to process GNSS receiver data and to get an accurate time tag based on GPS time and the CPU clock. Due to the higher frame rate, the frame-to-frame displacement on the image plane is smoother than for Database A. Like in the previous subsection, some statistics to summarize the content of this database are collected in Table 2. A small set of images from the two FTs with a zoom on the target is collected in Figure 12 to provide examples of the target appearance.

Table 2. Statistics of the FTs for database B.

FT	N	N_{IN}	N_{OUT}	ρ (m)		$\Delta u/\Delta v$ (pixel)	
				Mean	Std	Mean	Std
1	1330	1275	75	168.9	86.5	4/2	9/4
2	735	735	0	20.5	1.7	4/7	6/9

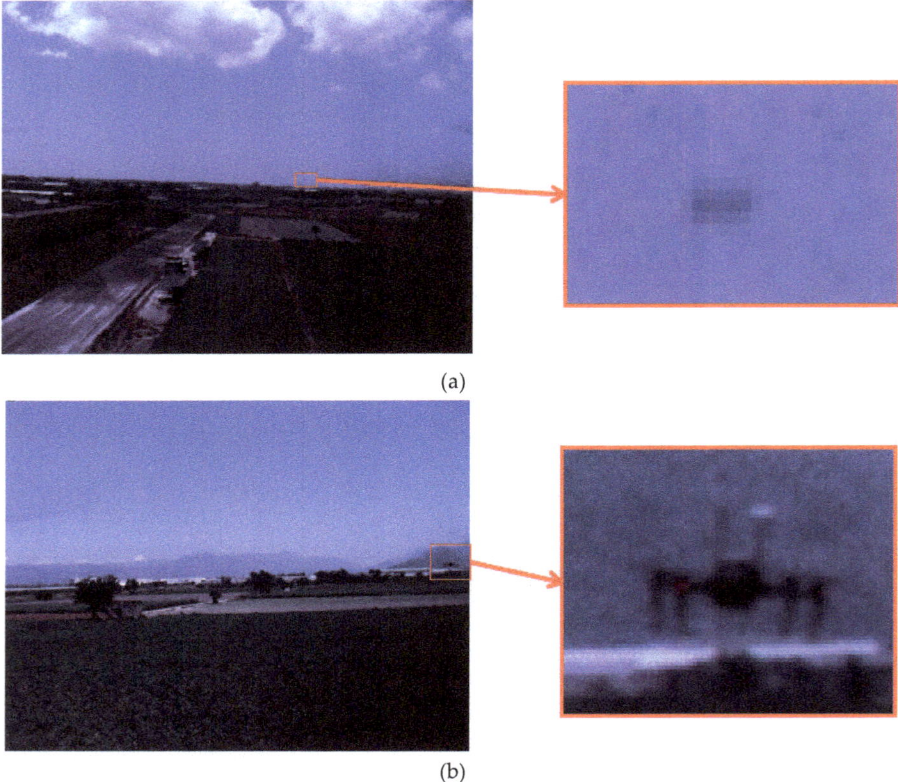

(a)

(b)

Figure 12. Example of images from FT1-B (**a**) and FT2-B (**b**).

5. Results

The YOLO v2 object detector is trained using a sub-set of 650 images selected from FT1-A and FT2-A. As anticipated in Section 3.1.2, these images are randomly cropped around the target with 150-by-150-pixel windows and labelled by a human operator to obtain the ground truth (supervised approach). The resulting neural network is first used within the *BB detection* block to test the performance of the proposed architecture on FT3-A. Then, the generalizing capabilities of the network are evaluated by testing the proposed architecture with images obtained during FT1-B and FT2-B which are different in terms of target UAV, camera installed on board the tracker and interval of target-tracker relative distance. Also, these flights are carried out on a different flight field and in a different period of the year.

Before entering the details of the achieved results, it is necessary to provide a concise definition of the performance parameters which are adopted within the next sub-sections. Specifically, the detection and tracking performance is analyzed in terms of four performance parameters.

- *Percentage of Correct Detections* (CD), i.e., the ratio between the number of frames in which the target is correctly declared to be either inside the image plane (less than 10-pixel error both horizontally and vertically) or outside the image plane and the total number of frames.
- *Percentage of Wrong Detections* (WD), i.e., the ratio between the number of frames in which the target is correctly declared to be inside the image plane but the error is larger than 10 pixels either horizontally or vertically, and the total number of frames.

- *Percentage of Missed Detections* (MD), i.e., the ratio between the number of frames in which the target is wrongly declared to be outside the image plane (it is not detected even if it is present in the image) and the total number of frames.
- *Percentage of False Alarms* (FA), i.e., the ratio between the number of frames in which the target is wrongly declared to be inside the image plane (it is detected even if it is not present in the image) and the total number of frames.

5.1. Detector Performance on FT3-A

A detailed analysis of the performance of the first step of the proposed visual-based architecture is carried out. Specifically, the DL-based detector is applied to each image acquired during FT3-A and the previously-defined performance metric are evaluated as a function of τ_{det}, i.e., the minimum value of the maximum score among the search windows (S_{max}) for which the detected BB is confirmed. These results are collected in Figure 13a. To highlight the advantages gained by exploiting the navigation data of both the tracker and the target to limit the search area, the performance metrics are also evaluated applying the DL-based detector without the *Target UAV prediction* block. In this case the YOLO v2 object detector is applied to a set of squared search windows composed of 150 × 150 pixels defined to cover the entire image plane. These results are collected in Figure 13b. As explained in Section 3.1.2, d_u is set to 100 pixels to obtain a partial overlap between consecutive windows. Clearly, a similar parameter can be defined in the vertical direction for the full-image analysis which is also set to 100 pixels.

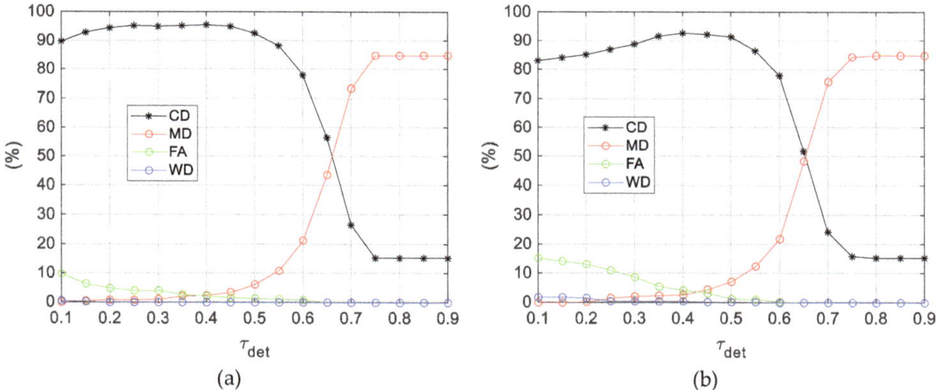

Figure 13. DL-based detector performance as a function of τ_{det}. FT3-A composed of 381 frames. (**a**) *Target UAV prediction* enabled (d_u = 100 pixels, N_w = 7 search windows). (**b**) *Target UAV prediction* disabled ($d_u = d_v$ = 100 pixels, N_w = 28 search windows).

By enabling the *Target UAV prediction* block, the average computational time gets 75% lower than for the full-image analysis. Besides this expected advantage in terms of computational efficiency due to the reduced number of search windows, the results above show that the use of target and tracker navigation data allows improving the detector performance for any considered metric and for any value of τ_{det}. Clearly, the most critical advantage is given by the significant reduction in FA, i.e., 53% on average in the τ_{det}-interval (0.10, 0.60). This phenomenon can be intuitively explained considering that the possibility that the YOLO v2 object detector provides a BB with a relatively-high score (containing something with the same appearance of the target UAV, like a bush on ground or a bird in the sky) increases if the target search is carried out in the full image plane, and especially if a large portion of the scene is below the horizon.

Thus, focusing on the performance level achieved by the full DL-based detector, as depicted in Figure 13a, some interesting considerations can be made:

- WD is always 0% (except for a couple of wrong detections occurring if τ_{det} is set below 0.20), which implies that, when the target is inside the FOV, it is always detected, and its LOS is estimated with a relatively-high accuracy level.
- CD and MD reach an asymptotic value (15.2% and 84.8%, respectively) when τ_{det} gets larger than 0.70. This occurs since the selection of large values of τ_{det} prevents false alarms for the 58 images (see N_{OUT} in Table 1) where the target is outside the FOV (or is not visible).
- The best performance in term of CD (95.5%) is obtained setting τ_{det} to 0.40. However, this choice leads to a residual FA of 2.1%. As τ_{det} increases, FA goes to 0% but MD also increases to the detriment of the number of correct detections. In general, the trade-off in the choice of τ_{det}, i.e., the problem of finding the best compromise between CD, FA and MD, depends on the mission task. For instance, cooperative UAVs can be used to improve navigation performance in GNSS-challenging areas [13]. In this case it is more important to limit the occurrence of false alarms, thus choosing larger values of τ_{det}.

Another interesting analysis regards the effectiveness of the *BB refinement* block described in Section 3.1.3 aiming at improving the accuracy achieved in the estimation of the target projection on the image plane, and consequently, on the corresponding LOS. This accuracy is evaluated as the error between the true and estimated azimuth (*Az*) and elevation (*El*) angles which can be obtained from the true and estimated target LOS, respectively. The relations between the target LOS in CRF ($\underline{\rho}_{c,los}$) and the two angles are given by Equations (3) and (4):

$$Az = \tan^{-1}\left(\frac{\rho_x}{\rho_z}\right) \tag{3}$$

$$El = \tan^{-1}\left(-\frac{\rho_y}{\rho_z}\cos(Az)\right) \tag{4}$$

where ρ_x, ρ_y and ρ_z are the components of $\underline{\rho}_{c,los}$ in CRF. A statistical analysis of the azimuth and elevation accuracy achieved either enabling the *BB refinement* block or not, and setting τ_{det} to 0.40, is presented in Table 3.

Table 3. Statistical analysis of the accuracy in the estimated target LOS. DL-based detector applied on the images (381) from FT3-A (τ_{det} = 0.40).

	BB Refinement Enabled		*BB Refinement Not Enabled*	
	Az Error (°)	*El Error (°)*	*Az Error (°)*	*El Error (°)*
Mean	−0.046	0.057	−0.049	0.063
Std	0.023	0.016	0.046	0.051
Rms	0.051	0.059	0.067	0.081

The introduction of the *BB refinement* block determines an improvement in all the statistical parameters both for the azimuth and elevation angles. In particular, the uncertainty (std) is reduced of around 50% in *Az*, and 68% in *El*, and a slight reduction is also observed in the mean error. Even including this systematic error, the rms value shows that the use of the *BB refinement* block allows achieving an accuracy of the order of the camera IFOV (0.05°) in both *Az* and *El*. Further error reduction is hindered by the uncertainty of the supervised approach used to mark the ground truth, especially at far target-tracker range, when the target is extremely small and blurred on the image plane. Finally, a comparison is presented in Table 4 between the best detection performance achieved on this FT using the proposed DL-based approach (τ_{det} = 0.40) against a state-of-the-art technique based on standard machine vision tools (i.e., template matching and morphological filtering) proposed by the authors in [48].

Table 4. Comparison between the detector proposed in this paper (DL-based) and a state-of-the-art approach [48], based on template matching and morphological filtering. FT3-A (381 images).

Detector	*CD*	*MD*	*FA + WD*
DL-based	95.5%	2.4%	2.1%
[48]	78%	18%	4%

5.2. Detector and Tracker Performance on FT3-A

The performance achieved by the full visual-based architecture proposed for detection and tracking of cooperative UAVs is now evaluated, considering the effects determined by the choice of the two main setting parameters, i.e., τ_{det} and τ_{tr}. Based on the results presented in the previous section in Figure 13a, τ_{det} is first set to 0.40 (which provided the best performance in terms of CD) while τ_{tr} is varied between 0.40 and 0.10.

The choice to set τ_{tr} to values less than or equal to τ_{det} is in line with the idea implemented by the DL-based tracker to limit the target search to a single window centered at the predicted target projection where the confidence to find the target is higher than in other regions of the image plane. The result of these tests obtained setting $d_{u,tr}$ and $d_{v,tr}$ to 150 pixels, are collected in Table 5.

Table 5. DL-based detector vs. DL-based detector and tracker performance as a function of τ_{tr} (τ_{det} = 0.40). FT3-A composed of 381 frames.

	τ_{det}	τ_{tr}	MD	FA	CD	WD
DL-based detector	0.40	/	2.4%	2.1%	95.5%	0.0%
DL-based detector and tracker	0.40	0.40	3.2%	1.8%	95.0%	0.0%
	0.40	0.30	1.8%	2.1%	96.1%	0.0%
	0.40	0.20	1.3%	2.4%	96.3%	0.0%
	0.40	0.10	0.5%	3.7%	95.5%	0.3%

The advantage of using the full detection and tracking architecture with respect to the DL-based detector only, is twofold. Clearly, it is mainly related to the improvement in the computational efficiency, since the YOLO v2 object detector is applied to a single search window. However, it is also possible to reduce the number of false alarms (setting τ_{tr} to 0.40) or to optimize the ratio between CD and MD (lowering τ_{tr} to 0.20).

To better highlight the advantage in terms of the performance metrics, gained using the DL-based tracker, and, consequently, optimizing the performance of the visual-based architecture, it is convenient to increase the value of τ_{det}. Indeed, it is reasonable to have less confidence in the target projection identified at the detection stage due to the lower amount of available a-priori information. Looking again at the results in Figure 13a, τ_{det} is now set to 0.65, which is the minimum value which allows nullifying FA (at the expense of an increase in MD). Instead, τ_{tr} is varied from 0.65 down to 0.20 (lower values are not considered as they tend to produce more false alarms without improvement in CD). The result of these tests (setting again $d_{u,tr}$ and $d_{v,tr}$ to 150 pixels), are collected in Table 6.

Table 6. DL-based detector vs. DL-based detector and tracker performance as a function of τ_{tr} (τ_{det} = 0.65). FT3-A composed of 381 frames.

	τ_{det}	τ_{tr}	MD	FA	CD	WD
DL-based detector	0.65	/	43.6%	0%	56.4%	0%
DL-based detector and tracker	0.65	0.65	40.4%	0%	59.6%	0%
	0.65	0.40	11.0%	0%	89.0%	0%
	0.65	0.20	7.35%	0%	92.65%	0%

Thanks to the selection of a larger detection threshold, the absence of false alarms is confirmed even if the tracking threshold is significantly reduced (down to 0.20), which simultaneously allows reaching a peak level for CD. Indeed, the remaining missed detections correspond to image frames in which the DL-based detector correctly determines the target position, but the score is lower than 0.65. With this configuration for the setting parameters (τ_{det} = 0.65 and τ_{tr} = 0.20), the accuracy level in the estimated LOS of the target is characterized by a rms error of 0.050° and 0.060° in azimuth and elevation, respectively.

5.3. Detector and Tracker Performance on FT1-B and FT2-B

The proposed algorithmic architecture for visual-based detection and tracking of cooperative UAV is now tested using a dataset of images acquired by a different tracker UAV equipped with a camera characterized by a better spatial resolution (1600 × 1200 pixels) at a higher frame-rate (see Section 4.2 for details). This latter aspect makes more effective the strategy of exploiting target and tracker navigation data to predict the target projection on the image plane during both the detection and tracking phases. Since also the target UAV is different from the previously-analyzed FTs, these tests allow preliminary assessing the generalizing capability of the YOLO v2 object detection system trained for the *BB detection* block within the DL-based detector and tracker.

Starting from FT1-B which presents more challenging conditions due to the extremely wide variation of the target-tracker range (from 25 m to 300 m), the performance metrics are evaluated considering different values of τ_{det} and τ_{tr}. First, τ_{det} is set to 0.50 and τ_{tr} is varied from 0.50 and 0.075. The results of these tests obtained setting $d_{u,det}$ to 100 (N_w = 15), $d_{u,tr}$ and $d_{v,tr}$ to 150 pixels, are collected in Table 7.

Table 7. DL-based detector and tracker performance as a function of τ_{tr} (τ_{det} = 0.50). FT1-B composed of 1330 frames.

	τ_{det}	τ_{tr}	MD	FA	CD	WD
DL-based detector and tracker	0.50	0.50	71.65%	0.08%	28.27%	0%
	0.50	0.30	61.73%	0.15%	38.12%	0%
	0.50	0.15	58.2%	0.15%	41.65%	0%
	0.50	0.075	42.8%	0.3%	56%	0.9%

Though this configuration of the setting parameters produces very low values of FA (up to 4 false alarms if τ_{tr} is 0.075), it limits the number of correct detections. This phenomenon can be explained considering that, due to the difference in the flight-test scenario, and, consequently, in the images collected during FT1-B with respect to the training dataset (obtained from FT1-A and FT2-A), the average score characterizing the BBs provided by the YOLO v2 object detector is lower than in the previous tests carried out on FT3-A. Specifically, there is a correlation between the target-tracker relative distance and the value of S_{max}, as shown in Figure 14a. As it could be expected, larger values of S_{max} (up to 0.8) are obtained when the target-chaser relative distance is in the interval covered by the training dataset (53–135 m). A histogram providing the distribution of the range characterizing the training images can be found in Figure 14b.

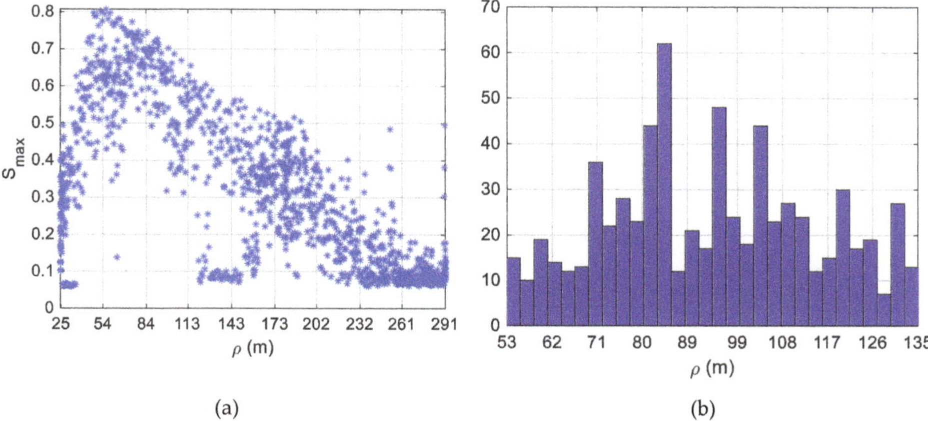

Figure 14. (**a**) Detection and tracking test on FT1-B (1330 images). Distribution of S_{max} as a function of the target-tracker relative distance. (**b**) Histogram providing the distribution of the target-tracker range characterizing the 650 images selected from FT1-A and FT2-A.

By looking at the significant increase in CD obtained by varying τ_{tr} from 0.50 (28.27%) to 0.075 (56%), it is clear that lower values of τ_{det} must be selected to try further increasing the number of correct detections. Hence, the result of additional tests obtained setting $d_{u,det}$ to 100 (N_w = 15), $d_{u,tr}$ and $d_{v,tr}$ to 150 pixels, are collected in Table 8.

Table 8. DL-based detector and tracker performance as a function of τ_{tr} (τ_{det} = 0.20). FT1-B composed of 1330 frames.

	τ_{det}	τ_{tr}	MD	FA	CD	WD
DL-based detector and tracker	0.20	0.20	39.62%	1.05%	57.59%	1.73%
	0.20	0.15	36.54%	1.05%	60.68%	1.73%
	0.20	0.10	31.65%	1.20%	65.64%	1.50%
	0.20	0.075	29.55%	1.43%	67.59%	1.43%

By setting τ_{det} to 0.20 and τ_{tr} to 0.075, it is possible to get a peak value of CD (67.59%) while keeping the number of false alarms and wrong detections below 40 out of 1330 tested frames. Looking at the still relatively large value of MD, the achieved performance could appear poor if compared with the results shown regarding FT3-A. However, the missed detections are related exclusively to a limit in the detection range. To clarify this point, Figure 15 shows the variation of the target-tracker relative distance during FT1-B, while highlighting the instances of correct detection.

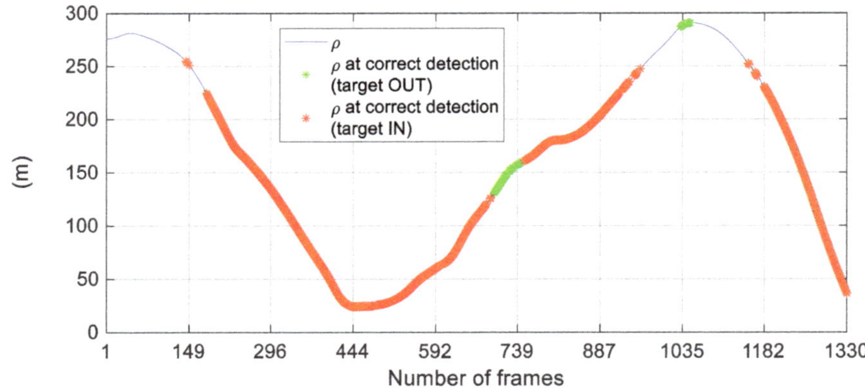

Figure 15. Variation of the target-chaser relative distance (blue line) during the FT1-B (1330 images). The DL-based detector and tracker are applied setting τ_{det} to 0.20 and τ_{tr} to 0.075. The frames where the algorithmic architecture provides correct detections are highlighted with red (target inside the FOV) and green (target outside the FOV) stars.

The proposed architecture can detect and track the target during the entire flight test. A relatively-large number of missed detections (which determines the percentage values of MD in Table 8) occurs only when the target-tracker relative distance gets larger than 225 m, which is well beyond the maximum target-tracker range characterizing the set of training images (≈135 m). Indeed, at such far range, the target not only occupies very few pixels on the image plane, but it is also blurred (its appearance is characterized by very low contrast with respect to the local background even above the horizon). Finally, Table 9 presents a statistical analysis of the accuracy level attained in the estimated LOS of the target for different values of the setting parameters.

Table 9. Statistical analysis of the accuracy in the estimated target LOS. DL-based detector and tracker applied on the images (1330) from FT1-B.

	τ_{det} = 0.20 and τ_{tr} = 0.075		τ_{det} = 0.50 and τ_{tr} = 0.15	
	Az Error (°)	*El Error (°)*	*Az Error (°)*	*El Error (°)*
Mean	−0.016	0.040	−0.025	0.036
Std	0.062	0.034	0.035	0.023
Rms	0.064	0.053	0.043	0.043

These results show that the best setting in terms of CD (τ_{det} = 0.20 and τ_{tr} = 0.075) leads to a rms error slightly larger (lower) than twice the camera IFOV (0.03°) for the *Az* (*El*) angle. Instead, if the best setting in terms of FA is selected (τ_{det} = 0.50 and τ_{tr} = 0.15), the rms error is around 0.04° both horizontally and vertically. This can be explained considering that this latter configuration of the setting parameters allows removing a larger number of correct detections but characterized by lower values of S_{max}, and, consequently, by a lower overlap with the real target position on the image plane.

An additional test is now carried out applying the proposed detection and tracking architecture to the images collected during FT2-B which is a coordinated flight at close range, i.e., the target-tracker relative distance varies in the interval (17 m, 23 m). During this image sequence the target is always inside the FOV so false alarms cannot occur. Since the target occupies a larger portion of the image plane with respect to the previously analyzed flight tests, the size of the search window is increased becoming a square of 300 × 300 pixels. This solution is adopted to avoid risking that the target is partially cut out of the search area either during the detection or the tracking process. Based on the results on FT1-B, τ_{tr} is set to a low value (0.10) while different values of τ_{det} are considered. Results of these tests are collected in Table 10.

Table 10. DL-based detector and tracker performance as a function of τ_{det} (τ_{tr} = 0.10). FT2-B composed of 735 frames.

	τ_{det}	MD	CD	WD
DL-based detector and tracker	0.50	23.3%	75.9%	0.8%
	0.30	5.3%	93.6%	1.1%
	0.20	4.9%	94%	1.1%

The results show that it is possible to significantly increase CD by reducing the value of τ_{det} down to 0.20. This solution allows correctly detecting and tracking the target during almost the entire flight, without causing an increase in the number of wrong detections However, the choice of accepting detections characterized by a low value of S_{max} leads to a slight worsening of the accuracy in the estimated LOS of the target. Indeed, if τ_{det} and τ_{tr} are set to 0.20 and 0.10 respectively, the rms error is around 0.08° in both *Az* and *El*.

6. Conclusions

This paper proposed an original algorithmic architecture for visual-based detection and tracking of cooperative UAVs. This architecture relies on a state-of-the-art Deep Learning neural network (YOLO v2) as main processing block, but it exploits navigation data of both the tracker and target UAV (also available thanks to the cooperative nature of the formation) to obtain a reliable prediction of the position of the target projection on the image plane, thus limiting the area to which the object detector must be applied. This strategy allows not only the computational time to be reduced, but it also extremely limits the number of false alarms which tend to occur when the target search is carried out in the entire image plane. Another innovative point is the development of an image processing technique to refine the position of the bounding box detected by YOLO, thus increasing the angular accuracy in the estimate of the target line-of-sight.

An experimental campaign of flight tests involving formations of two multirotor UAVs was carried out to build a database of images to both train the neural network and assess the performance of the proposed architecture. The flight tests were carried out with different UAVs as both tracker and target vehicles and using cameras with different resolution. The developed detection and tracking algorithms were tested over multiple datasets, thus being able to analyze the reliability and robustness against challenging conditions. In particular, the testing datasets were characterized by wide variability of the target appearance in terms of clutter level in the background, illumination condition and scale. Regarding this latter aspect, the generalizing capability of the trained neural network was evaluated due to the extremely wide variability of the target-tracker relative distance (from 17 m to 300 m) in the analyzed flight tests. The proposed architecture showed satisfying performance (more than 90% of correct detections) in all the analyzed cases while being able to estimate the target line-of-sight with an accuracy of the order of the camera IFOV (from 0.04° to 0.08° in the different cases).

Further investigations are still needed to improve the capability to correctly identify the target at far range below the horizon, where the target appearance may have very low contrast with respect to the local background. In fact, most of the wrong detections and false alarms noted analyzing the detection and tracking performance of the proposed architecture over the two datasets, occur below the horizon where the local background is particularly cluttered. Indeed, the closer the background is with respect to the camera, the lower the local contrast produced by the target gets and, consequently, the more difficult the detection process becomes. This problem could be addressed by training different DL-based neural network to operate above and below the horizon, respectively. Clearly, this solution would foresee the implementation of a reliable processing strategy to extract the horizon in order to decide which detector is more convenient depending on the position of the target prediction. Another open point for investigation regards the possibility to take advantage of the knowledge of the target-tracker relative distance also during the training process. This would allow entrusting the detection task to different neural networks each one purposely trained in a different interval of range, with the aim of increasing both reliability and accuracy. Indeed, the results presented in terms of the generalization capability of the neural network trained in this work, show that it is challenging to apply the same DL-based detector when the target-tracker distance gets too large or short with respect to the interval of ranges characterizing the training dataset. Overall, the possibility to investigate these open points leads to the need to generate a larger database of visual data by conducting additional flight tests. Finally, it is worth outlining that different DL-based object detector, other than YOLO v2, could be trained and tested within the proposed architecture without the need of further modifying the processing strategies.

Author Contributions: Conceptualization, R.O., G.F. and G.I. Data curation, R.O., G.F. and G.I. Formal analysis, R.O. Funding acquisition, G.F. Investigation, R.O. and G.I. Methodology, R.O. Project administration, G.F. Resources, G.F. Software, R.O. Supervision, G.F. Validation, R.O. and G.I. Writing, original draft, R.O. Writing, review and editing, R.O., G.F. and G.I.

Funding: This research was carried out within the framework of the "CREATEFORUAS"project, financially supported by the Italian Ministry for Education, University, and Research, within the PRIN Programme.

Conflicts of Interest: The authors declare no conflict of interest.

References

1. Se, S.; Firoozfam, P.; Goldstein, N.; Wu, L.; Dutkiewicz, M.; Pace, P.; Pierre Naud, J.L. Automated UAV-based mapping for airborne reconnaissance and video exploitation. In Proceedings of the SPIE 7307, Orlando, FL, USA, 28 April 2009; pp. 73070M-1–73070M-7.
2. Gonçalves, J.A.; Henriques, R. UAV photogrammetry for topographic monitoring of coastal areas. *ISPRS J. Photogramm. Remote Sens.* **2015**, *104*, 101–111. [CrossRef]
3. Ham, Y.; Han, K.K.; Lin, J.J.; Golparvar-Fard, M. Visual monitoring of civil infrastructure systems via camera-equipped Unmanned Aerial Vehicles (UAVs): A review of related works. *Vis. Eng.* **2016**, *4*, 1–8. [CrossRef]
4. Qi, J.; Song, D.; Shang, H.; Wang, N.; Hua, C.; Wu, C.; Qi, X.; Han, J. Search and rescue rotary-wing uav and its application to the lushan ms 7.0 earthquake. *J. Field Rob.* **2016**, *33*, 290–321. [CrossRef]
5. Maes, W.H.; Steppe, K. Perspectives for remote sensing with unmanned aerial vehicles in precision agriculture. *Trends Plant Sci.* **2019**, *24*, 152–164. [CrossRef]
6. Floreano, D.; Wood, R.J. Science, technology and the future of small autonomous drones. *Nature* **2015**, *521*, 460–466. [CrossRef]
7. Bürkle, A.; Segor, F.; Kollmann, M. Towards autonomous micro uav swarms. *J. Intell. Rob. Syst.* **2011**, *61*, 339–353. [CrossRef]
8. Sahingoz, O.K. Networking models in flying ad-hoc networks (FANETs): Concepts and challenges. *J. Intell. Rob. Syst.* **2014**, *74*, 513–527. [CrossRef]
9. Barrientos, A.; Colorado, J.; Cerro, J.D.; Martinez, A.; Rossi, C.; Sanz, D.; Valente, J. Aerial remote sensing in agriculture: A practical approach to area coverage and path planning for fleets of mini aerial robots. *J. Field Rob.* **2011**, *28*, 667–689. [CrossRef]
10. Messous, M.A.; Senouci, S.M.; Sedjelmaci, H. Network connectivity and area coverage for UAV fleet mobility model with energy constraint. In Proceedings of the 2016 IEEE Wireless Communications and Networking Conference, Doha, Qatar, 3–6 April 2016.
11. Vetrella, A.R.; Causa, F.; Renga, A.; Fasano, G.; Accardo, D.; Grassi, M. Multi-UAV Carrier Phase Differential GPS and Vision-based Sensing for High Accuracy Attitude Estimation. *J. Intell. Rob. Syst.* **2019**, *93*, 245–260. [CrossRef]
12. Vetrella, A.R.; Fasano, G.; Accardo, D. Attitude estimation for cooperating UAVs based on tight integration of GNSS and vision measurements. *Aerosp. Sci. Technol.* **2019**, *84*, 966–979. [CrossRef]
13. Vetrella, A.R.; Opromolla, R.; Fasano, G.; Accardo, D.; Grassi, M. Autonomous Flight in GPS-Challenging Environments Exploiting Multi-UAV Cooperation and Vision-aided Navigation. AIAA Information Systems-AIAA Infotech@Aerospace 2017, Grapevine, TX, USA, 5 January 2017.

14. Stoven-Dubois, A.; Jospin, L.; Cucci, D.A. Cooperative Navigation for an UAV Tandem in GNSS Denied Environments. In Proceedings of the 31st International Technical Meeting of The Satellite Division of the Institute of Navigation (ION GNSS+ 2018), Miami, FL, USA, 24–28 September 2018; pp. 2332–2339.
15. Maza, I.; Ollero, A.; Casado, E.; Scarlatti, D. Classification of multi-UAV Architectures. In *Handbook of Unmanned Aerial Vehicles*, 1st ed.; Springer: Dordrecht, The Netherlands, 2015; pp. 953–975.
16. Goel, S.; Kealy, A.; Lohani, B. Development and experimental evaluation of a low-cost cooperative UAV localization network prototype. *J. Sens. Actuator Networks* **2018**, *7*, 42. [CrossRef]
17. Yanmaz, E.; Yahyanejad, S.; Rinner, B.; Hellwagner, H.; Bettstetter, C. Drone networks: Communications, coordination, and sensing. *Ad Hoc Networks* **2018**, *68*, 1–15. [CrossRef]
18. Militaru, G.; Popescu, D.; Ichim, L. UAV-to-UAV Communication Options for Civilian Applications. In Proceedings of the 2018 IEEE 26th Telecommunications Forum (TELFOR), Belgrade, Serbia, 20–21 November 2018; pp. 1–4.
19. Chmaj, G.; Selvaraj, H. Distributed Processing Applications for UAV/drones: A Survey. In *Progress in Systems Engineering*; Advances in Intelligent Systems and Computing; Springer: Berlin, Germany, 2015; Volume 366, pp. 449–454.
20. He, L.; Bai, P.; Liang, X.; Zhang, J.; Wang, W. Feedback formation control of UAV swarm with multiple implicit leaders. *Aerosp. Sci. Technol.* **2018**, *72*, 327–334. [CrossRef]
21. Trujillo, J.C.; Munguia, R.; Guerra, E.; Grau, A. Cooperative monocular-based SLAM for multi-UAV systems in GPS-denied environments. *Sensors* **2018**, *18*, 1351. [CrossRef] [PubMed]
22. Basiri, M.; Schill, F.; Lima, P.; Floreano, D. On-board relative bearing estimation for teams of drones using sound. *IEEE Rob. Autom Lett.* **2016**, *1*, 820–827. [CrossRef]
23. Kohlbacher, A.; Eliasson, J.; Acres, K.; Chung, H.; Barca, J.C. A low cost omnidirectional relative localization sensor for swarm applications. In Proceedings of the 2018 IEEE 4th World Forum on Internet of Things (WF-IoT), Singapore, 5–8 February 2018; pp. 694–699.
24. Pugh, J.; Martinoli, A. Relative localization and communication module for small-scale multi-robot systems. In Proceedings of the 2006 IEEE International Conference on Robotics and Automation ICRA 2006, Orlando, FL, USA, 15–19 May 2006; pp. 188–193.
25. Walter, V.; Saska, M.; Franchi, A. Fast mutual relative localization of uavs using ultraviolet led markers. In Proceedings of the 2018 IEEE International Conference on Unmanned Aircraft Systems (ICUAS), Dallas, TX, USA, 12–15 June 2018; pp. 1217–1226.
26. LeCun, Y.; Bengio, Y.; Hinton, G. Deep learning. *Nature* **2015**, *521*, 436–444. [CrossRef] [PubMed]
27. Wang, C.; Wang, J.; Shen, Y.; Zhang, X. Autonomous Navigation of UAVs in Large-Scale Complex Environments: A Deep Reinforcement Learning Approach. *IEEE Trans. Veh. Technol.* **2019**, *68*, 2124–2136. [CrossRef]
28. Redmon, J.; Divvala, S.; Girshick, R.; Farhadi, A. You only look once: Unified, real-time object detection. In Proceedings of the 2016 IEEE conference on computer vision and pattern recognition, Las Vegas, NV, USA, 26 June–1 July 2016; pp. 779–788.
29. Fasano, G.; Accardo, D.; Moccia, A.; Maroney, D. Sense and avoid for unmanned aircraft systems. *IEEE Aerosp. Electron. Syst. Mag.* **2016**, *31*, 82–110. [CrossRef]
30. Lai, J.; Ford, J.J.; Mejias, L.; O'Shea, P. Characterization of Sky-region Morphological-temporal Airborne Collision Detection. *J. Field Rob.* **2013**, *30*, 171–193. [CrossRef]
31. Fasano, G.; Accardo, D.; Tirri, A.E.; Moccia, A.; De Lellis, E. Sky region obstacle detection and tracking for vision-based UAS sense and avoid. *J. Intell. Rob. Syst.* **2016**, *84*, 121–144. [CrossRef]
32. Huh, S.; Cho, S.; Jung, Y.; Shim, D. Vision-based sense-and-avoid framework for unmanned aerial vehicles. *IEEE Trans. Aerosp. Electron. Syst.* **2015**, *51*, 3427–3439. [CrossRef]
33. James, J.; Ford, J.J.; Molloy, T. Learning to Detect Aircraft for Long-Range Vision-Based Sense-and-Avoid Systems. *IEEE Rob. Autom. Lett.* **2018**, *3*, 4383–4390. [CrossRef]
34. Unlu, E.; Zenou, E.; Riviere, N. Using shape descriptors for UAV detection. *Electron. Imaging* **2018**, *9*, 1–5. [CrossRef]
35. Unlu, E.; Zenou, E.; Riviere, N.; Dupouy, P.E. Deep learning-based strategies for the detection and tracking of drones using several cameras. *IPSJ Trans. Comput. Vision Appl.* **2019**, *11*, 1–13. [CrossRef]
36. Martínez, C.; Campoy, P.; Mondragón, I.F.; Sánchez-Lopez, J.L.; Olivares-Méndez, M.A. HMPMR strategy for real-time tracking in aerial images, using direct methods. *Mach. Vision Appl.* **2014**, *25*, 1283–1308. [CrossRef]

37. Gökçe, F.; Üçoluk, G.; Şahin, E.; Kalkan, S. Vision-based detection and distance estimation of micro unmanned aerial vehicles. *Sensors* **2015**, *15*, 23805–23846.
38. Li, P.; Wang, D.; Wang, L.; Lu, H. Deep visual tracking: Review and experimental comparison. *Pattern Recognit.* **2018**, *76*, 323–338. [CrossRef]
39. Li, J.; Ye, D.H.; Chung, T.; Kolsch, M.; Wachs, J.; Bouman, C. Multi-target detection and tracking from a single camera in Unmanned Aerial Vehicles (UAVs). In Proceedings of the 2016 IEEE/RSJ International Conference on Intelligent Robots and Systems (IROS), Daejeon, Korea, 9–14 October 2016; pp. 4992–4997.
40. Rozantsev, A.; Lepetit, V.; Fua, P. Detecting flying objects using a single moving camera. *IEEE Trans. Pattern Anal. Mach. Intell.* **2017**, *39*, 879–892. [CrossRef]
41. Wu, Y.; Sui, Y.; Wang, G. Vision-based real-time aerial object localization and tracking for UAV sensing system. *IEEE Access* **2017**, *5*, 23969–23978. [CrossRef]
42. Sun, S.; Yin, Y.; Wang, X.; Xu, D. Robust Visual Detection and Tracking Strategies for Autonomous Aerial Refueling of UAVs. *IEEE Transactions on Instrumentation and Measurement* **2019**, in press. [CrossRef]
43. Cledat, E.; Cucci, D.A. Mapping Gnss Restricted Environments with a Drone Tandem and Indirect Position Control. *ISPRS Annals* **2017**, *4*, 1–7. [CrossRef]
44. Krajnik, T.; Nitsche, M.; Faigl, J.; Vanek, P.; Saska, M.; Preucil, L.; Duckett, T.; Mejail, M.A. A practical multirobot localization system. *J. Intell. Rob. Syst.* **2014**, *76*, 539–562. [CrossRef]
45. Olivares-Mendez, M.A.; Mondragon, I.; Cervera, P.C.; Mejias, L.; Martinez, C. Aerial object following using visual fuzzy servoing. In Proceedings of the 1st Workshop on Research, Development and Education on Unmanned Aerial Systems (RED-UAS), Sevilla, Spain, 30 November–1 December 2011.
46. Duan, H.; Xin, L.; Chen, S. Robust Cooperative Target Detection for a Vision-Based UAVs Autonomous Aerial Refueling Platform via the Contrast Sensitivity Mechanism of Eagle's Eye. *IEEE Aerosp. Electron. Syst. Mag.* **2019**, *34*, 18–30. [CrossRef]
47. Opromolla, R.; Vetrella, A.R.; Fasano, G.; Accardo, D. Airborne Visual Tracking for Cooperative UAV Swarms. In Proceedings of the 2018 AIAA Information Systems-AIAA Infotech@ Aerospace, Kissimmee, FL, USA, 8–12 January 2018.
48. Opromolla, R.; Fasano, G.; Accardo, D. A Vision-Based Approach to UAV Detection and Tracking in Cooperative Applications. *Sensors* **2018**, *18*, 3391. [CrossRef] [PubMed]
49. The Kalibr Visual-Inertial Calibration Toolbox. Available online: https://github.com/ethz-asl/kalibr (accessed on 1 September 2019).
50. Yang, Z.; Shen, S. Monocular visual–inertial state estimation with online initialization and camera–imu extrinsic calibration. *IEEE Trans. Autom. Sci. Eng.* **2017**, *14*, 39–51. [CrossRef]
51. Camera Calibration Toolbox for Matlab. Available online: http://www.vision.caltech.edu/bouguetj/calib_doc/ (accessed on 1 September 2019).
52. Guo, Y.; Liu, Y.; Oerlemans, A.; Lao, S.; Wu, S.; Lew, M.S. Deep learning for visual understanding: A review. *Neurocomputing* **2016**, *187*, 27–48. [CrossRef]
53. Liu, W.; Anguelov, D.; Erhan, D.; Szegedy, C.; Reed, S.; Fu, C.Y.; Berg, A.C. Ssd: Single shot multibox detector. Computer Vision—ECCV 2016. ECCV 2016. In *Lecture Notes in Computer Science*; Leibe, B., Matas, J., Sebe, N., Welling, M., Eds.; Springer: Cham, Switzerland, October 2016.
54. Ren, S.; He, K.; Girshick, R.; Sun, J. Faster r-cnn: Towards real-time object detection with region proposal networks. In Proceedings of the Advances in Neural Information Processing Systems, Montreal, Canada, 7–12 December 2015; pp. 91–99.
55. Zhang, S.; Wen, L.; Bian, X.; Lei, Z.; Li, S.Z. Single-shot refinement neural network for object detection. In Proceedings of the 2018 IEEE Conference on Computer Vision and Pattern Recognition, Salt Lake City, UT, USA, 19–21 July 2018; pp. 4203–4212.
56. Deng, L.; Yang, M.; Li, T.; He, Y.; Wang, C. RFBNet: Deep Multimodal Networks with Residual Fusion Blocks for RGB-D Semantic Segmentation. *arXiv preprint*, 2019; arXiv:1907.00135.
57. Redmon, J.; Farhadi, A. YOLO9000: Better, faster, stronger. In Proceedings of the 2017 IEEE Conference on Computer Vision and Pattern Recognition, Honolulu, HI, USA, 21–26 July 2017; pp. 7263–7271.
58. Redmon, J.; Farhadi, A. Yolov3: An incremental improvement. *arXiv preprint*, 2018; arXiv:1804.02767.
59. Liu, L.; Ouyang, W.; Wang, X.; Fieguth, P.; Chen, J.; Liu, X.; Pietikäinen, M. Deep learning for generic object detection: A survey. *arXiv preprint*, 2018; arXiv:1809.02165.

60. Zhao, Z.Q.; Zheng, P.; Xu, S.T.; Wu, X. Object detection with deep learning: A review. *IEEE Trans. Neural Netw. Learn. Syst.* **2019**, in press. [CrossRef]
61. Deep Learning Toolbox. Available online: https://it.mathworks.com/help/deeplearning/index.html (accessed on 1 September 2019).
62. He, K.; Zhang, X.; Ren, S.; Sun, J. Deep residual learning for image recognition. In Proceedings of the 2016 IEEE Conference on Computer Vision and Pattern Recognition, Las Vegas, NV, USA, 27–30 June 2016; pp. 770–778.
63. Brown, A.K. Test results of a GPS/inertial navigation system using a low cost MEMS IMU. In Proceedings of the 11th Annual Saint Petersburg International Conference on Integrated Navigation System, Saint Petersburg, Russia, 24–26 May 2004.
64. Nuske, S.T.; Dille, M.; Grocholsky, B.; Singh, S. Representing substantial heading uncertainty for accurate geolocation by small UAVs. In Proceedings of the AIAA Guidance, Navigation, and Control Conference, Toronto, ON, Canada, 2–5 August 2010.
65. Eichler, S. Performance Evaluation of the IEEE 802.11p WAVE Communication Standard. In Proceedings of the 2007 IEEE 66th Vehicular Technology Conference, Baltimore, MD, USA, 30 September–3 October 2007; pp. 2199–2203.

Article

A Framework for Multi-Agent UAV Exploration and Target-Finding in GPS-Denied and Partially Observable Environments

Ory Walker *,†,‡, Fernando Vanegas ‡ and Felipe Gonzalez ‡

Queensland University of Technology, Brisbane City, QLD 4000, Australia; f.vanegasalvarez@qut.edu.au (F.V.); felipe.gonzalez@qut.edu.au (F.G.)

* Correspondence: ory.walker@hdr.qut.edu.au; Tel.: +61-409-833-395

† Current address: 2 George Street, Brisbane City, QLD 4000, Australia.

‡ These authors contributed equally to this work.

Received: 30 June 2020; Accepted: 18 August 2020; Published: 21 August 2020

Abstract: The problem of multi-agent remote sensing for the purposes of finding survivors or surveying points of interest in GPS-denied and partially observable environments remains a challenge. This paper presents a framework for multi-agent target-finding using a combination of online POMDP based planning and Deep Reinforcement Learning based control. The framework is implemented considering planning and control as two separate problems. The planning problem is defined as a decentralised multi-agent graph search problem and is solved using a modern online POMDP solver. The control problem is defined as a local continuous-environment exploration problem and is solved using modern Deep Reinforcement Learning techniques. The proposed framework combines the solution to both of these problems and testing shows that it enables multiple agents to find a target within large, simulated test environments in the presence of unknown obstacles and obstructions. The proposed approach could also be extended or adapted to a number of time sensitive remote-sensing problems, from searching for multiple survivors during a disaster to surveying points of interest in a hazardous environment by adjusting the individual model definitions.

Keywords: POMDP; Deep Reinforcement-Learning; UAV; multi-agent; search

---

1. Introduction

In recent years the use of Unmanned Aerial Vehicles (UAVs) has been broadly explored for a number of applications, both in consumer and industrial operation environments. Many tasks require the capacity for autonomous searching or surveying of a known or unknown environment in the presence of hazards. Such applications include the broad field of search and action tasks, from search and rescue [1], environmental sampling and data collection [2–4], pursuit of targets in complex environments [5,6], even underground mining and surveying of confined spaces [7].

Often-times exploration tasks are time-sensitive and require the use of multiple agents to carry out the mission objective. However, coordinating multiple UAV agents over complex environments in the presence of partially observable obstacles remains a challenge. There are multiple works involving searching or navigating an unknown environment using single and multiple agent configurations [8]. Solutions in the past have not allowed for operation of the UAV under uncertainty [9] or operation in continuous action spaces, using a discrete list of actions to control the operation of the UAV [10].

Modelling of many robotic control problems involving uncertainty is usually performed with the Partially Observable Markov Decision Process (POMDP) [11] framework. The main reason being that POMDPs are useful for modelling uncertainty within a system. In the past POMDPs were most effectively solved by classical algorithms and solvers [12–14]. However, these approaches have a number of limitations, including discrete action spaces [12] or difficulty modelling continuous action spaces [12,13], pauses in operation to calculate the next optimal trajectory [12,13] in the case of online planners, or long pre-planning times for each new environment. While performance of traditional solvers is excellent for solving discrete time-independent planning problems [15], their limitations could be detrimental when applied to low-level continuous control.

In recent years however there has been an explosion in the application of machine learning techniques to solving POMDP tasks as a result of advancements made to deep-reinforcement learning techniques. They have shown that they are exceptional when solving well defined MDP and POMDP control tasks, from simulated robotic control tasks [16,17], to atari video games [18], with deep reinforcement learning based approaches even out-performing human experts in the games of Go [19,20], Dota 2 [21] and Starcraft 2 [22].

Included in the tasks that have seen an increase in the use of machine learning techniques are UAV control tasks [23]. Recently a deep Q-network approach was applied to the exploration and navigation of outdoor environments [24], and a multi-agent deep deterministic policy gradient method was applied to multi-agent target assignment and path planning in the presence of threat areas [25]. However, the application of such techniques to control the exploration of multiple UAV agents in partially observable and hazardous environments remains unrepresented.

This paper maintains a focus on the problem of target-finding using multiple UAV agents in partially observable environments.The solution presented in this paper can be expanded and adapted as necessary to a number of remote sensing problem spaces, from searching for survivors in a variety of environments such as disaster zones, buildings, caves systems, and open or forested areas, to surveying potentially hazardous or difficult to reach points of interest. We present the multi-agent UAV target-finding problem as two problems; a decentralised multi-agent planning problem, and a local control problem. The UAVs must be capable of searching local continuous environments containing unknown hazards, while also being directed between local environments within the global environment.

This paper presents a multi-agent target-finding framework, using the Robotic Operation System 2 (ROS2) platform [26], to search partially observable occupancy map style environments using multiple simulated UAV agents. The Adaptive Belief Tree (ABT) solver [12] is used for validating the planning component of the framework, while a Proximal Policy Optimization (PPO2) [27] algorithm is used for producing a solution for and validating the local control component of the framework.

The main contribution of this paper is a hierarchical framework that combines decentralised POMDP based planning and deep reinforcement learning to enable multiple UAV agents to search GPS-denied, partially observable occupancy map style environments for a target while adapting to unknown obstacles in real-time. Secondary contributions include modelling of both the multi-agent planning problem for use with the TAPIR POMDP software package and the local control model in the form of an open-ai gym environment that uses occupancy-grid style maps.

2. Background

The POMDP framework is suitable for modelling sequential decision making for robotic agents operating in the real world, as robotic agents rarely if ever have access to measurements and sensing free from uncertainty.

Formally, the POMDP framework can be defined by an 8-tuple (S, A, T, O, Ω, R, b0, γ). Where:

- S is a set of states,
- A is a set of potential actions,
- T is the transition function, defining the probability of transition between states,
- O is a set of observations, Ω is the observation function, defining the probability of observing o from state s after taking action a,
- R is the reward function,

- b_0 is the initial belief,
- and γ is the discount factor.

An agent (UAV) in a POMDP formulation does not have knowledge of the true state of the environment. The agent instead maintains a continuous belief $b(s)$ over the potential states. It updates this belief after taking an action a and making an observation o. Consider this belief a distribution of all possible states for the agent given the transition and observation functions T and Ω, with less certain functions (modelled for less certain problems) resulting in a larger belief space, and more uncertainty in the true state of the agent and environment.

The objective of a POMDP solver is to identify a sequence of actions that maximize the total expected discounted return of the model. Rather than calculating for a known true state s however, the POMDP solver uses an initial belief of potential states b0.

The solution a POMDP solver provides is a policy π that maps belief space to action space and identifies which action will maximize the expected discounted reward given a belief b(s). The optimal policy is then represented by π^* which yields the maximum reward value for each belief state, which is denoted by V^*. V^* (b) is therefore the value function that accounts for the maximum reward value for any given belief b.

Adaptive Belief Tree (ABT) [12] is a recent online POMDP solving alogrithm implemented in the TAPIR software toolkit [28] capable of solving complex POMDP problems with significant state-spaces despite potential changes in the problem space by efficiently adapting previously generated policies during run-time to incorporate new information and produce near-optimal solutions.

In this work the ABT solver is applied to solving the global planning problem modelled as a POMDP. Further work would extend this problem model to include dynamic responses to a changing environment and planning in a completely unknown environment.

Solving the POMDP problem of local UAV exploration control involves optimizing a policy with respect to a future discounted reward. The primary goal of deep reinforcement learning within our framework is to learn a near-optimal policy for the local exploration control problem. Deep Reinforcement Learning leverages the properties of deep neural networks and reinforcement learning algorithms to produce policies that maximise a reward function for potentially complex problems. For validation of the framework we opted to use the Proximal Policy Optimization (PPO2) algorithm that is implemented in the Stable Baselines [29] project. However, the framework is designed in a modular fashion which permits the use of alternative algorithms to solve the problem.

A key benefit of applying Deep Reinforcement Learning to the problem is the reduced control overhead required when implementing a learned policy and the near-instantaneous response to new information available to the UAV at any given time, without the need for costly re-calculation as more information is acquired about the environment. The PPO2 algorithm has been shown to learn complex policies for a variety of continuous control tasks, and is well suited to the control task within this framework.

3. Problem Definition

The problem that the proposed framework aims to solve is that of searching a gps-denied, partially observable environment for a target using multiple UAV agents. The problem assumes the following:

- A rough map of the environment is known prior to operation. This map helps define the connections between the search regions of the environment. Such a map might exist in the form of a floor-plan for a building or a pre-existing map for a cave system.
- Obstacles, obstructions and minor changes to the shape of the local environments are unknown to the agents prior to operation. Figure 1 shows the difference between the known and unknown environment information.
- Agents are capable of some form of Simultaneous Localisation and Mapping (SLAM).
- Local agent localisation is perfect or near-perfect.

- The target does not move.
- There is an equal chance of the target being at any location within the environment.
- The search occurs on a single 2D-plane—i.e., the agent doesn't benefit from changing its altitude during the search as the obstacles are floor to ceiling.

The implementation presented in this paper assumes that UAV agents are equipped with a front facing sensor (a camera) with a range of seven (7) metres and a horizontal field of view of ninety (90) degrees. In practice, the kind of sensor the UAV is equipped with only matters during training of the control policy outlined in Section 6. During operation the Planning and Control (PAC) framework proposed only requires that the SLAM system used by an agent outputs a grid-map of the environment and the agent's pose and velocity within that grid map. The decoupling of SLAM from the PAC framework means that it can readily be expanded for platforms using other sensors and implementations of SLAM. This also means that the sensor dynamics and sensor noise are not considered by the PAC Framework during operation. A simplified approximation of a SLAM system is simulated in order to validate the PAC framework.

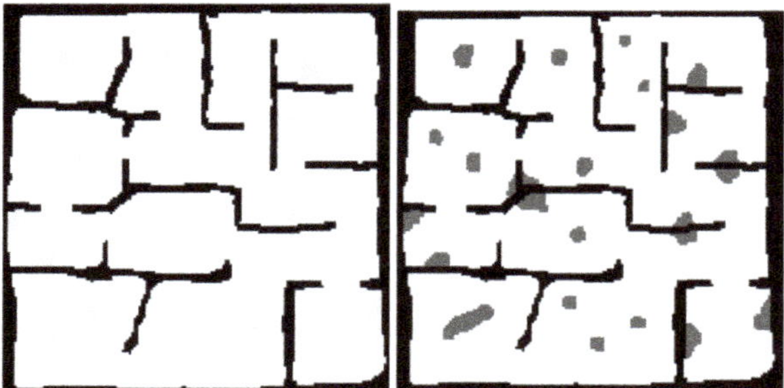

Figure 1. Prior Knowledge of the Environment vs. True State of the Environment.

4. Framework Definition

Developing the framework required the problem to be broken down into two main problem components:

- Planning for multiple agents over potentially large maps to find a target with an unknown location.
- Individual agent exploration control within local continuous partially observable environments.

To enable the framework to scale to large map sizes while also enabling local control, it was decided that the framework would use occupancy grid maps with a high level graph-map definition of the environment. The local controller would use local mapping and position information to control the agent, while the global planner considered the environment as a discrete graph map. The information contained within the map and how an environment is defined are outlined further in Section 4.1. It was decided for the global planner to use a decentralised approach, with each agent having its own planner and communicating the relevant information as necessary to all other agents. Each agent also has a local controller that handles communication between agents, receives macro-actions from the global planner, processes the pose, velocity and mapping information into an abstraction that can be used by the control policy, and outputs actions from the control policy to the flight controller. Figure 2 shows the resulting two layer-framework architecture for n-agents.

4.1. Environment Definition

Proper definition of the environment and the information available to the agent is instrumental to the problem definition and by extension the operation of the proposed framework. The environments used by this framework are occupancy style grid-maps and are composed of four main types of cells:

- Obstacles known prior to operation. Defined as a value of −3 in the occupancy-map.
- Obstacles unknown prior to operation. Defined as a value of −2 in the occupancy-map.
- Explored empty space. Defined as a value of −1 in the occupancy-map.
- Unexplored empty space. Defined as a value of >= 0 in the occupancy-map.

The map is further broken down at a higher level into search regions. These are differentiated via the use of different positive integers within the occupancy-map and their shape and size is generally defined by the natural borders created by known obstacles. The known obstacles are assumed prior knowledge within the problem space, and are defined from a pre-existing floorplan or map if one exists. Figure 3 shows the first steps in the creation of one of the test maps that were used for validation of the framework. First the known obstacles are defined, and then the search regions are filled into the empty space and unknown obstacles are added. The test map shown has existing known structures as a building or man made structure might have.

Once the regions are defined, the connections between the regions can be used to define a graph map of the environment. An example of this is shown Figure 4.

To finish creation of the map locations for the agents and targets to spawn are added. A complete map is shown in Figure 5.

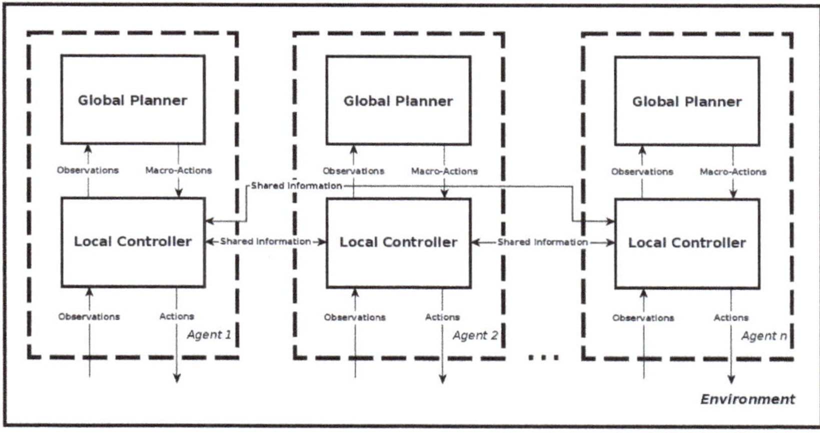

Figure 2. Framework Architecture

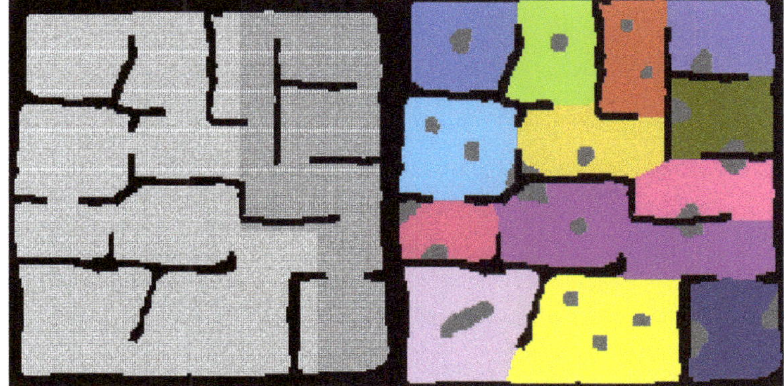

Figure 3. Map Creation: Defining the Known Obstacles, Search Regions and Unknown Obstacles. **Black Cells**: Known Obstacle Cells, **White Cells**: Explored Empty Cells, **Grey Cells**: Unknown Obstacle Cells, **Coloured Cells**: Unexplored Empty Cells (Search Regions).

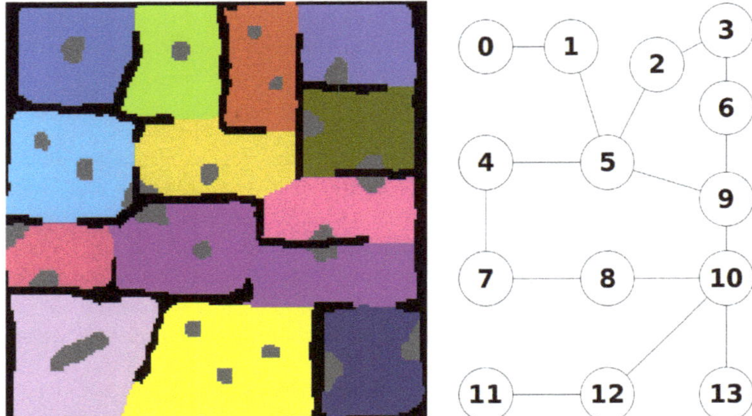

Figure 4. Graph Map Definition From Search Regions. **Black Cells**: Known Obstacle Cells, **White Cells**: Explored Empty Cells, **Grey Cells**: Unknown Obstacle Cells, **Coloured Cells**: Unexplored Empty Cells (Search Regions).

Figure 5. Complete Map. **Black Cells**: Known Obstacle Cells, **White Cells**: Explored Empty Cells, **Coloured Cells**: Unexplored Empty Cells (Search Regions), **Grey Cells**: Unknown Obstacle Cells, **Triangles**: Agent Spawn Locations, **Circles**: Target Spawn Locations.

5. Decentralised Multi-Agent Planner

Within the framework, each agent has their own global planner that is responsible for generating macro-actions for that agent. These macro-actions direct the UAV toward connected nodes within the graph-map according to a policy generated by the planner. The goal is to find a target within the environment in as few moves as possible. In its simplest form the planner directs the agent to search the graph-map of the environment until the agent is found or the environment is completely searched. For this paper the planner operates under the assumption that the target distribution is uniform across the environment. However, it would be simple to add the functionality for weighted distributions in the future. The positions of all agents are considered fully observable for the duration of the task, as localisation and communication are assumed, however the partially observable target location means that the problem must be modelled as a POMDP.

The policy for the POMDP model is generated using the online POMDP solver platform TAPIR, using the ABT algorithm. The implementation allows for the definiton of a move action for all nodes, however only those actions that are legal are considered during rollout, i.e., only actions that would result in the UAV moving to a connected node from its current node are considered during planning. The model also assumes that any node that the UAV visits will be searched and a penalty is given to an agent for occupying the same space as another agent in order to reduce the likelihood of the planners trying to have two agents occupy the same node, as this functionality is not available in the local controller at this time.

Due to the decentralised nature of the approach, coordination between agents is achieved in a loose fashion. Each agent is aware of the location of all other agents. During rollout each planner considers locations occupied by other agents as searched, just as it would consider a location occupied by its own agent. A planner also assumes during policy generation that other agents are equally likely to move to new locations at each time step, subsequently searching those locations. This causes the generation of a variety of future beliefs during rollout which collapse into a single belief whenever the positions of each agent are observed. By building these assumptions into the model, the planners produce individual agent policies that attempt to maximize a shared reward by finding the target in as few steps as possible, enabling decentralised collaboration. For instance, agents won't navigate toward regions occupied by other agents, as they assume that those areas will be searched by the agents occupying those regions. The problem is formulated as a Decentralised POMDP, considering only the actions for a single agent based on information received from all agents. It is modelled as shown in Table 1.

Table 1. Global Planner POMDP Formulation.

State-Space (S)	UAV Locations
	Target Location
Observations (O)	Location of UAV Agents
	Target Seen Status for each Agent
Actions (A)	Move to a Connected Node
Rewards (R)	Cost for each step
	Cost for occupying same node as other agent
	Reward when the target is found

6. Local Control Policy

The local controller component of the framework is composed of a single deep reinforcement learning control policy to generate actions and the ancillary components required for communication and observation generation for the local control policy and global planner. The following section details how the local control policy is generated. The goal of the Local Control Policy is to generate actions from observations such that the UAV navigates a local continuous portion of the global environment, exploring unexplored areas and avoiding obstacles as it discovers them. The actions that the policy generates are velocity commands to increase or decrease the velocity of the agent in the x, y, and yaw axes. To generate this control policy, a custom OpenAI gym environment was created, and then an agent was trained on that environment using the PPO2 implementation within Stable Baselines.

6.1. Open AI Gym Environment Definition

The Open AI Gym environment used to train the agent needed to be a good proxy for the simulated operation of an agent within the framework. Ideally it also needed to be lightweight in order to increase the speed at which the agent was trained. Because the agent was required to search environments with unknown obstacles, the problem also needed to be modelled as a POMDP. The problem was modelled as shown in Table 2.

Definition of the environment aspect of the state space for the problem was relatively simple. The gym environment uses one or more occupancy-maps, as defined in Section 4.1, as the training environments for the agent, each with only a single search region and no graph-map, as those features are only required for the global planner aspect of the framework. The map changes as the UAV explores, replacing unexplored cells with explored cells according to simulated sensing. The UAV location is simply the agents x, y and yaw value within the map, and is updated according to the agents velocity at the beginning of each time step.

Table 2. Local Control Environment Formulation.

State-Space (S)	UAV Pose
	UAV Velocity
	Environment Map
Observations (O)	Map Observations
	Agent Velocities (x, y, yaw)
Actions (A)	Change velocity in any of the operational axes (x, y, yaw)
Rewards (R)	Reward for each cell explored
	Cost for collision equal to sum of cumulative reward

The action space for the agent is a continuous space between −1 and 1 for each of the control axes of x, y, and yaw. The policy outputs a vector [x, y, yaw] at each time step that changes the agents velocity.

With respect to the observation space, the agent's velocity is fully observable. However the control policy can only receive information about the explored environment, and it does so via the use of thirty-two (32) distance readings and a corresponding type index. These readings are an abstraction of the UAVs pose within the local environment and the state of that local environment. They are generated irrespective of the sensors used by the agents. At each time step according to the agents pose within the environment, thirty two (32) distance readings spaced equally around the agent are projected to the first unexplored cell or obstacle in line with that reading. The type index is then updated to reflect whether the distance reading indicates an obstacle at that location or not. Figure 6 shows the agent within an environment, the cone of vision of the assumed front-facing sensor and the distance readings that are generated at each time step according the the UAV's pose and knowledge of the environment.

The environment is simulated at 10 hz i.e., observations are made and actions are generated every 0.1 s.

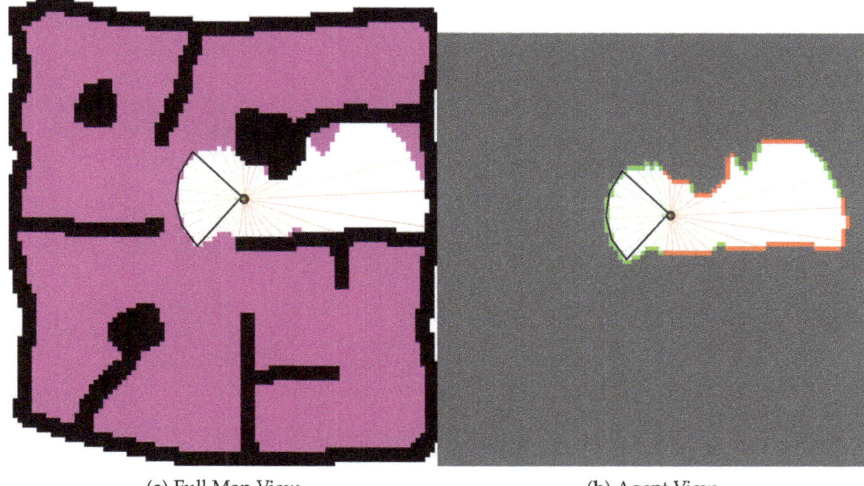

(a) Full Map View (b) Agent View

Figure 6. Full Map View vs. Agent View. **Black Cells**: Obstacle Cells, **White Cells**: Explored Empty Space Cells, **Purple Cells**: Unexplored Empty Cells, **Grey Cells**: Unexplored Area, **Red Cells**: Observed Obstacle Cells, **Green Cells**: Safe Edges of Unexplored Region, **Black Cone**: Field of View of Agent's Sensor, **Red Lines**: Obstacle Distance Readings, **Green Lines**: Safe Distance Readings, **Circle**: UAV.

6.2. Defining The Simulated Agent

The above Open Ai Gym training environment can be used to produce a control policy for different kinds of agents by changing the model of the agent within the environment. As a result the gym environment developed could be used to train a large sluggish UAV equipped with a LIDAR sensor, or a small agile UAV equipped with a front facing camera. In this case, a generic simulated agent with the characteristics shown in Table 3 was trained in the environment. Note that the characteristics listed are not the extent to which an agent might be modelled, and are only what were considered necessary for the generic agent model.

Table 3. Generic Simulated Agent Definition.

Radius (m)	0.5 metres
Linear Velocity Limit (m/s)	1.5
Angular Velocity Limit (ω/s)	$\pi/4$
Response Type	Linear
Linear Action Scale (m/s)	0.25
Angular Action Scale (ω/s)	0.125
Sensor Type	Front Facing Camera
Sensor Range (m)	7
Sensor FOV (° Horizontal)	90

The radius defines the size of a UAV and how close it can come to obstacles without triggering a collision and incurring a penalty.

The Linear and Angular Velocity Limits denote the max possible velocity in the linear and angular axes of control (x, y and yaw). The Response Type characteristic is where the simulated agent deviates largely from real world UAVs. In this case the simulated agent is modelled using a Linear Response.

For instance, if the simulated agent has a velocity of [0, 0, 0] and receives a control vector of [1, 0, 0], after one time step the agents velocity will be [1, 0, 0] with the average velocity of the agent over that time step being [0.5, 0, 0]. The Linear and Angular Action Scale parameters define the maximum change in velocity that can be requested by the controller at any time, i.e., the largest velocity delta for the agent over a single time step is capped at [0.25 m/s, 0.25 m/s, 0.125 ω/s]. Essentially they are a gain applied to the actions produced by the policy, which have a continuous range of [−1, 1] for each axis. The values of these parameters were selected because they seemed reasonable for the desired simulated agent.

The sensor type of the simulated agent was selected to be a front facing camera with a Sensor Range of seven (7) metres, and a horizontal field of view (FOV) of 90°. This sensor type was then modelled into the environment so that the agent could make observations during training, and the policy could learn how to search using that sensor. Training the agent is the only time when a model of the sensors used for making observations are required by the framework. The training step uses the model, which can be an approximation, of the sensor to simulate mapping of environments, such that the control policy learns the most optimal way to search unknown environments using that sensor while avoiding collisions. For instance, by simulating a LIDAR sensor, instead of a front facing camera, the control policy produced would be optimized for use with LIDAR based systems, instead of a system with a limited front-facing sensor.

Changing any of these parameters and how the agent is defined and modelled within the environment would change the learnt policy accordingly and enable control of a variety of UAV platforms. And as can be seen in Section 8.1, even a policy trained using an agent with a perfect linear response can be adapted to control physically simulated UAV platforms by changing the gain (action scales) applied to the output of the control policy.

6.3. Training and Using the Policy

The agent was trained on the environment detailed in Section 6.1 using the PPO2 algorithm implementation in the Stable Baselines project. Table 4 outlines the relevant training parameters used to produce the policy used in validation of the framework. The training was undertaken on a High End Desktop (HEDT), with a 32 Core (64 Thread) Processor (Threadripper 3970x) and took approximately five (5) hours to train. Training was split across sixty-four (64) parallel environments to improve the training time.

Thirty two (32) different training maps were created and used to increase the domain randomisation of the training, with agents being randomly spawned into a safe location in one of those maps during each training run. Eight (8) of these training maps can be seen in Figure 7. By increasing the variety of features contained within these maps the control policy produced can be used on a variety of local search environments without the need for retraining, as is the case with the policy tested in this paper. Only abstraction of the environment and inference using the trained policy need to be conducted on-board the UAV for the local control stack, resulting in a very computationally efficient controller.

Furthermore, if a particular environment type is expected to be the only type of environment faced by the agents, that type of environment could be weighted more heavily in the training. For instance if the expected operation environments were only caves and tunnels, you would only need to train the policy on cave and tunnel style environments.

The policy naturally learns to avoid observed obstacles as collisions impart a penalty during training, and reduce the total reward by ending searches prematurely.

Use of the policy in the framework requires the local controller node to generate the correct observations for the control policy. If the agent is not in the desired target region, as dictated by the global planner, the local controller node considers both its current region and target region when generating observations. However, once the UAV transitions into its target region the controller node treats all other regions as obstacles during the local search. This prevents the UAV from exploring outside the designated target region until the region is searched and the global planner gives the agent another action.

The implementation of the local control policy also checks if the agent is about to enter a region occupied by another agent. If it does, the agent holds position until the region becomes clear. Combined with the global-planners cost penalty for navigating into spaces occupied by another agent, this ensures the agents are never within the local space of each other, preventing any inter-agent collisions.

Table 4. Training Parameters.

Policy	PPO2-MLP
Layers	2
Neurons per Layer	64
Normalised	Yes
Environments	64
Mini Batches	32
Learning Rate	0.0002
Gamma	0.99
Clip Range	0.2
NoPtEpochs	10
Lam	0.925
Max Grad Norm	0.5
Ent Coef	0.0
Steps	1800

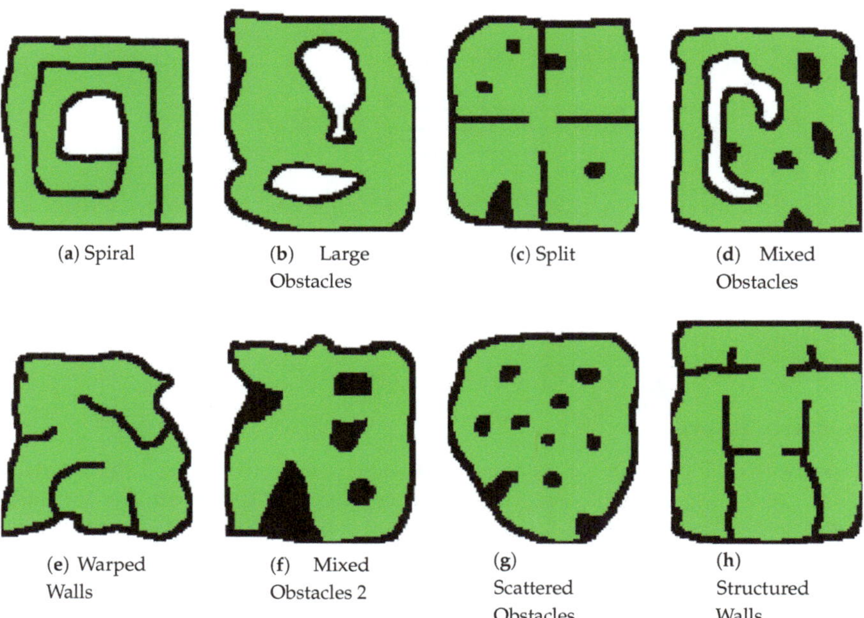

(**a**) Spiral (**b**) Large Obstacles (**c**) Split (**d**) Mixed Obstacles (**e**) Warped Walls (**f**) Mixed Obstacles 2 (**g**) Scattered Obstacles (**h**) Structured Walls

Figure 7. Training Maps One through Eight. **Black Cells**: Obstacle Cells, **Green Cells**: Traversable Cells.

7. Software Architecture

The components of the Framework were combined through the use of the Robotic Operation System 2 (ROS2) and executed on an Ubuntu 18.04 system. The software breakdown for an agent can be seen in Figure 8. The local control policy and global planner required the use of the stable-baselines python package and TAPIR ABT Implementation respectively, with custom models required for each. The full framework, along with a setup guide, is available at the following link: https://github.com/OryWalker/Multi-Agent-Target-Finding.

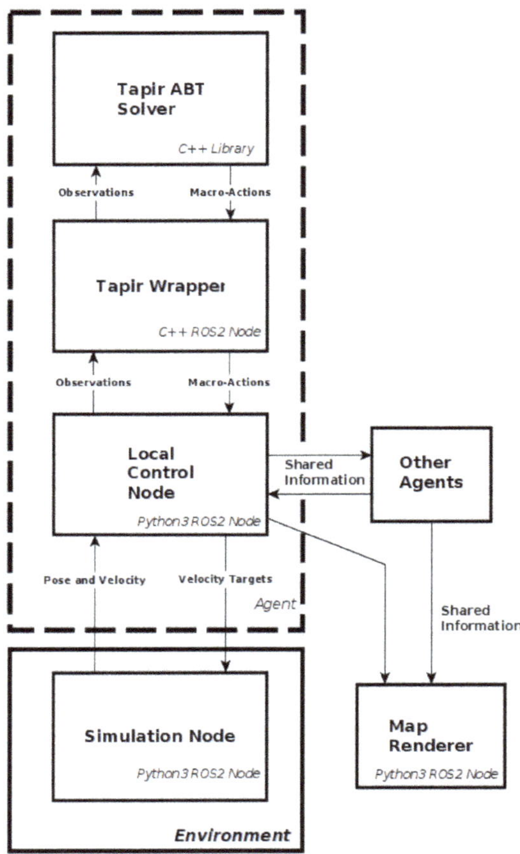

Figure 8. Software Architecture.

8. Experimental Results

Framework validation occured in two stages. First the performance of the Local Controller was confirmed. Once it was shown that the Local Controller was capable of controlling both ideal and physically simulated agents in an unseen environment, the full framework was tested using a varying number of ideal simulated agents on multi-region environments to validate overall performance. The following sections outline the testing undertaken.

8.1. Testing the Local Controller

The local control policy and local controller were validated across three separate UAV platforms; the generic simulated agent that the policy was trained on as a baseline, and two physically simulated

UAV platforms within the Gazebo simulation environment, the 3DR Iris and 3DR Solo. All agents were required to search the test environment show in Figure 9 to ninety-five (95) percent completion. The flyable area of the test environment was approximately one thousand five hundred (1500) square meters and the test environment was not part of the training environments. The baseline test using the default action scales ([0.25, 0.125]) and ideal simulated agent was run ten (10) times to obtain an idea of the time to search the environment. The IRIS and SOLO agents were then tested using a variety of action scales, with a total five (5) runs per configuration. The time to crash or finish the test was recorded. Table 5 shows the average times for the tests undertaken. If the agent crashed for the majority of the tests the time to crash is listed, while if the agent finished the test the majority of the time, the average time to finish is shown. For each test the number of crashes is listed, with five (5) meaning all runs failed for the physically simulated configurations.

It can be seen that the performance at the default action scales was undesirable. This is due to the sluggishness of the response of the IRIS and SOLO platforms when compared to the trained ideal agent. Increasing the action scale in increments of 0.125 for both the linear and angular action scales up to a maximum of 1 shows that the performance increases such that the physically simulated agents perform almost on par with the ideal agent the policy was trained on.

Figure 10 shows a snapshot of a 3DR Solo test using action scales of one (1.0). The full video of that test can be found at https://youtu.be/u2I5xYWlPuM.

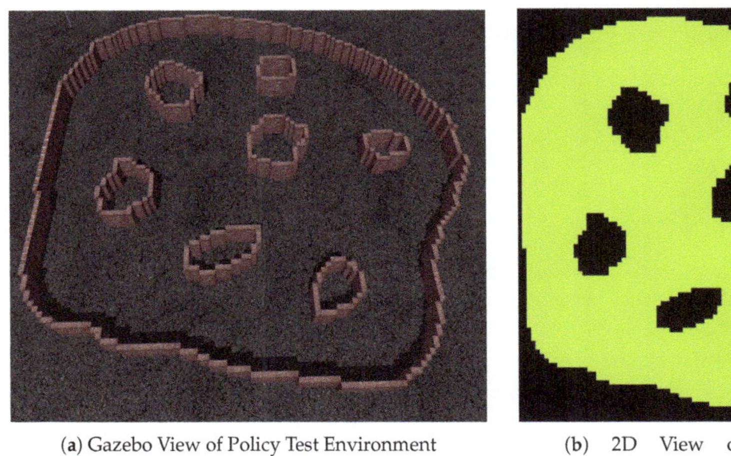

(a) Gazebo View of Policy Test Environment

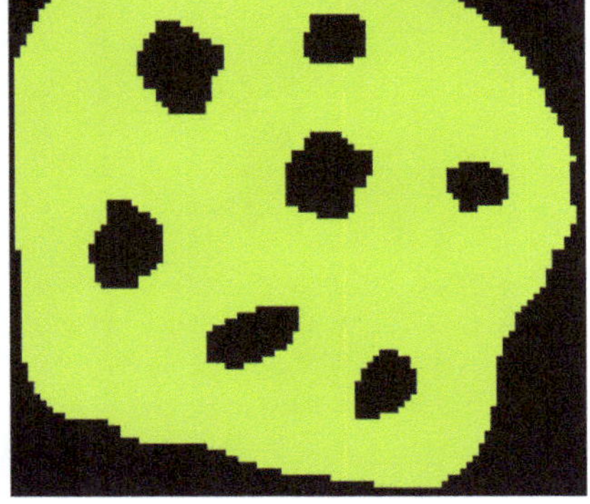

(b) 2D View of Policy Test Environment

Figure 9. Policy Test Environment.

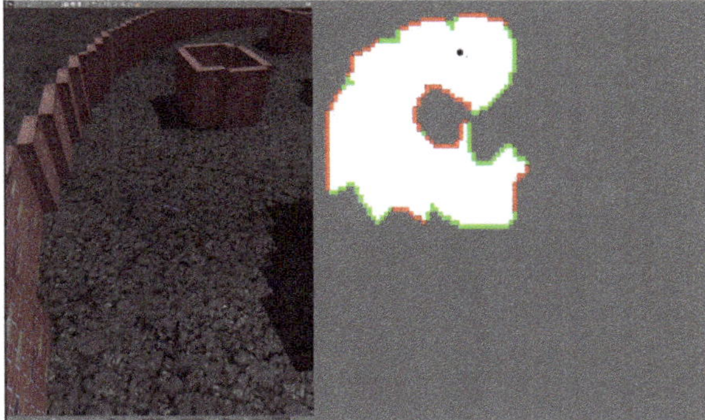

Figure 10. 3DR Solo Policy Test Snapshot.

Table 5. Local Controller Test Results.

Action Scales	Average Time (sec)	Outcome	No. of Crashes
	Ideal Agent		
0.25 // 0.125	127.62	**FINISH**	0
	3DR Solo		
0.25 // 0.125	144.27	**CRASH**	5
0.25 // 0.25	126.52	**CRASH**	4
0.375 // 0.375	133.65	**CRASH**	4
0.5 // 0.5	163.42	**FINISH**	2
0.625 // 0.625	153.94	**FINISH**	0
0.75 // 0.75	171.86	**FINISH**	0
0.875 // 0.875	161.52	**FINISH**	0
1.0 // 1.0	145.08	**FINISH**	0
	3DR Iris		
0.25 // 0.125	71.04	**CRASH**	5
0.25 // 0.25	45.83	**CRASH**	4
0.375 // 0.375	69.28	**CRASH**	4
0.5 // 0.5	153.33	**FINISH**	1
0.625 // 0.625	105.15	**FINISH**	4
0.75 // 0.75	122.82	**FINISH**	0
0.875 // 0.875	140.95	**FINISH**	1
1.0 // 1.0	138.32	**FINISH**	0

8.2. Testing the Full Framework

The combined performance of the framework was validated with generic simulated agents searching two test maps; a medium sized, fourteen (14) region environment, and a large twenty-five (25) region environment. Figure 11 shows both of the complete environments used for testing.

Testing was conducted using a variety of agent and target configurations. Test environment one was tested using one (1), two (2), three (3), and four (4) agent configurations. For each agent configuration five (5) tests were run for four (4) target locations, for a total of fifteen (20) test configurations for each agent configuration and a total of eighty (80) test runs for the first test environment.

The second test environment was also run eighty (80) times, using one (1), two (2), three (3), and four (4) agent configurations. No collisions with the environment were recorded for any of the test runs.

The numerical results can be seen notated in Table 6 and visualised in Figures 12–15.

Figure 12 outlines the individual target tests for the first environment, and it can be seen that on the whole, the times improved as the number of agents increased. This is further supported by the combined results shown in Figure 13. There are however a few anomolous results such as in Figure 12c, which displays the results for the Test Environment One, Target Location Three tests. After increasing the number of agents to two, the test results remain mostly consistent, with the system performing the same with two agents as with three, and four.

This is a result of the target location and the spawn of the second agent. The spawn of the second agent causes the first agent to consistently route almost directly toward the target location in an attempt to avoid planning conflicts. The route followed can be seen in Figure 16. These kinds of routing changes when adding agents are also responsible for other test case anomolies. Such as the Target Location Two test for Environment Two, shown in Figure 14b. In this case, it isn't until the fourth agent is added that the agents route themselves in such a way that the target is found in half the time compared with the previous tests.

(a) Test Environment One

(b) Test Environment Two

Figure 11. Complete Test Environments. **Black Cells**: Known Obstacle Cells, **White Cells**: Explored Empty Space Cells, **Coloured Cells**: Unexplored Empty Cells (Search Regions), **Grey Cells**: Unknown Obstacle Cells, **Triangles**: Agent Spawn Locations, **Circles**: Target Spawn Locations.

Table 6. Framework Test Results.

Test Environment One				
Agents	**One**	**Two**	**Three**	**Four**
Min Time (Sec)	210.07	103.08	87.64	46.05
Max Time (Sec)	485.31	459.75	213.66	142.61
Mean Time (Sec)	295.70	214.46	135.96	101.92
Test Environment Two				
Agents	**One**	**Two**	**Three**	**Four**
Min Time (Sec)	102.70	110.98	60.28	67.67
Max Time (Sec)	469.45	349.25	242.69	176.83
Mean Time (Sec)	215.22	178.34	134.61	121.60

It can be seen from the test results that there was a general trend of improved search time and consistency as the number of agents increased.

Figure 17 shows the progress and completion of a four (4) agent test on test environment one (1). The video at https://youtu.be/jh0dn33Ji0k shows a four (4) agent search of test environment one (1) at two (2) times playback speed.

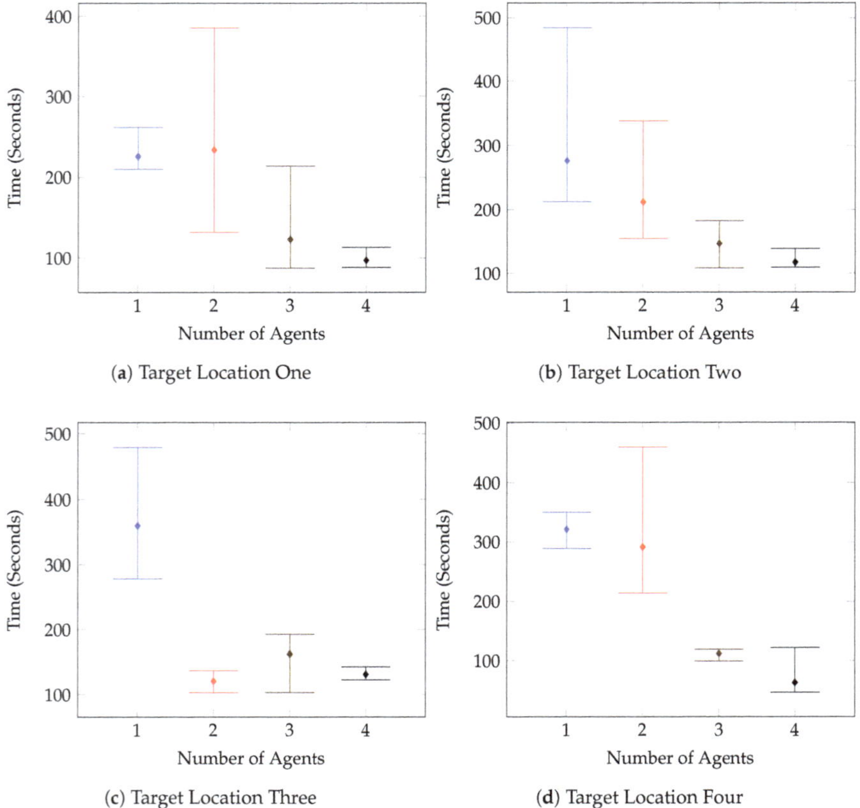

Figure 12. Number of Agents vs. Maximum, Mean and Minimum Times (Seconds)-Test Environment One.

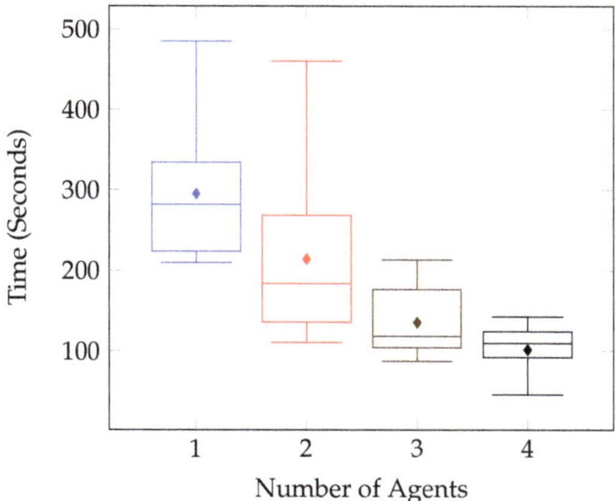

Figure 13. Number of Agents vs. Time (Seconds)-All Target Locations in Test Environment One.

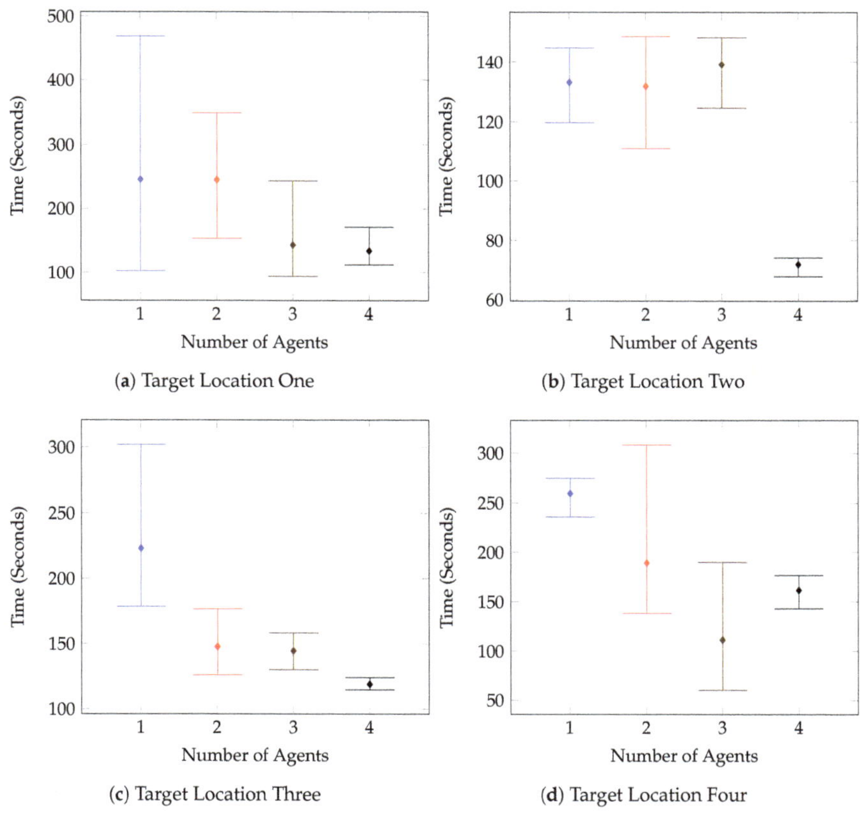

Figure 14. Number of Agents vs. Maximum, Mean and Minimum Times (Seconds)-Test Environment Two.

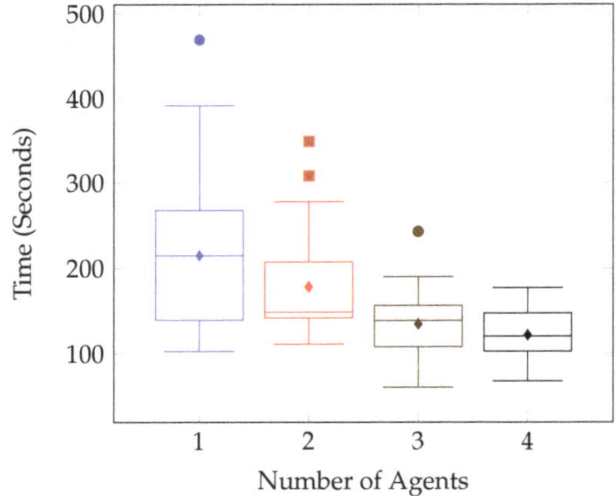

Figure 15. Number of Agents vs. Time (Seconds)-All Target Locations in Test Environment Two.

Figure 16. Target Three Test Example.

(**a**) Start Conditions

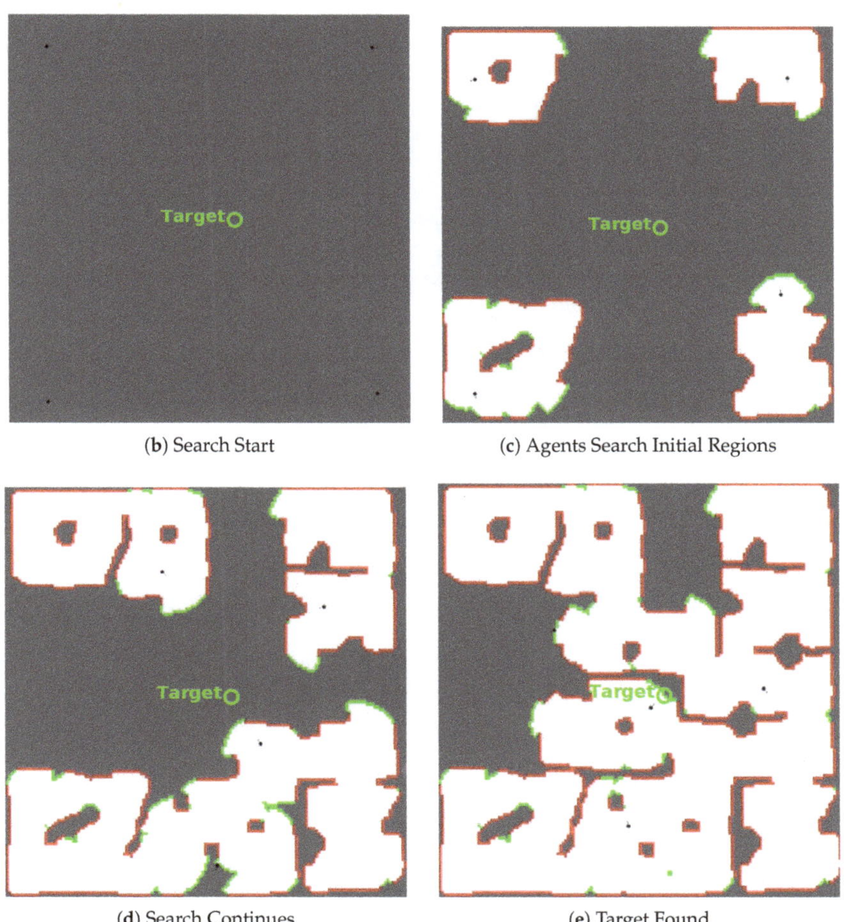

(**b**) Search Start (**c**) Agents Search Initial Regions

(**d**) Search Continues (**e**) Target Found

Figure 17. Four Agent Search of Test Environment One. **Grey Cells**: Unexplored Area, **Red Cells**: Observed Obstacle Cells, **Green Cells**: Safe Edges of Unexplored Region, **White Dotted Circle**: Target Location.

9. Conclusions

This paper has shown that using the proposed local control training environment, a small two layer neural net is capable of learning to control a generic UAV agent to explore a two dimensional (2D) occupancy-map of an environment while avoiding previously unknown obstacles. It has also shown that such a policy can be extended from use with the trained generic simulated agent to two different physically simulated platforms (3DR Iris, and 3DR Solo) simply by changing the gain for the control policy output.

Furthermore, the completed framework detailed by this paper has been shown to enable simulated UAV agents to search arbitrarily shaped, GPS-denied and partially observable environments using the combination of POMDP based planning and Deep Reinforcement Learning based control, under the assumption of accurate SLAM.

Given the performance of the physically simulated UAVs within Gazebo, using the PX4 software stack, the framework in its current form could be applied to a real-world agent with an accurate SLAM system and a map type that could be converted to the necessary 2D grid style environment that this framework uses. However, future work aims to integrate the use of existing map types such as the octo-map format, 3-D environments, and to model imperfect SLAM and positional noise within the control environment. This would be done in an attempt to increase the applicability of the framework and reduce the work required to enable this framework on a variety of platforms.

While the current framework prevents agents from interacting in a local environment, producing a control policy that enables inter-agent collision avoidance and cooperation at a local level is a target for future work. Additionally, while this paper does not consider strategies for optimizing the swarm configuration, this is also a goal for future works. Finally other future work could also include: improving the global planner to respond to large changes in map structure and lack of prior information, integration and validation of the framework on real-world hardware and in real-world environments, and development of additional problem definitions such as point of interest surveying and multi-target finding and pursuit.

Author Contributions: Conceptualization, O.W., F.V., and F.G.; Methodology, O.W., F.V., and F.G.; Software, O.W. and F.V.; Supervision, F.V. and F.G.; Validation, O.W.; Writing-Original Draft Preparation, O.W.; Writing-Review and Editing, F.V. and F.G. All authors have read and agreed to the published version of the manuscript.

Funding: This research received no external funding.

Conflicts of Interest: The authors declare no conflict of interest.

Abbreviations

The following abbreviations are used in this manuscript:

MDP	Markov Decision Process
POMDP	Partially Observable Markov Decision Process
UAV	Unmanned Aerial Vehicle
ABT	Adaptive Belief Tree
PPO	Proximal Policy Optimisation

References

1. Tomic, T.; Schmid, K.; Lutz, P.; Domel, A.; Kassecker, M.; Mair, E.; Grixa, I.L.; Ruess, F.; Suppa, M.; Burschka, D. Toward a fully autonomous UAV: Research platform for indoor and outdoor urban search and rescue. *IEEE Robot. Autom. Mag.* **2012**, *19*, 46–56. [CrossRef]
2. Hodgson, A.; Kelly, N.; Peel, D. Unmanned aerial vehicles (UAVs) for surveying marine fauna: A dugong case study. *PLoS ONE* **2013**, *8*, e79556. [CrossRef] [PubMed]
3. Rojas, A.J.; Gonzalez, L.F.; Motta, N.; Villa, T.F. Design and flight testing of an integrated solar powered UAV and WSN for remote gas sensing. In Proceedings of the 2015 IEEE Aerospace Conference, Big Sky, MT, USA, 7–14 March 2015; pp. 1–10.

4. Kersnovski, T.; Gonzalez, F.; Morton, K. A UAV system for autonomous target detection and gas sensing. In Proceedings of the 2017 IEEE aerospace conference, Big Sky, MT, USA, 4–11 March 2017; pp. 1–12.
5. Borie, R.; Tovey, C.; Koenig, S. Algorithms and complexity results for graph-based pursuit evasion. *Auton. Robot.* **2011**, *31*, 317. [CrossRef]
6. Ward, S.; Hensler, J.; Alsalam, B.; Gonzalez, L.F. Autonomous UAVs wildlife detection using thermal imaging, predictive navigation and computer vision. In Proceedings of the 2016 IEEE Aerospace Conference, Big Sky, MT, USA, 5–12 March 2016; pp. 1–8.
7. Gohl, P.; Burri, M.; Omari, S.; Rehder, J.; Nikolic, J.; Achtelik, M.; Siegwart, R. Towards autonomous mine inspection. In Proceedings of the 2014 3rd International Conference on Applied Robotics for the Power Industry, Foz do Iguassu, Brazil, 14–16 October 2014; pp. 1–6.
8. Berger, J.; Lo, N. An innovative multi-agent search-and-rescue path planning approach. *Comput. Oper. Res.* **2015**, *53*, 24–31. [CrossRef]
9. He, Y.; Zeng, Q.; Liu, J.; Xu, G.; Deng, X. Path planning for indoor UAV based on Ant Colony Optimization. In Proceedings of the 2013 25th Chinese Control and Decision Conference (CCDC) IEEE, Guiyang, China, 25–27 May 2013; pp. 2919–2923.
10. Vanegas, F.; Campbell, D.; Eich, M.; Gonzalez, F. UAV based target finding and tracking in GPS-denied and cluttered environments. In Proceedings of the 2016 IEEE/RSJ International Conference on Intelligent Robots and Systems (IROS) IEEE, Daejeon, Korea, 9–14 October 2016; pp. 2307–2313.
11. Sondik, E.J. The optimal control of partially observable Markov processes over the infinite horizon: Discounted costs. *Oper. Res.* **1978**, *26*, 282–304. [CrossRef]
12. Kurniawati, H.; Yadav, V. An online POMDP solver for uncertainty planning in dynamic environment. In *Robotics Research;* Springer: Cham, Switzerland, 2016; pp. 611–629.
13. Silver, D.; Veness, J. Monte-Carlo planning in large POMDPs. In Proceedings of the Advances in Neural Information Processing Systems, Vancouver, BC, Canada, 6–11 December 2010; pp. 2164–2172.
14. He, R.; Brunskill, E.; Roy, N. PUMA: Planning Under Uncertainty with Macro-Actions. In Proceedings of the Association for the Advancement of Artificial Intelligence, Atlanta, GA, USA, 11 July 2010; p. 7.
15. Al-Sabban, W.H.; Gonzalez, L.F.; Smith, R.N.; Wyeth, G.F. Wind-energy based path planning for electric unmanned aerial vehicles using markov decision processes. In Proceedings of the IEEE/RSJ International Conference on Intelligent Robots and Systems, Vilamoura-Algarve, Portugal, 7–12 October 2012.
16. Ebert, F.; Finn, C.; Dasari, S.; Xie, A.; Lee, A.; Levine, S. Visual foresight: Model-based deep reinforcement learning for vision-based robotic control. *arXiv* **2018**, arXiv:1812.00568.
17. Lillicrap, T.P.; Hunt, J.J.; Pritzel, A.; Heess, N.; Erez, T.; Tassa, Y.; Silver, D.; Wierstra, D. Continuous control with deep reinforcement learning. *arXiv* **2015**, arXiv:1509.02971.
18. Badia, A.P.; Piot, B.; Kapturowski, S.; Sprechmann, P.; Vitvitskyi, A.; Guo, D.; Blundell, C. Agent57: Outperforming the atari human benchmark. *arXiv* **2020**, arXiv:2003.13350.
19. Silver, D.; Huang, A.; Maddison, C.J.; Guez, A.; Sifre, L.; Van Den Driessche, G.; Schrittwieser, J.; Antonoglou, I.; Panneershelvam, V.; Lanctot, M.; et al. Mastering the game of Go with deep neural networks and tree search. *Nature* **2016**, *529*, 484. [CrossRef] [PubMed]
20. Silver, D.; Schrittwieser, J.; Simonyan, K.; Antonoglou, I.; Huang, A.; Guez, A.; Hubert, T.; Baker, L.; Lai, M.; Bolton, A.; et al. Mastering the game of go without human knowledge. *Nature* **2017**, *550*, 354–359. [CrossRef] [PubMed]
21. Berner, C.; Brockman, G.; Chan, B.; Cheung, V.; Dębiak, P.; Dennison, C.; Farhi, D.; Fischer, Q.; Hashme, S.; Hesse, C.; et al. Dota 2 with Large Scale Deep Reinforcement Learning. *arXiv* **2019**, arXiv:1912.06680.
22. Vinyals, O.; Babuschkin, I.; Czarnecki, W.M.; Mathieu, M.; Dudzik, A.; Chung, J.; Choi, D.H.; Powell, R.; Ewalds, T.; Georgiev, P.; et al. Grandmaster level in StarCraft II using multi-agent reinforcement learning. *Nature* **2019**, *575*, 350–354. [CrossRef] [PubMed]
23. Lopes, G.C.; Ferreira, M.; da Silva Simões, A.; Colombini, E.L. Intelligent control of a quadrotor with proximal policy optimization reinforcement learning. In Proceedings of the 2018 Latin American Robotic Symposium, 2018 Brazilian Symposium on Robotics (SBR) and 2018 Workshop on Robotics in Education (WRE), Joao Pessoa, Brazil, 6–10 November 2018; pp. 503–508.
24. Maciel-Pearson, B.G.; Marchegiani, L.; Akcay, S.; Atapour-Abarghouei, A.; Garforth, J.; Breckon, T.P. Online Deep Reinforcement Learning for Autonomous UAV Navigation and Exploration of Outdoor Environments. *arXiv* **2019**, arXiv:1912.05684.

25. Qie, H.; Shi, D.; Shen, T.; Xu, X.; Li, Y.; Wang, L. Joint optimization of multi-UAV target assignment and path planning based on multi-agent reinforcement learning. *IEEE Access* **2019**, *7*, 146264–146272. [CrossRef]
26. Maruyama, Y.; Kato, S.; Azumi, T. Exploring the performance of ROS2. In Proceedings of the 13th International Conference on Embedded Software, Pittsburgh, PA, USA, 1–7 October 2016; pp. 1–10.
27. Schulman, J.; Wolski, F.; Dhariwal, P.; Radford, A.; Klimov, O. Proximal policy optimization algorithms. *arXiv* **2017**, arXiv:1707.06347.
28. Klimenko, D.; Song, J.; Kurniawati, H. Tapir: A software toolkit for approximating and adapting pomdp solutions online. In Proceedings of the Australasian Conference on Robotics and Automation, Melbourne, Australia, 2–4 December 2014; Volume 24.
29. Hill, A.; Raffin, A.; Ernestus, M.; Gleave, A.; Kanervisto, A.; Traore, R.; Dhariwal, P.; Hesse, C.; Klimov, O.; Nichol, A.; et al. Stable Baselines. 2018. Available online: https://github.com/hill-a/stable-baselines (accessed on 15 December 2019).

Article

Evaluation of the Georeferencing Accuracy of a Photogrammetric Model Using a Quadrocopter with Onboard GNSS RTK

Martin Štroner *, Rudolf Urban, Tomáš Reindl, Jan Seidl and Josef Brouček

Department of Special Geodesy, Faculty of Civil Engineering, Czech Technical University in Prague, Thákurova 7, 166 29 Prague, Czech Republic; rudolf.urban@fsv.cvut.cz (R.U.); tomas.reindl@fsv.cvut.cz (T.R.); jan.seidl@fsv.cvut.cz (J.S.); josef.broucek@fsv.cvut.cz (J.B.)

* Correspondence: martin.stroner@fsv.cvut.cz

Received: 29 March 2020; Accepted: 16 April 2020; Published: 18 April 2020

Abstract: Using a GNSS RTK (Global Navigation Satellite System Real Time Kinematic) -equipped unmanned aerial vehicle (UAV) could greatly simplify the construction of highly accurate digital models through SfM (Structure from Motion) photogrammetry, possibly even avoiding the need for ground control points (GCPs). As previous studies on this topic were mostly performed using fixed-wing UAVs, this study aimed to investigate the results achievable by a quadrocopter (DJI Phantom 4 RTK). Three image acquisition flights were performed for two sites of a different character (urban and rural) along with three calculation variants for each flight: georeferencing using ground-surveyed GCPs only, onboard GNSS RTK only, and a combination thereof. The combined and GNSS RTK methods provided the best results (at the expected level of accuracy of 1–2 GSD (Ground Sample Distance)) for both the vertical and horizontal components. The horizontal positioning was also accurate when georeferencing directly based on the onboard GNSS RTK; the vertical component, however, can be (especially where the terrain is difficult for SfM evaluation) burdened with relatively high systematic errors. This problem was caused by the incorrect identification of the interior orientation parameters calculated, as is customary for non-metric cameras, together with bundle adjustment. This problem could be resolved by using a small number of GCPs (at least one) or quality camera pre-calibration.

Keywords: accuracy; onboard GNSS RTK; UAV

---

1. Introduction

Photogrammetry is presently a widely used method, particularly in combination with the Structure from Motion (SfM) technique. This combination allows the capture of surfaces in the form of point clouds (and derived products such as meshes, orthophotos, etc.), facilitating a relatively high degree of automation. The accuracy of multi-rotor UAV (Unmanned Aerial Vehicle) products is very important and is addressed in several studies [1–6].

Fixed-wing unmanned aerial vehicles (UAVs), such as eBee, mounted with a GNSS RTK (Global Navigation Satellite System, Real-Time Kinematic) module capable of determining the position of the drone with an accuracy of a few centimeters, have been available for some time now. Since last year, however, moderately priced UAV quadrocopters with GNSS RTK technology have come to the market. The pioneer in this area is DJI Phantom 4 RTK (approximately 7500 Eur).

From the perspective of workflow effectiveness, the idea of relying purely on GNSS RTK technology without the need for geodetic surveys (i.e., calculation of the photogrammetric model using only the imagery coordinates for georeferencing) is very attractive. Given the principle of the calculation where many variables (some of them mutually dependent) are used, it is, however, necessary to verify if georeferencing using only internal data can yield results of sufficient accuracy or whether the use of ground control points (GCPs) is necessary. The declared accuracy of the GNSS RTK receiver in the Czech Republic is approximately 0.025 m in the horizontal plane, while the vertical coordination is approximately 50% higher (0.04 m).

In geodesy, the use of GCPs is required to verify the accuracy of measurements. However, alternative appropriate solutions are preferable to reduce human effort. It is possible to find optimistic popular science articles dedicated to this topic on the internet (e.g., [7]). The use of such a simplified process would be welcome in many applications where it is difficult to stabilize GCPs, such as monitoring changes in the morphology of a volcano [8], landslides ([9,10]), dam and riverbed erosion [11–13], slow landslides [14], the risks associated with surface mining [15], slope stability in the vicinity of railways [16], the speed of glacier movement [17,18], or documenting rock outcrops [19]. This issue has been investigated by several authors who tested the accuracy of fixed-wing UAVs [20] with a higher travelling speed and working altitude and large custom-made UAVs, among others [21]. The novel testing of custom-made UAVs was published in [22,23] utilizing differential GPS (Global Positioning System) positions with the accuracy in decimeters. In [24], a digital surface model (DSM) was derived from the data acquired by a fixed-wing UAV SenseFly eBee-RTK, equipped with a compact camera Sony Cyber-Shot DSC-WX220 (18.2 Mpixel; focal length 4.5 mm). The latter study revealed that processing without the use of GCPs is problematic, particularly for the vertical component of the resulting model; thus, it is advisable to use at least one GCP set in the center of the area. In [25], the authors described a method for direct georeferencing without GCPs via the joint processing of data acquired by a quadrocopter flying at lower altitudes with data from a GNSS RTK-equipped fixed-wing UAV (senseFly eBee). The authors reported that their method was functional, although the results were again poorer for the vertical component and improved by adding at least a small number of GCPs. They also used pre-calibration for determining the interior orientation parameters directly on site, which, when used, improved the accuracy of (in particular) the vertical component (an approximately 1.5 (Ground sampling distance) GSD, compared to the original of approximately 3 GSD).

On the other hand, the authors in [26] stated that "the RTK/PPK method of the georeferencing can provide data with comparable or even higher accuracy compared to the GCP approaches, independently on the terrestrial measurements". The image acquisition was, similar to the previous study, performed using the senseFly eBee RTK, and all results derived solely through RTK georeferencing were better than the results derived using GCPs. In [27], the authors evaluated the performance of three software solutions (Agisoft PhotoScan, Pix4D, and MicMac) for the SfM processing of senseFly eBee RTK-acquired imagery, including testing the effect of the number of GCPs and their locations. The use of GCPs led to somewhat better results in the horizontal and approximately 50% improvement in the vertical components compared to the purely RTK-georeferenced data. The influence of the interior orientation parameters was also investigated, with the authors concluding that the "on-the-job calibration is likely in our opinion to be the best operational compromise, with the awareness that unless at least one GCP is provided, there might be biases in object coordinates, especially in elevation."

In [28], the effect of the trajectory of the data acquisition flight on the quality of the resulting DSM was studied along with the effect of the on-board GNSS RTK receiver recording camera positions. The data were acquired using a Mavinci Sirius Pro fixed-wing UAV mounted with a Panasonic LumixGX1-Pancake14mm-PRO camera. The authors demonstrated that a cross-flight pattern is beneficial for the quality of the resulting models and that the use of an on-board RTK receiver allows for accuracy similar to that achieved using GCPs but with a caveat regarding the need for a reasonable verification of the results.

A similar experiment was performed in [29], where the georeferencing of a topographical model using the on-board (GNSS) receiver and an inertial measurement unit without GCPs was assessed. This study compared photogrammetric models georeferenced using (a) a UAV equipped with a GNSS RTK receiver, (b) the same UAV mounted with a GNSS receiver only, and (c) independently measured GCPs only. The authors concluded that the horizontal accuracy of the RTK UAV data processed by direct georeferencing was equivalent to the horizontal accuracy of the non-RTK UAV data processed with GCPs, but the vertical error of the DSM from the RTK UAV data was 2 to 3 times greater than the DSM from the non-RTK data with GCPs.

Direct georeferencing using GNSS RTK was also studied, for example, in [30], where no problems or systematic shifts/errors were recorded, and in [31], where, conversely, the risk of developing a systematic shift in the horizontal direction was reported.

The results reported so far are, therefore, not entirely consistent and usually concern fixed-wing UAV data that move different from rotary-wing UAVs (in their speed, motion character, and stability). Further, residual systematic shifts due to the synchronization between GNSS RTK and camera records can occur, as noted in [28]. In this study, we tested the first suitable low-cost complex commercial solution for rotary-wing UAVs—the Phantom 4 RTK. Given the results reported for this UAV (e.g., in [25,27,32,33]), we particularly focused on the vertical component.

This experiment aimed to determine if it is possible to use only coordinates acquired by the UAV-mounted RTK receiver for processing and georeferencing (and if so, what the accuracy is) and what algorithms or processing methods can be used to improve the results. Two sites of different land types were used for testing—a mostly monochromatic homogenous area partially covered by tall woody vegetations (Rural) and a colorful broken (Urban) terrain with many buildings. Both of these areas were subject to three independent recordings to enable an assessment of the differences between individual image acquisitions for individual areas.

2. Materials and Methods

2.1. Data Acquisition

For the experiment, two areas of different characters from the perspective of tie point identification were used, each of them approximately 300 m × 300 m (90,000 m^2) in size. The expected flight altitude was 110 m. The extent, flight altitude, and trajectory were designed to allow image acquisition of the whole area during a single flight (limited by battery capacity) and to yield the expected resulting point cloud accuracy comparable to the GNSS RTK measurement accuracy (approximately 0.03 m in the horizontal and 0.04 m in the vertical direction).

According to [34], a horizontal accuracy of 0.5–1.0 GSD and a vertical accuracy of 1.5–2 GSD are generally achievable, corresponding to a flight altitude of approximately 100 m and a pixel size of approximately 0.03 m. On the other hand, the accuracy evaluation of a DEM (digital elevation model) for problematic terrain in [35] revealed significantly inferior vertical accuracy.

The first area of interest, the "Urban" area, is a built-up area with roads, houses, and greenery, featuring a surface with varied elevations and colors. This facilitates high-quality identification of tie points and, therefore, generally provides more reliable results than can be expected in the alternative area (see below; Figure 1). For the alternative area, an area with difficult-to-identify tie points was intentionally chosen (Figure 2). This area consists of homogenous surfaces—a field with low vegetation, grass, and a partial representation of tall and dense broadleaf trees (Rural).

For image acquisition, a UAV DJI Phantom 4 RTK mounted with a camera with a FC6310R lens (f = 8.8 mm), a resolution of 4864 × 3648 pixels, and a pixel size of 2.61 × 2.61 μm (price approximately 6000 Eur) was used. Figure 3 shows the trajectory of the image acquisition flights, including the dimensions; the image acquisition axis was always vertical (perpendicular to the flight axis).

Three independent flights using the same flight/image acquisition settings were performed (the same trajectory with a 75% forward overlap and a 75% sidelap at the ground level) at various times of the day (with approximately 2 h intervals; for the dates and times of the flights, see Table 1). Since the experiment sought to identify the problems of direct georeferencing using GNSS RTK, it was necessary to ensure that the impact of other circumstances of measurement was minimized. Agisoft Metashape has a special "Reduce overlap" function to determine the needed overlap of images. This was, in our case, 60% (using the high overlap option) in both directions. A 75% overlap in both directions was used to maximize the reliability of the results. The number of images was always 400 per flight. The GNSS RTK receiver was connected into the CZEPOS network of the permanent reference stations.

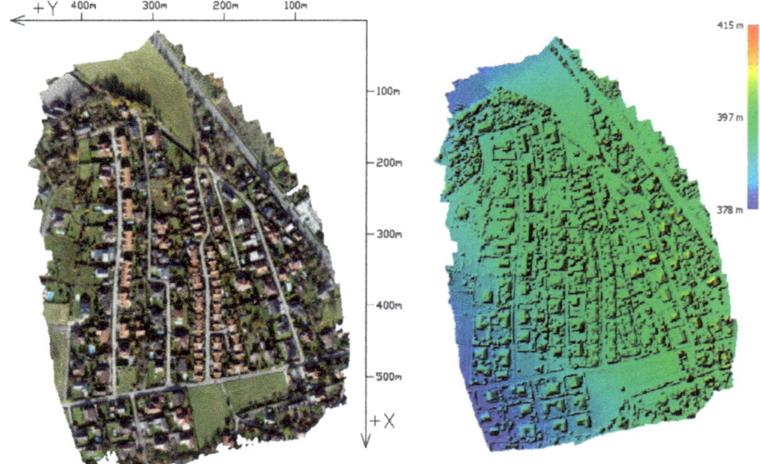

Figure 1. Urban area: orthophoto and DEM (Digital Elevation Model).

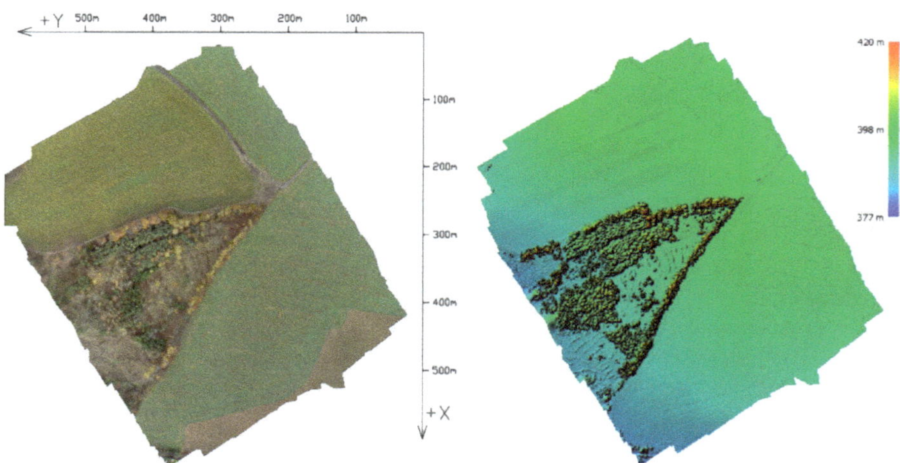

Figure 2. Rural area: orthophoto and DEM.

Table 1. Dates and times of the image acquisition flights.

Flight	Urban-7.11.2019	Rural-22.11.2019
1	11:08 to 11:27 h	11:15 to 11:34 h
2	13:11 to 13:29 h	13:28 to 13:46 h
3	15:02 to 15:20 h	15:22 to 15:41 h

In addition to the image acquisition itself, a geodetic survey of ground control points (GCPs) and checkpoints was performed using a GNSS RTK Trimble Geo XR receiver with a Zephyr 2 antenna connected into the CZEPOS network of the permanent reference stations (czepos.cuzk.cz). The ground control points were marked using portable black and white targets (stabilized using a single ten-centimeter-long nail) with a diameter of 80 cm or a white painted cross 80 cm in size on the tarmac.

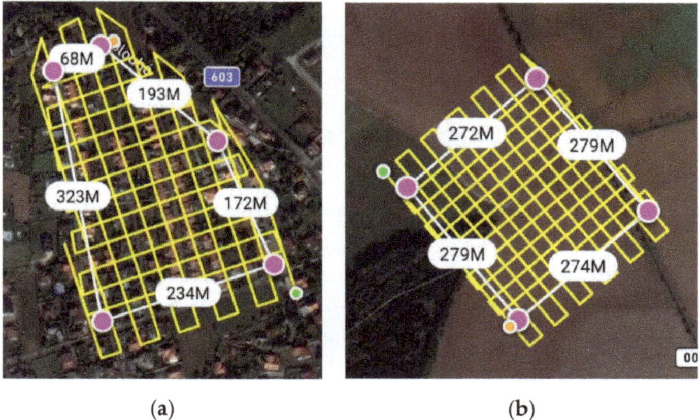

(a) (b)

Figure 3. Flight trajectory for the Urban (**a**) and Rural (**b**) areas including distances.

Two types of checkpoints (CPs) were used—one for verification of the horizontal accuracy and one for verification of the altitudinal accuracy. In the Urban area, the CPs present in the landscape that were clearly identifiable on the resulting georeferenced orthophoto were selected (considering the expected image resolution of 0.03 m, the checkpoint size had to be at least 15 cm long to display the image on at least five pixels). In the Rural area, the targets had to be used as checkpoints as there were practically no naturally occurring and easily identifiable points in the area. The CPs for assessment of the vertical accuracy were, on the other hand, placed in a way that ensured they would be located on a flatter surface, so the assessment could be unambiguous and less dependent on the fact that the elevations of individual CPs were compared to the surfaces constructed from the individual points of the point cloud. The number of individual points in both study areas is detailed in Table 2, for the location of the points see the Figure 4.

(a) (b)

Figure 4. Placement of the ground control points (ground control points (GCPs), blue) and checkpoints (horizontal—black/yellow; vertical—red) for both sites – Urban (**a**), Rural (**b**).

Table 2. The numbers of GCPs and control points (CPs) in the individual study areas.

	GCPs	CPs–Horizontal	CPs–Vertical
Urban	28	38	36
Rural	11	8	15

2.2. Data Processing

The GNSS receiver measurements were exported from the WGS 84 coordinate system (latitude, longitude, ellipsoidal height). The spatial position in the same coordinate system was also extracted from the images (containing GNSS RTK data) using the Exiftool utility. All data were converted into the Czech national coordinate positioning system (S-JTSK, System of Unified Trigonometric Cadastral Network) and the Bpv; vertical datum (Balt after adjustment) using the EasyTransform software (http://adjustsolutions.cz/easytransform/) to ensure the same algorithm was used on all data and thereby eliminate potential systematic errors that could occur as a result of different transformation algorithms.

The image processing was performed via the Agisoft Metashape 1.6.1 software with the Structure from Motion calculation method (SfM) using the non-default settings listed in Table 3:

Table 3. Agisoft Metashape software settings used for calculations.

Setting	Value
Align Photos	
ccuracy	high
Key point limit	40,000
Tie point limit	4000
Optimize Camera Alignment	fit all constants (f, cx, cy, k1–k4, p1–p4)
Build Dense Cloud	
Quality	High
Depth filtering	Moderate
DEM	
Projection	Geographic
Parameters	
Source data	Dense Cloud
Interpolation	Enabled
Advanced	
Resolution	2.8 cm/pix (implicit)
(Settings not detailed above were kept at default).	

As the UAV was not equipped with a professional metric camera, the interior orientation parameters were determined via calculations in the usual way. Although pre-calibration is generally recommended, the stability of the parameters of the UAV-mounted camera were not fully ascertained. Other studies also reported that the use of laboratory calibration can be even less [36] or equally [37] accurate in comparison to the method used in this study.

For each study area, the sparse point cloud was calculated in Agisoft Metashape, followed by the dense point cloud, and a DEM; based on this, an orthophoto with georeferencing was created.

As three flights were performed on each site, three sets of measurements were produced for each area of interest. For each of set, three variants of the calculations were performed: the "RTK" variant, "GCP" variant, and "Combined" variant. In the RTK variant, the calculations were performed using only the UAV-recorded camera coordinates with a 0.03 m accuracy setting. In the GCP variant, the calculation used the GCPs with the preset 0.03 m accuracy; the camera coordinates were kept in the calculation, but their accuracy was set to 104 m. This ensured that their impact on the calculation was negligible but that the differences between the measured RTK positions and the calculated positions could still be derived. The Combined variant used both the GCPs and camera coordinates with an accuracy of 0.03 m (the 0.03 m accuracy was used because this number represents the root mean square of the standard deviations of two horizontal and one vertical component of the GNSS RTK).

The results of the calculations, therefore, always represent areas of 91,600 m^2 with approximately 58 mil. points for the Urban area and 82,800 m^2 with approximately 39 mil points for the Rural area. The originally calculated point cloud was always cropped to frame the area with GCPs to prevent undesirable deformation in the model edges. These data were subsequently ground filtered (using the CSF plugin in CloudCompare software with the resolution set to 0.5 m and the classification threshold set to 0.5), and only ground points were used for further comparisons. Figures 5 and 6 show the individual variants and the coverage of the area by the imagery.

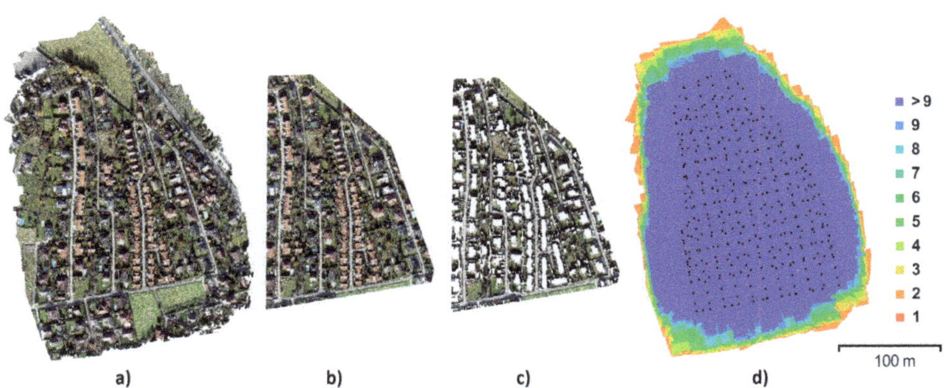

Figure 5. Urban data (**a**) original; (**b**) after cropping; (**c**) after ground filtering; (**d**) image coverage, black dots represent camera positions, and color hypsometry shows the number of overlapping frames.

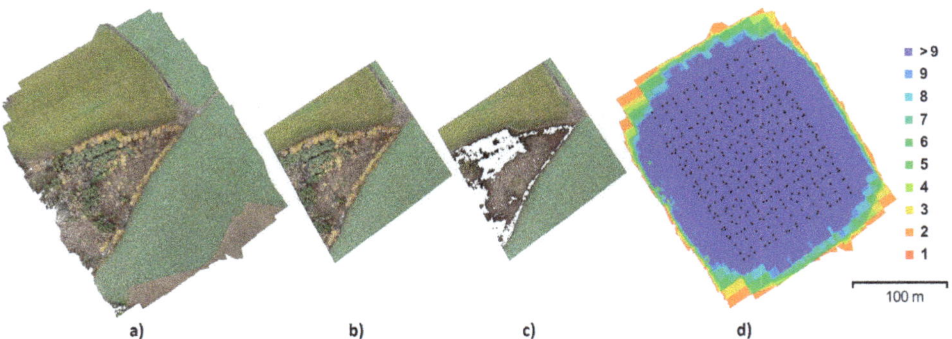

Figure 6. Rural data (**a**) original; (**b**) after cropping; (**c**) after ground filtering; (**d**) image coverage, black dots represent camera positions, and color hypsometry shows the number of overlapping frames.

2.3. Accuracy Assessment

The data accuracy was tested in various ways. First, the data from the checkpoints and GCPs using ground-survey GNSS receivers were evaluated, and the accuracy from three repeats of the measurements was assessed. Secondly, the accuracy characteristics available directly in the Agisoft software, providing information on the internal accuracy of the model, were analyzed. Subsequently, the results were compared to independent measurements, and, lastly, the resulting point clouds were mutually compared to identify local deformations. To preserve the clarity of the text (i.e., to maintain the link between individual methods of calculation and their results), the details of individual analyses are provided together with the results of these methods.

3. Results

3.1. The Accuracy of the GNSS RTK Geodetic Survey

The accuracy of the measurements of GCPs and CPs was determined with a geodetic GNSS RTK receiver. Each point was measured three times at approximately 2 h intervals (similar to the flights). Table 4 details the standard deviations from the repeated measurements and confirms the consistent accuracy in both areas. It also shows that the achieved accuracies of the individual components were better than those expected before the experiment.

Table 4. Standard deviations of the coordinates of the GCPs.

Area	S_x [m]	S_y [m]	S_z [m]
Urban	0.008	0.008	0.013
Rural	0.009	0.008	0.014

3.2. Internal Model Accuracy

While processing, Agisoft Metashape displays data for evaluation that can be used to analyze the internal calculation accuracy, such as the degree to which the coordinates acquired based on GCPs and those based on the UAV-mounted GNSS RTK mutually correspond. Without an additional (independent) verification, this is, in principle, the only information typically available to the user (however, this information does not provide the actual accuracy values).

Table 5 summarizes the accuracy characteristics calculated in Agisoft Metashape (root mean square deviations, RMSDs) and those based on its results (mean differences).

Table 5. Results for the Urban area—RMSDs and the mean differences between the calculated and measured camera/GCP positions.

Variant	Flight	Diff. on	RMSD			Mean Differences		
			X[m]	Y[m]	Z[m]	X[m]	Y[m]	Z[m]
GCP	1	Cameras	0.049	0.044	**0.246**	−0.000	−0.006	**0.245**
		GCPs	0.009	0.011	0.021	0	0	0
	2	Cameras	**0.086**	0.050	**0.222**	**−0.066**	0.016	**0.220**
		GCPs	0.009	0.010	0.025	0	0	0
	3	Cameras	0.055	0.051	**0.086**	−0.004	−0.016	**0.084**
		GCPs	0.009	0.010	0.020	0	0	0
RTK	1	Cameras	0.008	0.007	0.012	0	0	0
		GCPs	-	-	-	-	-	-
	2	Cameras	0.007	0.007	0.008	0	0	0
		GCPs	-	-	-	-	-	-
	3	Cameras	0.007	0.007	0.008	0	0	0
		GCPs	-	-	-	-	-	-
Combined	1	Cameras	0.017	0.018	0.013	0.001	0.000	0.000
		GCPs	0.012	0.012	0.021	−0.008	−0.001	−0.003
	2	Cameras	0.018	0.016	0.012	−0.005	0.001	0.000
		GCPs	0.064	0.020	0.016	0.064	−0.017	0.000
	3	Cameras	0.018	0.019	0.010	0.000	−0.001	0.000
		GCPs	0.009	0.013	0.017	−0.003	0.008	−0.002

For the detection of systematic errors, the arithmetic mean differences between the measured and calculated positions were also determined (if no systematic shift and only random errors are present, the mean difference should be equal to zero). The RMSD of the GCP positions in the GCP calculation variant represents the internal measurement accuracy. The deviations are significantly higher in the vertical component and are worse than the expected accuracy of the UAV-mounted GNSS RTK receiver. It is also necessary to mention the deviation in the X coordinate in Flight 2 in the Urban area, the deviation of which is also approximately two times higher than that of the other flights. In the "GCP" calculation variant, the GCP mean differences are obviously always equal to zero (as the model is always placed in the center of the GCPs). The values of these mean differences for cameras represent, in this calculation, variants of the systematic shifts of all camera positions. The fact that these values are close to those of the RMSD implies that a major part of the RMSD is caused by the systematic shift of the camera positions.

The RTK calculation variant (based solely on the RTK camera positions) showed similar results. That the GCP deviations could not have been calculated as a simple test shows us that the presence (i.e., tagging) of GCPs in the images in the Agisoft software affects—albeit to a small degree—the results (even if tagged only as checkpoints with small accuracy). Therefore, the GCPs were not present at all in the calculations when processing this variant. The derived RMSD values are very small and do not indicate any problems.

In the Combined calculation variant, the same accuracy (and therefore weight) was assigned to both the GCP and camera position coordinates. All resulting values show that although there were differences within the previous calculation variants, the measurements of both RTK camera positions and GCPs are correct, and the internal consistency of this model is better than the expected accuracy of the GNSS RTK measurement. The only exception is represented by the X coordinates of the second flight, where a major systematic shift between the on-board GNSS RTK and the terrestrial GNSS RTK receiver was observed. Its sizes are similar to those of the GCP variant. The presence of a shift in both calculations indicates that this systematic error was most likely caused by the GNSS RTK of the UAV receiver.

Table 6 shows the results of the identical calculations for the other area (Rural), the character of which entails a more difficult identification of tie points. For evaluation, it was also necessary to set a lower tie point accuracy (from the default value of 1 to 3.0 for the first flight, 3.5 for the second, and 4.5 for the third), as the coordinate differences after bundle adjustment were extremely high when using default values. Namely, the Total Error of the camera positions in the first flight was 2.129 m (values of 0.127 m; 0.113 m; 2.122 m for individual coordinates of X, Y, Z respectively), in the second flight, it was 0.535 m (0.109 m; 0.124 m; 0.509 m), and in the third flight, the Total Error was 0.632 m (0.096 m; 0.104; 0.613 m).

Table 6. Results for the Rural area—RMSDs and mean differences.

Variant	Flight	Diff. on	RMSD			Mean Difference		
			X [m]	Y [m]	Z [m]	X [m]	Y [m]	Z [m]
GCP	1	Cameras	0.035	0.026	**0.335**	0.010	0.006	**0.334**
		GCPs	0.003	0.004	0.015	0	0	0
	2	Cameras	0.028	0.031	**0.303**	0.013	0.007	**0.302**
		GCPs	0.005	0.004	0.011	0	0	0
	3	Cameras	0.026	0.029	0.016	0.006	−0.011	−0.006
		GCPs	0.005	0.006	0.007	0	0	0
RTK	1	Cameras	0.006	0.006	0.012	0	0	0
		GCPs	-	-	-	-	-	-
	2	Cameras	0.004	0.004	0.016	0	0	0
		GCPs	-	-	-	-	-	-
	3	Cameras	0.001	0.002	0.004	0	0	0
		GCPs	-	-	-	-	-	-
Combined	1	Cameras	0.012	0.012	0.013	0	0	0
		GCPs	0.008	0.007	0.010	0.003	0.003	−0.001
	2	Cameras	0.011	0.011	0.017	0	0	0
		GCPs	0.015	0.013	0.011	−0.009	−0.002	0.000
	3	Cameras	0.019	0.020	0.014	0	0	0
		GCPs	0.008	0.011	0.014	−0.004	0.009	0.005

Comparing the GCP" calculation variant with the Urban area, it is obvious that the camera position differences in the Rural area are greater than those in the Urban area (as high as tens of centimeters in the vertical component). The arithmetic mean differences indicate that they are to a major degree caused by systematic shifts in all camera positions.

The calculations of the RTK variant revealed only small deviations not exceeding the expected measurement accuracy. As the GCPs were removed from the analysis, just like for the Urban area, their accuracy could not be determined. The mean shifts in camera coordinates are always equal to zero, as the model was fit to them. The "Combined" calculation variant with equal weights for GCPs and camera positions again shows very good consistency, which suggests that the measurements themselves are correct.

Overall, the accuracy is clearly better in the Urban area where the conditions for the identification of tie points are much better than those in the Rural area. This accuracy is on par with the GNSS accuracy, with the only exception represented by Flight 2 in the Urban area, which was burdened with a systematic X coordinate shift. Although the corrections of the recorded camera positions in the GCP calculation variant can be relatively large (compared to the assumed accuracy), these corrections may not necessarily be caused by measurement errors (as is obvious from the results of the Combined calculation). In this case, the probable reason is instead computational instability (see the Discussion).

3.3. Verification Using Independent Checkpoints—Horizontal

To determine the absolute accuracy of the model, a comparison with an independent measurement was performed.

The positions of the horizontal checkpoints were derived from the georeferenced orthophoto in the Bentley Microstation software (ver. 8.11) and were compared to the coordinates determined by a ground survey to acquire the accuracy characteristics shown in Table 7, namely the mean error, standard deviation, and RMSD for the X and Y coordinates. Table 7 also presents the mean values for the individual calculation methods, individual areas, and an overall mean including both study areas. The numbers in brackets, where shown, indicate the means after leaving out Flight 2 in the Urban area with the obvious systematic shift in the axis.

Table 7. Horizontal checkpoint accuracy indicators.

Calculation Variant	Area	Flight	Mean Deviation		Standard Deviation		RMSD	
			X [m]	Y [m]	X [m]	Y [m]	X [m]	Y [m]
RTK	Urban	1	0.020	0.005	0.019	0.019	0.021	0.020
		2	0.082	−0.023	0.016	0.015	0.083	0.027
		3	0.005	0.006	0.015	0.016	0.016	0.017
	Rural	1	0.004	0.006	0.020	0.023	0.019	0.023
		2	0.001	0.002	0.022	0.025	0.020	0.024
		3	0.009	0.006	0.019	0.024	0.020	0.024
	Mean		**0.020 (0.008)**	**0.000 (0.005)**	**0.019 (0.019)**	**0.021 (0.022)**	**0.038 (0.019)**	**0.023 (0.024)**
GCPs	Urban	1	0.005	0.004	0.018	0.022	0.018	0.022
		2	0.012	−0.001	0.016	0.019	0.020	0.018
		3	0.008	0.004	0.020	0.023	0.021	0.023
	Rural	1	0.006	0.009	0.017	0.020	0.017	0.021
		2	0.007	0.012	0.022	0.026	0.022	0.027
		3	0.011	0.010	0.014	0.023	0.018	0.024
	Mean		**0.008**	**0.006**	**0.018**	**0.022**	**0.019**	**0.023**
Combined	Urban	1	−0.006	0.009	0.018	0.022	0.018	0.023
		2	0.072	−0.017	0.019	0.015	0.074	0.023
		3	0.006	0.008	0.021	0.020	0.021	0.021
	Rural	1	0.008	0.006	0.017	0.021	0.018	0.021
		2	0.002	0.007	0.025	0.024	0.024	0.024
		3	0.000	0.014	0.019	0.023	0.018	0.026
	Mean		**0.014 (0.002)**	**0.005 (0.009)**	**0.020 (0.020)**	**0.021 (0.022)**	**0.035 (0.020)**	**0.023 (0.023)**

The verification of the horizontal accuracy shows equal accuracy for all calculation methods and confirms the systematic error of 0.08 in the RTK measurement of the UAV-mounted GNSS receiver in Flight 2 in the Urban area (see Table 7). This error did not occur in any of the other flights. However, regardless of whether it was caused by an error in the UAV itself, by the reference station network, or by any other source, it confirms the need to perform independent verification as is usual in geodesy (one measurement is no measurement), without which such an error would pass undetected.

3.4. Verification Using Independent Checkpoints—Vertical

The vertical checkpoints were compared with the point clouds of the respective calculation variants using only the vertical component of the distance. This comparison was performed in the CloudCompare software (www.cloudcompare.org).

The point clouds were directly compared with the checkpoints by determining the minimum distance of each point from the point cloud to an irregular triangular network formed between the nearest twelve points of the SfM cloud (function Cloud to Cloud, tab Local modelling, and Local model option 2D1/2 Triangulation). The average vertical differences (systematic shift) and the standard deviations of the differences were also calculated and are shown in Table 8.

In the vertical component, the RTK calculation variant shows substantial errors, namely a systematic vertical shift of the entire model by up to 0.14 m in the Rural area. In the Urban area (when leaving out the problematic second flight), the results are better, and the standard deviations meet expectations. The results of the "GCP" and "Combined" variants are similar and correspond to expectations (considering the accuracy of the checkpoint measurement and pixel size).

Table 8. Results of the accuracy assessment of the vertical checkpoints.

Variant	Flight	Urban Mean Deviation [m]	Urban Standard Deviation [m]	Urban RMSD [m]	Rural Mean Deviation [m]	Rural Standard Deviation [m]	Rural RMSD [m]
RTK	1	0.029	0.021	0.036	**0.144**	0.032	**0.147**
	2	**−0.090**	**0.021**	**0.093**	0.041	0.028	0.049
	3	0.014	0.023	0.027	0.086	0.025	0.090
	Mean	**−0.016 (0.022)**	**0.022 (0.022)**	**0.060 (0.032)**	**0.090**	**0.028**	**0.103**
GCP	1	−0.010	0.028	0.030	−0.034	0.028	0.043
	2	−0.024	0.030	0.039	−0.029	0.028	0.040
	3	−0.014	0.029	0.032	−0.012	0.023	0.025
	Mean	**−0.016**	**0.029**	**0.034**	**−0.025**	**0.026**	**0.037**
Combined	1	−0.009	0.022	0.024	−0.023	0.030	0.037
	2	−0.023	0.024	0.034	−0.017	0.027	0.031
	3	−0.013	0.024	0.027	−0.014	0.027	0.030
	Mean	**−0.015**	**0.023**	**0.029**	**−0.018**	**0.028**	**0.033**

3.5. Comparison of Dense Clouds

When evaluating the results, it is also necessary to mutually compare the point clouds resulting from individual methods of calculation as the previous verifications were performed using isolated points only. Due to the point density, the comparison of dense clouds can be practically considered as continuous. Only the vertical component of ground filtered point clouds was compared here.

An interesting and important fact is that the standard deviation is practically very similar in all variants and comparisons; the mean standard deviation is 0.22 m. The data (i.e., the entire photogrammetric models) are, therefore, mutually shifted solely in the vertical direction. Figures 7 and 8 show the point clouds with a color scale indicating the sizes of the deviations and histograms showing the same color scale for Flight 1 in the Urban and Rural areas. The figures provide a comparison of the Combined variant (as the most accurate) with the RTK only variant (the one we were most interested in). The histograms show both the error distribution and systematic shift, confirming the conclusions from the checkpoints.

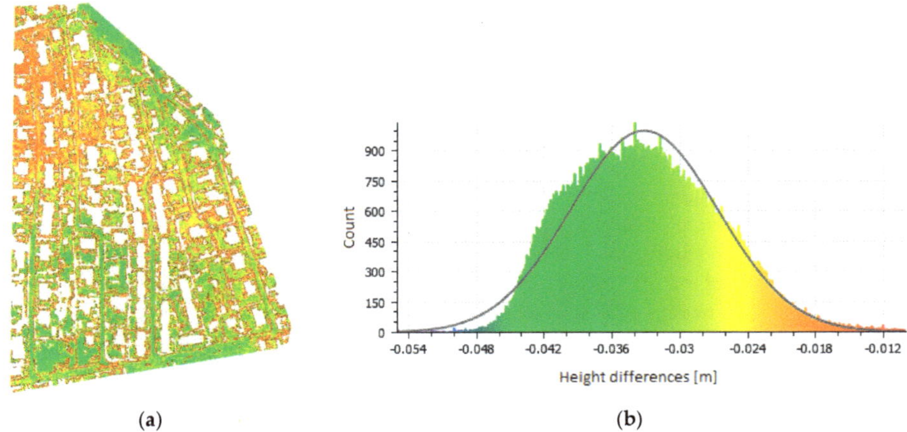

(**a**) (**b**)

Figure 7. Point cloud comparison (**a**) (Combined vs. RTK, Urban area, Flight 1, and vertical deviations) and the histogram of errors (**b**).

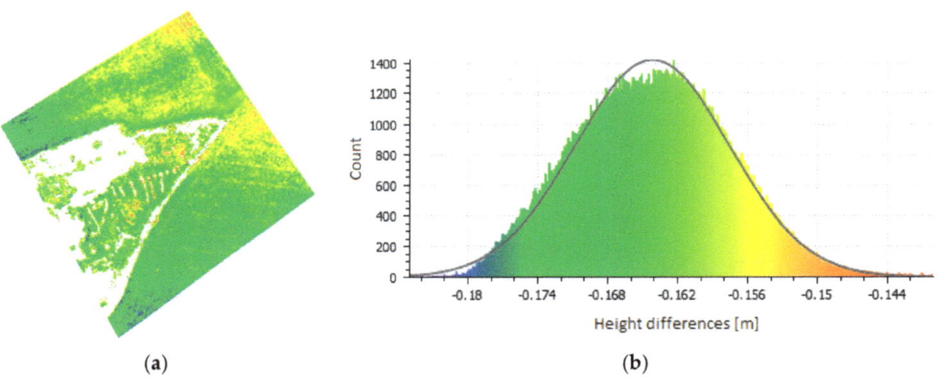

(**a**) (**b**)

Figure 8. Point cloud comparison (**a**) (Combined vs. RTK, Rural area, Flight 1, and vertical deviations) and the histogram of errors (**b**).

To compare the accuracy of the GCP and RTK, these variants were compared with the Combined variant (which was considered to be the most accurate) using a direct point cloud comparison (Table 9).

Table 9. Comparison of the GCP and RTK variants with the Combined variant.

Flight	Variant Compared		Urban Mean Deviation [m]	Urban Standard Deviation [m]	Rural Mean Deviation [m]	Rural Standard Deviation [m]
1	Combined	GCP	0.003	0.020	0.004	0.015
		RTK	**−0.033**	0.012	**−0.163**	0.019
2	Combined	GCP	0.003	0.020	0.006	0.020
		RTK	**0.060**	0.019	**−0.056**	0.015
3	Combined	GCP	0.003	0.017	−0.002	0.020
		RTK	**−0.025**	0.012	**−0.093**	0.018

Clearly, the RTK variant is always systematically shifted against both the Combined and GCP variants. In the Urban area, the systematic shift is approximately at the level of the ground sample distance (GSD; with the exception of the systematically shifted Flight 2); however, in the Rural area, it is significantly higher. When comparing identical calculation methods between individual flights (Table 10), the results are similar. In the RTK variant, the results for the Rural area are inferior to those for the Urban area. Otherwise, the standard deviations are practically equal to the GSD (ground sample distance), and the systematic errors are negligible when compared to the GSD.

Table 10. Mutual comparison of the calculation variants.

Variant	Flight Compared		Urban Mean Deviation [m]	Urban Standard Deviation [m]	Rural Mean Deviation [m]	Rural Standard Deviation [m]
Combined	1	2	0.006	0.025	0.006	0.017
		3	0.003	0.027	−0.011	0.019
GCP	1	2	0.006	0.026	−0.005	0.017
		3	0.003	0.026	−0.016	0.021
RTK	1	2	**0.100**	0.037	**0.099**	0.021
		3	0.010	0.027	0.058	0.020

4. Discussion

SfM-derived point clouds were constructed through multiple steps, including several decision and computation processes. It is, therefore, not possible to perform a simple estimation of the distribution of the measurement errors throughout computations or to derive the resulting accuracy only from the known measurement accuracy. Here, we used various data and processed them in several ways to show the differences that can result from different methods of calculation.

To simplify the process of cloud point generation and, in effect, improve its speed and reduce the price, it would be ideal if the camera coordinates acquired from the on-board GNSS RTK receiver could yield a result of sufficient quality. Of course, the term "sufficient quality" is relative and depends on the application. Nevertheless, when using a UAV for modelling a relatively small area, it is reasonable to assume an accuracy of approximately 1–2 GSD. In our experiment, this was not the case. In the Urban area, the second flight was systematically shifted for unknown reasons, although that area was suitable for automatic SfM processing. This alone emphasizes the need for independent verification that can easily detect such major errors. A shift can also occur in the vertical component of the model as shown in the Rural area. In that case, however, the reason was not a serious systematic error like with the previous issue; instead, this error had its roots in the imperfect automatic optimization of a principally correct calculation. Tables 5 and 6 show that without the use of independent verification (i.e., if only the RTK calculation based solely on the camera positions is used), there are no indications that anything is wrong. As the model is not deformed in any way in the RTKvariant (see Figures 7 and 8) but only vertically shifted (these results are different from those reported in [38]), we also tested the influence of a small number of GCPs on the results. In all cases, even a single GCP removed any systematic shift.

Flight 1 in the Rural area with a systematic vertical shift of 0.14 m (see Table 8) can serve as an example. After adding a single GCP, even in an area unsuitable from a configuration perspective, (i.e., in the lower right corner of the study area), the model was positioned more-or-less correctly (the systematic vertical shift dropped to 0.020 m, with a standard deviation of 0.026 m). We tested various positions for the individual GCPs, achieving practically identical results for all. Of course, we must bear in mind the principles of geodesy and always recommend multiple ground control points, at least in the four corners of the captured area.

The lack of accuracy in the vertical component in the rural area was likely due to the camera's self-calibration process, which may have led to the miscalculation of internal camera parameters, as observed in [25] and [27]. When utilising non-metric cameras, bundle adjustment is usually calculated together with a calculation of the internal orientation elements, which, in most cases, yields quality results when using GCPs. Here, for the Rural area (difficult for SfM processing from the perspective of identification of tie points), this method is not suitable, as the bundle adjustment did not yield correct interior orientation parameters, which led to a vertical shift of the model.

Table 11 shows that the calibrated focal lengths in the Combined calculation are very consistent while those of the RTK variant and provide results with much higher differences. We must, however, bear in mind that these values cannot be used as an unambiguous indicator as the interior orientation parameters are mutually dependent and cannot be evaluated separately. Further, the calibrated focal lengths are not consistent in the GCP variant, which otherwise provides correct results. The most reliable results (for the Combined calculation) are in bold, while italics denote the flight using a systematic shift. The underlined italics highlight the most outlying results (Rural RTK calculations).

For these reasons, we also performed an RTK calculation variant using calibration parameters derived from the Combined variant from the same flight, and the results were practically of the same quality as those from the Combined calculation (the mean vertical shifts were similar to those yielded using the Combined variant). This was true for all flights in both areas, with the exception of Flight 2 in the Urban area (the one with the systematic X-axis shift). This proves that the inferior results in the Rural area were likely caused by the incorrect determination of the interior orientation parameters during the calculation. In other words, inferior quality tie points make problematize the correct optimization of the interior parameters, which leads to inferior results. In the UAVs equipped with metric cameras where the stability of the interior orientation parameters can be expected, quality pre-calibration should, therefore, be able to overcome this problem and allow the production of quality models even without the use of GCPs. However, we still recommend to use some independent verification (e.g., a small number of "checkpoints") as there is always a risk of an error, such as that for Flight 2 of the Urban area, where the GNSS RTK receiver systematically determined the X coordinate erroneously in all images, no matter the cause.

Table 11. Interior orientation parameters determined for the individual calculation variants and flights (ME–Mean error in the vertical direction).

Area	Flight	Variant	F/pix	Cx/pix	Cy/pix	ME/m
Urban	1	Combined	**3683.1015**	**7.9537**	**31.7125**	0.024
	2		*3687.0770*	*7.0952*	*26.9476*	0.034
	3		**3683.8230**	**7.9741**	**31.6299**	0.027
	1	RTK	3684.2895	7.6251	30.8493	0.036
	2		3684.5967	6.7025	27.5137	0.093
	3		3684.5744	7.6400	30.8406	0.027
	1	GCP	3691.6372	9.1263	29.8708	0.030
	2		3694.8683	8.5679	25.1348	0.039
	3		3687.0244	9.2974	31.7190	0.032
Rural	1	Combined	**3683.3930**	**7.1279**	**30.7098**	0.037
	2		**3682.1743**	**7.5854**	**29.3024**	0.031
	3		**3684.4672**	**7.9307**	**31.1895**	0.030
	1	RTK	*3691.0187*	*7.1598*	*28.3578*	*0.147*
	2		*3686.3828*	*7.5746*	*27.8308*	*0.049*
	3		*3688.1387*	*7.7183*	*29.8556*	*0.090*
	1	GCP	3693.9528	7.7744	26.9313	0.043
	2		3692.3433	8.0277	25.2495	0.040
	3		3684.2539	8.2464	31.1109	0.025

Comparing our results achieved using a quadrocopter and those reported by [24], who used a fixed-wing UAV, we conclude that the expected improvement of the data accuracy based solely on the onboard GNSS RTK data (expected due to the lower travelling speed and better stability) was not confirmed.

5. Conclusions

Affordable GNSS-RTK equipped multicopters have only recently become available. Their lower traveling speed and higher stability (compared to fixed-wing UAVs) suggests that they could be more suitable for the direct processing of acquired imagery without the need for GCPs; this was the principal hypothesis of this study. Our results, however, prove that the use of a GNSS RTK receiver in a multicopter UAV without external verification is potentially very dangerous—a danger manifested predominantly as a systematic shift of the entire point cloud in the vertical components. The horizontal components (unless a systematic error of the UAV positioning system occurred) were practically unaffected.

To the best of our knowledge, we are the first to use two separate testing sites with different characteristics (Urban and Rural) for the evaluation of the suitability of GNSS RTK-equipped UAVs (fixed-wing or multicopter) for the production of models without GCPs. This allowed us to perform an in-depth comparison of the results and to analyse the reasons for their differences. The finding that the vertical error increases with the declining suitability of the terrain for SfM processing (even when using a GNSS-RTK-mounted multicopter) is not surprising but remains important for the practical application of this type of UAV.

Further testing, however, showed that this problem can be overcome in two ways. The first involves using a small number of ground control points (in a small area, such as those used in our study, as even a single GCP anywhere in the area was sufficient to rectify the problem). The other method is potentially more important for future practice. We showed that when using an optimum set of calibration parameters for a non-metric camera (reversely derived from the best calculation variant), one can achieve extremely accurate results, even when using GNSS RTK data only. This suggests that the use of metric cameras that can be perfectly pre-calibrated could also provide excellent results, even without the use of GCPs for the production of the model (note, however, that this model still needs to be verified externally afterwards to prevent shifts such as that in Flight 2 in the Urban area).

We also conclude that if GCPs are used, the horizontal and vertical accuracy will be practically identical to the GSD (here it is 0.028 m; it is also necessary to consider the accuracy of the verification method). This provides the safest method for data acquisition and processing. Moreover, the best model resulted from a combination of both GCPs and GNSS RTK UAV coordinates for the calculations.

Author Contributions: Conceptualization: M.Š.; methodology: M.Š.; ground data acquisition: T.R. and J.S.; UAV data acquisition: J.B.; data processing of UAV imagery: T.R. and J.S.; analysis and interpretation of results: M.Š. and R.U.; manuscript drafting: M.Š. and R.U. All authors have read and agreed to the published version of the manuscript.

Funding: This research was funded by the Grant Agency of CTU in Prague—grant number SGS20/052/OHK1/1T/11.

Conflicts of Interest: The authors declare no conflict of interest. The funders had no role in the design of the study; in the collection, analyses, or interpretation of data; in the writing of the manuscript, or in the decision to publish the results.

References

1. Hung, I.-K.; Unger, D.; Kulhavy, D.; Zhang, Y. Positional Precision Analysis of Orthomosaics Derived from Drone Captured Aerial Imagery. *Drones* **2019**, *3*, 46. [CrossRef]
2. Jaud, M.; Passot, S.; Le Bivic, R.; Delacourt, C.; Grandjean, P.; Le Dantec, N. Assessing the Accuracy of High Resolution Digital Surface Models Computed by PhotoScan® and MicMac® in Sub-Optimal Survey Conditions. *Remote Sens.* **2016**, *8*, 465. [CrossRef]
3. Gindraux, S.; Boesch, R.; Farinotti, D. Accuracy Assessment of Digital Surface Models from Unmanned Aerial Vehicles' Imagery on Glaciers. *Remote Sens.* **2017**, *9*, 186. [CrossRef]
4. Tonkin, T.N.; Midgley, N.G. Ground-Control Networks for Image Based Surface Reconstruction: An Investigation of Optimum Survey Designs Using UAV Derived Imagery and Structure-from-Motion Photogrammetry. *Remote Sens.* **2016**, *8*, 786. [CrossRef]

5. Nesbit, P.R.; Hugenholtz, C.H. Enhancing UAV–SfM 3D Model Accuracy in High-Relief Landscapes by Incorporating Oblique Images. *Remote Sens.* **2019**, *11*, 239. [CrossRef]
6. Puniach, E.; Bieda, A.; Ćwiąkała, P.; Kwartnik-Pruc, A.; Parzych, P. Use of Unmanned Aerial Vehicles (UAVs) for Updating Farmland Cadastral Data in Areas Subject to Landslides. *ISPRS Int. J. Geo-Inf.* **2018**, *7*, 331. [CrossRef]
7. Available online: https://www.heliguy.com/blog/2019/01/24/is-rtk-the-future-of-drone-mapping/ (accessed on 29 March 2020).
8. Derrien, A.; Villeneuve, N.; Peltier, A.; Beauducel, F. Retrieving 65 years of volcano summit deformation from multitemporal structure from motion: The case of Piton de la Fournaise (La Réunion Island). *Geophys. Res. Lett.* **2015**, *42*, 6959–6966. [CrossRef]
9. Jovančević, S.D.; Peranić, J.; Ružić, I.; Arbanas, Ž. Analysis of a historical landslide in the Rječina River Valley, Croatia. *Geoenviron. Disasters* **2016**, *3*, 26. [CrossRef]
10. Rossi, G.; Tanteri, L.; Tofani, V.; Vannocci, P. Multitemporal UAV surveys for landslide mapping and characterization. *Landslides* **2018**, *15*, 1045. [CrossRef]
11. Ridolfi, E.; Buffi, G.; Venturi, S.; Manciola, P. accuracy analysis of a dam model from drone surveys. *Sensors* **2017**, *17*, 1777. [CrossRef]
12. Buffi, G.; Manciola, P.; Grassi, S.; Barberini, M.; Gambi, A. Survey of the Ridracoli Dam: UAV–based photogrammetry and traditional topographic techniques in the inspection of vertical structures. *Geomat. Nat. Hazards Risk* **2017**, *8*, 1562–1579. [CrossRef]
13. Duró, G.; Crosato, A.; Kleinhans, M.G.; Uijttewaal, W.S.J. Bank erosion processes measured with UAV-SfM along complex banklines of a straight mid-sized river reach. *Earth Surf. Dyn.* **2018**, *6*, 933–953. [CrossRef]
14. Peppa, M.V.; Mills, J.P.; Moore, P.; Miller, P.E.; Chambers, J.E. Brief communication: Landslide motion from cross correlation of UAV-derived morphological attributes. *Nat. Hazards Earth Syst. Sci.* **2017**, *17*, 2143–2150. [CrossRef]
15. Salvini, R.; Mastrorocco, G.; Esposito, G.; Di Bartolo, S.; Coggan, J.; Vanneschi, C. Use of a remotely piloted aircraft system for hazard assessment in a rocky mining area (Lucca, Italy). *Nat. Hazards Earth Syst. Sci.* **2018**, *18*, 287–302. [CrossRef]
16. Kovacevic, M.S.; Car, M.; Bacic, M.; Stipanovic, I.; Gavin, K.; Noren-Cosgriff, K.; Kaynia, A. Report on the Use of Remote Monitoring for Slope Stability Assessments; H2020-MG 2014–2015 Innovations and Networks Executive Agency. Available online: http://www.destinationrail.eu/documents (accessed on 20 January 2020).
17. Kaufmann, V.; Seier, G.; Sulzer, W.; Wecht, M.; Liu, Q.; Lauk, G.; Maurer, M. Rock Glacier Monitoring Using Aerial Photographs: Conventional vs. UAV-Based Mapping—A Comparative Study. *Int. Arch. Photogram. Remote Sens. Spat. Inf. Sci.* **2018**, *XLII-1*, 239–246. [CrossRef]
18. Vivero, S.; Lambiel, C.H. Monitoring the crisis of a rock glacier with repeated UAV surveys. *Geogr. Helv.* **2019**, *74*, 59–69. [CrossRef]
19. Blišťan, P.; Kovanič, Ľ.; Zelizňaková, V.; Palková, J. Using UAV photogrammetry to document rock outcrops. *Acta Montan. Slovaca* **2016**, *21*, 154–161.
20. Moudrý, V.; Urban, R.; Štroner, M.; Komárek, J.; Brouček, J.; Prošek, J. Comparison of a commercial and home-assembled fixed-wing UAV for terrain mapping of a post-mining site under leafoff conditions. *Int. J. Remote Sens.* **2019**, *40*, 555–572. [CrossRef]
21. Koska, B.; Jirka, V.; Urban, R.; Křemen, T.; Hesslerová, P.; Jon, J.; Pospíšil, J.; Fogl, M. Suitability, characteristics, and comparison of an airship UAV with lidar for middle size area mapping. *Int. J. Remote Sens.* **2017**, *38*, 2973–2990. [CrossRef]
22. Bláha, M.; Eisenbeiss, H.; Grimm, D.; Limpach, P. Direct Georeferencing of UAVs. In Proceedings of the Conference on Unmanned Aerial Vehicle in Geomatics, Zurich, Switzerland, 14–16 September 2011; Volume 38, pp. 1–6.
23. Turner, D.; Lucieer, A.; Wallace, L. Direct georeferencing of ultrahigh-resolution UAV imagery. *IEEE Trans. Geosci. Remote Sens.* **2014**, *52*, 2738–2745. [CrossRef]
24. Forlani, G.; Dall'Asta, E.; Diotri, F.; Cella, U.M.; Roncella, R.; Santise, M. Quality Assessment of DSMs Produced from UAV Flights Georeferenced with On-Board RTK Positioning. *Remote Sens.* **2018**, *10*, 311. [CrossRef]
25. Forlani, G.; Diotri, F.; Cella, U.M.; Roncella, R. Indirect UAV Strip Georeferencing by On-Board GNSS Data under Poor Satellite Coverage. *Remote Sens.* **2019**, *11*, 1765. [CrossRef]

26. Tomaštík, J.; Mokroš, M.; Surový, P.; Grznárová, A.; Merganič, J. UAV RTK/PPK Method—An Optimal Solution for Mapping Inaccessible Forested Areas? *Remote Sens.* **2019**, *11*, 721. [CrossRef]
27. Benassi, F.; Dall'Asta, E.; Diotri, F.; Forlani, G.; Morra di Cella, U.; Roncella, R.; Santise, M. Testing Accuracy and Repeatability of UAV Blocks Oriented with GNSS-Supported Aerial Triangulation. *Remote Sens.* **2017**, *9*, 172. [CrossRef]
28. Gerke, M.; Przybilla, H.J. Accuracy Analysis of Photogrammetric UAV Image Blocks: Influence of Onboard RTK-GNSS and Cross Flight Patterns. *Photogramm.-Fernerkund.-Geoinf.* **2016**, *2016*, 17–30. [CrossRef]
29. Hugenholtz, C.; Brown, O.; Walker, J.; Barchyn, T.; Nesbit, P.; Kucharczyk, M.; Myshak, S. Spatial Accuracy of UAV-Derived Orthoimagery and Topography: Comparing Photogrammetric Models Processed with Direct Geo-Referencing and Ground Control Points. *Geomatica* **2016**, *70*, 21–30. [CrossRef]
30. Padró, J.-C.; Muñoz, F.-J.; Planas, J.; Pons, X. Comparison of four UAV georeferencing methods for environmental monitoring purposes focusing on the combined use with airborne and satellite remote sensing platforms. *Int. J. Appl. Earth Obs. Geoinf.* **2019**, *75*, 130–140. [CrossRef]
31. Stöcker, C.; Nex, F.; Koeva, M.; Gerke, M. Quality Assessment of Combined Imu/Gnss Data for Direct Georeferencing in the Context of UAV-Based Mapping. *ISPRS-Int. Arch. Photogramm. Remote Sens. Spat. Inf. Sci.* **2017**, *XLII-2/W6*, 355–361. [CrossRef]
32. Peppa, M.V.; Hall, J.; Goodyear, J.; Mills, J.P. Photogrammetric Assessment and Comparison of DJI Phantom 4 Pro and Phantom 4 RTK Small Unmanned Aircraft Systems. *Int. Arch. Photogramm. Remote Sens. Spat. Inf. Sci.* **2019**, *XLII-2/W13*, 503–509. [CrossRef]
33. Taddia, Y.; Stecchi, F.; Pellegrinelli, A. Coastal Mapping using DJI Phantom 4 RTK in Post-Processing Kinematic Mode. *Drones* **2020**, *4*, 9. [CrossRef]
34. Santise, M.; Fornari, M.; Forlani, G.; Roncella, R. Evaluation of DEM generation accuracy from UAS imagery. *Int. Arch. Photogramm. Remote Sens. Spat. Inf. Sci.* **2014**, *XL-5*, 529–536. [CrossRef]
35. Kršák, B.; Blišťan, P.; Pauliková, A.; Puškárová, V.; Kovanič, Ľ.; Palková, J.; Zelizňaková, V. Use of low-cost UAV photogrammetry to analyze the accuracy of a digital elevation model in a case study. *Measurement* **2016**, *91*, 276–287. [CrossRef]
36. Cramer, M.; Przybilla, H.-J.; Zurhorst, A. UAV cameras: Overview and geometric calibration benchmark. *Int. Arch. Photogramm. Remote Sens. Spat. Inf. Sci.* **2017**, *XLII-2/W6*, 85–92. [CrossRef]
37. Harwin, S.; Lucieer, A.; Osborn, J. The Impact of the Calibration Method on the Accuracy of Point Clouds Derived Using Unmanned Aerial Vehicle Multi-View Stereopsis. *Remote Sens.* **2015**, *7*, 11933–11953. [CrossRef]
38. James, M.R.; Robson, S. Mitigating systematic error in topographic models derived from UAV and ground-based image networks. *Earth Surf. Proc. Landf.* **2014**, *39*, 1413–1420. [CrossRef]

Article

A Method for Evaluating and Selecting Suitable Hardware for Deployment of Embedded System on UAVs

Nicolas Mandel [1,2,*], Michael Milford [1,2] and Felipe Gonzalez [1,2]

[1] Australian Centre of Excellence for Robotic Vision, Queensland University of Technology, Brisbane QLD 4000, Australia; michael.milford@qut.edu.au (M.M.); felipe.gonzalez@qut.edu.au (F.G.)
[2] QUT Centre for Robotics, Queensland University of Technology, Brisbane QLD 4000, Australia
* Correspondence: nicolasjohann.mandel@hdr.qut.edu.au

Received: 30 June 2020; Accepted: 5 August 2020; Published: 7 August 2020

Abstract: The use of UAVs for remote sensing is increasing. In this paper, we demonstrate a method for evaluating and selecting suitable hardware to be used for deployment of algorithms for UAV-based remote sensing under considerations of *Size*, *Weight*, *Power*, and *Computational* constraints. These constraints hinder the deployment of rapidly evolving computer vision and robotics algorithms on UAVs, because they require intricate knowledge about the system and architecture to allow for effective implementation. We propose integrating computational monitoring techniques—profiling—with an industry standard specifying software quality—ISO 25000—and fusing both in a decision-making model—the analytic hierarchy process—to provide an informed decision basis for deploying embedded systems in the context of UAV-based remote sensing. One software package is combined in three software–hardware alternatives, which are profiled in hardware-in-the-loop simulations. Three objectives are used as inputs for the decision-making process. A Monte Carlo simulation provides insights into which decision-making parameters lead to which preferred alternative. Results indicate that local weights significantly influence the preference of an alternative. The approach enables relating complex parameters, leading to informed decisions about which hardware is deemed suitable for deployment in which case.

Keywords: UAV; computer architecture; decision making; navigation; semantics

---

1. Introduction

Unmanned Aerial Vehicles (UAVs) are playing an increasingly important role in modern society. The past decade has seen an increased use of UAVs, with applications ranging from wildlife and agricultural monitoring [1–3], over racing [4], industrial monitoring [5], packet delivery [6], and search and rescue [7] to planetary exploration [8]. Solutions to computer vision problems, such as object detection [9], and robotics problems, such as localization [10], have been developed and disseminated at a rapid pace, with promising results on benchmarks [11]. Despite experiments on real robots being the common mode of evaluation for robotics applications [11], reproducing results requires a considerable effort [12]. Furthermore, computational performance has traditionally been considered a secondary factor [10], assuming that hardware progress will eliminate this issue.

At the same time, Wirth's law states that the increasing complexity in software outgrows the increase in computational power. Additionally, UAVs are heavily constrained through their *Size*, *Weight*, *Power*, and *Computation* [6,13–15]. The main limitation for small multirotor UAVs is their flight time, which is bound by the battery capacity, which is in turn limited by the *Size* and *Weight* of the UAVs. Over 90% of *Power* is consumed by the motors supplying the thrust; however, the influence of computational capabilities on *Power* is complex. While increasing computational load uses more

Power directly, faster computation can greatly reduce total power consumption for the same task due to reduced time spent at hover and lower accelerations, to up to 6 times faster completion of tasks [6].

Using the maximum computation available by offloading parts of the perception—planning—control loop over a network connection is the most common solution [6,16–20]. This method is highly effective, however it introduces networking effects and drop-outs, with potentially detrimental consequences when used in real-time cycles [21,22]. Furthermore, several remote-sensing applications cannot rely on network connections, making onboard computation the final goal. Some research has focused on the onboard application of algorithms [4,23–25], however most are the result of empirical processes. Only a finite amount of research is concerned with detailed evaluation of software and hardware performance [23,26,27]. Using a powerful laptop for computation is a possible solution, however *Size* and *Weight* do not always allow for such a payload. In this paper, the difference between onboard and real-time as illustrated in Figure 1 is termed *configuration*, while the combination of a software *configuration* and a specific hardware as illustrated in Figure 2 is termed *alternative*.

The question as to which hardware is considered as a suitable solution is therefore highly complex and requires consideration of mechanical, electrical, and computational constraints imposed on small UAVs. Boroujerdian et al. [6] uncovered the influence of computation on power consumption, while Krishnan et al. [27] connected mechanical and computational performance in a "roofline" model. However, to the best of our knowledge, no method exists that holistically combines and analyses all four of these factors together. To this end, a decision-making approach from mathematical economics, the analytic hierarchy process (*AHP*) [28], is proposed.

Additionally, the question as to which indicators underline the quality of software is posed. While fast computation reduces flight time in some cases, in others the benefit is marginal, especially when tasks do not require compute-intensive dynamic recalculation steps [6]. On the other hand, some UAVs may have supplementary tasks to fulfill, such as manipulation [29,30], with different needs, such as the capacity to accommodate other software. These diverging specifications call for a framework with a balance between flexibility and rigidity in definitions of quality and associated metrics. The use of the ISO 25000 family "Systems and Software Quality Requirements and Evaluation" (*SQUARE*) [31–33] is proposed to meet these requirements. Software development is an intricate process, especially for robotic systems, which are commonly evaluated in experiments and therefore require implementation on realistic hardware [11]. Standardized frameworks such as the Robot Operating System (*ROS*) are universal and alleviate issues with interoperability, however, reproducing results remains an issue [12]. Pinpointing bottlenecks when implementing a software on a hardware system is a difficult task and requires detailed profiling [34].

In this work, we demonstrate the methods using a software-in-the-loop (SITL) ROS package developed and validated by the authors [35]. Figure 3 shows the simulation in the process. The left image shows the Gazebo simulation environment, the top panel shows the UAV view with a marker visible, and the right panel shows the reconstructed map.

The main driver of the subsequent hardware-in-the-loop (HITL) development is the preference of evaluation over experimentation as a factor of successful developments as put forward by Corke et al. [11]. Two embedded devices—a nVidia Jetson Nano and a Raspberry Pi 3A+—with deviating properties are evaluated. The Nano is computationally powerful, while incorporating a large volume and being heavier. The Pi is smaller, with less powerful computation.

Profiling techniques are used to inspect the computational performance of the software–hardware combination and are put into relation with one another using the *SQUARE* family. The computational performance is combined with electrical and mechanical factors through the AHP to provide numerical values which *alternative* is considered best.

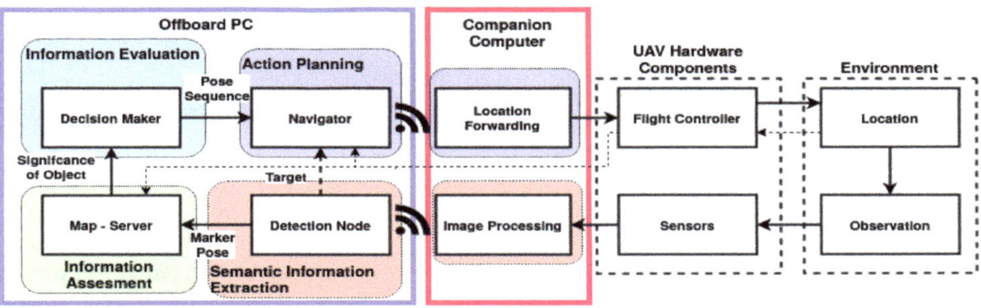

Figure 1. Modules developed in [35], onboard (**top**) and a real-time (**bottom**) *configuration*, with the system load distributed across a PC (purple) and an onboard computer (magenta). The thin dashed line indicates a serial connection and the wireless symbol a network connection. Differences in software set-up are called *configurations* in this work.

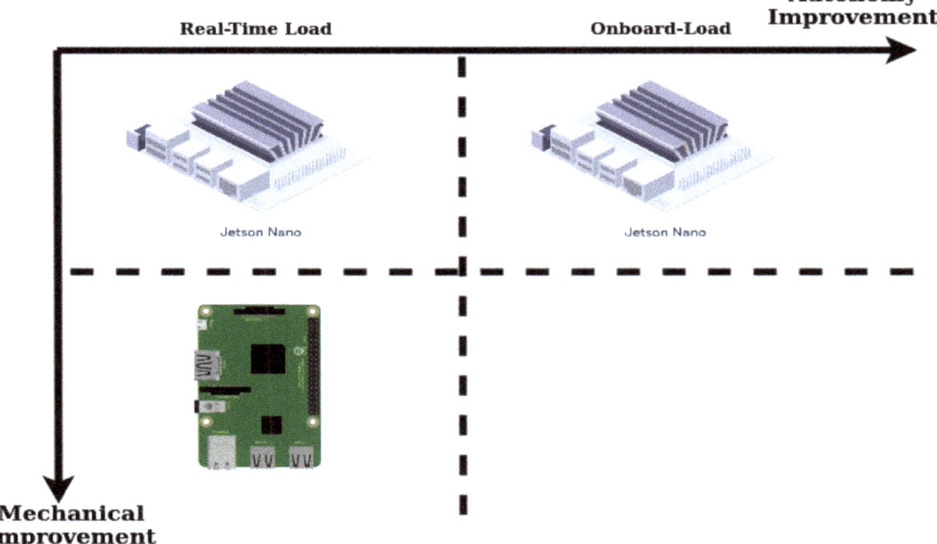

Figure 2. The three *alternatives* evaluated in this work. Differences in software *configurations* as shown in Figure 1 are shown on the *x*-axis. The *y*-axis denotes an additional mechanical variation through changing the companion computer.

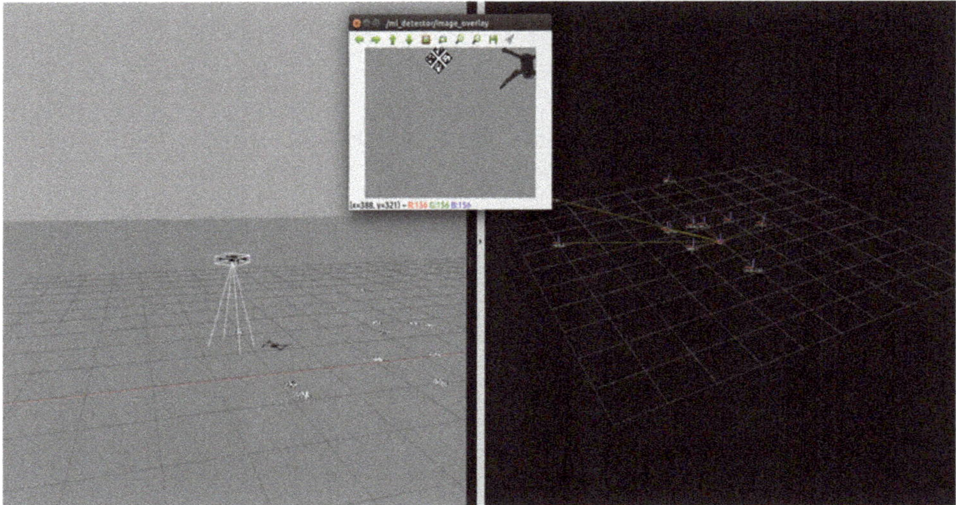

Figure 3. A software-in-the-loop (SITL) simulation in the Gazebo Environment is shown on the left. The UAV is conducting an autonomous remote-sensing search mission for a target marker. The top panel shows the view of the UAV; the right panel shows the map of detected markers.

Three case studies with different input weights for the AHP are proposed to verify the approach. The recorded parameters from the HITL simulation are used to show how the results change with varying the weights. Additionally, a Monte Carlo (MC) simulation is used over the admissible input space, which allows to compare the effects of different relative weights to give preference to another *alternative*.

Figure 1 presents one example of the differences between an onboard and a real-time system in the context of UAV onboard decision-making [35]. The split of the nodes is dependent on the frequency and the degree of closeness to hardware required by hardware components.

In this work, a methodical approach to HITL simulations is employed. The transition steps from SITL to HITL and their effects down the line of development are shown. A framework for evaluating the quality of software is proposed. Additionally, the AHP is drawn on for combining requirements along different dimensions, mainly mechanical and computational. Furthermore, to the best of our knowledge, the amount of research concerned with a systematic transition from simulation to experimental deployment is limited and we aim to augment the literature by providing an evaluation method.

2. Background

Evaluating system performance can be a slow, methodical task [34]. However, it is necessary to prevent detrimental effects down the line of development and deployment of UAVs. These can have catastrophic consequences when considering the deployment of robotic systems in critical conditions, such as near urban areas or oil rig inspection, where hardware failure may result in injuries to personnel on the ground or damage worth millions of dollars and years of work.

Models to evaluate the performance and potential of software on specific hardware are readily available. One example is the commonly known "roofline" model [36]. However, improvements implied by these models are gained through hardware-near programming techniques requiring profound knowledge of memory and CPU access. This is difficult to achieve for programmers working in a meta-environment such as ROS, which employs a complex variety of different tools in a software stack [12,37]. Models for evaluating the performance of GPUs in the context of contemporary deep learning approaches inspected the timeliness and performance [38] and the potential to be used with common embedded platforms [26]. The applicability of one of these algorithms on embedded platforms in the context of UAVs was evaluated by Kyrkou et al. [23] and subsequently a dedicated architecture solution was proposed [39].

The constraints on the computational capabilities of small multirotor UAVs caused by reduced availability of *Size*, *Weight*, and *Power* require a careful balance between computer hardware and algorithm efficiency [6,13–15]. Krishnan et al. [27] tied the computational performance together with mechanical properties of a UAV in the derivation of a modified "roofline" model [36] to provide hard limits on flying velocity. However, their model only considers obstacle avoidance as a task and neglects higher level navigation tasks [40]. While *Size*, *Weight*, and *Power* are relatively straightforward to approximate, *Computation* is more difficult to put into numbers. Tools to passively record the computational performance—commonly called "profiling" tools—are available in the scientific literature [22,23,26,27,36], as well as in industry [31,34,41]. However, the information gathered from these systems is either used to identify computational bottlenecks [34] or not used to make decisions at all. The use of the ISO 25000 SQUARE software model is proposed to relate computational performance indicators extracted from profiling.

2.1. The ISO SQUARE Family

The SQUARE family was published in the early 2010s [31]. It provides eight characteristics, each with varying sub-characteristics, on which to evaluate the quality of a software, and explicitly extends to embedded hardware–software systems. Which characteristics to use and at what point to define what is to be evaluated (e.g., at function calls, at server accesses, etc.) is up to the discretion of the developer, however it requires explicit definition. ISO 25021 defines *quality measure elements*, while ISO 25023 proposes some metrics on which to evaluate the sub-characteristics, as well as how to develop and combine these [33]. The analysis in this work is reduced to the following characteristics.

1. Performance Efficiency
2. Compatibility
3. Reliability

The remaining criteria of the ISO are *Functional Suitability*, *Usability*, *Security*, *Maintainability*, and *Portability*. ROS as a standardized framework that handles *Maintainability* and *Portability* [12]. *Functional Suitability*, *Usability*, and *Security* are equal across all *configurations* for the evaluated software and are therefore not considered in this work. If *alternatives* would display differences across these characteristics, they would require integration into the hierarchy and thus the analysis.

CPU load and memory are recorded to evaluate *Performance Efficiency* and *Compatibility*, as these are commonly considered the main determining performance factors [36]. *Resource Utilization* is the sub-characteristic of *Performance Efficiency* targeted in this research. *Time behavior* is not evaluated due to its inherent issues discussed in the literature [21,22]. GPU usage has been broadly evaluated in the literature [23,26,27], and as the evaluated application does not target it [35], it is omitted. *Capacity* is deemed not adequate to evaluate, because the system load only changes minutely over time and inputs cannot be scaled infinitely.

For *Compatibility*, the sub-characteristic *Interoperability* is equated through ROS. This research focuses on *Co-Existence*, which refers to the property of not negatively influencing other software running on the same system.

Fault Tolerance and *Recoverability* are sub-characteristics of *Reliability*, which are governed by the flight controller. *Availability* is a network property not considered adequate for the evaluation of the system; however, *Maturity* is evaluated through the number of faults.

Commonly, profiling is used as an approach to identify bottlenecks in the system [6,34]. However, bottlenecks reduce issues to a single part of the system and neither consider the impact on the entire system nor other dimensions beyond computation. The use of the AHP is proposed to alleviate this issue, which was developed by Thomas Saaty in the late twentieth century [28] and has been applied to a diverse set of decisions systems [42], including, but not limited to, medical [43], mining [44], flood susceptibility [45], and transfer learning [46] analyses.

3. Materials and Methods

3.1. Analytic Hierarchy Process

In the AHP, a hierarchy of criteria, termed levels in the following sections, is developed. The criteria at each level are compared against one another and assigned a weight between 1 and 9 to describe their relative importance [42]. The weights are commonly assigned subjectively by stakeholders, for example, through consensus voting [28] or sampling from expert opinions [45]. While other methods are available, the choice of weights is beyond the scope of this paper and discussed in [28,42,47]. To illustrate, *Power* is considered twice as important as *Weight* and *Weight* three times more important than *Size*. A matrix is constructed with these weights and its normalized maximal Eigenvector is calculated, which yields the local priority of each criterion. Only matrices with consistencies as calculated by Equation (1),

$$CI = \frac{\lambda_{max} - n}{n - 1} \tag{1}$$

with λ_{max} as the maximum Eigenvalue and n the dimension of the matrix, are allowed, as some constellations are considered unrealistic. For example, if *Weight* is twice as important as *Size* and *Power* is three times as important as *Weight*, *Size* cannot be more important than *Power*, as this would violate transitivity. However, small inconsistencies, which are bound to arise due to the constrained scale, are allowed and therefore the Consistency Index CI is divided by an empirically calculated R, $CR = \frac{CI}{R}$, to result in a value which should be below 0.1 [28,43]. This procedure is repeated for each hierarchical level, where the global priority is calculated through multiplication with the priority of the subsuming characteristic from the next higher level, thus resulting in a hierarchy of priorities.

The *alternatives* are compared pairwise with respect to each of the criteria and result in a relative preference for either one of the *alternatives*. For example, if one *alternative* is twice as heavy as another one, its preference for weight will be 1/2. These preferences are then multiplied with the priorities to result in a priority for each criterion for each *alternative*.

This method allows the hierarchical characteristics and sub-characteristics of the SQUARE family to be related to one another, as well as put into context with the criteria *Power*, *Weight*, and *Size* commonly mentioned in the literature [6,13,14]. It furthermore allows to evaluate discrete *alternatives* of embedded systems and put values into context with one another. Figure 4 shows the hierarchy developed and used in the following sections.

3.1.1. Case Studies

The global preference of the available *alternatives* largely depends on the weights assigned to each of the criteria at the different levels of the hierarchy. Three cases inspired from remote sensing scenarios are used to derive weights for the levels of the AHP. These weights are used to calculate different priorities, which are then multiplied with the same preference values for the *alternatives* resulting from profiling as described in Section 4.2. Without loss of generality, the preference values from the same experiments are used to illustrate the influence of differing weights on final priority. The following three tasks are chosen.

1. A fully autonomous UAV performing oil rig inspection [5]
2. A UAV in a controlled environment with a suspended arm [29,30]
3. An additional module on a UAV to improve performance by integrating semantics into its navigation [35]

These cases lead to assumptions concerning the assignment of weights, which require revision until the matrices can be considered consistent as defined by Equation (1). The weight matrices are explained in detail in Section 4.5.

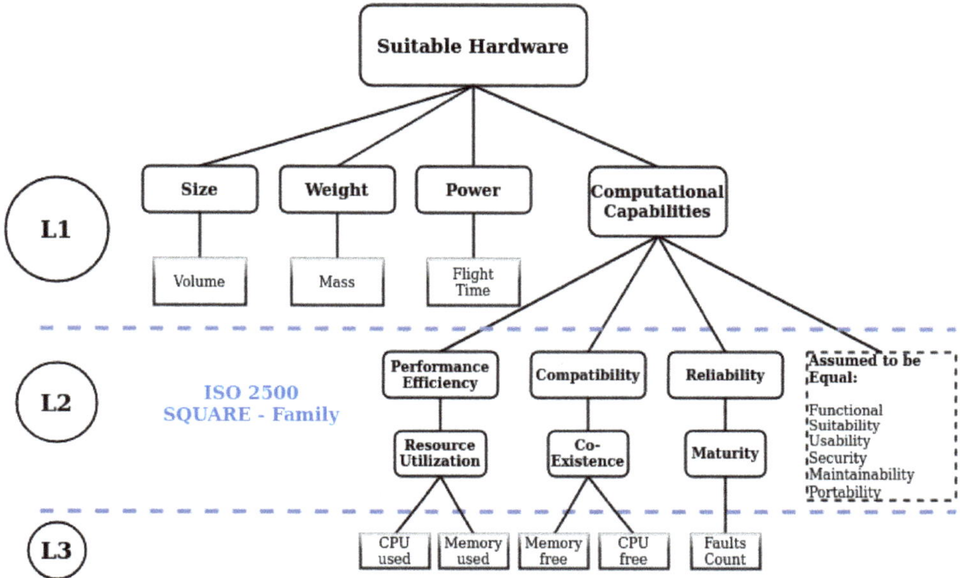

Figure 4. The final analytic hierarchy process (AHP) hierarchy used in the evaluation steps. The first level (L1) depicts the parameters common to literature [6]. The second level (L2) includes the parameters that are derived from the *SQUARE* family [31] and the third (L3) includes common computational parameters [33,36].

3.1.2. Monte-Carlo Simulation

Each of the *alternatives* (presented in Figure 2) can result as the best *alternative* given certain input weights. A MC simulation is conducted to find these weights. The space of consistent weight matrices is searched for the most dominant results, and the relative weights are presented. The results collected from the initial experiments—detailed in Section 4.5 and shown in Table 1—are used as inputs for the MC simulation. Ten-thousand iterations are run to generate values for the upper right triangular matrix from the original range from 1 to 9 and their inverses $f : S \mapsto X, f(s) = 1/s$, with $D = X \cup S$, $\underline{D} = 17$ admissible values [28]. Values from D are chosen at random to construct the upper right triangular matrix M, with the lower triangular matrix filled with the inverse, so $m_{j,i} = \frac{1}{m_{i,j}} \forall j > i$. The resulting matrix M is checked for consistency using Equation (1) and added to the set of admissible matrices only if it is consistent.

To clarify, for $rnk(M) = 2$ this leads to 17 possible combinations, because there is only one degree of freedom, $m_{1,2}$, which can take all 17 values from the admissible set D. For 3 and 4, 1228 and 479 consistent matrices are saved, respectively. The set of applicable matrices is denoted with Y_i, $i \in 2, 3, 4$ Algorithm 1 details the process on calculating the final score for each *alternative*. M_i denotes a matrix from the set of consistent matrices. $\hat{v}_j$ the global priority after multiplication with the higher level $j-1$. P denotes the matrix of recorded values derived from Table 1 and Section 4.4 and $\hat{p}$ the final priority of the *alternative*. The 10,000 resulting global priorities $\hat{p}$ are stored along with their respective weight matrices, M. In a final step, the constellation from the global priorities which would have led to the clearest preference of a given *alternative* are calculated and the weight matrices for that option are extracted for qualitative inspection.

Table 1. Profiling results for the three alternative cases presented in Figure 2.

Device	Nano		Pi	
Load	RT	Onb	RT	Unit
CPU Load Max	82.9	91.3	94.5	%
Used Mem Max	547	966	218	MB
Avail Mem Min	3249	2830	154	MB
Current	3.981	4.233	3.689	A
Faults	1	0	310	#
Volume	0.4464	0.4464	0.0672	1
Weight	0.252	0.252	0.044	kg

Algorithm 1 Monte Carlo Simulation

for 1 **to** 10,000 **do**
$M \leftarrow M_j \in Y_i$
$v_j \leftarrow$ max. Eigenvector of M_j
$\hat{v}_j \leftarrow v_j \cdot \hat{v}_{j-1}$
$\hat{p} \leftarrow P \cdot \hat{v}_j$
return $\hat{p}, M$
end for

4. Experimental Design

This work is targeting the HITL development step shown in Figure 5. Algorithm development, numerical simulations, and SITL—shown in Figure 3—were conducted in previous research and the performance of the developed package was verified [35]. The current work focuses on translating the simulation with the package into a distributed set-up, with the simulation running on a host PC—connected via Ethernet to the router—and parts of the package—in variations shown in Figure 1—running on the embedded system. The flight controller is connected to each device with an independent serial connection. This way it communicates with the simulator as well as the embedded computer.

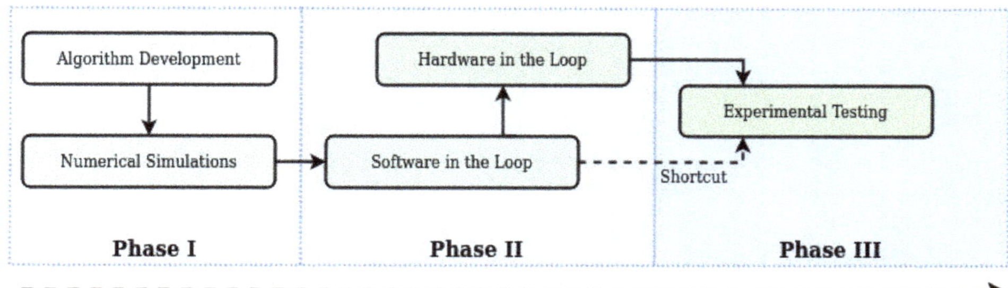

Figure 5. A generic autonomy development cycle, adapted from Ladosz et al. [48]. The arrow depicts the line of development, with the shading indicating the completeness. This work focuses on the transition included in Phase II.

4.1. System Specifications

All development procedures are undertaken on a Dell Notebook with an nVidia GeForce GTX 1060 Max-Q graphics card and an Intel i9-8950HK Processor with 16 GB of RAM and 24 GB of Swap. Ubuntu 18.04 with ROS melodic is used alongside Gazebo-9 for the simulation and development.

A mix of C++ and Python nodes are used during development. An open source node developed in-house is used for the image detection, while all other nodes were verified previously [35]. A vanilla version 10.1 of the Px4 Firmware on a PixHawk 4 is used. OpenCV version 3.4.6 with contrib modules is used for image processing. The network setup is run over a dedicated router within 1 m in line-of-sight, operating in the 5 to 2.4 GHz range to minimize network effects. The implementation details to enable reproduction [12] are detailed in Appendix A.

4.2. Development Steps

The simulation environment is shown in Figure 3. An UAV is exploring its environment and uses the camera feed to look for markers. If a certain type of marker is detected, the position is recorded and the path recalculated under consideration of known and unknown areas, as well as location of markers. Ten different marker placements, which were used for SITL [35], are used to record values according to Section 4.4 for each *alternative* shown in Figure 2.

Translating from SITL to HITL requires consideration of non-local communication. Packages are downloaded to the more powerful embedded device, the Jetson Nano. In a first step, the device is connected via Ethernet to the router, as shown in the first row of Figure 6, and one simulation is run with the real-time *configuration* of the package, as shown in Figure 1. Upon successful completion without notable errors or warnings, one simulation in onboard *configuration* is run, shown in the last column of Figure 6. When this test passes without notable errors, the network connection is switched to wireless and the test is repeated with both *configurations*, shown in the second row of Figure 6. When all tests are passed, the package is translated to the less powerful embedded device, the Raspberry Pi, as shown in the last column of Figure 6.

Each of these steps has an impact on the performance of the system, which may manifest itself in different ways and can be counteracted through adaptive software or hardware measures before continuing with the next step. The impact of these measures down the line of development should not be neglected, and therefore these steps are kept separate. Each measure requires re-synchronizing the software across the devices.

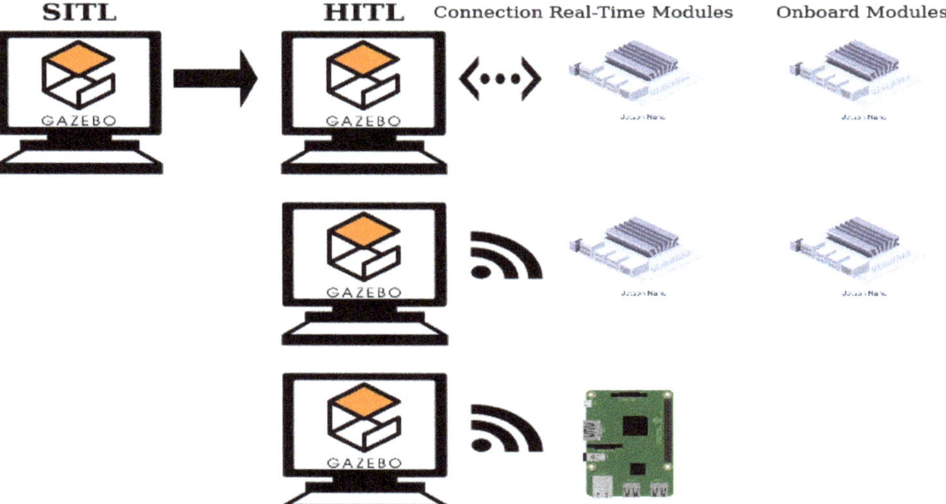

Figure 6. Hardware-in-the-loop set-ups. Each row shows changes in the hardware set-up and each column on the right changes in the software. The first row shows the transition from SITL to HITL using an Ethernet connection. The first column shows a real-time *configuration* as shown in Figure 1 and the the second column shows an onboard *configuration*. The second row shows the transition to a wireless connection. The third row shows the transition to the second embedded device.

4.3. Hardware and Software Alternatives

The preliminary development steps and checks lead to three software–hardware *alternatives* as candidates for the final evaluation detailed in Section 4.4. The *configurations* are chosen such that they represent opposite ends of the spectrum: the Nano is a powerful computing platform, with a selection of connectors and a large heat sink, therefore being larger and heavier than the Raspberry Pi. The Pi is a fraction of the *Size* and *Weight* of the Nano, however the decreased *Size* comes at the expense of connectors and processing power. The *alternatives* are shown in Figure 2.

The baseline *alternative* is the Nano real-time *alternative*, shown in the top left of Figure 2. The nodes running on the Nano are framed in pink in the bottom row of Figure 1, while all other nodes run on the host PC. This is equivalent to a distributed system where hardware-near and higher-frequency processing is done on the onboard computer, while higher-level, lower-frequency computational tasks are offloaded [6,40].

The high computational capacity of the Nano makes the system overpowered for these reduced tasks, therefore the other two *alternatives* shown in Figure 2 are evaluated to explore diverging specifications:

1. Can the computational load of the onboard computer be increased, allowing for a higher level of autonomy, without a detrimental impact on its performance [6]?
2. Can a smaller computer be used, allowing for a lighter payload and improved flight mechanics [27]?

These three *alternatives* are evaluated with the same experiments detailed in Section 4.4, after the development steps of Figure 6 are completed.

4.4. Profiling

Profiling—passive recording of system performance parameters—is undertaken with an independent ROS package. The official rosprofiler package [41] is updated to fit current psutil methods and message fields are adapted. Specifically, power logging, if available on the host, is included; CPU load is consolidated across all cores; virtual memory (used and available) as well as swap (used and available) are included. The package queries the host OS through psutil at a frequency of 5 Hz to collect samples. The number of samples, the average, the standard deviation, and the maximum or minimum—whichever is more conservative—over the samples is published at a frequency of 0.5 Hz. A client node is added, which is running on the notebook and records the values matching a predefined list of hosts and nodes. It writes the values to spreadsheet file at the end of each simulation run.

Postprocessing

The first and last five recorded values (equivalent to ten seconds each) are omitted during post-processing to reduce the influence of start-up and shutdown processes. Following this step, the number of samples per message is checked. Due to the publishing and querying frequencies, the number of samples should be ~10. If it is less than 8, a fault is recorded and all values of the sample omitted from further processing. This fault indicator has become evident during the development steps detailed in Figure 6. The current draw of the Raspberry Pi cannot be recorded through the OS, and therefore the nominal current draw of the device is multiplied with the CPU load as an approximation.

Using extreme values is preferred over average values in computational evaluation and aerospace systems, to account for worst case scenarios. The values of interest are therefore the maximum CPU load, the maximum used memory, as well as the minimum available memory. The 75th percentile value is extracted, to introduce robustness against outliers. These factors were determined through the inspection of intermittent development steps as shown in Figure 6.

The values recorded need to be consolidated into singular values and compared pairwise to map between 0 and 9 [28,42], depending on what is considered to be a priority. For *Size* comparisons, the volume of an, approximately square, 3D printed enclosure is measured: $V = l \cdot w \cdot h$. Both devices are weighed including their cases. An UAV, including a battery and excluding an onboard computer, is weighed: $w_{UAV} = 1322$ g. The relative increase in weight in percent, $\%_w = \frac{w_{onb}}{w_{UAV}+w_{onb}}$, is calculated. *Power* is commonly used as a proxy for flight time [6] and nominal values are used to calculate a baseline power usage P_0:

$$P_0 = \frac{U_0 \cdot (I \cdot t) \cdot c}{t_0} \tag{2}$$

With values U_0 being the average voltage rating, calculated through $U_0 = n_{cells} * U_{cell}$, with $n_{cells} = 3$ and $U_{cell} = 3.7\ V$, common for UAV LIPO batteries. c is a safety factor for battery drainage, for which 0.8 is used. $(I \cdot t)$ is a capacity rating in Amph for batteries, for which 4.0 is used. t_0 is a nominal flight time at hover, for which $t_0 = 1/3$, equal to 20 min of flight time is used. As the decrease in flight time caused by the power consumption, without additional performance effects [6], is of interest, the new flight time is calculated using the additional electrical power consumption by the onboard computer calculated through $P_e = V \cdot I$, evaluating to

$$t_f = \frac{U_0 \cdot (I \cdot t)}{P_0 + P_e} \tag{3}$$

The influence of additional *Weight* on the flight time is omitted, because its influence is considered marginal in comparison to the relative decrease in flight time through additional power consumption $\Delta t = \frac{t_0 - t_f}{t_0}$, which serves as a reasonable first approximation. The power draw of the device is either calculated through multiplication of the nominal voltage with the recorded current.

Computational performance is determined through the evaluation of *Resource Utilization*, *Co-Existence*, and *Maturity*. For *Resource Utilization*, CPU load, and memory usage are recorded.

For *Co-Existence*, available memory is recorded and CPU capacity calculated through 1−CPU load. Memory values are recorded as percentages of the nominal system capacity, with the capacity of the Nano being 4096 MB and the Raspberry Pi being 512 MB. These values are used for the comparison.

If the faults recorded are either 0 or 1, their value is considered equivalent. Otherwise, the logarithms are used as inputs for relative importance.

Figure 4 depicts the final hierarchy, which is to be used for the evaluation. The first level is derived from values common across literature. The second level is based on the *SQUARE* family [31] and the third level are common values from computational architectures [36].

4.5. Case Studies

4.5.1. Case 1—Oil Rig Inspection

The weights for the case of oil rig inspection are shown in Figure 7. The following assumptions lead to these weights; *Computation* would be most important, with *Power* following, due to the relationship uncovered by Boroujerdian et al. [6]. *Size* should be less important, albeit more important than *Weight* due to limited space. Reliability would be most important, as failures could have catastrophic consequences. Memory should be considered more important than CPU for both cases, because this would likely be the constraining factor for system updates in the distant future.

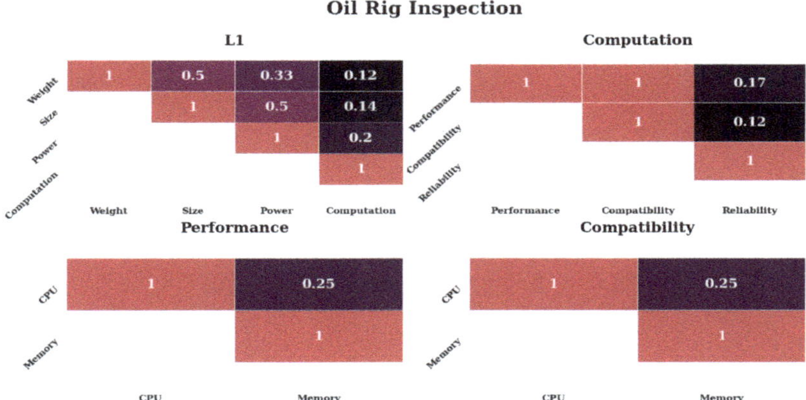

Figure 7. Weights put into the analysis for the case of an oil rig inspection. The top row shows the first two levels, the bottom row the third level. Only the upper diagonal weights are shown for simplicity. Darker colors indicate less preference, lighter colors more preference.

4.5.2. Case 2—Suspended Arm

The weights for the assumed case of a suspended arm in a controlled environment are shown in Figure 8. *Size* and *Weight* should be considered most important, because the suspended arm should yield a large increase in *Weight*, coming close to legal limits. *Power* should be considered least important, as the controlled environment would allow for quick exchange of batteries. On the computational level, compatibility should be considered most important, because the control of the arm would require implementation space. Reliability should be considered least important, as motion capture systems could serve as a fallback. Memory should be considered more important than CPU, as the control of the arm could require additional software tools and slower CPU cycles could be compensated through time spent at hover.

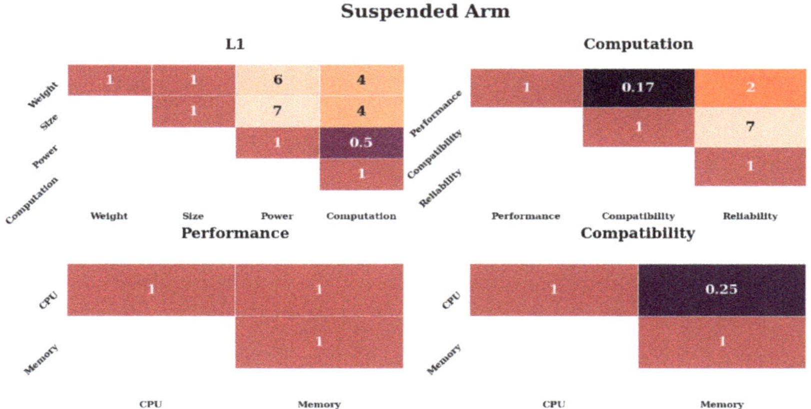

Figure 8. Weights put into the analysis for the case of a suspended arm. The top row shows the first two levels, the bottom row the third level. Only the upper diagonal weights are shown for simplicity. Darker colors indicate less preference, lighter colors more preference.

4.5.3. Case 3—Additional Module

Figure 9 shows the weights for the case of an additional module placed on an UAV. The assumptions are made that a module would have strict requirements on *Size* and *Weight*, because it would add to the total load. *Power* would be moderately important. Performance would be more important than compatibility, since the hardware would be entirely dedicated to the software module. Therefore, requirements on free space would be secondary, as would reliability, as the module would be considered nonessential. CPU performance would be more important than memory, as CPU load would likely be the first bottleneck.

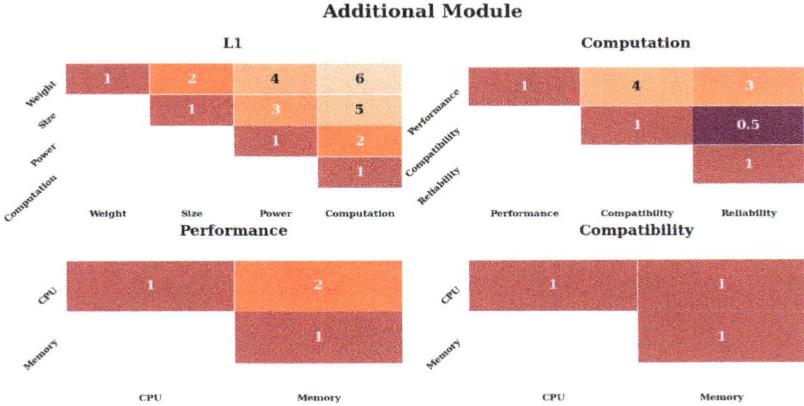

Figure 9. Weights put into the analysis for the case of an additional developed module. The top row shows the first two levels, the bottom row the third level. Only the upper diagonal weights are shown for simplicity. Darker colors indicate less preference, lighter colors more preference.

5. Results

Preliminary development steps, detailed in Section 4.2 and shown in Figure 6, are the reason for the following changes compared to SITL. First, the image type is changed to a compressed representation due to networking effects [21,22], which causes an increase in CPU load. Second, the MAVROS node responsible for translating between ROS and Mavlink is required to run on the onboard computer, increasing its load. Third, buffer sizes are increased for all messages except imaging. Fourth, the serial port speed is increased to 921,600 to allow for background messaging, increasing susceptibility to transmission errors. Fifth, the system load background usage through OS services is decreased. The final *configuration* for experiments are documented in Table A1 in Appendix A.

5.1. Profiling

Table 1 shows the values collected across the ten simulation runs exemplified by Figure 3. The values are collected for each of the three variations shown in Figure 2 and postprocessed according to Section 4.4.

The computational load is near the maximum across all devices, while the used memory is scaled according to the capacity of the device. The available memory of the Pi is much lower, in absolute as well as relative values. These effects can be explained by OS scheduling and paging techniques beyond the scope of this work. Current draw appears to be near the limits as well, which were recorded on the Nano and approximated on the Pi. The number of faults show a great variation. The Pi displays a substantial increase in number of faults compared with the Nano, while its *Size* and *Weight* are one order of magnitude lower than the Nano's.

5.2. Case Studies

Figure 10 and Table 2 show the results of the case study. The first case of oil rig inspection leads to the result that the Nano in a real-time *configuration* is preferred over an onboard *configuration*. This can be explained through the omission of networking effects, which are addressed in the literature [21,22]. The second and third case show that the Pi can be a highly preferred *alternative* under relative priorities, despite the large count of faults. The Pi is a better *alternative* when mechanical requirements prevail, as described in Section 4.5.

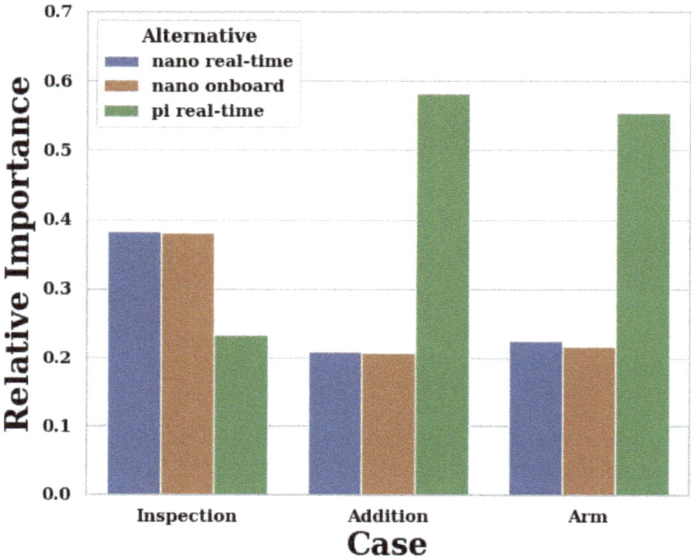

Figure 10. Relative preference for the case studies. The y-axis shows the relative preference, the x-axis a bar for each *alternative*: green for the Pi-real-time, blue for the Nanoreal-time and orange for the Nano-onboard *alternative*.

Table 2. Relative preferences for each *alternative* for the three case studies. Higher values indicate higher preference. Bold numbers indicate which alternative is preferred/the best for each column.

Device	Nano		Pi
Load	RT	Onb	RT
Inspection	**0.3832**	0.3818	0.2350
Suspended Arm	0.2267	0.2187	**0.5546**
Add. Module	0.2095	0.2087	**0.5828**

The first case, as shown in Figure 7, places an emphasis on *Power* and *Computation*, specifically reliability. These are the characteristics where the Nano has better values. Furthermore, free memory space leads to higher preference. The almost identical values are explained with the only slight difference in CPU and memory values (see Table 1), which only impacts the third level.

The second case as shown in Figure 8 emphasizes the importance of *Size* and *Weight*. This focus on values where the Pi is on another order of magnitude than the Nano renders the increased computational performance of the Nano almost negligible.

The same can be said about the third case, shown in Figure 9, albeit a little less pronounced. The increase in importance of performance on the second level and subsequent focus on CPU performance are rendered irrelevant through the high-level focus place on *Size* and *Weight*.

5.3. Monte Carlo Simulation

For each of the subsystem combinations listed in Table 1 and shown in Figure 2, there will be a case where the weights result in that combination being the most suitable *alternative*. The tabulated values for each of these alternative cases can be found in Appendix B. These cases are presented as follows.

5.3.1. Nano-Onboard

The first four heat maps of Figure 11 show the weights required to lead to a preference of the Nano-onboard *alternative*. The preference value is 0.37926, slightly above a third. The values indicate that the Nano is preferred over the Pi, however only a slight preference was given to the full load on the Nano over the partial load. The first level shows that *Computation* receives the highest preference compared to all others. The second level indicates that reliability is considered most important, with performance being the second most important. The third level shows that memory usage is very much preferred over CPU usage and memory space a little over CPU.

These values indicate that when mechanical parameters are not as important as computational and reliability plays a significant role, the Nano dominates the decision, as expected.

5.3.2. Nano-Real-Time

The preference value of the Nano in a real-time setting is 0.42533. This result indicates that the Nano in a real-time setting has quite a large preference over the other *alternatives*. The weights that lead to this preference are shown in the second set of Figure 11. The first level shows that *Computation* has the highest priority, followed by *Power*, while the other two are considered less important. The second level demonstrates the benefit of available space, which has a significantly higher value than reliability and lastly performance.

5.3.3. Pi-Real-Time

The last set of heatmaps in Figure 11 shows the weights leading to the highest preference for the Raspberry Pi-real-time *alternative*. The relative preference value is 0.64268, more than three times as high as the other two *alternatives* as shown in Appendix B. Inspecting the first level reveals that *Weight* is considered most important, and *Size* second most, while *Power* and *Computation* are considered relatively less important. Computational parameters have little significance through this; however, performance largely outweighs compatibility and reliability. On the third level, memory is preferred over CPU.

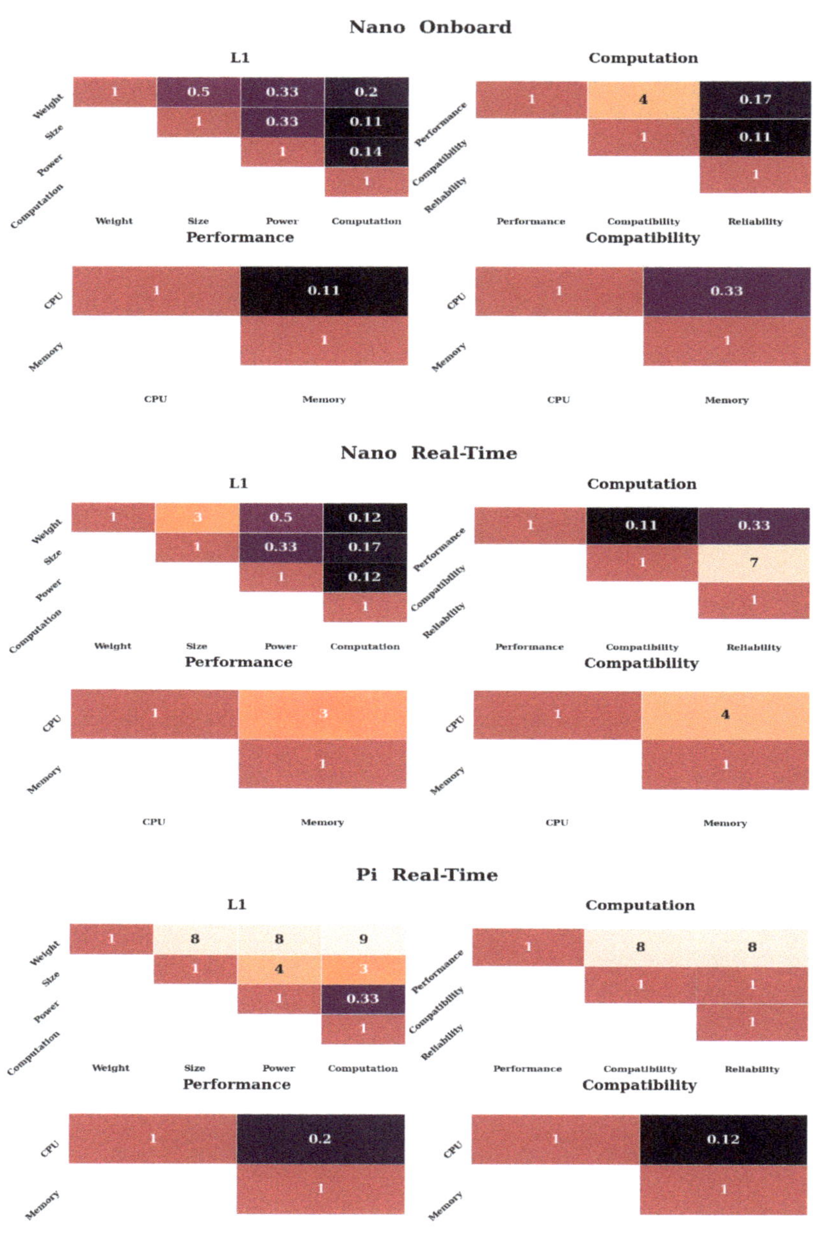

Figure 11. The weight matrices leading to the highest preferences for each *alternative*. The first heat maps show the weights for the Nano-onboard-*alternative*, the second set of heatmaps the weights for the Nano-real-time-*alternative*, and the third row the weights for the Pi-real-time *alternative*. The top row shows the first two levels, the bottom row the third level. Only the upper diagonal weights are shown for simplicity. Darker colors indicate less preference, lighter colors more preference.

6. Discussion

6.1. The AHP

The AHP is a model mainly used for high-level decision-making and has both benefits and drawbacks. On one hand it is constrained by its scale and the required consistencies [47], as well as the subjectivity of the choice of weights by stakeholders. On the other hand, the scales and consistencies force users to put their criteria into numerical values by comparing pairwise and checks whether the criteria can be considered consistent. Users are required to re-evaluate their criteria if these are deemed inconsistent and increased dimensions allow for higher degrees of inconsistencies, which are more likely to arise.

The hierarchy allows to be extended or reduced. For example, *Size* could have included an additional parameter for the form factor (for example, cubic as bad, but flat as a good form factor). Alternatively, other characteristics of the ISO, as mentioned in Section 2.1, would require consideration if *alternatives* would differ with respect to these characteristics, for example, if a solution not based on ROS would be provided, the equalities mentioned in Section 2.1 would not hold. However, the hierarchy reduces the influence of lower levels on the higher levels, which is indicated through the limited influence of reliability, despite the evident shortcoming of the Pi. Shifting values between levels may result in different outcomes, however the model was chosen in compliance with contemporary literature and the ISO model. The question whether to place reliability on a higher level should be evaluated with respect to the decomposition of structurally similar indicators as described by Saaty [28], as well as the inclusion into the ISO [31]. Weights on the third level are diluted through the higher levels and have partial co-dependence, as indicated by the inverse relation of CPU used and CPU free. Additionally, the use of ROS limits the freedom of software optimizations [36] for developers, paying the price of flexibility for reduced complexity.

The complex structure of the problem [6,27] would render common optimization techniques difficult to be applicable and is beyond the scope of this work. Unlike convex optimization or gradient-based methods, the AHP does not yield absolute performance optima; however, it provides relative preferences on readily available *alternatives* that can be compared with each other. Furthermore, the evaluation disregards hardware settings that might have an influence, such as cable shielding or memory card wear.

The research could also be extended to further *alternatives*, as only three, with two comprising of the same hardware, were evaluated. However, with an increase in *alternatives*, final preferences become less pronounced due to the mathematical normalization inherent to the AHP. Furthermore, the *alternatives* were chosen such that they represent opposite ends of the spectrum: the Pi as a mechanically lightweight *alternative*, with little computing power and the Nano as a comparably large and powerful module. The variation between the onboard and real-time *configurations* is used to illustrate differences in software load, albeit the effects are less pronounced. This can be explained through smart OS priority and assignment schemes for memory and CPU, leading to marginal differences for these aspects.

6.2. Case Studies

The case studies, inspired by remote-sensing scenarios, indicate that varying requirements lead to different results which *alternatives* are preferred. While the step of decreasing mechanical *Size* or increasing *Computation* may trigger a logical choice in experienced engineers, evaluating the benefits may be difficult for other users.

Networking effects are omitted, because these were addressed in the literature [21,22] and with the deployment of ROS2. However, network settings hold the potential to impact decisions, as they have an influence in UAV remote sensing applications.

The case studies are used to illustrate the influence of relative weights and how these can differ on a case-by-case basis, however the tasks presented do not necessarily correlate with the task defined in the original work [35]. Nonetheless, the methodology serves its purpose of illustrating the effect that different weights have on a final preference.

6.3. Monte Carlo Simulation

The MC simulation compares a large variety of different weights as inputs, combined with the same results from profiling, as shown in Table 1. This exposes the relative weights, which could have led to the preference of an *alternative*. While these results are indicative of the relative weights which could have to the preference of an *alternative*, the AHP is not used to select weights. This would contradict the paradigm of the AHP that relative weights require assignment, and not *alternatives*. The method also depends on results from profiling and therefore a variation in recording performance values may lead to a change in preference. It also poses the questions whether shifting around certain aspects to other levels, such as reliability to the topmost level, should not be considered to ensure an increased focus.

6.4. Comparing Case Studies and Monte Carlo Simulation

The case studies are inspired by three remote-sensing scenarios and result in different preferences, as indicated in Table 2 and Figure 10, despite being run with the same performance indicators as shown in Table 1. Figures 7–9 indicate that the distinctiveness of the first level has a significant influence on the outcome. The mechanical properties *Size* and *Weight* are considered more important for the case of an additional module (Figure 9) as well as the case of a suspended arm (Figure 8), which are also the cases which give preference to the Pi, as shown in Appendix B.

The dominance of the Pi under consideration of mechanical properties is underlined by the MC results as shown in Figure 11 and Appendix B, where the first level of the Pi real-time result is predominantly focused on *Weight*. The other two alternatives are almost equal, as their mechanical properties are equal and the influence of *Computation* is diminished. When *Computation* is considered important, the Nano prevails, albeit not by as large a margin as the Pi (see Appendix B), due to the reduced benefits arising from the lower levels.

Altogether the Nano shows considerable benefits when *Computation*, and foremost reliabilitity, is important, while the Pi prevails when *Size* and *Weight* are required to be small. However, the inclusion of reliability into the computational level may lead to be misleading for fully autonomous missions and may require researchers to ensure other safety measures are in place. Generally the weights tend to shift towards mechanical importance when the configuration is not considered mission-critical and safety can be guaranteed through other measures. If basic flight depends on the configuration, computational factors gain influence, which is in agreement with contemporary research [6,27].

7. Conclusions

This research details a method for evaluating and selecting suitable hardware for the hardware-in-the-loop development step for deployment of remote-sensing UAVs. While computational performance is often considered secondary, it has a significant influence on which choice of software–hardware combination should be considered as suitable for a given task. A HITL software package was tested under three different software–hardware setups and profiled to collect performance indicators. The indicators are used in a decision-making model alongside subjective preference values to show how changes in priorities lead to preference of different software–hardware combinations. A Monte Carlo simulation indicates weights under which one of the *alternatives* shown in Figure 6 would be preferred.

The results provide insights into the complex decisions required to deploy autonomous algorithms on hardware and require careful selection of the preference weights as they rely on subjective assignment. However, preference weights only require ordinal-scaled definition and are checked for mathematical consistency, therefore allowing awareness for priorities of potentially conflicting *alternatives*.

The methodology presented in this research can enable remote-sensing researchers to evaluate whether their choice of software–hardware combination is sustainable by running hardware-in-the-loop simulations. This can lead to a reduced failures, improve performance consistencies, and reduce development time, improving deployment of sophisticated UAV-based remote sensing technology.

Author Contributions: N.M. conceptualization, methodology, software, analysis, data curation, writing—original draft preparation, and visualization; M.M. conceptualization and writing—review and editing, supervision; F.G. conceptualization, resources, writing—review and editing, supervision, and project administration. All authors have read and agreed to the published version of the manuscript.

Funding: This research was conducted by the Australian Research Council project number CE140100016, and supported by the QUT Centre for Robotics and ARC grant FT140101229.

Acknowledgments: N.M. acknowledges F. Vanegas, K. Morton, and G. Suddrey for their support during the development of this work. The authors acknowledge continued support from the Queensland University of Technology (QUT) through the Centre for Robotics.

Conflicts of Interest: The authors declare no conflicts of interest.

Appendix A. Hardware Implementation Details

Table A1. Configurations of the two embedded hardware devices following initial experiments.

Device	Nano		Pi	
Software	**Type**	**Installation**	**Type**	**Installation**
OS	Jetpack Ubuntu 18.04	Etcher	Ubuntu Mate 18.04	Cross-compilation Pi 3B
Swap	Swap	2048 MB	Zram/Swap	256/1052 MB
ROS	Melodic	Package Manager	Melodic	Package Manager
MavRos	0.33.4	Source	0.33.4	Source
OpenCV	3.4.6.	Source + Contrib	3.4.6	Source + Contrib Cross-compilation
Timesync	NTP	ntp.qut	NTP	ntp.qut
Proprietary Changes	Power Mode	0	Serial Clock speed	Elevated-mavros
	HDMI	disabled	HDMI	disabled
	WiFi	Powersaving disabled	WiFi	Powersaving disabled
	Power Supply	Barrel Jack 4A	Display Service	disabled
	SD-Card	32 GB SanDisk SDHC U3	SD-Card	32 GB SanDisk SDHC U1
	Serial Port Config	921600 8N1	Serial Port Config	921600 8N1
			Bluetooth	disabled-Serial Port

Appendix B. Monte-Carlo Simulation Results

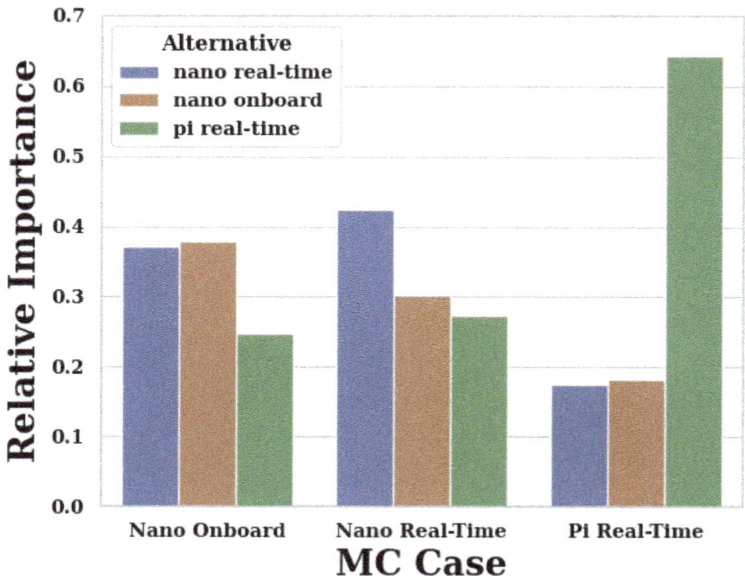

Figure A1. Relative preference for the Monte Carlo simulation. The y-axis shows the relative preference, the x-axis a bar for each *alternative*: green for the Pi-real-time, blue for the Nano-real-time and orange for the Nano-onboard *alternative*. Each MC results in one of the alternatives being preferred.

Table A2. Results from the Monte-Carlo Simulation. Each row shows the global preference of the alternatives. The rows indicate the case for which these values were the highest absolute preference. Bold numbers indicate which alternative is preferred/the best for each column.

Device	Nano		Pi
Load	RT	Onb	RT
Nano-RT	**0.42533**	0.30175	0.27292
Nano-Onb	0.37235	**0.37926**	0.24840
Pi-RT	0.17570	0.18163	**0.64268**

References

1. Gonzalez, L.F.; Montes, G.A.; Puig, E.; Johnson, S.; Mengersen, K.; Gaston, K.J. Unmanned Aerial Vehicles (UAVs) and Artificial Intelligence Revolutionizing Wildlife Monitoring and Conservation. *Sensors* **2016**, *16*, 97. [CrossRef] [PubMed]
2. Vanegas, F.; Bratanov, D.; Powell, K.; Weiss, J.; Gonzalez, F. A Novel Methodology for Improving Plant Pest Surveillance in Vineyards and Crops Using UAV-Based Hyperspectral and Spatial Data. *Sensors* **2018**, *18*, 260. [CrossRef] [PubMed]
3. Villa, T.F.; Gonzalez, F.; Miljievic, B.; Ristovski, Z.D.; Morawska, L. An Overview of Small Unmanned Aerial Vehicles for Air Quality Measurements: Present Applications and Future Prospectives. *Sensors* **2016**, *16*, 1072. [CrossRef] [PubMed]
4. Kaufmann, E.; Loquercio, A.; Ranftl, R.; Dosovitskiy, A.; Koltun, V.; Scaramuzza, D. Deep Drone Racing: Learning Agile Flight in Dynamic Environments. *arXiv* **2018**, arXiv:1806.08548.

5. Welburn, E.; Khalili, H.H.; Gupta, A.; Watson, S.; Carrasco, J. A Navigational System for Quadcopter Remote Inspection of Offshore Substations. In Proceedings of the Fifteenth International Conference on Autonomic and Autonomous Systems, Athens, Greece, 2–6 June 2019.
6. Boroujerdian, B.; Genc, H.; Krishnan, S.; Cui, W.; Faust, A.; Reddi, V. MAVBench: Micro Aerial Vehicle Benchmarking. In Proceedings of the 2018 51st Annual IEEE/ACM International Symposium on Microarchitecture (MICRO), Fukuoka, Japan, 20–24 October 2018; pp. 894–907. [CrossRef]
7. Vanegas, F.; Gonzalez, F. Uncertainty based online planning for UAV target finding in cluttered and GPS-denied environments. In Proceedings of the 2016 IEEE Aerospace Conference, Big Sky, MT, USA, 5–12 March 2016; pp. 1–9. [CrossRef]
8. Galvez Serna, J.; Vanegas Alvarez, F.; Gonzalez, F.; Flannery, D. A review of current approaches for UAV autonomous mission planning for Mars biosignatures detection. In *IEEE Aerospace Conference*; IEEE: Piscataway, NJ, USA, 2020; in press.
9. Liu, L.; Ouyang, W.; Wang, X.; Fieguth, P.; Chen, J.; Liu, X.; Pietikäinen, M. Deep Learning for Generic Object Detection: A Survey. *Int. J. Comput. Vis.* **2019**. [CrossRef]
10. Cadena, C.; Carlone, L.; Carrillo, H.; Latif, Y.; Scaramuzza, D.; Neira, J.; Reid, I.; Leonard, J.J. Past, Present, and Future of Simultaneous Localization and Mapping: Toward the Robust-Perception Age. *IEEE Trans. Robot.* **2016**, *32*, 1309–1332. [CrossRef]
11. Corke, P.; Dayoub, F.; Hall, D.; Skinner, J.; Sünderhauf, N. What can robotics research learn from computer vision research? *arXiv* **2020**, arXiv:2001.02366.
12. Cervera, E. Try to Start It! The Challenge of Reusing Code in Robotics Research. *IEEE Robot. Autom. Lett.* **2019**, *4*, 49–56. [CrossRef]
13. Zhao, Y.; Zheng, Z.; Liu, Y. Survey on computational-intelligence-based UAV path planning. *Knowl.-Based Syst.* **2018**, *158*, 54–64. [CrossRef]
14. Lu, Y.; Xue, Z.; Xia, G.S.; Zhang, L. A survey on vision-based UAV navigation. *Geo Spat. Inf. Sci.* **2018**, *21*, 21–32. [CrossRef]
15. Kang, K.; Belkhale, S.; Kahn, G.; Abbeel, P.; Levine, S. Generalization through Simulation: Integrating Simulated and Real Data into Deep Reinforcement Learning for Vision-Based Autonomous Flight. In Proceedings of the 2019 International Conference on Robotics and Automation (ICRA), Montreal, QC, Canada, 20–24 May 2019; pp. 6008–6014. [CrossRef]
16. Toudeshki, A.G.; Shamshirdar, F.; Vaughan, R. Robust UAV Visual Teach and Repeat Using Only Sparse Semantic Object Features. In Proceedings of the 2018 15th Conference on Computer and Robot Vision (CRV), Toronto, ON, Canada, 8–10 May 2018; pp. 182–189. [CrossRef]
17. Loquercio, A.; Maqueda, A.I.; del Blanco, C.R.; Scaramuzza, D. DroNet: Learning to Fly by Driving. *IEEE Robot. Autom. Lett.* **2018**, *3*, 1088–1095. [CrossRef]
18. Kouris, A.; Bouganis, C. Learning to Fly by MySelf: A Self-Supervised CNN-Based Approach for Autonomous Navigation. In Proceedings of the 2018 IEEE/RSJ International Conference on Intelligent Robots and Systems (IROS), Madrid, Spain, 1–5 October 2018; pp. 1–9. [CrossRef]
19. Kim, D.K.; Chen, T. Deep Neural Network for Real-Time Autonomous Indoor Navigation. *arXiv* **2015**, arXiv:1511.04668.
20. Ross, S.; Melik-Barkhudarov, N.; Shankar, K.S.; Wendel, A.; Dey, D.; Bagnell, J.A.; Hebert, M. Learning monocular reactive UAV control in cluttered natural environments. In Proceedings of the 2013 IEEE International Conference on Robotics and Automation, Karlsruhe, Germany, 6–10 May 2013; pp. 1765–1772. [CrossRef]
21. Tardioli, D.; Parasuraman, R.; Ögren, P. Pound: A multi-master ROS node for reducing delay and jitter in wireless multi-robot networks. *Robot. Auton. Syst.* **2019**, *111*, 73–87. [CrossRef]
22. Bihlmaier, A.; Hadlich, M.; Wörn, H. Advanced ROS Network Introspection (ARNI). In *Robot Operating System (ROS): The Complete Reference*; Koubaa, A., Ed.; Studies in Computational Intelligence; Springer International Publishing: Cham, Switzerland, 2016; Volume 1, pp. 651–670. [CrossRef]
23. Kyrkou, C.; Plastiras, G.; Theocharides, T.; Venieris, S.I.; Bouganis, C. DroNet: Efficient convolutional neural network detector for real-time UAV applications. In Proceedings of the 2018 Design, Automation Test in Europe Conference Exhibition (DATE), Dresden, Germany, 19–23 March 2018; pp. 967–972. [CrossRef]

24. Dang, T.; Papachristos, C.; Alexis, K. Autonomous exploration and simultaneous object search using aerial robots. In Proceedings of the 2018 IEEE Aerospace Conference, Big Sky, MT, USA, 3–10 March 2018; pp. 1–7. [CrossRef]
25. Papachristos, C.; Kamel, M.; Popović, M.; Khattak, S.; Bircher, A.; Oleynikova, H.; Dang, T.; Mascarich, F.; Alexis, K.; Siegwart, R. Autonomous Exploration and Inspection Path Planning for Aerial Robots Using the Robot Operating System. In *Robot Operating System (ROS): The Complete Reference*; Koubaa, A., Ed.; Studies in Computational Intelligence; Springer International Publishing: Cham, Switzerland, 2019; Volume 3, pp. 67–111. [CrossRef]
26. Modasshir, M.; Li, A.Q.; Rekleitis, I. Deep Neural Networks: A Comparison on Different Computing Platforms. In Proceedings of the 2018 15th Conference on Computer and Robot Vision (CRV), Toronto, ON, Canada, 8–10 May 2018; pp. 383–389. [CrossRef]
27. Krishnan, S.; Wan, Z.; Bhardwaj, K.; Whatmough, P.; Faust, A.; Wei, G.Y.; Brooks, D.; Reddi, V.J. The Sky Is Not the Limit: A Visual Performance Model for Cyber-Physical Co-Design in Autonomous Machines. *IEEE Comput. Archit. Lett.* **2020**, *19*, 38–42. [CrossRef]
28. Saaty, R.W. The analytic hierarchy process—What it is and how it is used. *Math. Model.* **1987**, *9*, 161–176. [CrossRef]
29. Suarez, A.; Vega, V.M.; Fernandez, M.; Heredia, G.; Ollero, A. Benchmarks for Aerial Manipulation. *IEEE Robot. Autom. Lett.* **2020**, *5*, 2650–2657. [CrossRef]
30. Morton, K.; Toro, L.F.G. Development of a robust framework for an outdoor mobile manipulation UAV. In Proceedings of the 2016 IEEE Aerospace Conference, Big Sky, MT, USA, 5–12 March 2016; pp. 1–8. [CrossRef]
31. ISO; IEC. *25010:2011-Systems and Software Engineering-Systems and software Quality Requirements and Evaluation (SQuaRE)-System and Software Quality Models*; Standard 25010:2011; ISO/IEC: Geneva, Switzerland, 2011.
32. ISO; IEC. *25021:2012-Systems and Software Engineering-Systems and Software Quality Requirements and Evaluation (SQuaRE)-Quality Measure Elements*; Standard 25021:2012; revised 2019; ISO/IEC: Geneva, Switzerland, 2012.
33. ISO; IEC. *25023:2016-Systems and Software Engineering-Systems and Software Quality Requirements and Evaluation (SQuaRE)-Measurement of System and Software Product Quality*; Standard 25023:2016; ISO/IEC: Geneva, Switzerland, 2016.
34. Simmonds, C. *Mastering Embedded Linux Programming*; Packt Publishing Ltd.: Birmingham, UK, 2017.
35. Mandel, N.; Vanegas, F.; Milford, M.; Gonzalez, F. Towards Simulating Semantic Onboard UAV Navigation. In *IEEE Aerospace Conference*; IEEE: Big Sky, MT, USA, 2020.
36. Williams, S.; Waterman, A.; Patterson, D. Roofline: An insightful visual performance model for multicore architectures. *Commun. ACM* **2009**, *52*, 65–76. [CrossRef]
37. Cervera, E.; Del Pobil, A.P. ROSLab: Sharing ROS Code Interactively With Docker and JupyterLab. *IEEE Robot. Autom. Mag.* **2019**, *26*, 64–69. [CrossRef]
38. Huang, J.; Rathod, V.; Sun, C.; Zhu, M.; Korattikara, A.; Fathi, A.; Fischer, I.; Wojna, Z.; Song, Y.; Guadarrama, S.; et al. Speed/accuracy trade-offs for modern convolutional object detectors. In Proceedings of the IEEE Conference on Computer Vision and Pattern Recognition, Honolulu, HI, USA, 21–26 July 2017.
39. Kouris, A.; Venieris, S.I.; Bouganis, C. Towards Efficient On-Board Deployment of DNNs on Intelligent Autonomous Systems. In Proceedings of the 2019 IEEE Computer Society Annual Symposium on VLSI (ISVLSI), Miami, FL, USA, 15–17 July 2019; pp. 568–573. [CrossRef]
40. Siegwart, R. *Introduction to Autonomous Mobile Robots*, 2nd ed.; Intelligent Robotics and Autonomous Agents Series; MIT Press: Cambridge, MA, USA, 2011.
41. Brooks, D. Rosprofiler-ROS Wiki. Available online: http://wiki.ros.org/rosprofiler (accessed on 7 August 2020).
42. Subramanian, N.; Ramanathan, R. A review of applications of Analytic Hierarchy Process in operations management. *Int. J. Prod. Econ.* **2012**, *138*, 215–241. [CrossRef]
43. Mühlbacher, A.C.; Kaczynski, A. Der Analytic Hierarchy Process (AHP): Eine Methode zur Entscheidungsunterstützung im Gesundheitswesen. *Pharmacoecon. Ger. Res. Artic.* **2013**, *11*, 119–132. [CrossRef]
44. Ataei, M.; Shahsavany, H.; Mikaeil, R. Monte Carlo Analytic Hierarchy Process (MAHP) approach to selection of optimum mining method. *Int. J. Min. Sci. Technol.* **2013**, *23*, 573–578. [CrossRef]

45. Dahri, N.; Abida, H. Monte Carlo simulation-aided analytical hierarchy process (AHP) for flood susceptibility mapping in Gabes Basin (southeastern Tunisia). *Environ. Earth Sci.* **2017**, *76*, 302. [CrossRef]
46. Zamir, A.; Sax, A.; Shen, W.; Guibas, L.; Malik, J.; Savarese, S. Taskonomy: Disentangling Task Transfer Learning. *arXiv* **2018**, arXiv:1804.08328.
47. Franek, J.; Kresta, A. Judgment Scales and Consistency Measure in AHP. *Procedia Econ. Financ.* **2014**, *12*, 164–173. [CrossRef]
48. Ladosz, P.; Coombes, M.; Smith, J.; Hutchinson, M. A Generic ROS Based System for Rapid Development and Testing of Algorithms for Autonomous Ground and Aerial Vehicles. In *Robot Operating System (ROS): The Complete Reference*; Koubaa, A., Ed.; Studies in Computational Intelligence; Springer International Publishing: Cham, Switzerland, 2019; Volume 3, pp. 113–153. [CrossRef]

Article

UAV Autonomous Localization Using Macro-Features Matching with a CAD Model

Akkas Haque [1], Ahmed Elsaharti [1], Tarek Elderini [2], Mohamed Atef Elsaharty [1,*] and Jeremiah Neubert [1]

[1] Department of Mechanical Engineering, University of North Dakota (UND), Upson II Room 266, 243 Centennial Drive, Stop 8359, Grand Forks, ND 58202, USA; akkasuddin.haque@und.edu (A.H.); ahmed.elsaharti@und.edu (A.E); jeremiah.neubert@und.edu (J.N.)

[2] School of Electrical Engineering and Computer Science, University of North Dakota (UND), Upson II Room 369. 243 Centennial Drive, Stop 7165, Grand Forks, ND 58202, USA; tarek.elderini@und.edu

* Correspondence: mohamed.elsaharty@und.edu; Tel.: +1-701-777-5996

Received: 9 January 2020; Accepted: 27 January 2020; Published: 29 January 2020

Abstract: Research in the field of autonomous Unmanned Aerial Vehicles (UAVs) has significantly advanced in recent years, mainly due to their relevance in a large variety of commercial, industrial, and military applications. However, UAV navigation in GPS-denied environments continues to be a challenging problem that has been tackled in recent research through sensor-based approaches. This paper presents a novel offline, portable, real-time in-door UAV localization technique that relies on macro-feature detection and matching. The proposed system leverages the support of machine learning, traditional computer vision techniques, and pre-existing knowledge of the environment. The main contribution of this work is the real-time creation of a macro-feature description vector from the UAV captured images which are simultaneously matched with an offline pre-existing vector from a Computer-Aided Design (CAD) model. This results in a quick UAV localization within the CAD model. The effectiveness and accuracy of the proposed system were evaluated through simulations and experimental prototype implementation. Final results reveal the algorithm's low computational burden as well as its ease of deployment in GPS-denied environments.

Keywords: autonomous localization; 3D registration; UAV; GPS-denied environment; real-time

---

1. Introduction

Due to the proliferation of Unmanned Aerial Vehicles (UAV) applications in the past decade, vision-based localization of UAVs has been an important and active field of research for several years. This is primarily due to the increase in applications where Global Positioning Systems (GPS) and Global Navigation Satellite System (GNSS) are infeasible [1,2]. Furthermore, auxiliary functions such as object recognition can be integrated into the vision system reducing overall complexity. In indoor applications, GPS signal can be weak, inaccurate or unavailable, thus preventing autonomous UAVs from localizing within buildings which gives rise to multiple complications that hamper navigation efforts [2,3].

Non-vision-based localization in GPS-denied environments either rely on wireless sensor network [4,5], Time of Flight, and Received Signal Strength Indicator (RSS) [6]. Such solutions require special building-wide setup consisting of sensors and/or transmitters to be constructed adequate functionality [5]. Furthermore, the increased cost as the setup footprint increases as well as synchronization between different nodes yields a major complexity in this type of setups [6].

As a resolution to such complications, vision-based localization approaches in GPS-denied environments have been widely discussed in literature. Detection and mapping to the corresponding location in the environment plays a key role in vision-based localization. Most of the approaches discussed in literature rely upon Simultaneous Localization and Mapping (SLAM) systems which create a 3D map of an unknown environment [7]. This helps in determining the relative location of the UAV at any instant within a GPS-denied environment. However, SLAM based systems create a heavy computational burden which yields a bottleneck for a real-time processing scenario [3].

To overcome such a challenge, captured image feature extraction and comparison with a preset model has been discussed in [8–10]. In [8], a localization method is proposed though extraction of lines from the environment using an omnidirectional camera. These lines are then matched with 3D line segments in a CAD model of the environment using a robust matching algorithm. Although this approach brings in numerous innovative features, it does require the robot to be moving only in Special Euclidean group (SE(2)). However, moving in SE(3) would exponentially increase the search space for the initial localization of the robot. Not to mention its need for a relatively bulky omnidirectional camera which could be problematic in some cases. This method has been utilized in [9] as well using the onboard RGB camera to determine the pose of the camera with respect to the environment. This in turn provides complementary drift update to the localization and mapping SLAM derived pose estimation. However, this has been implemented solely for camera pose estimation and never for localization of the UAV. A different approach in [10,11] proposes a method of unknown indoor environment exploration that utilizes the semi-dense nature of the map obtained from Line Segment Detector-SLAM (LSD-SLAM) to generate an octo-map. LSD-SLAM utilizes all the available image information such as lines and edges which leads to a robust and high accuracy denser 3D reconstruction of the environment. This is then used to locate spaces that a UAV can navigate while populating the unknown regions of the space around it. Although such approaches can be effective, the computational burden of these methodologies creates a bottleneck for future added features to the system.

This paper proposes a technique that starts by extracting macro-features [12] from stereo images of the environment (captured during flight) using a Convolutional Neural Network (CNN) [13]. These macro-features are then registered to a SLAM map that is generated during flight. Finally, the registered information is then used to find correspondences between the SLAM map and a 3D CAD model of the environment. This would enable the UAV to quickly and effectively create a transformation between the SLAM map (in-which it is localized by default) and the CAD model, thus localizing the UAV within the CAD model of the environment.

The main contribution of this paper is the construction of a novel feature vector made up of spatial relations between macro-features. This feature vector is to be leveraged to efficiently search for and correlate similarities in macro-features between the SLAM map and the 3D CAD model of the environment. With such a methodology, localization can be achieved in a computationally effective manner as compared to previously discussed approaches. Furthermore, the performance of the proposed system is evaluated via experimental implementation on a UAV that autonomously localizes itself within a building given a 3D CAD model of the building. After localization, the system is further tested by having the UAV autonomously navigate to a set goal position within the CAD model to evaluate the overall accuracy of the proposed localization method.

This paper is divided into five sections. Section 2 explains the utilized methodology from pre-processing to macro-features detection, extraction and matching. The hardware implementation used to test and evaluate the proposed method is discussed in Section 3. Section 4 reveals the results of the experiment and analyzed in terms of detection accuracy, feature vector effectiveness as well as localization accuracy. Finally, conclusions are presented in Section 5.

2. Methodology

The proposed localization technique will be utilized in an indoor environment where the macro-features of the environment are doors and windows. The preliminary function is to detect such features using a CNN discussed in Section 2.2. For each detected macro-feature, a unique binary description vector is created by encoding the distances and angles between the neighboring macro-features. These description vectors are compared (in Section 2.3) to the preprocessed vectors (in

Section 2.1) in a 3D CAD model. Once multiple matches are found, the UAV can localize itself within the CAD model and therefore would be able to follow a calculated trajectory to the goal position.

2.1. Pre-Processing

Prior to the UAV flight, a Building Information Model (BIM) [14] is loaded using Open Asset Import Library (Assimp) [15] which provides a unified method of accessing the data within. A BIM is a 3D CAD model typically produced during the planning phase of a construction project and includes information about macro-features as well as their locations as shown in Figure 1.

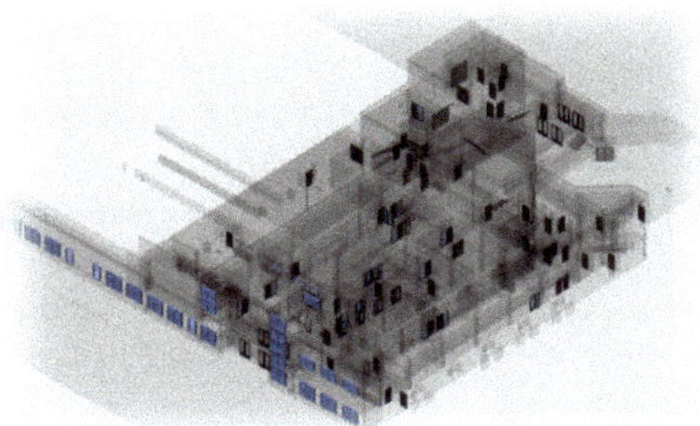

Figure 1. The Building Information Model: The required macro-features (doors and windows) are extracted and highlighted. Doors are depicted in black and windows in blue. All the other components are represented in translucent gray.

Macro-features are fixed features in the indoor environment such as doors, windows, and vent panels in which the proximity of each feature to the other can be used to identify a particular location on a 3D map. The macro-features from the CAD model are extracted and their centroids' locations are stored in a K-Dimensional Tree (k-d Tree) [16,17] of dimension 3. This facilitates retrieval of information about the nearest neighboring macro-feature of any given reference macro-feature which is required while generating the description vector within the CAD model.

Using the k-d Tree, vectors are computed for each macro-feature within the CAD model using an algorithm explained in Section 2.3. These are later used to detect correspondences between the detected macro-features during the UAV flight and those present within the CAD model.

2.2. Macro-Feature Detection and Extraction Method

The objective here is to device a method to detect doors and windows (macro-features) reliably and efficiently from observed data while simultaneously registering them to the generated SLAM map. A CNN YOLOv2 [13] that has been previously trained to detect macro-features would ensure a high level of accuracy while maintaining a quick computational speed. Since windows are harder to detect using traditional image processing techniques due to reflections and the presence of objects behind them. YOLOv2 performs real-time detection by applying a single neural network to the entire image. This network divides the image into regions and predicts bounding boxes and probabilities of predictions for each region.

2.2.1. Data Preparation and Training

The training data is collected using short video sequences containing doors and windows from the real datasets. The data is collected from a variety of different sources including the actual indoor flight environment. Care is taken to collect images in different lighting conditions as well as from different distances and angles to ensure training effectiveness. The images are annotated using LabelImg [18] tool. Doors and windows in each image were marked using rectangular boxes and the coordinates were stored in normalized coordinates by expressing them as a fraction of the length and width of the image. Slight perturbation in angle were introduced to the images during augmentation. This was done by applying a rotation to the input images given by,

$$R_z(\theta) = \begin{bmatrix} \cos\theta & -\sin\theta & 0 \\ \sin\theta & \cos\theta & 0 \\ 0 & 0 & 1 \end{bmatrix}, \tag{1}$$

where R_z is a rotation about an axis perpendicular to xy plane and θ is the angle of rotation. This angle is kept to a maximum of ±20 degrees in 5-degree increments to simulate the maximum possible roll. Furthermore, inverse warping is applied to get rid of holes in the warped images. These holes were artifacts of forward warping where pixels in the warped images are not painted in. This is due to the unavailability of one-to-one mapping from the source images to the destination images.

A total of 1540 images are collected and a total of 12,320 images are generated through augmentation. A total of 80% of the dataset is randomly selected as the training set and the remaining 20% are to validate the detection accuracy. This is done to compute the prevision-recall curve of the detection which provides information to set the cut-off threshold of the detection confidence. Detections are classified into two classes, one for the doors and the other for windows. In addition, to remove detections that looked like neither, a third class is added. This class is not used in the evaluation but leads to an increase in the neural network accuracy.

2.2.2. Detection Refinement

The YOLOv2 provides rectangular boxes bounding the proposed detection with the image space as shown in Figure 2a. Each of these bounding boxes corresponds to a single detection of either a door or a window. These detections are then translated into the actual macro-features in 3D space. To do this, lines are first detected in the area within each of the bounding boxes using Line Segment Detector (LSD) [19,20]. In practice, even though the algorithm does perform quite well line detection, the output is prone to generate broken lines even after rigorous parameter tuning as in Figure 2b. Further refinement is applied to such an issue by joining vertical side lines within a threshold distance from bounded box provided they meet certain criteria. The LSD outputs lines as point pairs denoting the endpoints of the line segments. These endpoints are used to compute the slopes of line segments. Line segments are joined when such line segments have a slope within a threshold angle as well as close endpoints are within a threshold distance. The threshold for the angle is set to 2 degrees and the threshold distance between close endpoints is set to 10 pixels. The longest lines closest to the two vertical sides of the bounds are then taken to denote the vertical sides of the macro-feature as illustrated in Figure 2c. This is similarly illustrated in Figure 2d,e on one and multiple doors, respectively.

2.2.3. 3D Reconstruction

Using stereo image pairs captured during flight, the 2D lines detected and denoted as the vertical edges of each macro-feature are projected into 3D space. The image pairs are used to compute the depth from disparity at each point in the image. This computation is highly parallelizable and is done on board the UAV. The sampled points are shown in Figure 3a along the 2D vertical side lines detected in Figure 2c and the depth of each point are projected into the SLAM map in Figure 3b. This is computed by:

$$\begin{bmatrix} P_{3x} \\ P_{3y} \\ P_{3z} \end{bmatrix} = d \begin{bmatrix} (P_{2x} - c_x)/f_x \\ \left(P_{2y} - c_y\right)/f_y \\ 1 \end{bmatrix}, \tag{2}$$

where P_{3x}, P_{3y}, and P_{3z} are the coordinates of the point in 3D with respect to the current pose of the camera in the world coordinate system. While P_{2x} and P_{2y} are the coordinates of the point in the 2D image coordinate system with a computed depth of d. Camera coordinates and focal lengths are represented by c_x, c_y and f_x, f_y, respectively.

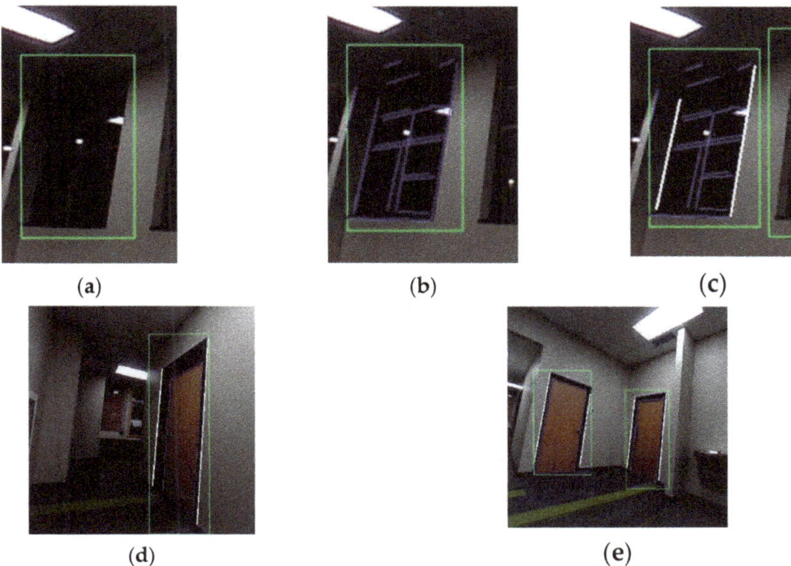

Figure 2. Macro-feature detection and refinement process from (**a**) YOLOv2 bounding box, (**b**) lines detection using Line Segment Detector (LSD) algorithm, (**c**) vertical sides of the object detected and marked, (**d**) vertical sides of single door detected and marked and (**e**) multiple doors detected and marked.

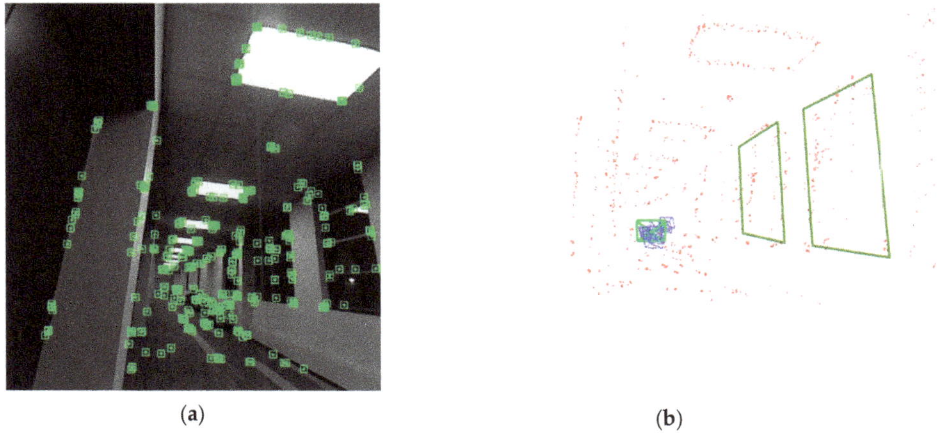

Figure 3. Reconstruction and registration of macro-features in 3D from (**a**) tracked features to (**b**) the generated Simultaneous Localization and Mapping (SLAM) map showing the 3D reconstructed windows and previously registered keyframes in blue quads.

The sampled points are used to compute the line in 3D using a least-squared line fitting algorithm relying on orthogonal distance. Outliers are then removed by removing points that are at a threshold distance from the computed line. The line is once again recomputed from the inliers using the orthogonal regression used previously. This two-step computation of the line provides a better estimate and is resistant to outliers in practice. The endpoints of the line are computed by finding the projection of the two endpoints of the line segment in 3D onto the fitted line. The four points that make up these line segments now represent the macro-feature in 3D in the view coordinates. Each of these macro-features detected are assigned to the keyframe in which they are first detected during flight.

Keyframes are frames in a graph-based SLAM that are tracked across multiple frames. They consist of the camera pose at an instant along with map points in view of the keyframe and contain the connection to other keyframes in the SLAM map that share closely correlated map points. Note that assigning the detected macro-features to the keyframes ensures spatial consistency in the event of a bundle adjustment step occurring before the localization step is complete.

A homogenous point in 3D is defined by:

$$\mathbf{P} = \begin{bmatrix} P_x \\ P_y \\ P_z \\ 1 \end{bmatrix}, \tag{3}$$

where P_x, P_y, and P_z are the three scalar coordinates. A transformation, $T \in SE(3)$ is represented by:

$$T = \begin{bmatrix} R & t \\ 0 & 1 \end{bmatrix}, \tag{4}$$

where $R \in SO(3)$ is a rotation matrix and t is a translation.

At any point in time, the locations of the macro-feature points in the world coordinates can be computed by,

$$\mathbf{P}_w = T_w^c \mathbf{P}_c, \tag{5}$$

where $\mathbf{P_w}$ are the homogeneous coordinates of the point in the world coordinate system, $\mathbf{P_c}$ are the homogeneous coordinates of the point in the view coordinate system, and T_w^c is the transformation between the camera and the world coordinate systems.

For the course registration phase that takes place after this step, each of the macro-features computed above are further reduced to their respective centroids in the world coordinate system. The description vector for these centroids are computed according to the following subsection.

2.3. Macro-Feature Registeration and Matching

In the proposed system, the UAV is already localized within the SLAM coordinate system. This means that the successful alignment of the CAD model with the SLAM map would result in localization of the UAV within the CAD model which is the main goal. This can be achieved by successfully matching the macro-features between both the CAD model and the SLAM map.

A SLAM algorithm that utilizes stereo images ensures that the SLAM map generated has an accurate estimation of scale. Since the two sets of data have the same scale, a rigid transformation is sufficient to align both sets of data. To compute this transformation, point-to-point correspondences is required between both datasets. Once the correspondences are found, a Random Sample Consensus (RANSAC) [21] based approach uses subsets of these correspondences to determine the best set of corresponding macro-features to utilize and the required transformation is computed.

To speed up the process of finding corresponding macro-features, a novel method is being devised that encodes each macro-feature into a binary macro-feature description vector that is orientation and position invariant. This is the main contribution of the paper and is explained in detail in the following subsections. It is to be noted that the use of a binary description vector ensures fast macro-feature matching due to the utilization of Hamming distance as the distance measurement unit. This is because the computation of a Hamming distances can be done in a single XOR operation between the two binary strings/vectors. The orientation and position invariance help in removing the dependence of a vector on the coordinate system.

To obtain matches between the macro-features in both datasets, the information within the vector needs to be decoupled from both coordinate systems and should mainly be dependent on the macro-features themselves. This is done by encoding the distances and angles between the closest 5 macro-features into the description vector making the data within the vector dependent on only the macro-features and the co-relations between them. Since the SLAM coordinate system has close to accurate scale, the relative distances and orientations between the closest macro-features stay the same in both coordinate systems. Figure 4 illustrates when the macro-features are reduced to their respective centroids, the relative distances and orientations between the macro-features stay the same.

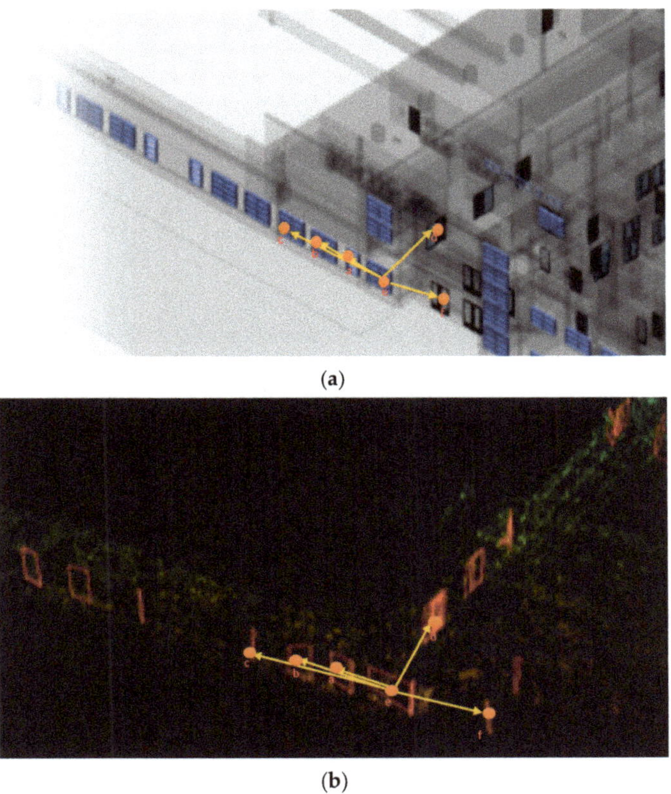

Figure 4. Orientation invariance of the macro-feature description vector. The images show the relative distances and orientation of a few selected macro-features, with 'e' being the microfeature of which the description vector is being calculated. Image (**a**) shows the macro-features in the 3D CAD model while image (**b**) shows the same macro-features in the SLAM map.

Figure 4a shows portion of the CAD model with a macro-feature whose vector is being calculated while Figure 4b shows the SLAM map generated with the detected macro-features and corresponding labels in both images. The description vector is computed by first retrieving the 5 macro-features closest to a macro-feature for which the vector is being computed. This is done efficiently by utilizing the k-d Tree to store the centroids of the detected macro-features. Next, using the macro-feature closest to the macro-feature for which the computation is occurring as a base, the angles to each of the 4 other macro-features are computed using the formula,

$$\alpha = \arccos\left(\frac{\mathbf{a}.\mathbf{b}}{||\mathbf{a}||.||\mathbf{b}||}\right). \tag{6}$$

The vector is clarified in Figure 5, where **a** is the vector between the current macro-feature and the closest (base) macro-feature, **b** is the vector between the current macro-feature and the macro-feature whose angle is to be calculated, and α is the angle between the two vectors.

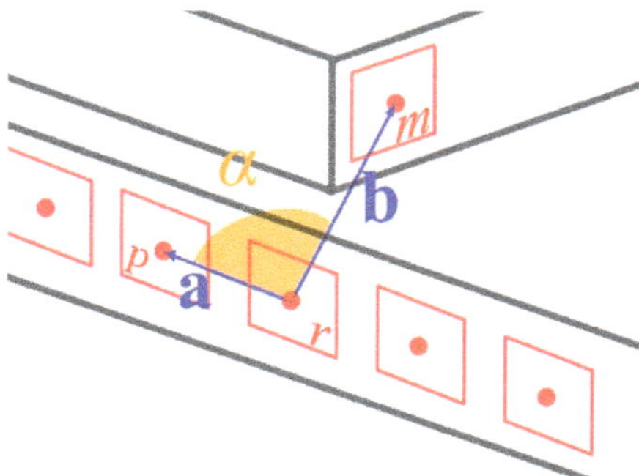

Figure 5. Angle calculation—In this figure, r is the macro-feature for which the description vector is being calculated, p is the closest macro-feature to r (taken as a base for the calculation), m is the macro-feature for which the angle relative to r is being calculated, and α is the angle between vectors **a** and **b** that connects r, p, and m.

The encoding of the relative orientations and distances to each of the macro-features is achieved through a lookup table. The lookup table ensures that the values of distances and angles that are close to local means are assigned the same value. This binning process enables the encoding of the distances and the angles into a compact form that can then be used to form the binary descriptive vector. This process is outlined in Algorithm 1. In addition to the distance and orientation information, the first bit of vector (*feature_type*) is used to denote the type of the macro-feature i.e., door (0) or window (1). This is further explained in the next subsection.

Algorithm 1: Computation of Feature Descriptor

Result: Descriptor vector, d

Extract closest 5 features from k-d Tree with distances;

$v \leftarrow all\_5\_points$

$x \leftarrow current\_pt$

$vec_0 \leftarrow v_0 - x$

forall i *in* [1, 2, 3, 4] **do**

  $vec_i \leftarrow v_i - x$

  $dist_i \leftarrow LUT_{dist}(|vec_i, vec_0|)$

  $\angle i \leftarrow LUT_{angle}(vec_i, vec_0)$

end

Descriptor vector, $d \leftarrow [feature\_type, [dist_1], [\angle 1], [dist_2], [\angle 2], [dist_3], [\angle 3], [dist_4], [\angle 4]$

2.3.1. Lookup Table and Vector Structure

The lookup table formed in Algorithm 1 is computed through a binning process. Two lookup tables are created, one for angles and the other for distances. These tables are used to group together values that are close to local means. The boundaries of these bins are computed using a Kernel Density Estimation (KDE) process [22].

First, all the possible values are computed during the pre-processing stage using the CAD model. These values are then sorted and grouped into clusters which is achieved by sorting all the values and then estimating the shape of a probability density function f (using a KDE) that provides a representation of the data. The Kernel Density Estimator, $\hat{f}_h$ is given by:

$$\hat{f}_h(x) = \frac{1}{n}\sum_{i=1}^{n} K_h(x - x_i) = \frac{1}{nh}\sum_{i=1}^{n}\left(\frac{x - x_i}{h}\right), \tag{7}$$

where K is the kernel (in this case a Gaussian kernel), K_h is the scaled kernel, h is a smoothing parameter and $x_1, x_2, \ldots, x_n$ are univariate, independent, and identically distributed samples drawn from the distribution function. The bins required for the lookup table are created by estimating and analyzing the shape of the function then dividing the distribution along its local minima as illustrated in Figure 6.

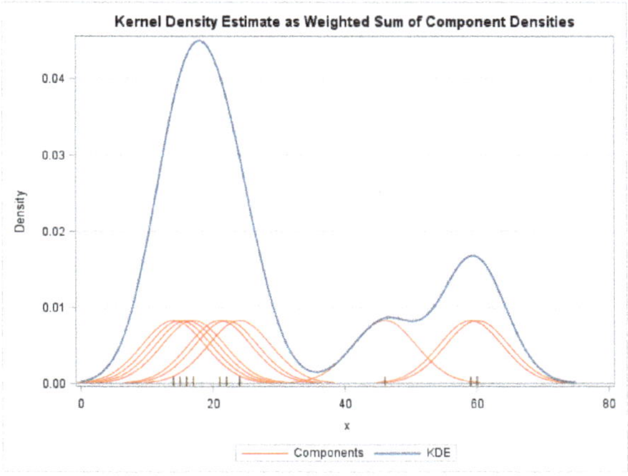

Figure 6. The Kernel Density Estimation used to create bins for the lookup table.

By considering the underlying density of this function to be Gaussian and $\hat{\sigma}$ as the standard deviation of values, the value of h can be estimated using Silverman's rule of thumb as follows:

$$h = \left(\frac{4\hat{\sigma}^5}{3n}\right)^{\frac{1}{5}} \approx 1.06\hat{\sigma}n^{-1/5}, \tag{8}$$

Each of the bins are then assigned a number between 0 to the total number of bins, the maximum being 255 to fit an 8-bit width in the binary representation of the number. This way, each neighboring macro-feature gives rise to two 8-bit values, one for the distance and one for the angle. As mentioned earlier that 4 neighboring macro-features are considered to define a specific macro-feature, the total size of the descriptive vector is thus 64 bits. The first bit in the 64-bit vector is usually zero as the number of bits used to represent the number of bins in the angles is less than 8. This enables the use of the first bit to store the type of the macro-feature while using the remaining 7 bits to store the first angle. The final structure of the macro-feature description vector is shown in Figure 7. In this study, the macro-feature type bit takes a 0 or 1 value depending on the feature being a door or window. If the macro-feature types match, the vector score is evaluated using the rest of the bits, if not, the maximum value for the distance is returned i.e., 64.

1bit	7bits	8bits	8bits	8bits	8bits	8bits	8bits	8bits
Feature Type	angle1	dist1	angle2	dist2	angle3	dist3	angle4	dist4

Figure 7. The 64-bit orientation invariant description vector with the first bit used to define the macro-feature type.

Since this is a binary vector, matching of the vector is done by finding the Hamming distance between two vectors. The vectors are matched on the basis of the lowest Hamming distance. Hamming distance will compute the number of positions of two equal strings where the corresponding values are different. This is computed very easily and efficiently using the XOR operator and then summing the total number of set bits in the result. Computers with newer hardware usually have support for counting the total number of set bits in CPU instructions. This results in a computational efficient matching mechanism. The complexity for finding matches between the observed macro-features and the CAD macro-features is thus, $O(mn)$, with m being the number of observed macro-features and n being the number of CAD macro-features.

2.3.2. Initial Registration

During the initial registration phase, macro-features are extracted using the methods outlined in Section 2.2. For each extracted macro-feature, the centroid of the macro-feature is stored in the k-d Tree and the vector is calculated when the macro-feature has more than a threshold number of neighbors. These vectors and the vectors computed from the CAD model during the pre-processing stage are then matches using the methods outlined in Section 2.3.1.

The matching macro-features are then used in a RANSAC which works by selecting the minimum number of parameters needed to estimate a model hypothesis. RANSAC is a robust iterative method that is used to estimate the parameters of a mathematical model from noisy data that has outliers. The generated model is then used to compute the total number of inliers for the model hypothesis. This process is repeated for a total of N times to generate N hypotheses and the hypothesis with the greatest number of inliers is selected as the transformation. The number N is a function of the desired probability of success (p). If ω is the probability of selecting an inlier for each selected point, and m is the number of points required to estimate the model, then $1-\omega^m$ will be the probability of finding at least one outlier among the m points. Meanwhile, $(1-\omega^m)^N$ is thus the probability of never selecting m points that are all inliers. Thus, the total number of hypothesis that need to be generated for a desired probability of success is:

$$N = \log(1-p) / \log(1-\omega^m). \tag{9}$$

Therefore, N different hypotheses are generated to estimate the model. In this study, the model is a rigid transformation between two 3D point sets. Since this requires at least four point-to-point correspondences, the value of m in this case is four, which are randomly selected from the matched macro-features. The rigid transformation is then computed using a linear least squares method based on Singular Value Decomposition (SVD).

Coplanarity or collinearity of the points give rise to degenerate conditions and must be avoided. Since collinear points are also coplanar it is sufficient to perform the check for coplanarity. The test for coplanarity can be done by computing the scalar triple product, which is given by:

$$\left|(x_3 - x_1).[(x_2 - x_1) \times (x_4 - x_1)]\right| < \varepsilon, \tag{10}$$

where $x_1,\ x_2,\ x_3,\ x_4\ \in \mathbb{R}^3$ are the four distinct points and $\varepsilon = 0.01$ is the margin of error. Any set of points giving rise to a value less than ε are considered coplanar. If the selected points pass the coplanarity test, they are rejected, and a new set of points are selected for the hypothesis. The hypothesis that generates the greatest number of inliers is then selected as the transformation.

The rigid transformation in SE(3) between the two point sets is computed using the following methodology. If $A := \{\mathbf{a}_i | i = 1, 2, \ldots, n, \mathbf{a}_i \in \mathbb{R}^3\}$ and $B := \{\mathbf{b}_i | i = 1, 2, \ldots, n, \mathbf{b}_i \in \mathbb{R}^3\}$ are two corresponding sets of points, A being the source and B being the destination set, the rigid transformation to be computed is found by minimizing the squared error of the transformed coordinates, which is given by:

$$(R, \mathbf{t}) = \underset{R \epsilon SO(3), \mathbf{t} \in \mathbb{R}^3}{\operatorname{argmin}} \sum_{i=1}^{n} \omega_i ||(R\mathbf{a}_i + \mathbf{t}) - \mathbf{b}_i||^2, \tag{11}$$

where R is the rotation in *SO(3)* and $\mathbf{t}$ in the translation while $\omega_i > 0$ are the weights assigned to the squared differences for each point pair. The weighted centroids on both sets are computed by:

$$\overline{\mathbf{a}} = \frac{\sum_{i=1}^{n} \omega_i \mathbf{a}_i}{\sum_{i=1}^{n} \omega_i}, \ \overline{\mathbf{b}} = \frac{\sum_{i=1}^{n} \omega_i \mathbf{b}_i}{\sum_{i=1}^{n} \omega_i}, \tag{12}$$

where $\overline{\mathbf{a}}$ and $\overline{\mathbf{b}}$ are the centroids of the two sets. Vectors are then computed using the point sets and the respective centroids by,

$$x_i := \mathbf{a}_i - \overline{\mathbf{a}}, \ y_i := \mathbf{b}_i - \overline{\mathbf{b}}, \ i = 1, 2, \ldots, n \tag{13}$$

where x_i and y_i are the corresponding vectors originating at the respective centroids. The 3×3 covariance matrix is then computed using:

$$S = XWY^T \tag{14}$$

where X and Y are the vectors of dimension 3 and $W = \operatorname{diag}(\omega_1, \omega_2, \ldots, \omega_n)$, the weight matrix. The singular value decomposition of

$$S = U\sigma V^T \tag{15}$$

is computed where, U and V are orthogonal unitary matrices, and σ is a diagonal matrix with non-negative real numbers along the diagonal which are the singular values of S. The rotation matrix R is then given by

$$R = VU^T \tag{16}$$

The translation can then be calculated as

$$\mathbf{t} = \overline{\mathbf{b}} - R\overline{\mathbf{a}} \tag{17}$$

After computing the transformation matrix using the least-squares based RANSAC method outlined above, the transformation is refined by using all the points from all the features. This means that all the four points that make up the corners of the macro-features (doors/windows) in the 3D space are used to compute the least-squares rigid transformation. This gives rise to a better estimation of the transformation.

3. Hardware Implementation

The hardware used to implement and test the proposed system comprised of Parrot's SLAM dunk module mounted on top of a Parrot Bebop 2 drone and a ground control station. The SLAM dunk utilizes NVIDIA's Jetson TK1 board along with integrated stereo camera, an Inertial Measurement Unit (IMU), a front facing ultrasonic sensor, a barometer, and a magnetometer as shown in Figure 8.

Figure 8. The testing hardware setup—A SLAM dunk mounted on a Parrot Bebop 2 drone.

The SLAM dunk module controls the drone by connecting to an access point hosted by the Bebop 2 via USB. On the other hand, the connection between the SLAM dunk and the ground control station is established via Wi-Fi using the Bebop's access point as an interface. This means that all the communication between the SLAM dunk and the ground control station is to be routed via the Parrot's server. This is done to make use of the powerful Wi-Fi hardware present on-board the Bebop 2. The complete network configuration of the setup is illustrated in Figure 9.

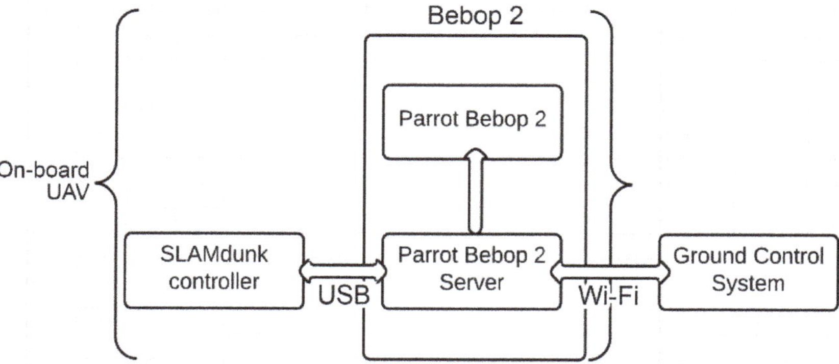

Figure 9. Communication setup between Ground Control, SLAM dunk, and Parrot Bebop 2.

The UAV is set to have two main modes of operation: exploration mode and localized mode. In the exploration mode, the UAV's operation is based completely on the SLAM map that is being generated during flight while the localizing sequence continuously attempts to localize the UAV with the CAD model. A flow chart of this mode of operation including the pre-processing steps discussed in Section 2.1 can be seen in Figure 10. Once localized, the UAV switches to localized mode where it is made to follow a set of calculated waypoints (based on its localized position) to a goal location.

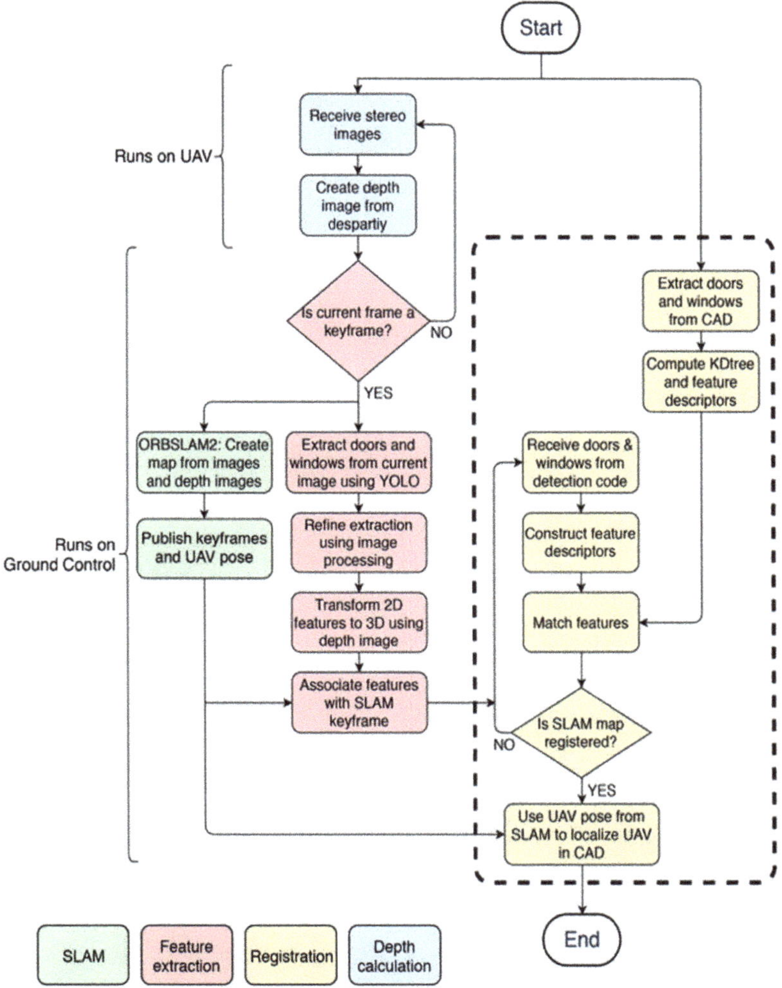

Figure 10. Pipeline process flow.

3.1. Preprocessing

As explained in Section 2.1 for the macro-feature vector matching to occur the BIM of the building needs to be pre-processed. This is done experimentally by loading the BIM into Assimp to be able to access the information within the model. Since Assimp is written in C++ it is easy to integrate into the process pipeline.

Since the BIM format already contains all the macro-features of the environment Figure 1, the next step was to extract the position of the centroids of the macro-features and store them in a k-d Tree which facilitated the creation of the macro-feature description vector for each of the macro-features as discussed explained previously using Algorithm 1. These vectors are later used to find the correspondences between the SLAM and the CAD maps.

3.2. Exploration Mode

In exploration mode, the left and right images from the stereo cameras were used to compute depth from disparity using GPU optimized Semi Global Block Matching (SGBM) [23] on-board the

SLAM dunk. The left image and the computed depth images were then sent via Wi-Fi to the ground control station at a rate of 30 fps in addition to IMU updates that were sent over at ~150 Hz. It is to be noted that all communication between the UAV and the ground control including image transfer are handled by Robot Operating System (ROS) [24] communication packets.

On the ground control unit, the system was divided into two main components: the SLAM module and the localization module. The SLAM module estimates the current UAV pose in a map generated using the images and IMU data as illustrated in Figure 11a. ORB-SLAM2 [25,26] was used as the SLAM algorithm and was used in the RGBD mode, which utilized the left and computed depth images from the UAV to construct a map consisting of a sparse set of 3D map points while also estimating the six-degree-of-freedom (6DOF) current UAV pose in it.

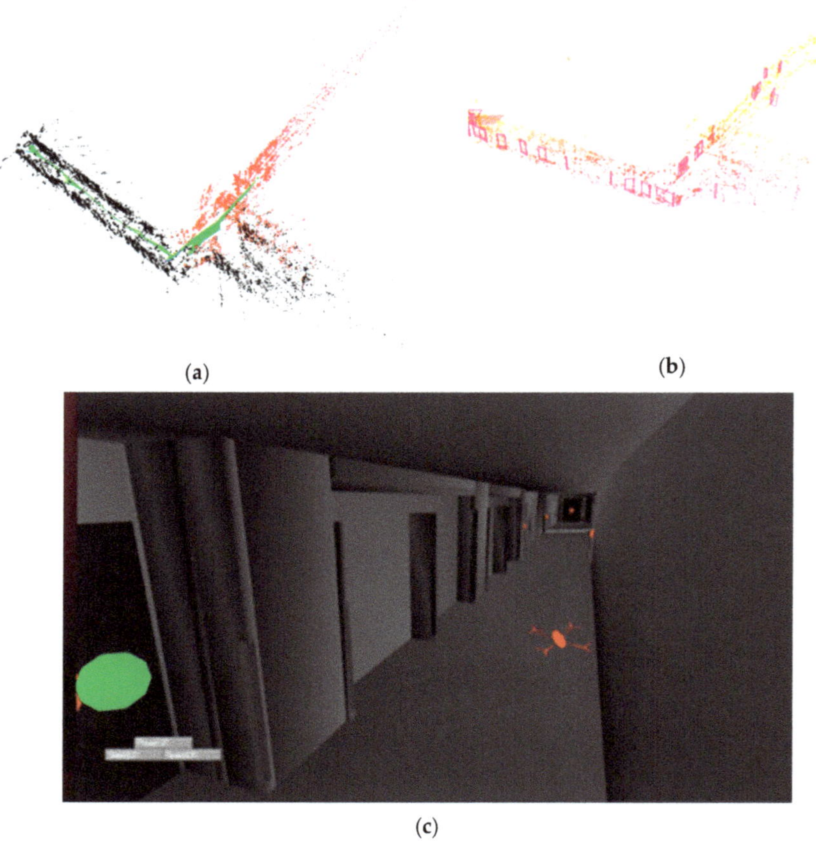

Figure 11. Exploration mode (**a**) SLAM generated map, (**b**) detected and reconstructed doors in the slam map, and (**c**) localized Unmanned Aerial Vehicles (UAV).

The localizer extracted the macro-features discussed earlier (Figure 11b) from the images and tried to localize the UAV in the CAD model by first forming correspondences between the environment's detected macro-features and those extracted from the CAD model. These are then used to estimate a rigid transformation in SE(3) between the two coordinate frames (SLAM and CAD model) using a robust transformation estimation algorithm (Section 2.3.2). The resulting localized UAV is illustrated in Figure 11c.

3.3. Localized Mode

Once the UAV was localized using the localizer module, the UAV began the process of traveling to the destination. The height of the UAV from the ground level was retrieved using an ultrasound sensor located beneath the Bebop 2 drone. A cross-section of the current floor, roughly at the UAV's flight height was generated from an octomap [27,28] representation of the building. The octomap provided a probabilistic 3D voxel representation of the environment/building that was updated in real-time with objects that did not appear in the original CAD model. The goal position and the current location of the UAV were then projected onto the 2D occupancy grid that is generated from the cross section generated earlier. An A* search algorithm [29] shown in Figure 12, to plan a path from the current location to the goal position was then run on this grid map. This generated a path of shortest route between the current location and the goal location. Dividing the floor into a 2D grid reduces the search space and consequently provided an efficient method to compute the shortest distance. Using the path generated, waypoints were sent to the UAV representing the centroids of the cells in the 2D grid which were connected by the computed path. Since the UAV did not undergo any fast motions, it was sufficient to supply the waypoints as a set of location coordinates with respect to the CAD coordinate system.

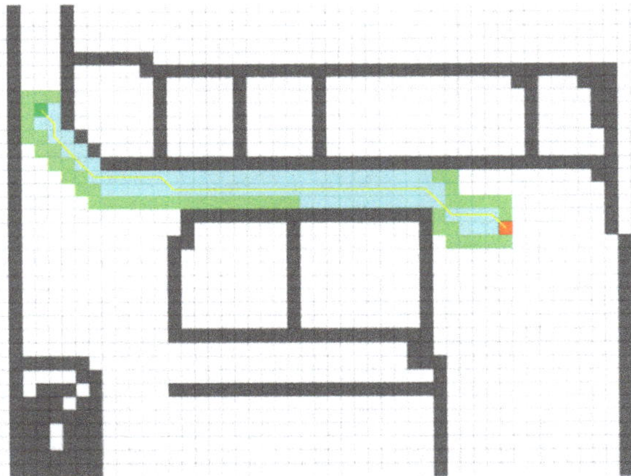

Figure 12. A* algorithm for path finding. A cross section of the floor is used to generate the 2D occupancy grid. The dark green cell is the start location and the red cell is the goal location. The gray cells are occupied, and the white cells are free. A path is computed by applying the A* algorithm and waypoints are generated using the centroids of the grid cells.

4. Experimental Results and Analysis

The following sections evaluates the performance and accuracy of the different algorithms used. Section 4.1 evaluates the performance of the feature descriptor extraction and matching. Sections 4.2 and 4.3 evaluate the performance of the entire system operation through localization as well as re-localization within the CAD model.

4.1. Object Vector Results and Accuracy

A basic method of selecting features based on distances to neighboring features was initially explored. In this method, lookup table of distances from each feature to every other feature was first computed from the CAD model in the preprocessing phase. Two features were then selected at random from the observed dataset and matched with distances computed from the CAD model. On finding a successful match, a new feature was then selected such that the distances to the two selected features was consistent with the distances in the CAD model. This process was repeated until a total of four correspondences were found. This method was found to be resource intensive and slow.

The histogram of the distances to the closest five features from each feature whose descriptor is to be calculated is shown in Figure 13a. The probability density function of the Kernel Density Estimator is shown in Figure 13b. The shape of the KDE closely resembles the histogram and thus represents the underlying data. Dividing the probability density function along the local minima yields 24 different bins for the lookup table.

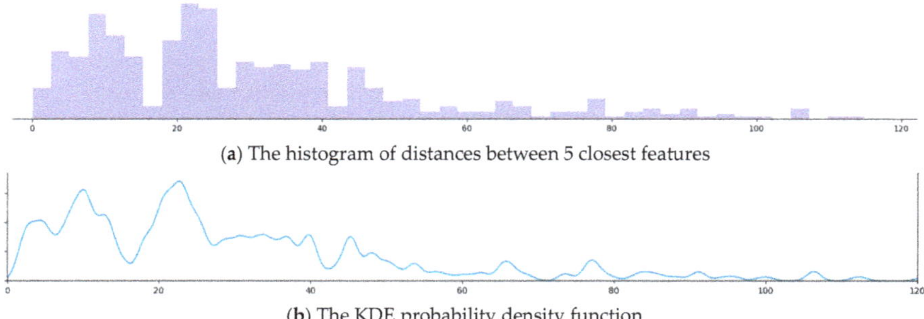

(**a**) The histogram of distances between 5 closest features

(**b**) The KDE probability density function

Figure 13. The histogram and Kernel Density Estimator (KDE) probability density function for distance. The horizontal axis represents the Hamming distance starting from 0. In (**a**), the vertical axis represents the number of features at a particular distance. In (**b**), the vertical axis represents the probability density of a particular distance.

The accuracy of the matches formed based on the descriptor created from this lookup table has been evaluated across 5 separate runs of the algorithm in different settings and has been shown in Table 1. Each of the matches are formed by finding the feature descriptor with the least Hamming distance from the features in the CAD model.

Table 1. Feature Descriptor Accuracy.

Run Id	Features Detected	Correct Matches	Incorrect Matched	Accuracy
Run 1	21	15	6	71.4%
Run 2	25	19	6	76%
Run 3	7	4	3	57.1%
Run 4	31	24	7	77.4%
Run 5	19	14	5	73.6%

It is seen that the matching accuracy increases with the number of features detected. This is expected since increasing the number of features results in a more complete picture of neighboring features which are used to build the feature vector. Lower number of features detected leads to greater chance of ambiguity in the feature vectors due to multiple features having similar distances. Run 3 is an example of such a scenario.

The extraction of the 5 nearest neighbors from the k-d Tree has been found to take an average of approximately 0.052ms on the test system consisting of a 6th generation core-i7 processor. The matching of descriptors using the brute-force matching technique has a complexity of $O(mn)$ and takes approximately 0.3ms for 25 descriptors in the observed dataset and 106 descriptors in the CAD dataset.

4.2. Localization within the CAD Model

The descriptors computed from both the CAD model and the observed data are used to match the two sets of features and calculate the SE(3) rigid transformation between the SLAM map and the CAD model. Figure 14 shows 5 runs of different lengths within the test setup. AprilTags [30,31] were placed at the goal positions for each run. AprilTags are fiducial markers that are not prone to the ambiguities inherent in other markers like Checkerboards or Circlegrids. To measure the accuracy of the localization, the distance between the final location of the UAV and the Apriltag located at the goal position was measured.

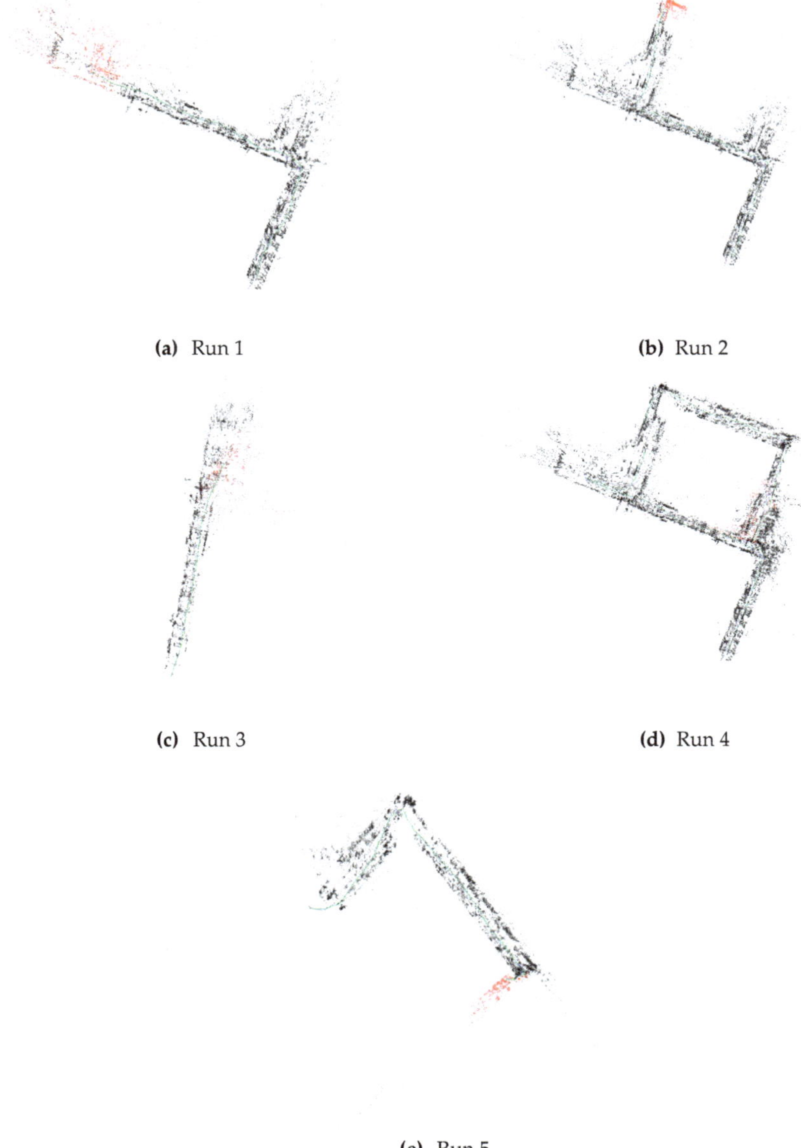

(a) Run 1 **(b)** Run 2

(c) Run 3 **(d)** Run 4

(e) Run 5

Figure 14. Runs of different lengths in the test setup. The top-down view of the generated SLAM map is shown here for each of the 5 runs of the system. The run in (d) has been set with multiple goal locations to generate a larger map for use in re-localization testing.

The localization time was calculated based on how long it took the UAV to successfully localize itself in the CAD model from the time the system was started. Table 2 shows the localization time and the error between the system goal position and the actual goal position for each of the runs depicted in Figure 14.

Table 2. Localization accuracy.

Run Id	Localization Time (s)	Error (m)	Error (% of Trajectory Length)
Run 1	16	0.25	4%
Run 2	17	0.14	3%
Run 3	13	0.23	3%
Run 4	15	0.15	1%
Run 5	13	0.26	2%

The UAV was able to successfully localize itself within the CAD model approximately 15 seconds of starting on average. Run 4 had been performed with multiple goal locations to generate the complete map of the floor. This map is then used in a run to test the re-localization mode of the UAV.

4.3. Re-Localization within the CAD Model

To test the re-localization capability of the system, a SLAM map was generated from a previous run and its transformation with respect to the CAD model was stored. These were then used when running the UAV a second time. The system was able to quickly localize the UAV within a second of the system initialization. This is possible because of ORB-SLAM2 uses a fast and efficient re-localization module. The initial SLAM map and the view from the perspective of the camera is shown in Figure 15. The real-time experimental results can be viewed in Video S1.

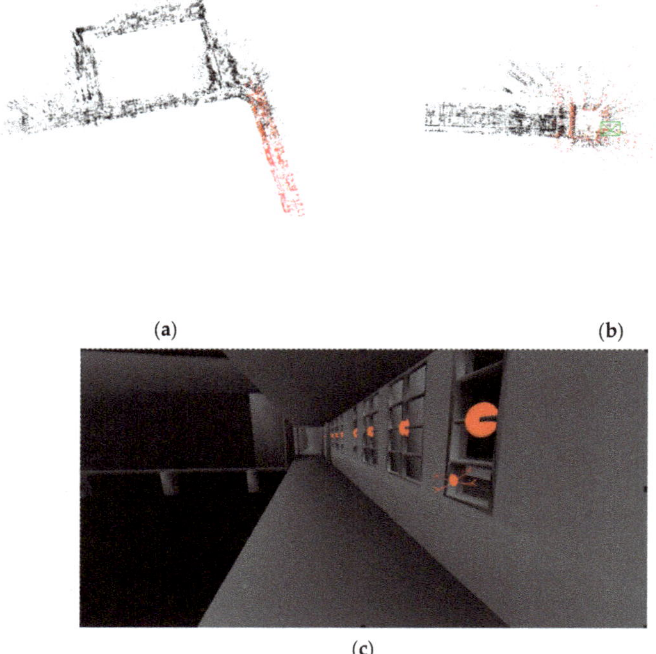

Figure 15. Re-localization of the UAV in previously constructed map. The previously constructed map is shown in (**a**). The UAV is shown localized in this map in (**b**) while (**c**) shows its representation in 3D CAD model.

5. Conclusions

The proposed system utilized a pre-existing scale accurate SLAM system using stereo images to estimate depth, and a pre-existing real-time Convolutional Neural Network for object detection coupled with a novel and computationally efficient method of registration of the SLAM map with the CAD model.

This paper proposed a novel pipeline combining a real-time CNN based object detection network, a SLAM system and a novel registration mechanism combined with image processing techniques to localize a UAV in a CAD model. Furthermore, the paper proposed a computationally efficient, orientation invariant feature descriptor to match features in the CAD model and the observed data, based solely on the spatial correlation between the features.

A major contribution of this work has been the fast and efficient computation of orientation invariant feature descriptors that were used to form correspondences between the observed features and the features in the CAD model. The use of a k-d Tree enabled the quick extraction neighboring features. The use of KDE to group clusters of discrete values together to reduce the time needed to match features had played a crucial part in improving the efficiency and performance of the algorithm.

The experimental validation shows that leveraging readily available 3D CAD models of buildings provided valuable information regarding the environment of the UAV. It provided the system with all the information needed to deduce the current location of the UAV within the CAD model, and enabled the UAV to find its pose without the use of unreliable GPS in indoor environments or other costly systems that emulate the GPS mechanism indoors. The use of a SLAM system registered to a CAD model also enabled the system to leverage the re-localization feature of the SLAM, thus enabling fast localization in previously visited locations.

Future Work

The system suffers from a few limitations that could be improved. The accuracy of the feature descriptor could be improved by applying a threshold on the computed Hamming distance. This could be calculated by saving all the computed descriptors and then studying the result of using different thresholds varying from 0 to 64 for the descriptor matches. All these matches could then be used to generate a Receiver Operating Characteristics (ROC) curve, which could be used to deduce the best possible value for the threshold distance.

The system performance could also be significantly improved by including other macro-features like drinking fountains, exit signs, posts, and other macro-features. Including more macro-features would enable the UAV to localize quickly due to the abundance of macro-features as well as the variations in the macro-feature types which would give rise to less ambiguity in the macro-feature descriptors. This would also alleviate problems arising from symmetries. Furthermore, since the system was developed for a single floor localization problem, added macro-features unique to each floor can further help the localization problem. Similar floors can be a challenging problem with the presented algorithm which is also in consideration in our future work in progress.

Supplementary Materials: The following are available online at https://youtu.be/OR3acLQsoBY, Video S1: Autonomous Localization of UAV in a CAD Model.

Author Contributions: Conceptualization, A.H. and J.N.; methodology, A.H.; software, A.H.; validation, A.H., writing—original draft preparation, A.H.; writing—review and editing, T.E., A.E., and M.A.E.; supervision, J.N. All authors have read and agreed to the published version of the manuscript.

Funding: This research was funded by State of North Dakota under Research ND grant.

Conflicts of Interest: The Authors declare no conflict of interest.

References

1. Lee, J.Y.; Chung, A.Y.; Shim, H.; Joe, C.; Park, S.; Kim, H. UAV Flight and Landing Guidance System for Emergency Situations. *Sensors* **2019**, *19*, 4468. [CrossRef] [PubMed]

2. Nguyen, D.D.; Elouardi, A.; Florez, S.A.R.; Bouaziz, S. HOOFR SLAM System: An Embedded Vision SLAM Algorithm and Its Hardware-Software Mapping-Based Intelligent Vehicles Applications. *IEEE Trans. Intell. Transp. Syst.* **2019**, *20*, 4103–4118. [CrossRef]
3. Padhy, R.P.; Verma, S.; Ahmad, S.; Choudhury, S.K.; Sa, P.K. Deep Neural Network for Autonomous UAV Navigation in Indoor Corridor Environments. *Procedia Comput. Sci.* **2018**, *133*, 643–650. [CrossRef]
4. Liu, H.; Darabi, H.; Banerjee, P.; Liu, J. Survey of Wireless Indoor Positioning Techniques and Systems. *IEEE Trans. Syst. Man Cybern. Part C Appl. Rev.* **2017**, *37*, 1067–1080. [CrossRef]
5. Montanha, A.; Polidorio, A.M.; Dominguez-Mayo, F.J.; Escalona, M.J. 2D Triangulation of Signals Source by Pole-Polar Geometric Models. *Sensors* **2019**, *19*, 1020. [CrossRef]
6. Oliveria, L.; Di Franco, C.; Aburdan, T.E.; Almeida, L. Fusing Time-of-Flight and Received Signal Strength for adaptive radio-frequency ranging. In Proceedings of the 16th International Conference on Advanced Robotics (ICAR), Montevideo, Uruguay, 25–29 November 2013; pp. 1–6.
7. Fraga-Lamas, P.; Ramos, L.; Mondéjar-Guerra, V.; Fernández-Caramés, T.M. A Review on IoT Deep Learning UAV Systems for Autonomous Obstacle Detection and Collision Avoidance. *Remote Sens.* **2019**, *11*, 2144. [CrossRef]
8. Koch, O.; Teller, S. Wide-Area Egomotion Estimation from Known 3D Structure. In Proceedings of the IEEE Conference on Computer Vision and Pattern Recognition, Minneapolis, MN, USA, 17–22 June 2007; pp. 1–8.
9. Li-Chee-Ming, J.; Armenakis, C. UAV navigation system using line-based sensor pose estimation. *Geo Spat. Inf. Sci.* **2018**, *21*, 2–11. [CrossRef]
10. Engel, J.; Schöps, T.; Cremers, D. LSD-SLAM: Large-Scale Direct Monocular SLAM. In Proceedings of the European Conference on Computer Vision, Zurich, Switzerland, 6–12 September 2014; pp. 834–849.
11. Hasan, A.; Qadir, A.; Nordeng, I.; Neubert, J. Construction Inspection through Spatial Database. In Proceedings of the 34rd International Association for Automation and Robotics in Construction, Taipei, Taiwan, 28 June–1 July 2017; pp. 840–847.
12. Pătrăucean, V.; Armeni, I.; Nahangi, M.; Yeung, J.; Brilakis, I.; Haas, C. State of research in automatic as-built modelling. *Adv. Eng. Inf.* **2015**, *29*, 162–171. [CrossRef]
13. Zhang, H.; Wu, J.; Liu, Y.; Yu, J. VaryBlock: A Novel Approach for Object Detection in Remote Sensed Images. *Sensors* **2019**, *19*, 5284. [CrossRef]
14. Li, S.; Niu, Y.; Feng, C.; Liu, H.; Zhang, D.; Qin, H. An Onsite Calibration Method for MEMS-IMU in Building Mapping Fields. *Sensors* **2019**, *19*, 4150. [CrossRef]
15. Assimp: Open Asset Import Library. Available online: http://www.assimp.org (accessed on 11 December 2019).
16. Arya, S.; Mount, D.M.; Netanyahu, N.S.; Silverman, R.; Wu, A.Y. An optimal algorithm for approximate nearest neighbor searching fixed dimensions. *J. ACM* **1998**, *45*, 891–923. [CrossRef]
17. Anzola, J.; Pascual, J.; Tarazona, G.; González Crespo, R. A Clustering WSN Routing Protocol Based on k-d Tree Algorithm. *Sensors* **2018**, *18*, 2899. [CrossRef] [PubMed]
18. Tzutalin. LabelImg. Git Code. 2015. Available online: https://github.com/tzutalin/labelImg (accessed on 11 December 2019).
19. Von Gioi, R.G.; Jakubowicz, J.; Morel, J.M.; Randall, G. LSD: A fast line segment detector with a false detection control. *IEEE Trans. Pattern Anal. Mach. Intell.* **2010**, *32*, 722–732. [CrossRef] [PubMed]
20. Sun, Y.; Wang, Q.; Tansey, K.; Ullah, S.; Liu, F.; Zhao, H.; Yan, L. Multi-Constrained Optimization Method of Line Segment Extraction Based on Multi-Scale Image Space. *ISPRS Int. J. Geo Inf.* **2019**, *8*, 183. [CrossRef]
21. Derpanis, K.G. Overview of the RANSAC Algorithm. *Image Rochester NY* **2010**, *4*, 2–3.
22. Elgammal, A.; Harwood, D.; Davis, L. Non-parametric Model for Background Subtraction. In Proceedings of the 6th European Conference on Computer Vision-Part II, Dublin, Ireland, 26 June–1 July 2000; pp. 751–767.
23. Hirschmuller, H. Stereo Processing by Semiglobal Matching and Mutual Information. *IEEE Trans. Pattern Anal. Mach. Intell.* **2007**, *30*, 328–341. [CrossRef]
24. Quigley, M.; Conley, K.; Gerkey, B.; Faust, J.; Foote, T.; Leibs, J.; Wheeler, R.; Ng, A.Y. ROS: An open-source Robot Operating System. In Proceedings of the ICRA Workshop on Open Source Software, Kobe, Japan, 17 May 2009; p. 5.
25. Mur-Artal, R.; Tardós, J.D. Orb-slam2: An open-source slam system for monocular, stereo, and rgb-d cameras. *IEEE Trans. Robot.* **2017**, *33*, 1255–1262. [CrossRef]

26. Qin, J.; Li, M.; Liao, X.; Zhong, J. Accumulative Errors Optimization for Visual Odometry of ORB-SLAM2 Based on RGB-D Cameras. *ISPRS Int. J. Geo Inf.* **2019**, *8*, 581. [CrossRef]
27. Wurm, K.M.; Hornung, A.; Bennewitz, M.; Stachniss, C.; Burgard, W. OctoMap: A probabilistic, flexible, and compact 3D map representation for robotic systems. In Proceedings of the ICRA 2010 Workshop on Best Practice in 3D Perception and Modeling for Mobile Manipulation, Anchorage, AK, USA, 3–7 May 2010; pp. 403–412.
28. Faria, M.; Ferreira, A.S.; Pérez-Leon, H.; Maza, I.; Viguria, A. Autonomous 3D Exploration of Large Structures Using an UAV Equipped with a 2D LIDAR. *Sensors* **2019**, *19*, 4849. [CrossRef]
29. Zhang, H.; Li, M.; Yang, L. Safe Path Planning of Mobile Robot Based on Improved A* Algorithm in Complex Terrains. *Algorithms* **2018**, *11*, 44. [CrossRef]
30. Olson, E. AprilTag: A robust and flexible visual fiducial system. In Proceedings of the 2011 IEEE International Conference on Robotics and Automation, Shanghai, China, 9–13 May 2011; pp. 3400–3407.
31. Abbas, S.M.; Aslam, S.; Berns, K.; Muhammad, A. Analysis and Improvements in AprilTag Based State Estimation. *Sensors* **2019**, *19*, 5480. [CrossRef] [PubMed]

Article

Fully-Actuated Aerial Manipulator for Infrastructure Contact Inspection: Design, Modeling, Localization, and Control

Pedro J. Sanchez-Cuevas [1,*], Antonio Gonzalez-Morgado [1], Nicolas Cortes [1], Diego B. Gayango [1], Antonio E. Jimenez-Cano [2], Aníbal Ollero [1] and Guillermo Heredia [1]

1 Robotics, Vision and Control Group, Universidad de Sevilla, Camino de los Descubrimientos s.n., 41092 Sevilla, Spain; antgonmor5@alum.us.es (A.G.-M.); niccorfer@alum.us.es (N.C.); diebengay@alum.us.es (D.B.G.); aollero@us.es (A.O.); guiller@us.es (G.H.)

2 Laboratory for Analysis and Architecture of Systems, Centre National de la Recherche Scientifique, 7 Avenue du Colonel Roche, 31400 Toulouse, France; aejimenezc@laas.fr

* Correspondence: psanchez16@us.es

Received: 30 June 2020; Accepted: 17 August 2020; Published: 20 August 2020

Abstract: This paper presents the design, modeling and control of a fully actuated aerial robot for infrastructure contact inspection as well as its localization system. Health assessment of transport infrastructure involves measurements with sensors in contact with the bridge and tunnel surfaces and the installation of monitoring sensing devices at specific points. The design of the aerial robot presented in the paper includes a 3DoF lightweight arm with a sensorized passive joint which can measure the contact force to regulate the force applied with the sensor on the structure. The aerial platform has been designed with tilted propellers to be fully actuated, achieving independent attitude and position control. It also mounts a "docking gear" to establish full contact with the infrastructure during the inspection, minimizing the measurement errors derived from the motion of the aerial platform and allowing full contact with the surface regardless of its condition (smooth, rough, ...). The localization system of the aerial robot uses multi-sensor fusion of the measurements of a topographic laser sensor on the ground and a tracking camera and inertial sensors on-board the aerial robot, to be able to fly under the bridge deck or close to the bridge pillars where GNSS satellite signals are not available. The paper also presents the modeling and control of the aerial robot. Validation experiments of the localization system and the control system, and with the aerial robot inspecting a real bridge are also included.

Keywords: aerial systems; applications, inspection robotics, bridge inspection with UAS

1. Introduction

In the last few years, the interest in Unmanned Aerial Vehicles (UAVs) has been continuously growing [1]. Especially during the last decade, the robotic community has paid attention to modeling, design, and control of multirotors [2], and then, due to their mechanical simplicity and availability of actuators, sensors, and electronics, their use has been multiplying in many areas with a huge variety of applications. The first cases of the use of these aerial robots were focused on perceiving, monitoring, or filming the environment. However, in recent years, the appearance of robotic aerial manipulation has widened the application range of these UAVs. Aerial manipulators [3,4] are aerial robots with arms or mechanical devices which are able to physically interact with the environment, performing tasks as object manipulation [5,6], operating devices [7,8], surface repairing [9], and taking measurements with sensors in contact with surfaces [10].

One of these applications in which the use of UAVs is being increasingly sought is the maintenance and safe operation of the existing civil infrastructure like bridges, viaducts, and tunnels, whose progressive deterioration requires in many cases inspection, assessment, and repair work.

Infrastructure inspection is done in most cases with non-destructive testing (NDT) methods. Visual inspection is one of the main methods used for inspection of the concrete walls of bridges and tunnels, which allow for detecting and identifying defects and the overall state of the infrastructure. On the other hand, sonic and ultrasonic sensors are used to measure crack depth and width and characterize the condition of the infrastructure concrete elements. Ultrasonic methods need the sensor be directly in contact with the surface of the bridge or tunnel and applying a constant force during the measurements.

Over the last decade, UAVs have been used for this purpose, and now there are prototypes [11,12] and commercially available drones for an initial, non-contact inspection of difficult to access areas of the infrastructure, mostly using imagery, videos [13–16], or three-dimensional point clouds [17]. In all these cases, though, the inspection is carried out remotely, and when a damage of concern is being detected, an in-depth inspection has to follow with experienced inspectors in need of hands-on-access to the infrastructure section.

Infrastructure inspection and maintenance has been one of the priority robotics research areas of the Horizon 2020 framework program. In particular, the H2020 AEROBI project has addressed successfully the problem of performing concrete bridge contact inspection with aerial robots and aerial manipulators [18]. The ongoing RESIST project is extending their use also for tunnel inspection and steel bridges [19], and the recently started PILOTING project aims at further developing these technologies to help them reach the commercialization stage [20].

Nowadays, there are three main approaches for infrastructure contact inspection. One of them consists of using a multirotor which can stick firmly to the bridge beams and place a reflector prism in contact with the bridge surface to measure beam deformation using a robotic Total Station on ground, which can measure the position of the prism with millimeter accuracy [21]. The main advantage of this approach is that the measurements can achieve the required accuracy for structural assessment since the robot is completely still while the inspection measurements are collected. However, the robot is not able to move over the bridge surface once it is touching it, making it very difficult or even impossible to reach the location of a specific defect in a second inspection. The second approach uses aerial manipulators with a robotic arm placed on the upper part of the multirotor to reach the inspection area while flying [22]. This solution opens the spectrum of applications that can be accomplished because the end effector of the manipulator can be designed for a specific use case, such as sensor installation [23], hammering tasks [24], deflection measurements [25], and others. The third approach is about using a morphing platform or a more complex robot with the capability of moving along the infrastructure, maintaining the contact with the surface [26,27]. Nevertheless, although the last two approaches seem to be more versatile, the required positioning accuracy of the robot and the requirements in the measurement collection process are, again, very difficult to reach while flying and sensor installation usually requires a controlled force while the glue is curing to grant the adhesion of the device to the surface. These are challenging applications in which the robot must fight the environmental disturbances, like wind and aerodynamic effects among others, while the end effector of the robotic arm is applying a constant and controlled force.

On the other hand, the aerial robot in [25] and most aerial manipulators [3–10] use standard multirotor configurations with 4, 6, or 8 co-planar propellers. They are, therefore, under-actuated robots that need to tilt to move or apply forces in the horizontal plane. Several designs have been developed in the literature to overcome this limitation and achieve full range of motion in the 6 DOF, which are generally known as multi-directional thrust multirotors. A first group have the motors fixed to the frame with different non-parallel axes, and can achieve full range multi-directional motion without any other actuation [28–31]. Other designs are able to rotate their rotors about the axis of the individual multirotor arm [32–35] or with bi-axial tilting mechanisms [36,37], or rotate frame sections [38], although they have the penalty of the added weight of the tilting mechanism and actuation.

The multi-directional multirotor designs have also been combined with a robotic arm with few additional degrees of freedom to reach the desired position with the end-effector for pipe inspection in oil and gas plants [39,40]. Although they have not been used before for infrastructure inspection, fully actuated platforms have good characteristics for accurately positioning the arm end-effector and applying forces in confined spaces as tunnels and below the bridge deck.

Aerodynamic effects can arise also in these kinds of scenarios and applications, in which the aerial robot is required to operate in the surrounding of the infrastructure. Previous works have shown how the aerodynamic ceiling effect can change the performance of a multirotor when it approaches a surface from the underside [41,42] and usually requires using a specific controller [21,43]. However, in this specific work, the system will operate far enough from the infrastructure not to be affected by this aerodynamic disturbance.

This paper presents the design of an aerial manipulator (see Figure 1) based on a fully actuated hexarotor with tilted rotors and a special arm placed at the top for contact inspection of bridges and tunnels. It describes a new prototype of aerial manipulator including the aerial platform, the robotic arm, and a specific "docking gear" focused on bridge inspection. The aerial platform uses a standard multirotor frame adapted to be fully actuated. The robotic arm has compliant joints that allow it to adapt and absorb contact forces, and has sensors to measure joint rotation and deformation. A custom sensing device has been included inside the arm to minimize its weight and size. Finally, the device called "docking gear" acts as a landing gear mounted on the top of the aerial platform which helps to establish the contact between the bridge and the aerial robot and maintain a security distance avoiding also being affected by the aerodynamic ceiling effect. In this way, it significantly increases the positioning accuracy of the end effector of the arm and facilitates its force control because the robot is not moving while inspecting. Moreover, the docking gear also improves the safety conditions preventing the propellers from hitting the structure and, thanks to their material in the contact interface, it helps to precisely maintain the position during the operation. These two aspects are considered by the end-users of this technology as mandatory and essential for proper commercial exploitation in bridge maintenance programs. Since the use case of this robot imposes flying very close to the infrastructure inspected, a localization system to operate in GNSS denied environments is also presented using a robotic total station and a commercial camera which is running a VSLAM algorithm [44]. Such algorithm has been optimized by Intel and it runs in an internal processor of the camera, fusing data from an IMU and stereoscopic vision [45]. The modeling and control techniques used in the robot are also described and validated through outdoors experiments.

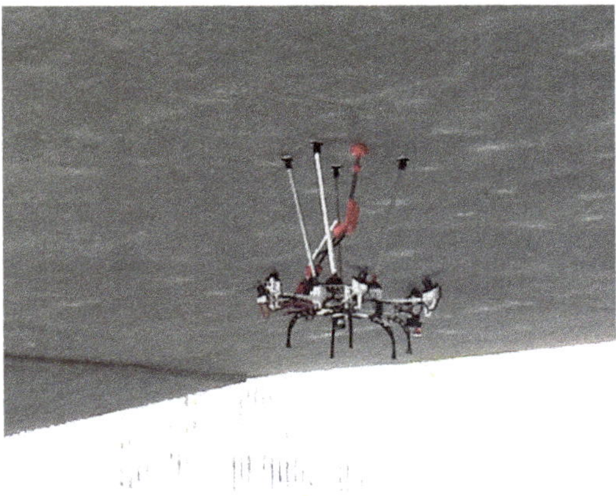

Figure 1. Fully-actuated aerial robot during an inspection by contact operation.

The rest of paper is organized as follows: Section 2 presents a description of the global system including details related to the hardware and software of the final solution and the decision criteria. Section 3 presents the localization system to operate in a GNSS denied environment and in a time consistent reference frame. Section 4 describes the dynamic model of both the multirotor and the robotic arm followed by the control techniques implemented in these subsystems. Section 5 shows the experiments used to validate the system and its capabilities and, finally, conclusions and future works are presented in Section 6.

2. System Description

The aerial part of the system proposed is composed of three main subsystems: the fully-actuated aerial platform, the robotic arm (both specifically designed for bridge inspection applications), and the localization system. On the other hand, the ground part consists of three subsystems: a ground control station running on a PC with MATLAB/Simulink, a robotic total station which is used as the inspection sensor, and it is in charge of providing the position estimation of the aerial platform in case of the flight is carried out in a GNSS-denied environment, a standard router to establish the link between the autopilot on-board and the ground control system (GCS) which is also the robotic total station server. All these components as well as the software/hardware architecture are described in detail in this section.

2.1. *Tilted Rotor Design*

As it was mentioned in the Introduction, standard hexarotors with co-planar propellers are under-actuated platforms in terms of the force that they can exert and the maneuvers that they can perform. However, both characteristics are essential in the field of aerial contact-inspection robotics and give a chance for the application of a fully-actuated platform. The simplest multirotor considered as a fully-actuated platform consists of a hexarotor with tilted propellers to exert a full vector of forces and torques including lateral forces, unlike a co-planar hexarotor which is only able to develop four control actions, the vertical force, and the three torques. In other words, a tilted hexarotor has not only more control capability, but also an enhanced performance followed by the ability to exert forces in the horizontal plane without tilting the platform, which has a priceless value in aerial manipulation applications.

The prototype presented in this work (see Figure 2a) has been built customizing a DJI F550, a commercial frame available in the market, and the selected motorization is the combination of the rotors DJI 2312E rotors with the DJI 9×4.5 in propellers and the XRotor 40A ESCs as propulsive system.

The main modifications to the commercial frame are: spacers to increase the space in the avionics bay to be able to integrate the autopilot board and supports to tilt the rotors and provide the ability to exert the above-mentioned horizontal forces. In order to facilitate the installation, these supports are formed by two separated pieces. The left side in Figure 2b shows an explodsion of both parts and the screws necessary for its union. The white part shown is the one in charge of making the inclination of the rotor. This piece is screwed with the frame using four screws and four nuts, which are placed in four side cavities of the piece. The black part is in charge of making the union between the motor and the white piece. To do this, first, the black piece is screwed to the motor, and finally the black piece to the white piece. Due to this frame being widely used in the research community, the authors have decided to share the designed parts to transform this platform in a multirotor with tilted propellers (Stl files in https://hdvirtual.us.es/discovirt/index.php/s/YPGbxt9Lz96MSQ3). This link also hosts the spacer part designed to increase the avionic bay of the DJI frame. All the modifications to the original design of the DJI frame have been mainly printed in ABS.

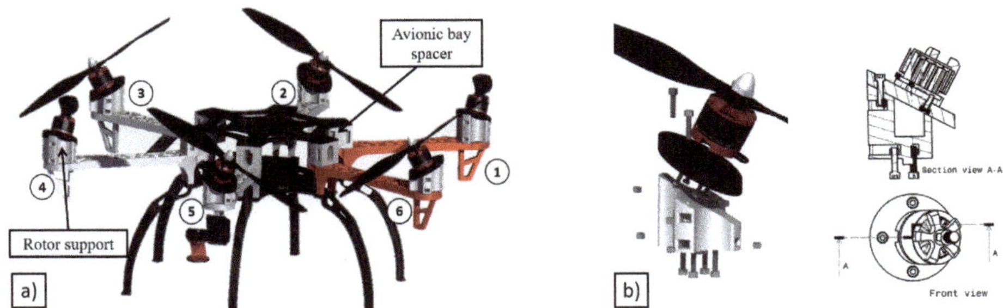

Figure 2. Fully-actuated platform. (**a**) 3D concept; (**b**) rotor support detail.

Following the process described in [46] and for the purpose of minimizing the control effort, the global tilt angle of the propellers has been decomposed in two rotations around a local reference frame linked to the rotor. The sequence of this rotation is shown in Figure 3 and consisting of a first rotation of an angle α_i around the X_i-axis, and a second one of β_i in the Y_i'-axis, where the index i represents the i-rotor being $i = 1$, the front left rotor and following a counter-clockwise direction for the others (see Figures 2a and 3). Moreover, in order to simplify the design, it has been imposed that the $\forall i \|\alpha_i\| = \alpha = \|\beta_i\| = \beta = 20°$, and their signs are such that $\alpha_i = -\alpha$ and $\beta_i = \beta$ when $i = 2n+1$ where $n = 0, 1, 2$ and $\alpha_i = \alpha$ and $\beta_i = -\beta$ when $i = 2n$ being $n = 1, 2, 3$, as it is summarized in Table 1. The value of 20° is justified due to this inclination does not compromise the lift of the multirotor and guarantees enough lateral forces for inspection by contact operations.

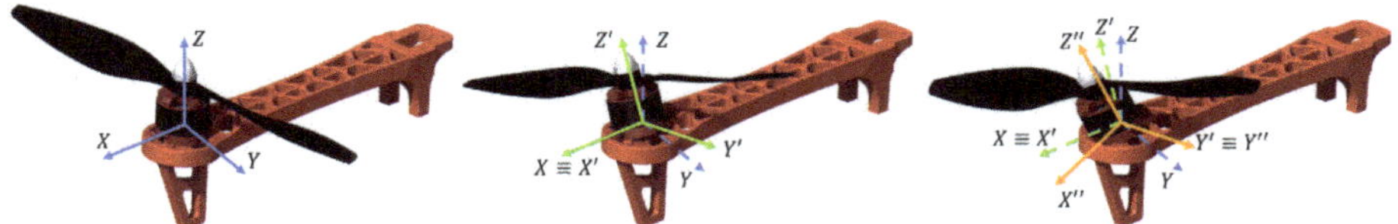

Figure 3. Sequence of the propeller rotation.

Table 1. Rotation angles.

i	1	2	3	4	5	6
α_i	−20°	20°	−20°	20°	−20°	20°
β_i	20°	−20°	20°	−20°	20°	−20°

In order to analyze the fully-actuation of the platform, we study the forces on the x-axis and y-axis that the platform is able to generate, without modifying its orientation, depending on the angles of inclination of the motors α and β. To obtain these work regions, the optimization problem (1) has been solved for different values of α and β, where F is the modulus of the force exerted in the direction $\vec{v}$, m is the mass of the platform, g is the acceleration of gravity, $M(\alpha, \beta)$ is the mixer matrix, presented in Section 4, and ω_{min} and ω_{max} are the maximum and minimum speeds of the motors:

$$\begin{aligned} \underset{\omega^2 \in \mathbb{R}^6}{\text{maximize}} \quad & F \\ \text{subject to} \quad & M(\alpha, \beta)\omega^2 = [F\vec{v} + mg\vec{z}; 0; 0; 0], \\ & \omega_{min}^2 < \omega^2 < \omega_{max}^2. \end{aligned} \tag{1}$$

Figure 4 shows the results of the optimization problem for the x-axis and y-axis. It can be seen how the forces can be generated whenever we are under the blue curve, for the y-axis, and the red curve for the x-axis. At the moment that we exceed these curves, we will not be able to generate these lateral forces maintaining a null inclination angle.

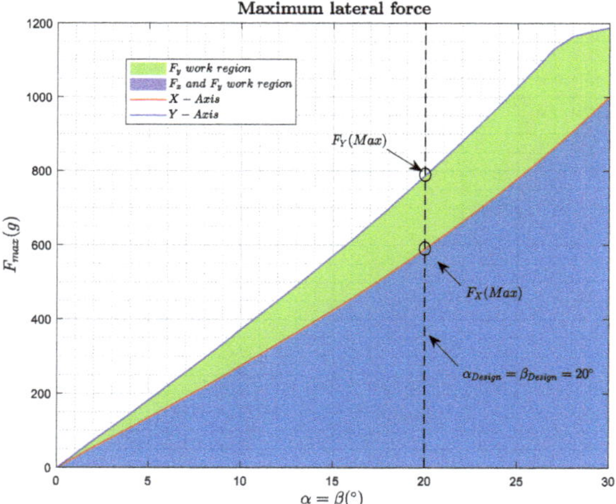

Figure 4. Maximum lateral force achievable by the platform.

In addition, to continue to see the limits of the platform, the maximum angle of inclination of the platform in hover has been studied. For this, the optimization problem (2) has been solved, where θ is the inclination angle around the $v - axis$, R_v is the rotation matrix around the v-axis, m is the mass of the platform, g is the acceleration of gravity, $M(\alpha, \beta)$ is the mixer matrix, and w_{min} and w_{max} are the maximum and minimum speeds of the motors:

$$\begin{aligned} \underset{\omega^2 \in \mathbb{R}^6}{\text{maximize}} \quad & \theta \\ \text{subject to} \quad & M(\alpha, \beta)\omega^2 = [R_v(\theta)mg\vec{z}; 0; 0; 0], \\ & \omega^2_{min} < \omega^2 < \omega^2_{max}. \end{aligned} \tag{2}$$

Figure 5 shows the maximum roll and pitch angles in a hovering maneuver. Below these curves is the working region, i.e., the region where our platform is still a full platform.

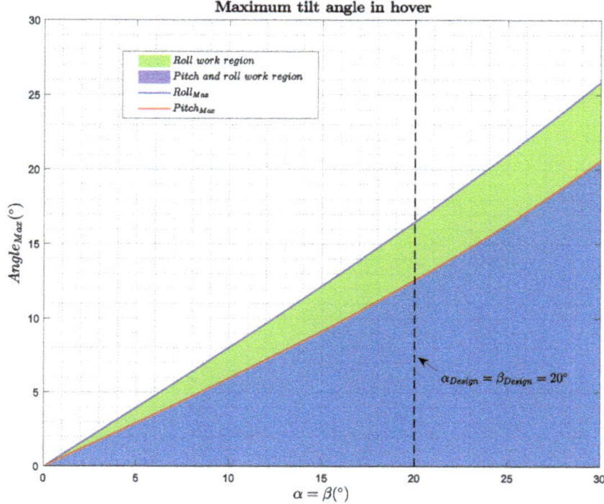

Figure 5. Maximum tilt angle in hover achievable by the platform.

The main features of the fully-actuated platform designed are shown in Table 2. Note that the weight presented in this table is the minimum take off weight including batteries and avionics and that the *Maximum achievable lateral force* are the maximum forces exerted by the platform limiting the thrust of the motors to the 90% of their maximum thrust.

Table 2. Main features of the fully-actuated platform.

Weight	2.00 kg
Payload	$\sim$ 600 g
Maximum achievable lateral forces	$F_x = 415$ g $F_y = 555$ g
Propellers	DJI 9 $\times$ 4.5 in
Battery	4S 5300 mAh
Rotation angles	$\alpha = 20°$ $\beta = 20°$

2.2. *Arm Design*

The aerial platform presented in this paper is equipped with a 3DoF lightweight robotic arm (see Figure 6a), which has been designed minimizing the inertia as much as possible. In order to achieve this objective, light materials like carbon fiber links, ABS, and aluminum have been used. The servo motors have been installed as close as possible to the base of the arm to minimize the changes in the mass distribution due to the movement of the arm links. Moreover, it is equipped with a force sensor to accomplish the inspection operations [47], such as installing a sensor or collecting data to monitor the bridge health. The sensor consists of a linear potentiometer and a compression spring installed in the third joint which is therefore a compliant joint (Figure 7). In this way, the arm is able to measure the contact force that it is available also to the autopilot which is an integrated controller that considers the dynamic behavior of both the aerial platform and the robotic arm.

Figure 6a shows the three degrees of freedom of the lightweight robotic arm, where γ_1 is the shoulder pitch angle, γ_2 is the elbow pitch angle, and γ_3 is the slider bar displacement.

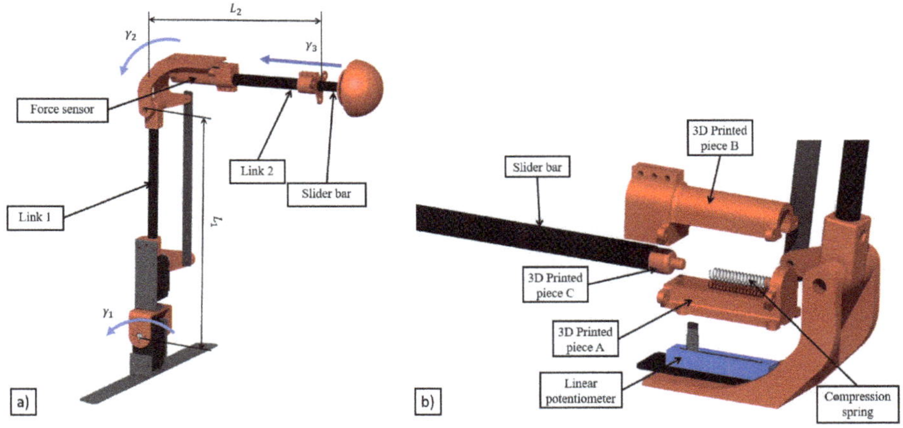

Figure 6. Lightweight robotic arm. (**a**) 3D concept; (**b**) force sensor detail.

The force sensor is composed of four components: a linear potentiometer, a compression spring, a carbon fiber bar, and 3D printed pieces (see Figure 6b). The linear potentiometer is fixed on the elbow of the arm by the plastic piece A. This part also holds the spring at one end. The plastic piece B encapsulates the whole sensor and holds the link B of the arm, inside which the carbon fiber bar slides. Finally, the carbon fiber bar has a plastic piece C at its end, which facilitates the contact with the spring.

The stiffness of the spring has been characterized experimentally. Different values of the compression forces have been applied to the spring through a variable weight, and the sensor value measured at the ADC port has been recorded. With all the data, a two-section piece-wise linear approximation has been done (Figure 7) and the parameters can be seen in Table 3. Finally, Table 4 shows the main features of the robotic arm.

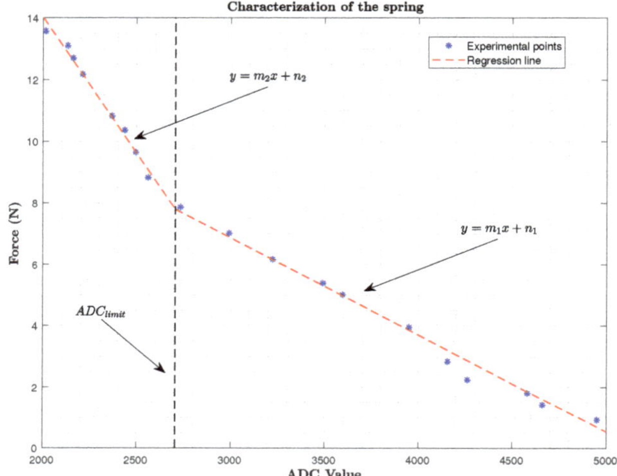

Figure 7. Spring experimental characterization.

Table 3. Parameters of the linear regression.

$m_1 = -0.0031661$	$n_1 = 16.368$
$m_2 = -0.0088239$	$n_2 = 31.674$
$ADC_{limit} = 2705$	

Table 4. Main features of the robotic arm.

Weight	0.370 kg
Dimensions	Link 1 = 25 cm Link 2 = 20 cm
Rotation range	Shoulder pitch = $[-100°, 100°]$ Elbow pitch = $[20°, 160°]$
Battery	2S 1300 mAh

2.3. Final Prototype Integration

The final integrated system (see Figure 8) with the aerial platform and the robotic arm has been completed with the installation of the aforementioned "docking gear", which acts as a landing gear, installed on the top of the aerial platform, made of aluminum and ABS parts. The main function of this device is to facilitate the inspection decoupling the controller of the aerial platform during the inspection phase from the inspection operation guaranteeing that the contact is safe and stable. The device has been lined with a rubbery material which prevents horizontal movements while inspecting. On the other hand, the robotic arm has been installed between two rotors in order to increase the arm workspace, using a part manufactured in aluminum. Although the robotic arm has not been mounted in the geometric center of the UAV, the batteries have been placed to guarantee that the center of the mass (CoM) of the entire system matches this point.

From the application point of view, it is important to remark that, although the system has already been tested inspecting bridge beams, it has also been designed with the capabilities of inspecting vertical structures such as pillars or walls. This is possible because the position of the robotic arm allows it to reach positions beyond the propeller's workspace.

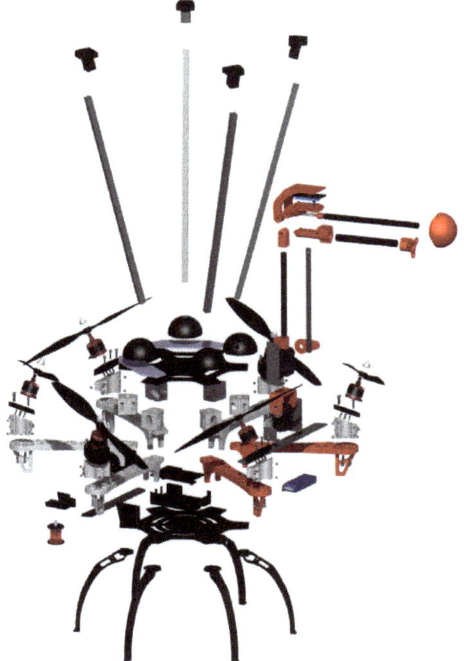

Figure 8. Exploded view of the system.

2.4. On-Board Avionics

The on-board avionics consists of a custom autopilot code running in a Raspberry Pi 3 Model B connected to the Emlid's sensors shield Navio2. The aerial platform and the Ground Control Station (GCS) are connected by Wi-Fi Technology. A range sensor model SF11/C has also been mounted on the platform. This prototype has been equipped with a multi-sensor localization system using two kinds of sensors, a laser sensor on the ground with a reflective target on-board, and a VSLAM sensor on-board. The first one consists of a Leica MS50 and a GRZ101 360° reflector prism on-board. This Leica MS50 is a robotic total station which was defined in [48] as "a robotized sensor used to measure distances with laser technology which can measure the distance to any surface or a reflector prism using a highly accurate laser system", in this case, we are using the aforementioned reflector prism to improve the measurements and facilitate the tracking. The second sensor on-board will be an Intel Realsense T265 tracking camera, Santa Clara, CA, USA, which is a stand-alone solution for simultaneous localization and mapping. The measurements of the total station are communicated by a Wi-Fi link with the autopilot and then those are integrated with the camera estimation using an extended Kalman filter running in the Raspberry Pi. After that, the output of this Kalman filter is used as feedback in the position control loop. In this way, it is possible to carry out autonomous flights in GNSS denied environments such as underneath bridges or tunnels. Section 3 explains in detail the fusion algorithm as well as the alignment method and data preparation stage, it also emphasizes the advantages of using this multi-sensor approach which mixes and absolute localization system (total station) with a local one (tracking camera). Finally, Figure 9 shows a scheme of the hardware system mounted in this prototype.

The aerial robot controller has been developed using our GRVC autopilot tool (https://github.com/AEnrique/GRVCAutopilot) [49]. This tool has been developed in the University of Seville and designed to be used inside a Raspberry Pi® 3B with a Navio®2 sensor shield. The autopilot allows the implementation of control strategies at low and high levels using MATLAB/Simulink, as well as the easy integration of new sensors and data monitoring.

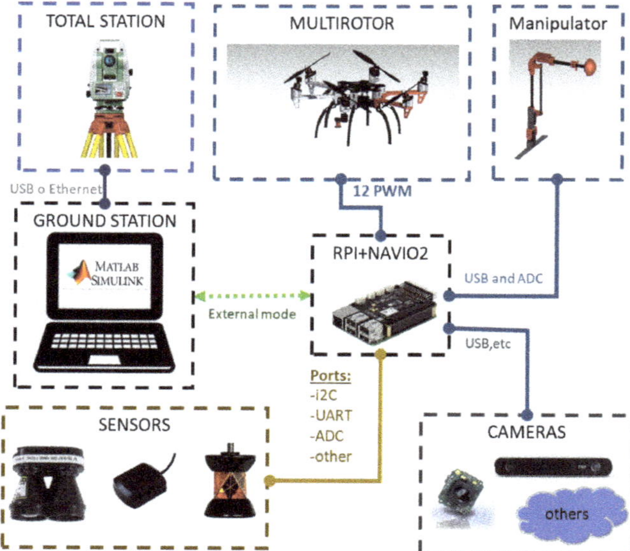

Figure 9. Hardware architecture.

3. Localization System

An aerial robot flying close to or under the bridge deck will have in general poor GNSS satellite visibility or no visibility at all. Thus, for this application, the aerial robot needs a reliable localization system that does not use the measurements of the GNSS receivers.

Moreover, although there are several solutions that aim to solve this localization problem in a GNSS-denied environment, those usually consist of algorithms that generate a local reference frame each time they are executed, like the classical VSLAM [44], LOAM [50], and other SLAM algorithms [51]. However, the use case of infrastructure inspection imposes other unavoidable requirements for the localization system. On the one hand, it is necessary to define a common time-consistent reference frame to be able to localize the defects once they are detected and reach their location in different flights and days to assess their evolution over time. This reference frame must be independent of the sensors on-board and, preferably, it should be linked to the physical infrastructure. In this way, the defects as well as the flight plan and others will be referenced in a common frame and will not depend on the initial position and orientation of the platform which may vary from one flight to another.

3.1. Multi-Sensor Localization for Bridge Inspection

The proposed solution consists of creating a reference frame using a Leica MS50 robotic total station to solve the time-consistent reference frame because, with the Leica software, it is possible to measure and store the position of non-deformable points of the terrain as a reference of this time-consistent reference frame. Then, in a second inspection which could be carried out deploying the Leica in a different location or even inspecting in a different day, those reference points could be measured again on a local axis, but, using the least squared method, and the total station triangulates its position and axes in the original time-consistent reference frame. However, despite in a previous work [48], this device has been used to directly feedback the position control loop in the autopilot, the total station has drawbacks when it is used as a stand-alone positioning sensor. First, the sample rate is variable from 0 up to 20 Hz. Second, it is necessary to keep the reflective target mounted on the platform in a direct line of sight with the total station. In addition, finally, since the total station must be deployed on the ground, it requires using a wireless link between the robot and the Leica server, and, if the communication fails, the robot will lose the positioning measurements and lead to a crash.

For those reasons, it was proposed to use a multi-sensor solution, which should be able to converge to the axis created by the total station with on-board sensors that do not depend on a wireless link and work in a more stable way. In this case, the measurements provided by the robotic total station will be combined with the ones provided by an Intel RealSense T265, a dual fish-eye lens camera that is running a VSLAM algorithm out of the box. In the same way, this solution could be applied to other SLAM or LOAM techniques. Those algorithms usually work creating their own local reference frame which depends on the initial position and orientation of the aerial platform, it being necessary to define a calibration protocol to align them to the common reference frame established by the robotic total station.

In this case, the calibration protocol consists of collecting data while moving the aerial platform manually or in a first security flight while the different algorithms are running in parallel. After that, it is necessary to calculate the deviation between the different frames and the physical frame, and, using a numerical method like the least square fitting algorithm, it is possible to obtain the transformation of each local inertial reference system to the common one.

Once we have the set of data, the first task to do is to make it uniform, that is to say, since the sensors have different sample rate, the length of the data vector will not be the same for each sensor, and it is necessary to be the same for the sensor fusion algorithm. We shall use linear interpolation presented in [52].

Now, we can introduce the alignment process. We define the following transformation equation, which transforms a point from the camera frame to the total station frame:

$$\hat{x} = Tx + P_0, \tag{3}$$

where x is the position given by the VSLAM algorithm of the camera, $\hat{x}$ the position given by the total station, T the rotation matrix from the camera frame to the total station frame, and P_0 the multirotor initial position. We have to use all the points prepared previously to conform an over-determined linear equation system and to obtain the nine components of T (t_{ij}) and the three components of P_0 (p_i). Finally, the prepared equation system results as follows:

$$\begin{pmatrix} \hat{x}_1 \\ \hat{x}_2 \\ \vdots \\ \hat{x}_n \end{pmatrix} = \left(\begin{array}{ccc|ccc|c|ccc} x_1 & & & x_2 & & & & x_n & & \\ & x_1 & & & x_2 & & \cdots & & x_n & \\ & & x_1 & & & x_2 & & & & x_n \\ \hline & I_3 & & & I_3 & & \cdots & & I_3 & \end{array} \right)^T u, \tag{4}$$

where I_3 is the three-dimensional identity matrix, $(\cdot)^T$ the transposition operator, and n the number of total points after interpolation. The blank spaces of the above matrix are supposed to be 0 and u is the unknowns vector

$$u = \begin{pmatrix} t_{11} & t_{12} & t_{13} & t_{21} & t_{22} & t_{23} & t_{31} & t_{32} & t_{33} & p_1 & p_2 & p_3 \end{pmatrix}^T. \tag{5}$$

This over-determined equation system can be solved using a least square fitting algorithm implemented in MATLAB or using any C++ library such as *Armadillo* (https://gitlab.com/conradsnicta/armadillo-code) to solve it on-board.

Figure 10 shows an example of how this method works adjusting the reference system of the RealSense T265 and the Leica MS50. The XY trajectory of a flying platform manually piloted has been represented. It has also been represented the trajectory seen by the Optitrack of the University of Seville, which is a high precision motion capture system that has been used as ground truth. In yellow, the raw trajectory calculated by the tracking camera is shown. It can be clearly observed that, by default, it suffers scale, alignment, and position errors. However, it is clearly shown that the aforementioned calibration protocol improves the results of the camera and automatically transforms their measurements for the absolute reference frame created by the total station. The results of the camera after performing the calibration are represented in purple.

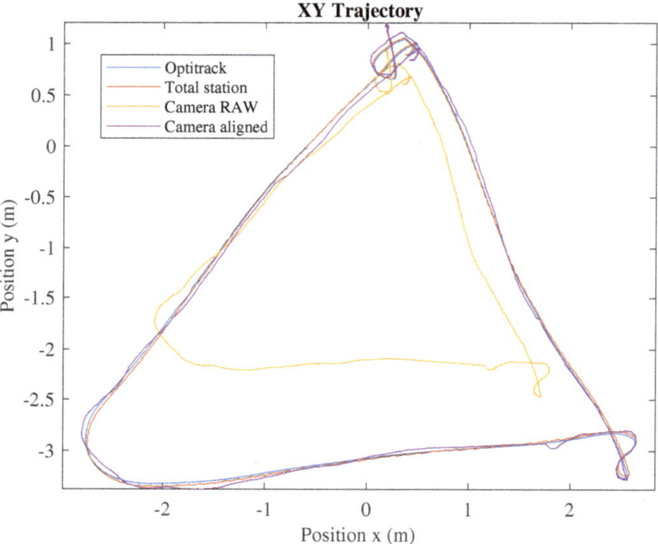

Figure 10. Camera alignment.

Now, it is possible to employ all the data to get an accurate estimate of the multirotor position using the Kalman filter [53,54]. We first define the following state vector at instant k, where the three first components are the vehicle position and the last three ones are its velocity

$$x_k = \begin{pmatrix} y_1 & y_2 & y_3 & \dot{y}_1 & \dot{y}_2 & \dot{y}_3 \end{pmatrix}_k^T . \tag{6}$$

The linear difference equation that governs the vehicle state is:

$$x_k = A x_{k-1} + w_{k-1}, \tag{7}$$

and a measurement is given by

$$z_k = H x_k + v_k. \tag{8}$$

Here, w and v are random variables representing the process and the sensors noise, respectively. They are assumed to be white and with normal probability distribution:

$$p(w) \sim N(0, Q), \tag{9}$$
$$p(v) \sim N(0, R). \tag{10}$$

Matrix A is built using kinematics equations, integrating velocity to obtain the position:

$$A = \left(\begin{array}{c|c} I_3 & T \cdot I_3 \\ \hline 0_{3\times3} & I_3 \end{array} \right), \tag{11}$$

where $0_{3\times3}$ is a 3 by 3 zero matrix and T the elapsed time from $k-1$ to k. H relates the state with the sensors measurements, since our sensors provide us with the position and velocity directly; this matrix is the identity matrix.

The first stage of Kalman filter is prediction. We will predict the state vector and its covariance matrix P

$$\hat{x}_k^- = A\hat{x}_{k-1}, \tag{12}$$
$$P_k^- = AP_{k-1}A^T + Q. \tag{13}$$

The second stage of Kalman filter is to correct the state and covariance matrix using sensors data

$$K_k = P_k^- H^T (H_k P_k^- H^T + R)^{-1}, \tag{14}$$
$$\hat{x}_k = \hat{x}_k^- + K_k(z_k - H\hat{x}_k^-), \tag{15}$$
$$P_k = (I - K_k H)P_k^-. \tag{16}$$

In order to tune the Kalman Filter, due to it being hard to estimate the measurements and process noise (R and Q), we employed a heuristic method, changing the parameters manually until the filter provided an acceptable response. For the sake of simplicity, the parameters were not supposed correlated, that is to say, the identity matrix multiplied by a certain gain. The resulting parameters are $Q = 2 \cdot 10^{-6} \times I_6$ (Process noise covariance), $R_{ts} = 5 \cdot 10^{-3} \times I_6$ (Total Station measurement covariance), and $R_c = 1 \cdot 10^{-2} \times I_6$ (Camera VSLAM measurement covariance).

We can summarize the whole process in the following steps:

1. Align sensors to the common reference frame:
   (a) Move the aerial platform manually or in a first safe flight and record the measurements from the sensors.
   (b) Interpolate the collected sensor data to have the same length.
   (c) Solve the over-determined equation system using a least square fitting algorithm to obtain T and P_0.
2. Apply the Kalman filter to obtain an accurate estimate of the platform position.

Figure 11 shows the architecture of the positioning system.

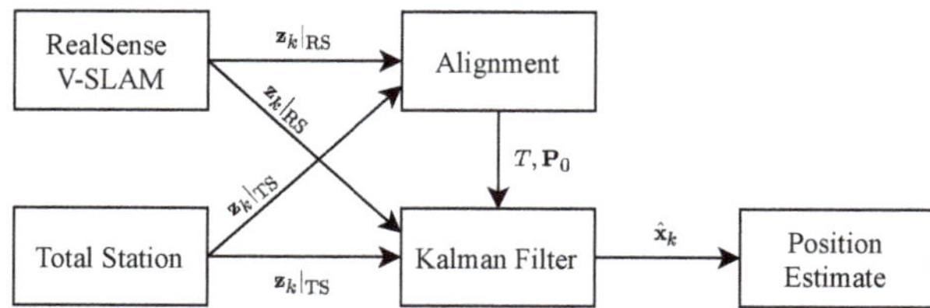

Figure 11. Positioning system architecture.

3.2. Experimental Results

Figure 12 shows the results of the localization system in an outdoor experiment. In this experiment, we could not use Optitrack as ground truth because it is not available in our outdoor experimental field; however, in Figure 10, we can see that Total Station has enough accuracy as ground truth and can be used as a valid reference system in outdoor applications. The UAV starts the flight at $(-1.5, -1)$. The pink line represents the outcome of the implemented Kalman filter algorithm, fusing the data from the aligned tracking camera and the robotic total station. In this test, a failure in the total station was simulated (red asterisk), even though it is shown that the algorithm continues estimating the position just using the tracking camera. When the total station is recovered at 73 s (blue asterisk), the accumulated error between the position estimated with the VSLAM algorithm and the real position in the Leica reference system is 20.5 cm, and, after that, the filter uses both the total station and tracking camera measurements again.

Finally, we conclude that, for getting the position of the aerial platform, we will use the method previously presented because it solves the sensor alignment problem, it provides a time-consistent reference frame, it is GNSS-free, and it fuses all the available sensor measurements, getting a precise localization algorithm that is robust against sensor failures. Additionally to this approach, the next efforts will be devoted to extend this methodology including a 3D LIDAR.

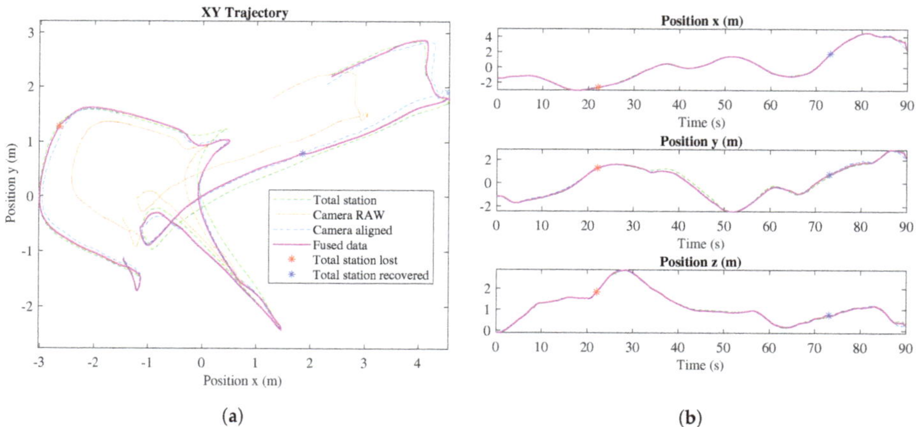

Figure 12. Fused trajectory using Kalman Filter (**a**) and separated coordinates x, y, and z (**b**).

4. Modeling and Control

4.1. Modeling

The dynamic model for a multirotor and a 3-link manipulator arm can be obtained using the Euler–Lagrange formulation. Defining the state vector as $\xi = [p\ \eta\ \gamma]^T$, where $p = [x\ y\ z]$ is the position of the multirotor center of gravity in Cartesian coordinates in the Earth fixed frame $\{E\}$, $\eta = [\phi\ \theta\ \psi]$ that represents the Euler angles of the aerial platform attitude, commonly known as roll, pitch, and yaw in the multirotor base or body frame $\{B\}$ that is attached to the IMU of the autopilot, $\gamma = [\gamma_1\ \gamma_2\ \gamma_3]$ is the vector which contains the joint angles of the robotic arm, γ_1 and γ_2 are the active links while γ_3 is the passive one which acts as a force sensor during the contact. These variables are illustrated in Figure 13.

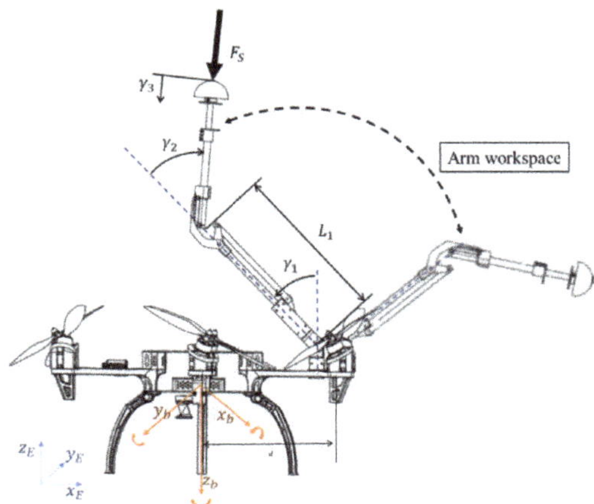

Figure 13. Joint angles definition.

Thus, the dynamic equations of the aerial manipulator can be obtained using the Euler–Lagrange equations:

$$\frac{d}{dt}\frac{\partial \mathcal{L}}{\partial \dot{\xi}_i} - \frac{\partial \mathcal{L}}{\partial \xi_i} = \mathcal{F} + \mathcal{F}_C \tag{17}$$

where $\mathcal{L}$ is the Lagrangian of the system defined as $\mathcal{L} = T - V$, where $T = 1/2\sum_{k=1}^{N} m_k v_k^2$ is the total kinetic energy of the system and V is the potential energy.

In the Euler–Lagrange equation, $\mathcal{F}$ is the vector of forces and torques generated by the actuators, $\mathcal{F}_C$ includes the contact forces that can be estimated with the sensor installed in the end effector and they are mainly a function of the passive joint, $\mathcal{F}_C \sim \mathcal{F}_C(\gamma_3)$. Thus, the dynamic model of the full system can be written in compact matrix form as:

$$M(\xi)\ddot{\xi} + C(\xi, \dot{\xi})\dot{\xi} + G(\xi) = F(\alpha, \beta) + F_C(\xi) + F_{ext} \tag{18}$$

where $M(\xi)$ is the inertia matrix, $C(\xi, \dot{\xi})$ represents the Coriolis and centrifugal terms, the gravitational forces that are represented by $G(\xi)$, which depends on the state of the system due to the movements of the aerial manipulator, change the position of the center of gravity of the platform. On the other hand, $F(\alpha, \beta)$ collects the forces and torques of the rotors and active joints. $F_C(\xi)$ models the contact forces and F_{ext} includes external forces and torques which are not known or modeled like the ones produced by the wind or the environment.

In summary, these forces can be expressed as:

$$F(\alpha, \beta) = \begin{bmatrix} F_w(\alpha, \beta) \\ F_a \end{bmatrix} \tag{19}$$

where $F_w(\alpha, \beta)$is the vector of forces generated by the rotors, and F_a are the torques generated by the arm joint servos, as follows: C_T being the thrust coefficient of the rotors, L the distance from the CoM to the rotor position, and C_f the drag coefficient.

Matrices $A_{3\times6}$ and $G_{3\times6}$ in (20) have the form:

$$F_w(\alpha,\beta) = N(\alpha,\beta)\begin{bmatrix}\omega_1^2 & \dots & \omega_6^2\end{bmatrix}^T = \begin{bmatrix}A_{3\times6}C_T \\ G_{3\times6}LC_T + H_{3\times6}C_f\end{bmatrix}$$

$$F_a = \begin{bmatrix}\tau_1 & \tau_2 & 0\end{bmatrix}^T \tag{20}$$

$$A_{3\times6} = \begin{bmatrix} \frac{\sqrt{3}}{2}s\beta - \frac{s\alpha c\beta}{2} & s\alpha c\beta & -\frac{\sqrt{3}}{2}s\beta - \frac{s\alpha c\beta}{2} & \frac{\sqrt{3}}{2}s\beta - \frac{s\alpha c\beta}{2} & s\alpha c\beta & -\frac{\sqrt{3}}{2}s\beta - \frac{s\alpha c\beta}{2} \\ \frac{s\beta}{2} + \frac{\sqrt{3}}{2}s\alpha c\beta & -s\beta & \frac{s\beta}{2} + \frac{\sqrt{3}}{2}s\alpha c\beta & \frac{s\beta}{2} + \frac{\sqrt{3}}{2}s\alpha c\beta & -s\beta & \frac{s\beta}{2} - \frac{\sqrt{3}}{2}s\alpha c\beta \\ c\alpha c\beta & c\alpha c\beta & c\alpha c\beta & c\alpha c\beta & c\alpha c\beta & c\alpha c\beta \end{bmatrix} \tag{21}$$

$$G_{3\times6} = \begin{bmatrix} \frac{c\alpha c\beta}{2} & c\alpha c\beta & \frac{c\alpha c\beta}{2} & -\frac{c\alpha c\beta}{2} & -c\alpha c\beta & -\frac{c\alpha c\beta}{2} \\ \frac{\sqrt{3}}{2}c\alpha s\beta & 0 & \frac{\sqrt{3}}{2}c\alpha s\beta & \frac{\sqrt{3}}{2}c\alpha s\beta & 0 & -\frac{\sqrt{3}}{2}c\alpha s\beta \\ s\alpha c\beta & -s\alpha c\beta & c\alpha c\beta & -s\alpha c\beta & c\alpha c\beta & -s\alpha c\beta \end{bmatrix} \tag{22}$$

In addition, the $H_{3\times6}$ matrix (20) can be obtained from $A_{3\times6}$, changing the sign of the first, third, and fifth columns.

Relative to the force measured through the sensor installed in the end-effector, F_C could be written as:

$$F_C(\xi) = \begin{bmatrix} R^{EB} & 0 \\ 0 & I \end{bmatrix} \tag{23}$$

$$\begin{bmatrix} -F_s \sin(\gamma_1 - \gamma_2) \\ 0 \\ -F_s \cos(\gamma_1 - \gamma_2) \\ 0 \\ F_s d \cos(\gamma_1 - \gamma_2) - F_s L_1 \sin(\gamma_2) \\ 0 \\ F_s L_1 \sin(\gamma_2) \\ 0 \\ F_s \end{bmatrix} \tag{24}$$

F_s being the force measured with the sensor which is assumed as an input data such that $F_s = K_s\gamma_3$, K_s is the spring stiffness constant and d is the distance from the CoM to the point in which the arm is attached, and $R^{EB} \in SO(3)$ is the rotation matrix which transforms a vector in body frame $\{B\}$ to the inertial earth fixed frame $\{E\}$. These models of forces and torques are essential to provide the fully actuated capabilities and also to take it into account to guarantee the safety and stability of the aerial platform during the contact phase. This model is validated by comparing model prediction with real flight data in Section 5.1.

4.2. Control Scheme

The design of the aerial manipulator controller combines the backstepping approach in [55] with an open-loop adaption of the inertia matrix to improve the performance of the aerial platform and the manipulator arm [56]. After estimating the six independent inertia values for each of the manipulator settings and its validation (See Section 5), the application of certainty-equivalence principle is immediate. The demonstration of its suitability for a standard multirotor with a robotic arm has been proved in [56,57]. In this section, the previous approach has been adapted to a fully actuated platform with a robotic sensorized arm.

One of the major advantages of this technique is that it allows the coupled control of the full system dynamics and the external forces induced by surface contact and the arm movements. Moreover, thanks to the linear potentiometer placed at the end effector (see Figure 6b), the external forces during the contact can be estimated and the torques applied in each arm joint, due to the contact force, can be calculated from (24). The estimation of these forces and torques allows the implementation of a controller to track trajectories on a structure surface applying the desired force needed to carry out measurements of deflections in a beam, the inspection of cracks using an ultrasonic sensor or sensor placement in the infrastructure. On the other hand, due to the servo actuators used in the robotic arm, the controller outputs, U_γ, must be mapped to desired joints inputs, γ_d. This mapping has been generated thanks to a standard procedure of system identification.

The full control scheme is shown in Figure 14, and the control equations of the aerial platform are presented from (25) to (31):

$$U_p = m[ge_3 + K_p e_p + K_v e_v + L_p X_p + \ddot{p}_d] + G_p + C_p + \tilde{N}_{p\ddot{\gamma}} - F_{C_p} \tag{25}$$

$$U_\eta = I_\eta[K_\eta e_\eta + K_{\dot{\eta}} e_{\dot{\eta}} + L_\eta \chi_\eta + \ddot{\eta}_d] + C_\eta + G_\eta + \tilde{N}_{\eta\ddot{\gamma}} - F_{C_\eta} \tag{26}$$

$$U_\gamma = J_E^T M_M e_\gamma + C_M + G_M - J^T F_d + M_{QM}^T \begin{bmatrix} \ddot{p} \\ \ddot{\eta} \end{bmatrix} - F_{c_\gamma} \tag{27}$$

$$e_p = p - p_d;\ e_v = k_p e_p + \lambda_p \chi_p + v_d - v \tag{28}$$

$$e_\eta = \eta_d - \eta;\ e_{\dot{\eta}} = k_\eta e_\eta + \lambda_\eta \chi_\eta + \dot{\eta}_d - \dot{\eta} \tag{29}$$

$$e_\gamma = I\ddot{X} + k_p(X_d - X) + k_v(\dot{X}_d - \dot{X}) \tag{30}$$

$$F_d = F_{s_d} + k_f(F_{s_d} - F_s) \tag{31}$$

U_p, U_η, and U_γ are the position, attitude, and arm joints control inputs, the values of K_i are defined such that $K_n = 1 - \lambda_n^2 + k_n$, $K_{\dot{n}} = k_n + k_{\dot{n}}$ and $L_n = -k_n\lambda_n$, are positive constant controller gains, e_p, e_v, e_η, $e_{\dot{\eta}}$, and e_γ are the tracking errors, χ_i are the integral tracking errors, J_E^T is the Jacobian of the manipulator, X represents the position of the end effector in the Earth fixed frame $\{E\}$ and p_d, v_d, η_d, X_d, and F_{s_d} are the commanded references. $\tilde{N}_{p\ddot{\gamma}}$ and $\tilde{N}_{\eta\ddot{\gamma}}$ introduce the estimation of the dynamic coupling, which is considered negligible due to the low inertial and lightweight manipulator design. I_η represents the submatrix of $M(\xi)$ that govern the attitude dynamic. The rest of the terms C_p, C_η, C_M, G_p, G_η, and G_M represent the terms of the Coriolis and the gravity matrix that affects each variable.

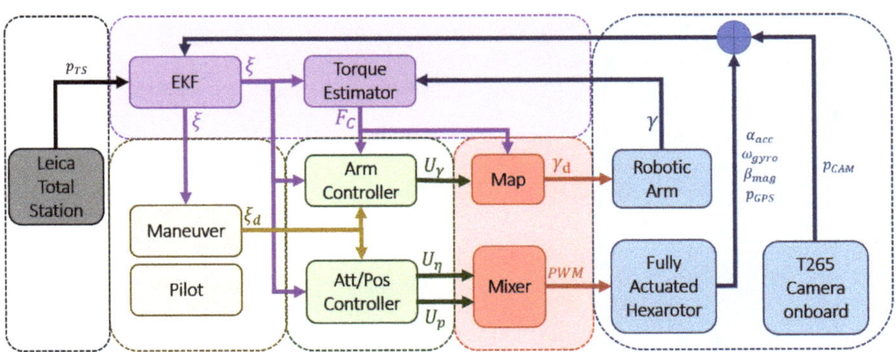

Figure 14. Control architecture.

For a better understanding of how the control law is obtained and how it achieves the origin being Globally Uniformly Asymptotically Stable (GUAS), the derivation of the roll tracking controller ϕ_d is briefly explained below.

Defining $z_1 = \phi_d - \phi$ as a tracking roll error, a candidate Lyapunov function is:

$$V_1 = \frac{1}{2}z_1^2 + \frac{1}{2}\lambda_1\chi_1^2 \tag{32}$$

where $\chi_1 = \int_0^t z_1(\tau)\,d\tau$ is the integral error. Now, we choose $\dot{\phi}$ as a virtual control, and find its value to stabilize V_1. Derivating (32) and making $\dot{V}_1$ non-positive, we have:

$$\dot{V}_1 = z_1(\dot{\phi}_d - \dot{\phi} + \lambda_1\chi_1) \tag{33}$$

Therefore, $\dot{\phi}$ would be $\dot{\phi} = \dot{\phi}_d + \lambda_1\chi_1 + k_1 z_1$, the derivate of Lyapunov function $\dot{V}_1$ would be non-positive, and V_1 would be bounded. Furthermore, applying LaSalle–Yoshizawa Lemma, it proves the regulation of z_1:

$$\dot{V}_1 = -k_1 z_1^2 \leq W(z_1) \tag{34}$$

$$\lim_{t\to+\infty} W(z_1) = 0 \tag{35}$$

Equation (35) implies that z_1 would tend to zero. However, $\dot{\phi}$ is only a state and not a control input. Therefore, we define the stabilizing function α and the error between this function and the virtual control as:

$$\alpha = \dot{\phi}_d + \lambda_1\chi_1 + k_1 z_1 \tag{36}$$

$$z_2 = \alpha - \dot{\phi} \tag{37}$$

Now, we construct a control Lyapunov function for the entire system, where the error between the stabilizing function and the virtual control be penalized. Therefore,

$$V_2 = \frac{1}{2}z_1^2 + \frac{1}{2}\lambda_1\chi_1^2 + \frac{1}{2}z_2^2 \tag{38}$$

Derivating (38), we have:

$$\dot{V}_2 = -k_1 z_1^2 + z_1 z_2 + z_2(\dot{\alpha} - \ddot{\phi}) \tag{39}$$

In (39), it can be seen that, if $\dot{\alpha} - \ddot{\phi} = -k_2 z_2 - z_1$, the derivate of the Lyapunov function $\dot{V}_2$ would be non-positive and, therefore, V_2 is bounded. In addition, applying LaSalle-Yoshizawa Lemma, the regulation is also proved:

$$\dot{V}_2 = -k_1 z_1^2 - k_2 z_2^2 \leq W(z_1, z_2) \tag{40}$$

$$\lim_{t\to+\infty} W(z_1, z_2) = 0 \tag{41}$$

To comply $\dot{\alpha} - \ddot{\phi} = -k_2 z_2 - z_1$, after some mathematical operations, the following expression can be obtained:

$$\ddot{\phi} = \ddot{\phi}_d + (1 - k_1^2 + \lambda_1)z_1 + (k_1 + k_2)z_2 - k_1\lambda_1\chi_1 \tag{42}$$

Proceeding in the same way with the pitch and yaw, similar expressions can be obtained:

$$\ddot{\theta} = \ddot{\theta}_d + (1 - k_3^2 + \lambda_2)z_3 + (k_3 + k_4)z_4 - k_3\lambda_2\chi_2 \tag{43}$$

$$\ddot{\psi} = \ddot{\psi}_d + (1 - k_5^2 + \lambda_3)z_5 + (k_5 + k_6)z_6 - k_5\lambda_3\chi_3 \tag{44}$$

By introducing the expressions (42), (43) and (44) in the dynamic Equation (18), the expression of the control law (26) is deduced.

5. Experimental Results

This section aims to validate the performance of the fully-actuated platform and its capability to carry out the inspection operations in an outdoors scenario. First, the experiment consisted of analyzing the model estimated to be used in the controller. Second, a test bench experiment shows the capabilities of the robotic arm in a controlled environment that simulate the use case. Third, the advantages of using a platform with tilted rotors are shown comparing the results obtained of this fully-actuated system with a similar platform that implements the standard cascade linear controller. Finally, the full system is deployed and validated carrying out an inspection by contact operation in a bridge outdoors.

The prototype used along these experiments is presented Figure 15, along with the final mass specifications.

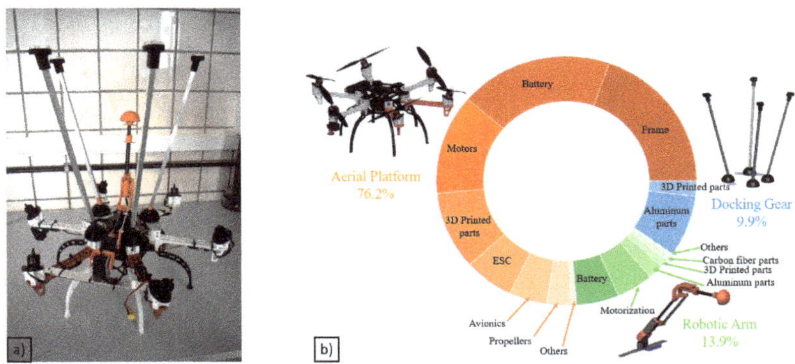

Figure 15. Fully actuated platform: (**a**) system integration; (**b**) mass distribution of the system.

5.1. Model Validation

Due to the controller implemented being based on a model estimation, the first step consisted of carrying out a model identification. To achieve this goal, all flight data such as control inputs, angular velocities, etc, were logged implementing a basic linear controller. During the flight, it was necessary to make aggressive maneuvers to excite the dynamics in such a way that the order of magnitude of any variable in the system is much greater than the standard deviation of the intrinsic noise to it.

After logging the data, a filtered version of the signals and the different derivatives that appear in the model were obtained, such as angular acceleration from the gyro data. Then, applying the equation of the dynamic model (18), the angular acceleration can be obtained for each sample time. Figure 16 shows that the real and estimated angular pitch acceleration match fairly well.

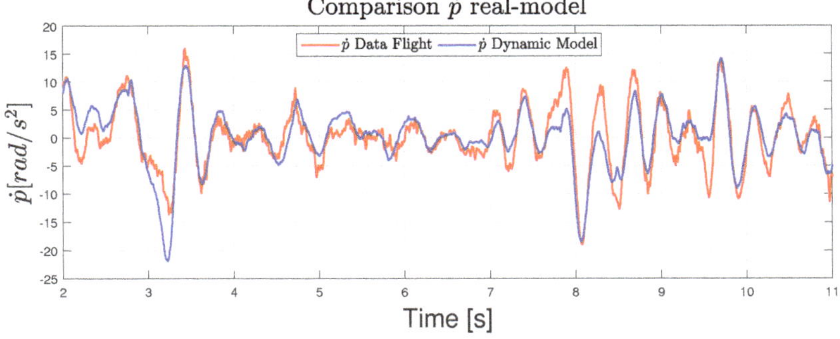

Figure 16. Comparison of experimental and dynamic model on roll attitude.

Repeating this procedure with several configuration of the arm angles γ_1 and γ_2, the results were as good as the ones shown in the previous figure. Therefore, the dynamic model and the parameters estimation were considered validated.

5.2. Inspection in the Test Bench

This section aims to demonstrate the inspection capability of the robotic arm, before its installation on the hexarotor. For this purpose, an inspection has been done on a test bench on which the arm has been installed (Figure 17).

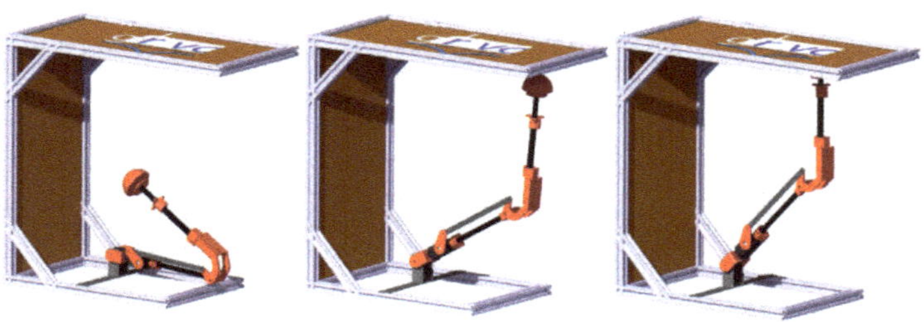

Figure 17. Test bench for the robotic arm.

The inspection process of a vertical surface consists of different phases. First, the robotic arm is in the rest position ($\gamma_1 = 95^\circ$ and $\gamma_2 = 152^\circ$). The robotic arm then moves to the inspection position ($\gamma_1 = 61^\circ$ and $\gamma_2 = 55^\circ$). In this position, the robotic arm is close to the surface but does not touch it. Finally, the inspection takes place, during which the robotic arm performs a force control.

Finally, Figure 18 shows the values of the articular variables, as well as the force demanded during an inspection of a vertical surface carried out on the test bench. This test simulates the force required to stick a sensor using a two-component glue.

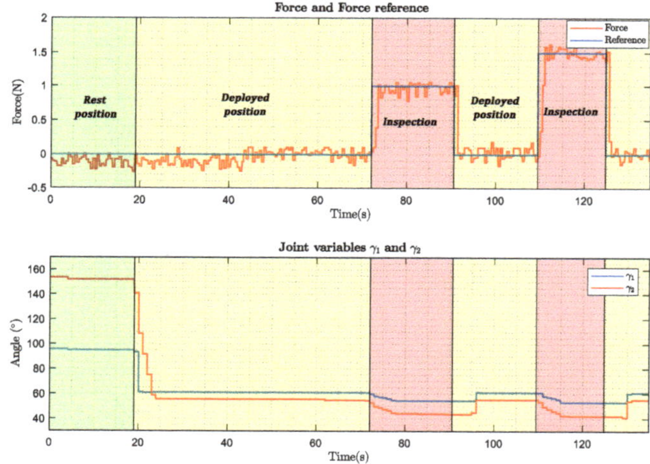

Figure 18. Inspection in the test bench.

5.3. Tilted Rotor Capabilities

In order to compare the differences between a fully-actuated platform and an under-actuated robot, three different flight modes have been implemented in the controller. In the first mode, called "standard stabilization", the multirotor has a standard cascade linear controller without taking into account the inclination of the propeller, and then without the capability of accomplishing horizontal and angular movements independently. In contrast, the "command position" and "command angle" mode has been implemented to be able to command specific positions or angles in an independent way.

Figure 19 collects some pictures of the maneuvers that the fully-actuated platform is able to perform. Moreover, telemetry results of these maneuvers are also shown in Figure 20. The green background is the aforementioned "standard stabilization" mode in which is not possible to decouple the angle and position dynamics, since there is not a position controller implemented in this mode, so the dX and dY results do not have to match the reference. Yellow and red backgrounds show the fully-actuated modes in which the hexarotor is able to perform maneuvers in position and angles in an independent way. In yellow, the controller is focused on controlling the position and velocity without changing the attitude ("command position"), and in red the aerial platform controls the attitude maintaining a constant position. These results show how the platform is able to tilt 10 degrees without changing the position thanks to the full-actuation capabilities.

Figure 19. Fully actuated platform maneuvers. (**a**) during a planar translation ("com. position" flight mode); (**b**) tilted without losing the position reference ("com. angle" flight mode).

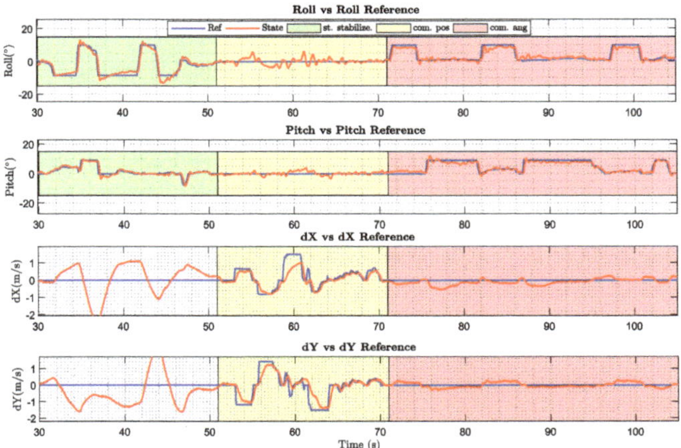

Figure 20. Performance of the fully-actuated platform in each flight mode implemented.

5.4. Inspection Experiments

This section aims to demonstrate the applicability that these kinds of platforms have in the field of bridge inspection by contact. The experiment carried out is illustrated in the sequence and scheme of Figures 21 and 22. The experiment is carried out in "command position" flight mode presented in the previous section. The aircraft is commanded to follow references in X, Y, and Z coordinates until it reaches the bridge surface. Once the docking gear is in contact with the bridge surface, the phase called "inspection phase" starts. During this phase, the robotic arm is commanded to exert a force on the surface, for example, to install a vibration sensor, take measurements of the bridge, or accomplish a maintenance operation. Finally, the arm is commanded to stop the operation and the platform lands on a safe place.

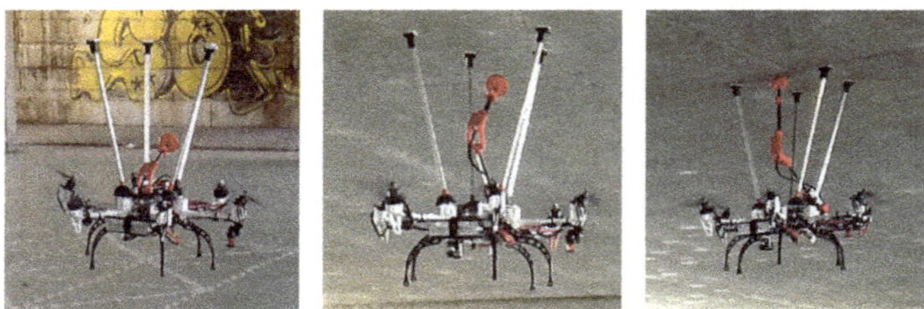

Figure 21. Sequence of a bridge inspection operation with the system proposed.

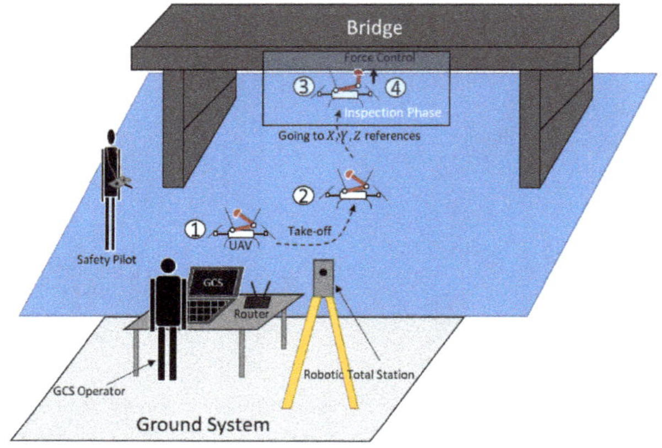

Figure 22. Scheme of the typical experiment.

Figure 23 illustrates the performance of the aerial manipulator in two different experiments. It shows the results in terms of how the controllers follow the values commanded in X, Y, Z, and F_C during two different experiments. The area colored yellow in each result represents the inspection phase which starts when the docking gear is in contact with the bridge. It clearly shows that the system proposed is very stable during the inspection operation; moreover, they validate the feasibility of applying the robotic system proposed to the field of the bridge inspection operation using UAVs. A video compilation with one of the experiments can be found in [58].

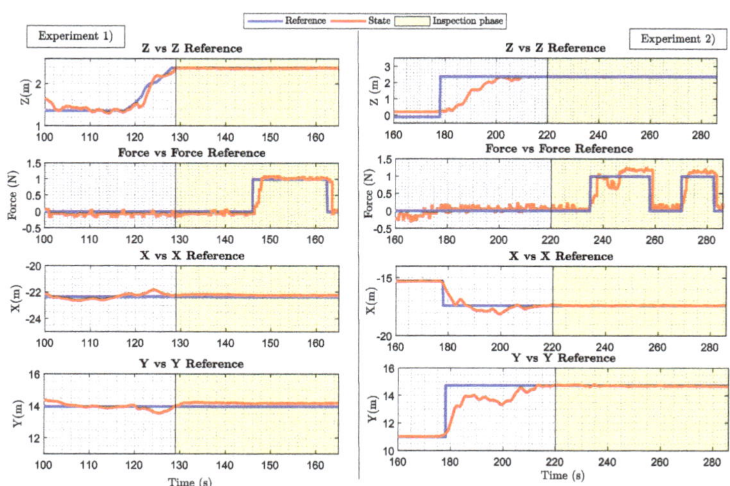

Figure 23. Experimental results.

6. Conclusions

This paper has presented a fully actuated aerial platform with a lightweight robotic arm for contact inspection of bridges and tunnels. The use of a multi-directional thrust aerial platform shows advantages over standard co-planar multirotor configurations for flying in the confined spaces under bridge decks and inside tunnels. Another advantage is the capability of applying forces in the horizontal plane of the presented prototype using its arm and the force sensor at the end effector, and without needing to tilt the platform for exerting lateral forces.

The presented prototype includes a new concept of docking gear which also provides an advantage during the inspection, getting the best accuracy during the contact operation, while facilitating the control strategy due to the decoupling between the flight and the inspection, and increasing safety during the operation. These aspects have been considered by the end-users as mandatory and are essential for proper commercial exploitation in bridge maintenance programs.

The localization system proposed is able to operate in GNSS-denied conditions and solves common problems that arise in several inspection applications. In those, it is essential to be able to visit the same points in different inspection operation to assess and trace how some specific defects evolve along the time.

The prototype as well as the localization system and the control technique implemented have been evaluated in a real bridge inspection which is a GNSS-denied environment and uses a robotic total station and a reflector prism on-board as position estimation to close the position control loop and inspection sensor.

This work demonstrates that it is possible to solve the time-consistent reference frame issue that affects most of the inspection applications using UAVs in a GPS denied environment. This has been considered as one of the most important requirements imposed by the need for following the evolution of the defects over time. It also shows the advantages of using a fully-actuated platform against a conventional co-planar multirotor to accomplish an inspection by contact application one more time. It is clear that those platforms have enhanced capabilities in applications that require exerting forces with a flying robot. Finally, it has been shown that using a docking gear can improve the measurements obtained thanks to the stability provided by maintaining a contact with the infrastructure.

As future work, the prototype will be enhanced to inspect pillars and tunnel walls, including an adjustable docking gear, and the multi-sensor localization system will increase the number of sensors and its technology.

Author Contributions: Conceptualization, P.J.S.-C. and G.H.; methodology, P.J.S.-C.; software, D.B.G. and A.E.J.-C.; validation, P.J.S.-C., A.G.-M., and D.B.G.; formal analysis, N.C. and A.E.J.-C.; investigation, A.G.-M., D.B.G., and N.C.; resources, G.H.; writing—original draft preparation, P.J.S.-C., A.G.-M., D.B., N.C., and A.E.J.-C.; writing—review and editing, G.H. and A.O.; supervision, G.H. and A.O.; project administration, G.H.; funding acquisition, G.H. and A.O. All authors have read and agreed to the published version of the manuscript.

Funding: This work has been supported by the ARTIC Project, funded by the Spanish Ministerio de Economía, Industria, y Competitividad (RTI2018-102224-B-I00), the H2020 RESIST (H2020-MG-2017-769066) and PILOTING (H2020-ICT-2019-2-871542) projects, funded by the European Commission, and by the FPU Program of the Spanish Ministerio de Educación, Cultura y Deporte.

Conflicts of Interest: The authors declare no conflict of interest.

References

1. Vachtsevanos, G.; Valavanis, K. *Handbook of Unmanned Aerial Vehicles*; Springer: Berlin/Heidelberg, Germany, 2015; pp. 93–103.
2. Mahony, R.; Kumar, V.; Corke, P. Multirotor Aerial Vehicles: Modeling, Estimation, and Control of Quadrotor. *IEEE Robot. Autom. Mag.* **2012**, *19*, 20–32. [CrossRef]
3. Ruggiero, F.; Lippiello, V.; Ollero, A. Aerial manipulation: A literature review. *IEEE Robot. Autom. Lett.* **2018**, *3*, 1957–1964. [CrossRef]
4. Ollero, A.; Siciliano, B. *Aerial Robotic Manipulation*; Springer: Cham, Switzerland, 2019.

5. Pounds, P.E.I.; Bersak, D.R.; Dollar, A.M. The Yale aerial manipulator: Grasping in flight. In Proceedings of the 2011 IEEE International Conference on Robotics and Automation (ICRA), Shanghai, China, 9–13 May 2011; pp. 2974–2975.
6. Seo, H.; Kim S.; Kim, H.J. Aerial grasping of cylindrical object using visual servoing based on stochastic model predictive control. In Proceedings of the 2017 IEEE International Conference on Robotics and Automation (ICRA), Singapore, 29 May–3 June 2017; pp. 6362–6368.
7. Korpela, C.; Orsag M.; Oh, P. Towards valve turning using a dual-arm aerial manipulator. In Proceedings of the 2014 IEEE/RSJ International Conference on Intelligent Robots and Systems (IROS), Chicago, IL, USA, 14–18 September 2014; pp. 3411–3416.
8. Shimahara, S.; Leewiwatwong, S.; Ladig, R.; Shimonomura, K. Aerial torsional manipulation employing multi-rotor flying robot. In Proceedings of the 2016 IEEE/RSJ International Conference on Intelligent Robots and Systems (IROS), Daejeon, Korea, 9–14 October 2016; pp. 1595–1600.
9. Chermprayong, P.; Zhang, K.; Xiao, F.; Kovac, M. An integrated Delta manipulator for aerial repair: A new aerial robotic system. *IEEE Robot. Autom. Mag.* **2019**, *26*, 54–66. [CrossRef]
10. Fumagalli, M.; Naldi, R.; Macchelli, A.; Forte, F.; Keemink, A.Q.L.; Stramigioli, S.; Carloni, R.; Marconi, L. Developing an aerial manipulator prototype: Physical interaction with the environment. *IEEE Robot. Autom. Mag.* **2014**, *21*, 41–50. [CrossRef]
11. Yang, C.H.; Wen, M.C.; Chen, Y.C.; Kang, S.C. An Optimized Unmanned Aerial System for Bridge Inspection. In Proceedings of the 32nd International Symposium on Automation and Robotics in Construction and Mining (ISARC), Oulu, Finland, 15–18 June 2015.
12. Recchiuto, C.T.; Sgorbissa, A. Post-disaster assessment with unmanned aerial vehicles: A survey on practical implementations and research approaches. *J. Field Robot.* **2018**, *35*, 459–490. [CrossRef]
13. Greenwood, W.W.; Lynch, J.P.; Zekkos, D. Applications of UAVs in Civil Infrastructure. *J. Infrastruct. Syst.* **2019**, *25*, 04019002. [CrossRef]
14. Jung, H.-J.; Lee, J.-H.; Yoon, S.; Kim, I.-H. Bridge Inspection and condition assessment using Unmanned Aerial Vehicles (UAVs): Major challenges and solutions from a practical perspective. *Smart Struct. Syst.* **2019**, *24*, 669–681.
15. Jung, S.; Song, S.; Kim, S.; Park, J.; Her, J.; Roh, K.; Myung, H. Toward Autonomous Bridge Inspection: A framework and experimental results. In Proceedings of the 2019 16th International Conference on Ubiquitous Robots (UR), Jeju, Korea, 24–27 June 2019; pp. 208–211. [CrossRef]
16. Hallermann, N.; Morgenthal, G. Visual inspection strategies for large bridges using Unmanned Aerial Vehicles (UAV). In Proceedings of the 7th International Conference on Bridge Maintenance, Safety and Management (IABMAS), Shanghai, China, 7–11 July 2014; pp. 661–667.
17. Chen, S.; Linh, T.-H.; Laefer, D.; Mangina, E. Automated bridge deck evaluation through UAV derived point cloud. In Proceedings of the 2018 Civil Engineering Research in Ireland Conference, Dublin, Ireland, 29–30 August 2018.
18. AEROBI Project (AErial RObotic System for In-Depth Bridge Inspection by Contact). Available online: http://www.aerobi.eu/ (accessed on 15 August 2020)
19. RESIST Project (Resilient Transport Infrastructure to Extreme Events). Available online: http://www.resistproject.eu/ (accessed on 15 August 2020).
20. PILOTING Project (PILOTs for Robotic INspection and Maintenance Grounded on Advanced Intelligent Platforms and Prototype Applications). Available online: https://piloting-project.eu/ (accessed on 15 August 2020)
21. Jimenez-Cano, A.E.; Sanchez-Cuevas, P.J.; Grau, P.; Ollero, A.; Heredia, G. Contact-Based Bridge Inspection Multirotors: Design, Modeling, and Control Considering the Ceiling Effect. *IEEE Robot. Autom. Lett.* **2019**, *4*, 3561–3568. [CrossRef]
22. Ikeda, T.; Yasui, S.; Fujihara, M.; Ohara, K.; Ashizawa, S.; Ichikawa, A.; Okino, A.; Oomichi, T.; Fukuda, T. Wall contact by octo-rotor UAV with one DoF manipulator for bridge inspection. In Proceedings of the 2017 IEEE/RSJ International Conference on Intelligent Robots and Systems (IROS), Vancouver, BC, Canada, 24–28 September 2017; pp. 5122–5127.
23. Hamaza, S.; Georgilas, I.; Fernandez, M.; Sanchez, P.; Richardson, T.; Heredia, G.; Ollero, A. Sensor Installation and Retrieval Operations Using an Unmanned Aerial Manipulator. *IEEE Robot. Autom. Lett.* **2019**, *4*, 2793–2800. [CrossRef]

24. Abe, Y.; Ichikawa, A.; Ikeda, T.; Fukuda, T. Study of hammering device to put on multi-copter targeted for bridge floor slabs. In Proceedings of the 2017 International Symposium on Micro-NanoMechatronics and Human Science (MHS), Nagoya, Japan , 3–6 December 2017; pp. 1–3. [CrossRef]
25. Jimenez-Cano, A.E.; Heredia, G.; Ollero, A. Aerial manipulator with a compliant arm for bridge inspection. In Proceedings of the 2017 International Conference on Unmanned Aircraft Systems (ICUAS), Miami, FL, USA, 13–16 Jun 2017; pp. 1217–1222.
26. Myeong, W.; Myung, H. Development of a Wall-Climbing Drone Capable of Vertical Soft Landing Using a Tilt-Rotor Mechanism. *IEEE Access* **2019**, *7*, 4868–4879. [CrossRef]
27. Yamada, M.; Nakao, M.; Hada, Y.; Sawasaki, N. Development and field test of novel two-wheeled UAV for bridge inspections. In Proceedings of the 2017 International Conference on Unmanned Aircraft Systems (ICUAS), Miami, FL, USA, 13–16 June 2017; pp. 1014–1021. [CrossRef]
28. Franchi, A.; Carli, R.; Bicego, D.; Ryll, M. Full-Pose Tracking Control for Aerial Robotic Systems with Laterally-Bounded Input Force. *IEEE Trans. Robot.* **2018**, *34*, 644271. [CrossRef]
29. Brescianini, D.; D'Andrea, R. Design, modeling and control of an omni-directional aerial vehicle. In Proceedings of the 2016 IEEE International Conference on Robotics and Automation (ICRA), Stockholm, Sweden, 16–20 May 2016; pp. 3261–3266. [CrossRef]
30. Park, S.; Lee, J.; Ahn, J.; Kim, M.; Her, J.; Yang, G.; Lee, D. ODAR: Aerial Manipulation Platform Enabling Omnidirectional Wrench Generation. *IEEE/ASME Trans. Mechatron.* **2018**, *23*, 1907–1918. [CrossRef]
31. Lee, J.Y.; Leang, K.K.; Yim, W. Design and control of a fully-actuated hexarotor for aerial manipulation applications. *J. Mech. Robot.* **2018**, *10*, 1–10. [CrossRef]
32. Kamel, M.; Verling, S.; Elkhatib, O.; Sprecher, C.; Wulkop, P.; Taylor, Z.; Siegwart, R.; Gilitschenski, I. The Voliro Omniorientational Hexacopter: An Agile and Maneuverable Tiltable-Rotor Aerial Vehicle. *IEEE Robot. Autom. Mag.* **2018**, *25*, 34–44. [CrossRef]
33. Long, Y.; Cappelleri, D. J. Linear control design, allocation, and implementation for the Omnicopter MAV. In Proceedings of the IEEE International Conference on Robotics and Automation, Karlsruhe, Germany, 6–10 May 2013; pp. 289–294.
34. Ryll, M.; Bulthoff, H.H.; Giordano, P.R. A Novel Overactuated Quadrotor Unmanned Aerial Vehicle: Modeling, Control and Experimental Validation. *IEEE Trans. Control Syst. Technol.* **2015**, *23*, 540–556. [CrossRef]
35. Ryll, M.; Bicego, D.; Franchi, A. Modeling and control of FAST-Hex: A fully-actuated by synchronized-tilting hexarotor. In Proceedings of the 2016 IEEE/RSJ International Conference on Intelligent Robots and Systems (IROS), Daejeon, Korea, 9–14 October 2016; pp. 1689–1694. [CrossRef]
36. Zheng, P.; Tan, X.; Kocer, B.B.; Yang, E.; Kovac, M. TiltDrone: A Fully-Actuated Tilting Quadrotor Platform. *IEEE Robot. Autom. Mag.* **2020**. [CrossRef]
37. Odelga, M.; Stegagno, P.; Bulthoff, H.H. A fully actuated quadrotor uav with a propeller tilting mechanism: Modeling and control. In Proceedings of the 2016 IEEE International Conference on Advanced Intelligent Mechatronics (AIM), Barcelona, Spain, 12–15 July 2016; pp. 306–311.
38. Zhao, M.; Kawasaki, K.; Anzai, T.; Chen, X.; Noda, S.; Shi, F.; Okada, K.; Inaba, M. Transformable multirotor with two-dimensional multilinks: Modeling, control, and whole-body aerial manipulation. *Int. J. Robot. Res.* **2018**, *37*, 1085–1112. [CrossRef]
39. Tognon, M.; Chávez, H.A.; Gasparin, E.; Sablé, Q.; Bicego, D.; Mallet, A.; Lany, M.; Santi, G.; Revaz, B.; Cortés, J.; et al. A Truly-Redundant Aerial Manipulator System With Application to Push-and-Slide Inspection in Industrial Plants. *IEEE Robot. Autom. Lett.* **2019**, *4*, 1846–1851. [CrossRef]
40. Ollero, A.; Heredia, G.; Franchi, A.; Antonelli, G.; Kondak, K.; Sanfeliu, A.; Viguria, A.; Martinez-de Dios, J.R.; Pierri, F.; Cortés, J.; et al. The AEROARMS Project: Aerial Robots with Advanced Manipulation Capabilities for Inspection and Maintenance. *IEEE Robot. Autom. Mag.* **2018**, *25*, 12–23. [CrossRef]
41. Sanchez-Cuevas, P. J.; Heredia, G.; Ollero, A. Multirotor UAS for bridge inspection by contact using the ceiling effect. In Proceedings of the 2017 International Conference on Unmanned Aircraft Systems (ICUAS), Miami, FL, USA, 13–16 Jun 2017; pp. 767–774. [CrossRef]
42. Hsiao, Y. H.; Chirarattananon, P. Ceiling Effects for Hybrid Aerial–Surface Locomotion of Small Rotorcraft. *IEEE/ASME Trans. Mechatron.* **2018**, *24*, 2316–2327. [CrossRef]

43. Kocer, B.B.; Tiryaki, M.E.; Pratama, M.; Tjahjowidodo, T.; Seet, G.G.L. Aerial robot control in close proximity to ceiling: A force estimation-based nonlinear mpc. In Proceedings of the 2019 IEEE/RSJ International Conference on Intelligent Robots and Systems (IROS), Macau, China, 4–8 November 2019; pp. 2813–2819. [CrossRef]
44. Taketomi, T.; Uchiyama, H.; Ikeda, S. Visual SLAM algorithms: A survey from 2010 to 2016. *IPSJ Trans. Comput. Vision Appl.* **2017**, *9*, 16. [CrossRef]
45. Robust Visual-Inertial Tracking from a Camera that Knows Where it's Going. Available online: https://www.intelrealsense.com/visual-inertial-tracking-case-study/ (accessed on 15 August 2020)
46. Rajappa, S.; Ryll, M.; Bülthoff, H.H.; Franchi, A. Modeling, control and design optimization for a fully-actuated hexarotor aerial vehicle with tilted propellers. In Proceedings of the 2015 IEEE International Conference on Robotics and Automation (ICRA), Seattle, WA, USA, 26–30 May 2015; pp. 4006–4013.
47. Suarez, A.; Heredia, G.; Ollero, A. Lightweight compliant arm with compliant finger for aerial manipulation and inspection. In Proceedings of the 2016 IEEE/RSJ International Conference on Intelligent Robots and Systems (IROS), Daejeon, Korea, 9–14 October 2016; pp. 4449–4454.
48. Sanchez-Cuevas, P.J.; Ramon-Soria, P.; Arrue, B.; Ollero, A.; Heredia, G. Robotic system for inspection by contact of bridge beams using UAVs. *Sensors* **2019**, *19*, 305. [CrossRef] [PubMed]
49. Jimenez-Cano, A.E. Modeling and Control of Aerial Manipulators. Ph.D. Thesis, University of Sevilla, Seville, Spain, 2019.
50. Zhang, J.; Singh, S. LOAM: Lidar Odometry and Mapping in Real-time. *Robot. Sci. Syst.* **2014**, *2*, 1–9.
51. Mur-Artal, R.; Montiel, J.M.M.; Tardos, J.D. ORB-SLAM: A versatile and accurate monocular SLAM system. *IEEE Trans. Robot.* **2015**, *31*, 1147–1163. [CrossRef]
52. Akima, H. A new method of interpolation and smooth curve fitting based on local procedures. *J. ACM* **1970**, *17*, 589–602. [CrossRef]
53. Kalman, R.E. A New Approach to Linear Filtering and Prediction Problems. *J. Basic Eng.* **1960**, *82*, 35–45. [CrossRef]
54. Welch, G.; Bishop, G. An Introduction to the Kalman Filter. 1995. Available online: http://citeseerx.ist.psu.edu/viewdoc/download?doi=10.1.1.336.5576&rep=rep1&type=pdf (accessed on 19 August 2020).
55. Bouabdallah, S.; Siegwart, R. Full control of a quadrotor. In Proceedings of the 2007 IEEE/RSJ International Conference on Intelligent Robots and Systems (IROS), San Diego, CA, USA, 29 October–2 November 2007; pp. 153–158.
56. Jimenez-Cano, A.E.; Braga, J.; Heredia, G.; Ollero, A. Aerial manipulator for structure inspection by contact from the underside. In Proceedings of the 2015 IEEE/RSJ International Conference on Intelligent Robots and Systems (IROS), Hamburg, Hamburg, 28 September–2 October 2015; pp. 1879–1884.
57. Heredia, G.; Cano, R.; Jimenez-Cano, A.E.; Ollero, A. Modeling and Design of Multirotors with Multi-joint Arms. In *Aerial Robotic Manipulation*; Ollero, A., Siciliano, B., Eds.; Springer: Cham, Switzerland, 2019.
58. Video of the Experiments. Available online: https://hdvirtual.us.es/discovirt/index.php/s/xpadr7AZEk8cbeE (accessed on 15 August 2020).

Article

A Monocular SLAM-based Controller for Multirotors with Sensor Faults under Ground Effect

Antonio Matus-Vargas [1], Gustavo Rodriguez-Gomez [1] and Jose Martinez-Carranza [1,2,*]

[1] Department of Computational Science, Instituto Nacional de Astrofísica, Óptica y Electrónica (INAOE), Puebla 72840, Mexico; matusv@inaoep.mx (A.M.-V.); grodrig@inaoep.mx (G.R.-G.);

[2] Department of Computer Science, University of Bristol, Bristol BS8 1UB, UK

* Correspondence: carranza@inaoep.mx

Received: 30 September 2019; Accepted: 8 November 2019; Published: 13 November 2019

Abstract: Multirotor micro air vehicles can operate in complex and confined environments that are otherwise inaccessible to larger drones. Operation in such environments results in airflow interactions between the propellers and proximate surfaces. The most common of these interactions is the ground effect. In addition to the increment in thrust efficiency, this effect disturbs the onboard sensors of the drone. In this paper, we present a fault-tolerant scheme for a multirotor with altitude sensor faults caused by the ground effect. We assume a hierarchical control structure for trajectory tracking. The structure consists of an external Proportional-Derivative controller and an internal Proportional-Integral controller. We consider that the sensor faults occur on the inner loop and counteract them in the outer loop. In a novel approach, we use a metric monocular Simultaneous Localization and Mapping algorithm for detecting internal faults. We design the fault diagnosis scheme as a logical process which depends on the weighted residual. Furthermore, we propose two control strategies for fault mitigation. The first combines the external PD controller and a function of the residual. The second treats the sensor fault as an actuator fault and compensates with a sliding mode action. In either case, we utilize onboard sensors only. Finally, we evaluate the effectiveness of the strategies in simulations and experiments.

Keywords: multirotor; ground effect; sensor faults

---

1. Introduction

The interest in Unmanned Aerial Vehicles (UAVs) has been growing in recent years. Different types of UAVs have been used in many applications such as photography, cinematography, surveillance, remote inspection and emergency response, to mention a few. Among the types of UAVs, rotary-wing vehicles are one of the most versatile platforms. Particularly, multirotors offer large payload, high mobility and simple construction.

Micro Air Vehicles (MAVs) equipped with onboard sensors are ideal platforms for autonomous navigation. Due to payload and energy restrictions, an MAV is equipped with fewer sensors. The minimal sensor suite for autonomous localization has been reported to be a monocular camera and an Inertial Measurement Unit (IMU) [1]. However, it is common that an MAV is also equipped with sensors to measure the altitude. For low-level flights, the altitude is obtained from a combination of a range sensor and a barometer; this combination is referred to as an altimeter. With a monocular camera and the altitude measurement, the autonomous flight of a quadrotor has been achieved for indoor scenarios [2].

Multirotor MAVs can operate in complex and confined environments that are not accessible to larger drones. Within such environments, a multirotor is likely to move close to horizontal and vertical surfaces. These situations result in airflow interactions between the propellers and proximate surfaces. The most common of these interactions is the one produced by the ground. This phenomenon, known as the ground effect, is more pronounced in rotorcraft operating in hover and at low speeds. In addition to the increment of thrust efficiency, it has been reported that the ground effect causes variations in the pressure altitude reported by the barometer [3].

The ground effect has been well researched for helicopters [4–6]. For multirotors, empirical models have been developed [7,8]. Furthermore, several control schemes have been proposed for compensating the ground effect in quadrotors such as adaptive [9], fuzzy [10], sliding mode [7] and multi-controller [11]. However, the studies above have only considered the case in which there are no faults in the multirotor. Those controllers are not suitable for operating the vehicle in the case of actuator or sensor faults.

When there is no fault, a hierarchical control structure, composed by an external Proportional-Derivative (PD) controller and an internal Proportional-Integral (PI) controller, is developed for trajectory tracking. In contrast to previous work, we do not use external sensors to observe system states such as the vehicle's pose. Instead, we employ a technique known as *Monocular Simultaneous Localization and Mapping* (Monocular SLAM) to observe the position and orientation of the multirotor. The measurement of these states are estimated with metric, thus enabling us to implement position feedback for the external PD controller; the internal PI controller receives measurements from the altimeter. Sensor faults occur on the inner loop and we counteract them in the outer loop. We design the fault detection unit as a logical process that depends on the weighted residual. The residual compares estimations from the SLAM system with faulty internal readings. When a sensor fault is identified, the system switches to a control sub-law. We derive two control sub-laws—the first combines the external PD controller and a function of the residual; the second treats the sensor fault as an actuator fault and compensates with a sliding mode action. To the authors' knowledge, this is the first time that sensor faults caused by the ground effect are tackled utilizing a state estimator based on monocular SLAM. The latter is a well-known technique in robotics, typically used to estimate a robot's pose, but it has never been used to address any issues related to ground effect in multirotors.

The organization of the rest of the paper is as follows. Section 2 presents the related literature considering the ground effect in multirotors. In Section 3, we propose the fault detection unit and the control sub-laws. Section 4 discusses the simulation and experimental results. Finally, conclusions are given in Section 5.

2. Related Work

For safety reasons, full-scale helicopters are not allowed to operate close to obstacles. However, operating close to the ground is unavoidable. This circumstance has motivated the study of aerodynamic interactions between the helicopter and the ground while ignoring other surfaces like walls and the ceiling. As a result, the most known of these interactions is the ground effect. Simplified models for one rotor in ground effect were proposed, such as the model of Cheeseman [4] and the model of Hayden [5]. The applicability of these models to micro-rotors has been questioned because they operate at significantly different Reynolds numbers [12]. On the other hand, the wake flow of a single micro-rotor in ground effect has been studied thoroughly [6,13,14]. For a small helicopter, an empirical formula for ground compensation has been proposed [7]. Though simple, the generalization of this formula to other vehicles is not clear.

The effect of the ground on multirotor MAVs has been acknowledged in several works. An earlier approach was to model this effect as a variable thrust coefficient, which could be adapted on-line [9,10,15]. Alternatively, a method for modeling the same effect on a quadrotor used visual feedback from streamers attached to the ground [8]. Here, the aggregate energy due to the quadrotor down-wash was calculated and combined with the current throttle to predict the ground effect. The limitation of this work is that it requires streamers on the ground. In Reference [16], the model of Cheeseman has been compared to empirical data collected with a micro quadrotor, showing that the ground effect manifested at a higher height than predicted by the model. Similarly, a series of experiments was conducted to juxtapose the same model and data obtained from a coaxial quadrotor [17]. Again, the results pointed to a stronger ground effect than predicted by the theory. Motivated by this, the model of Cheeseman was extended for a quadrotor in Reference [18]. The authors also introduced the partial ground effect, which appears when only some of the rotors experience the ground effect. Naturally, the aforementioned models have been combined with control techniques [19–21].

When a multirotor flies over an obstacle, it is influenced by the ground effect. Considering that a depth image of the obstacle is available, the impulse affecting a quadrotor that is passing over that obstacle has been predicted from prior experience [22]. It was demonstrated that the aerodynamic interactions between the vehicle and the local environment produce consistent effects. The prediction scheme was nevertheless prone to either overestimating or underestimating these effects. This scheme was later incorporated into the control loop [23]. With this approach, a significant improvement was achieved for seen obstacles. For unobserved objects, the disturbance was not properly compensated. A similar problem was attacked in Reference [18], where the form of the obstacle was restricted to a box.

A state observer incorporates a mathematical model of the quadrotor and can be used to estimate external disturbances. The estimation can be combined with a feedback controller. This combination has been reported to improve the control performance when external disturbances exist [24–27]. However, the observer's estimation deteriorates for small disturbances and large sensor noise. Filtering the inputs and outputs of the observer can help in this case. The tuning of the filters is intricate and the filters introduce delays in the estimation, which deteriorates the overall control performance. On the contrary, stochastic estimation algorithms are designed to handle process and sensor noise at the cost of increasing the computational burden. An unscented Kalman filter has been presented to estimate external force and torque for quadrotors [28]. Nevertheless, it is not clear if this approach can be applied to disturbance rejection.

In Reference [11], we have shown that the position control performance under the ground effect can be enhanced with a rapid switching control algorithm. The position feedback was provided by an external motion capture system at a high frequency. We proposed a multi-controller structure where a classical controller and an extra one with a firm variable action were combined. The latter was designed to take action when the vehicle was not responding to the former. We treated these situations as mere disturbances caused by the increment of thrust in the ground effect. However, we noticed that sometimes the vehicle was suddenly starting to descend without being commanded to do so. This observation motivated us to check for possible faults in the drone. Indeed, we found spurious altimeter faults in the ground effect (see Section 4.1). Even though our previous approach did improve the control performance, there was still room for improvement, especially in avoiding reliance on external sensors.

The literature mentioned above has only considered the case in which faults do not occur in the multirotor. In the case of actuator or sensor faults, conventional controllers are not suitable for operating the vehicle due to significant control errors. In Reference [29], the research on fault diagnosis and fault-tolerant control for quadrotors was reviewed. Little research has examined the fault diagnosis and control for multirotors with sensor faults. In particular, the literature comprises faults in sensors like accelerometers, magnetometers and rate gyros [30–32]. More recently, the work in Reference [33] has considered faults in the velocity measurement obtained from the Global Positioning System (GPS) and the Inertial Navigation System (INS). One observer was designed for detection and another to estimate the sensor fault. Under several constraints, the estimation error was proved to be uniformly ultimately bounded. The bound is directly proportional to the magnitude of the derivative of the sensor fault. This fact might limit the effectiveness of such an observer since the sensor faults under the ground effect tend to have rapid changes. Furthermore, this approach considered faults occurring in the same loop.

In summary, the literature is lacking in terms of the diagnosis and control of sensor faults for multirotors. Furthermore, this is the first time that sensor faults induced by the ground effect have been reported. Consequently, the first attempts at mitigating these faults are going to be presented in this paper.

3. Methods and Materials

This section starts with a summary of the multirotor dynamics model. Subsequently, a specific control scheme is assumed and the stability conditions for the fault-free system are shown analytically. Finally, the fault diagnosis scheme and two control sub-laws are proposed.

3.1. Multirotor Dynamics

The mathematical model of the multirotor is deduced by introducing a world-fixed coordinate system $\{W\}$ and a body-fixed coordinate system $\{B\}$ (see Figure 1). By using either the Newton-Euler equations or the Euler-Lagrange formalism, the model of the multirotor can be obtained [34–36]. We selected the following model to describe the multirotor unmanned helicopter:

$$\begin{aligned} \ddot{x}(t) &= u_z(t)(\sin\psi\sin\phi + \cos\psi\sin\theta\cos\phi)/m \\ \ddot{y}(t) &= u_z(t)(\sin\psi\sin\theta\cos\phi - \cos\psi\sin\phi)/m \\ \ddot{z}(t) &= u_z(t)\cos\theta\cos\phi/m - g \\ \ddot{\phi}(t) &= u_\phi(t)/J_x \\ \ddot{\theta}(t) &= u_\theta(t)/J_y \\ \ddot{\psi}(t) &= u_\psi(t)/J_z \end{aligned}, \tag{1}$$

where $[x, y, z]^T \in \mathbb{R}^3$ is the position vector from the origin of the body frame to the origin of the world frame, $[\phi, \theta, \psi]^T \in \mathbb{R}^3$ is the vector of Euler angles (roll, pitch, and yaw), $m \in \mathbb{R}$ is the mass of the vehicle, $\text{diag}(J_x, J_y, J_z) \in \mathbb{R}^{3\times 3}$ is the inertia matrix, $g \in \mathbb{R}$ is the acceleration of gravity and $[u_z, u_\phi, u_\theta, u_\psi]^T \in \mathbb{R}^4$ is the vector of control inputs.

Assuming that the rotorcraft moves around the hovering state, the model (1) reduces to:

$$\begin{aligned} \ddot{x}(t) &= \theta(t)g \\ \ddot{y}(t) &= -\phi(t)g \\ \ddot{z}(t) &= u_z(t)/m - g \\ \ddot{\phi}(t) &= u_\phi(t)/J_x \\ \ddot{\theta}(t) &= u_\theta(t)/J_y \\ \ddot{\psi}(t) &= u_\psi(t)/J_z \end{aligned}. \tag{2}$$

From (2), it is clear that the translational axes are decoupled. Thus, the z-axis dynamic model of the multirotor can be separated as follows:

$$\ddot{z}(t) = u_v(t). \tag{3}$$

This model implies that any possible gravity is subsumed under the u_v term:

$$\ddot{z}(t) = \frac{u_z(t)}{m} - g = u_v(t), \tag{4}$$

where $u_z(t) = m[\ddot{z}(t) + g]$ is the actual upward thrust generated by the rotorcraft in Newtons. In control system theory, it is common to compensate factors such as gravity and the mass into u_v, as the calculation of u_z is straightforward and does not depend on any state variables.

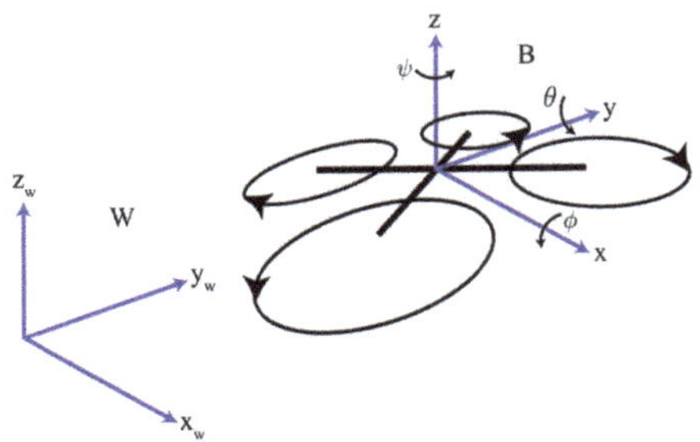

Figure 1. Coordinate frames definition.

3.2. Altitude Control

Before developing the fault-free control scheme, the following assumption is required. Define $z_{ref}(t) \in \mathbb{R}$ as the trajectory of reference. This trajectory satisfies that $\dot{z}_{ref}(t) \in \mathbb{R}$ and $\ddot{z}_{ref}(t) \in \mathbb{R}$ exist for all $t \geq 0$. This assumption is reasonable since, in practice, the trajectory that a multirotor can track is limited by the physical attributes of the multirotor.

We follow a cascade control scheme, where a low-level controller is present as the internal loop and a trajectory tracking controller is running as the external loop [37]. The inner loop controls the velocity at a high frequency and the outer loop controls the position at a low frequency. When there are no sensor faults, the formulation of the latter is:

$$u_p(t) = k_{p,1}[z_{ref}(t) - z(t)] + k_{d,1}\frac{d}{dt}[z_{ref}(t) - z(t)] + \dot{z}_{ref}(t), \tag{5}$$

where $k_{p,1} \in \mathbb{R}$ and $k_{d,1} \in \mathbb{R}$ are controller parameters to be designed. The reference velocity for the inner loop controller is obtained simply as $v_{ref} = u_p$. Then, the formulation of this loop's controller is: title = Active Fault-Tolerant Control for a Quadrotor with Sensor Faults,

$$u_v(t) = k_{p,2}[v_{ref}(t) - v_z(t)] + k_{i,2}\int_0^t [v_{ref}(t) - v_z(t)]d\tau + k_{d,2}\frac{d}{dt}[v_{ref}(t) - v_z(t)] \tag{6}$$

where $k_{p,2} \in \mathbb{R}$, $k_{i,2} \in \mathbb{R}$, and $k_{d,2} \in \mathbb{R}$ are controller parameters and $v_z(t) \in \mathbb{R}$ is the z-axis velocity of the robot.

Define the position and velocity errors as $e_z = z_{ref} - z$ and $e_v = v_{ref} - v_z$, respectively. The following two propositions provide the stability conditions for the outer and inner controllers.

Proposition 1. *The position error e_z, resulting from the application of the external controller (5) to the model (3), converges asymptotically to zero if $k_{p,1}/(1 + k_{d,1}) > 0$.*

Proof. Given that the internal loop is faster than the external one, we can regard the model from u_p to $\dot{z}$ as a proportion. Therefore, recalling the form of e_z, we have that:

$$\dot{e}_z(t) = \dot{z}_{ref}(t) - u_p = -k_{p,1}e_z - k_{d,1}\dot{e}_z = -\frac{k_{p,1}}{1 + k_{d,1}}e_z. \tag{7}$$

The characteristic equation of (7) is:

$$\lambda + \frac{k_{p,1}}{1 + k_{d,1}} = 0. \tag{8}$$

Since $k_{p,1}/(1 + k_{d,1}) > 0.$, the only root of (8) is negative, which implies that (7) is stable and $e_z(t)$ reaches zero asymptotically. □

Proposition 2. *The velocity error e_v, resulting from the application of the internal controller (6) to the model (3), converges asymptotically to zero if $k_{p,2}/(1+k_{d,2}) > 0$ and $k_{i,2}/(1+k_{d,2}) > 0$.*

Proof. For small time steps, it is reasonable to assume constant acceleration between time steps, which means that $\ddot{v}_{ref}(t) = 0$. Then, recalling the form of e_v, we have that:

$$\begin{aligned} \dot{e}_v(t) &= \dot{v}_{ref}(t) - \dot{v}_z(t) \\ \ddot{e}_v(t) &= \ddot{v}_{ref}(t) - \dot{u}_v(t) = -k_{p,2}\dot{e}_v(t) - k_{i,2}e_v(t) - k_{d,2}\ddot{e}_v(t) \end{aligned}. \tag{9}$$

Then, the system can be rewritten as:

$$\begin{bmatrix} \dot{e}_v(t) \\ \ddot{e}_v(t) \end{bmatrix} = \begin{bmatrix} 0 & 1 \\ -\frac{k_{i,2}}{1+k_{d,2}} & -\frac{k_{p,2}}{1+k_{d,2}} \end{bmatrix} \begin{bmatrix} e_v(t) \\ \dot{e}_v(t) \end{bmatrix}. \tag{10}$$

The characteristic equation of (10) is:

$$\lambda^2 + \frac{k_{p,2}}{1+k_{d,2}}\lambda + \frac{k_{i,2}}{1+k_{d,2}} = 0. \tag{11}$$

Since $k_{p,2}/(1+k_{d,2}) > 0$ and $k_{i,2}/(1+k_{d,2}) > 0$, the real parts of the roots of (11) are negative, which implies that the system (10) is stable and $e_v(t)$ reaches zero asymptotically. □

In addition to the conditions in Propositions 1 and 2, the parameters of the external and internal controllers should be determined according to the real circumstances of the multirotor under consideration. Particularly, it is common to use only PI control in the inner loop of multirotors [38]. This choice is justified in the stability analysis, where the derivative gain appears, dividing the others, meaning that it slows down the internal closed-loop dynamics.

3.3. Metric Monocular SLAM

External sensors for localization of the drone are not adequate for complex environments. To overcome this restriction, we can use a visual SLAM method. Given a monocular onboard camera, ORB-SLAM2 can be employed to obtain the camera pose and 3D point estimates without metric [39]. To address the scale problem, assuming planar ground and knowing the camera angle and distance to the ground, we can obtain a synthetic depth image by resolving the ray-ground intersection geometry. This synthetic image can be coupled with incoming RGB images and then fed to the RGB-D version of ORB-SLAM2, which generates metric pose estimates [2].

Figure 2 illustrates the side view of the ray-ground geometric configuration. Knowing the height above ground of the camera h and the camera angle α, we can define a vector n perpendicular to the ground. Therefore, a vector l departing from the camera's optical center (x_0, y_0) and passing through a pixel (x, y) will intersect the planar ground for some scalar d. The value of d is obtained with the ray-plane intersection equations:

$$\begin{aligned} l &= [(x_0 - x)/f, (y_0 - y)/f, 1]^T \\ n &= [0, -h\sin\alpha, h\cos\alpha]^T \\ d &= \frac{n^T \cdot n}{l^T \cdot n} \end{aligned}, \tag{12}$$

where f is the focal length. This approach has been evaluated using the Vicon motion capture system in indoor environments. It was found that the pose error is 2% on average [2]. Figure 3 shows examples of the system carrying out the metric mapping. In this figure, it is noticed that all map points (red and black, obtained from the synthetic depth map) are on a three dimensional plane, which corresponds to the ground.

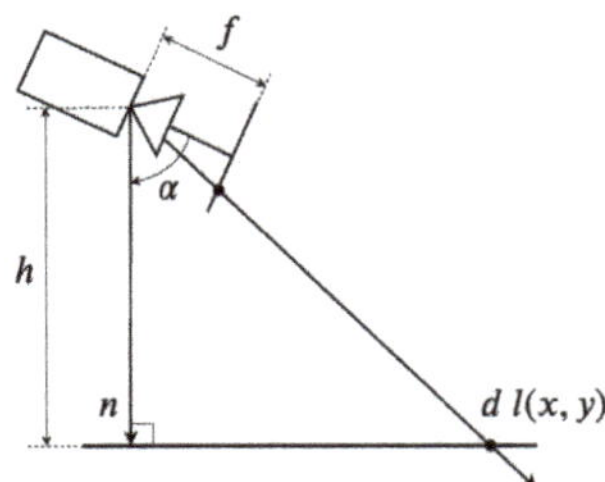

Figure 2. Geometric configuration to generate the synthetic depth image.

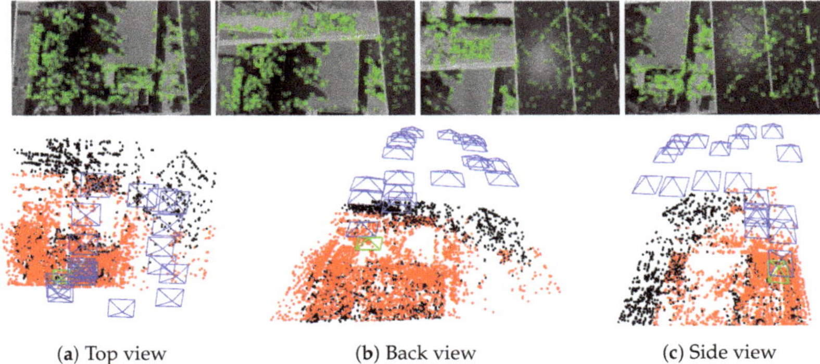

(**a**) Top view (**b**) Back view (**c**) Side view

Figure 3. Snapshots that illustrate the functioning of the metric monocular SLAM; the top row provides samples from camera frames with the tracked features (green markers); the bottom row displays three views of the map with keyframes (blue pyramids, the first one is green) and map points (red ones are being used for tracking, while black ones are ignored).

The height above ground of the camera can be obtained from the altimeter of the drone. The camera angle can be set a priori either via hardware or via software if its field of view can be foveated. We propose using the metric monocular SLAM system because it offers the following advantages. First, the metric monocular SLAM system is suitable for autonomous navigation, which means that such a system is likely to be already present in the control architecture. Second, if correctly initialized, it is robust against erroneous measurements of altitude (see Section 4.1). This property holds since the original SLAM algorithm at startup creates a keyframe with the first frame, sets its pose as the origin and creates an initial map from all keypoints with depth information [39]. The system optimizes the camera pose by finding features matches to the local map. It also optimizes the entire map by keeping the origin keyframe fixed. Therefore, the initial synthetic depth image should be generated with faultless altitude readings, which are easily obtained while hovering far from horizontal and vertical surfaces.

3.4. Fault-tolerant Control

We utilize a fault-tolerant control architecture. Figure 4 illustrates this architecture applied to our problem. First, we will describe the interior of the fault detection unit. We assume that there are available three measurements related to the altitude of the drone—range-based (y_r), inertial-based (y_i), and vision-based (y_v). The first measurement can be obtained from a combination of a range sensor (ultrasonic, infrared, laser) and a barometer. The second is commonly obtained from the IMU. The last is from the vision algorithms which use an onboard camera. These measurements relate to the state as:

$$\begin{array}{l} y_r(t) = z(t) + \delta(t) \\ y_i(t) = \dot{z}(t) \\ y_v(t) = z(t) \end{array}, \tag{13}$$

in which $\delta(t) \in \mathbb{R}$ is the sensor fault. Let us consider $r(t) \in \mathbb{R}^2$ as the residual vector defined as follows:

$$r(t) = \begin{bmatrix} r_1(t) \\ r_2(t) \end{bmatrix} = \begin{bmatrix} y_r(t) - y_v(t) \\ \dot{y}_r(t) - y_i(t) \end{bmatrix}. \tag{14}$$

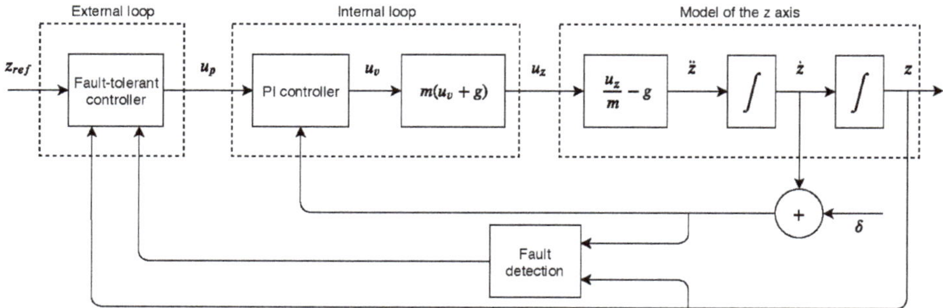

Figure 4. Fault-tolerant control schematic.

Next, we define the residual evaluation function $J_{eval}(t) \in \mathbb{R}$ with the formulation below:

$$J_{eval}(t) = [r^T(t)Wr(t)]^{1/2} = ||r(t)||_W, \tag{15}$$

where $W \in \mathbb{R}^{2\times2}$ is a real symmetric positive definite weighting matrix. In the literature, the fault diagnosis relies only on the norm of the residual [40]. We propose the weighting matrix to select the relative importance of the different components of the residual vector. Thus, the fault detection unit is operated by a logical process:

$$d_{lp} = \begin{cases} J_{eval}(t) \leq J_{th} & \text{normal case} \\ J_{eval}(t) > J_{th} & \text{fault case} \end{cases}, \tag{16}$$

where $J_{th} \in \mathbb{R}$ is the threshold of the fault detection unit and its value can be designed according to the experience of experts, considering uncertainties and disturbances present in the system.

So far, the detection or diagnosis scheme has been presented. In what follows, we will describe the general form of the fault-tolerant control law. Then, we will derive two control sub-laws that will be activated when a sensor fault is identified; the first deals with the sensor fault directly and the second deals with it as an actuator fault.

In general, the objective is to find a fault-tolerant control law for the external loop like the following equation:

$$u_p(t) = \begin{cases} u_p^n(t) & \text{normal case} \\ u_p^f(t) & \text{fault case} \end{cases}. \tag{17}$$

Inserting the measurement, u_p^n will always remain as:

$$u_p^n(t) = k_{p,1}[z_{ref}(t) - y_v(t)] + k_{d,1}\frac{d}{dt}[z_{ref}(t) - y_v(t)] + \dot{z}_{ref}(t). \tag{18}$$

The idea is to design u_p^f such that the multirotor stays close to the trajectory of the reference whenever a sensor fault is identified. If the magnitude of the sensor fault is small, the fault will not be detected by our fault detection unit. In this case, the control law (18) still guarantee the stability of the multirotor though the tracking error will not stay close to zero.

For the first control sub-law, suppose the inner loop uses a measurement with the form of the range sensor measurement in (13):

$$u_v(t) = k_{p,2}[v_{ref}(t) - \dot{y}_r(t)] + k_{i,2}\int_0^t [v_{ref}(t) - \dot{y}_r]d\tau. \tag{19}$$

For compensating disturbances produced by sensor faults of the inner loop, we will design an additive term $u_a \in \mathbb{R}$ for the outer loop. Taking into account the sensor fault and the additive term, u_v becomes:

$$u_v(t) = k_{p,2}e_v(t) + k_{i,2}\int_0^t e_v(t)d\tau + k_{p,2}[u_a - \dot{\delta}(t)] + k_{i,2}\int_0^t [u_a - \dot{\delta}(t)]d\tau. \tag{20}$$

From the previous equation, it is clear that u_a should be designed to be equal to $\dot{\delta}$. However, the time derivative of δ can be computed either with respect to the external loop or to the internal loop. To justify the selection, we will analyze the perturbation term $p = u_a - \dot{\delta}$. Let T_p and T_v be the sampling periods of the external and internal loops, respectively. We assume that $\mu = T_p/T_v \in \mathbb{Z}^+$. In discrete form, after one step of T_p, it yields:

$$p = \sum_{j=k-\mu+1}^{k}\left(u_a - \frac{\delta_j - \delta_{j-1}}{T_v}\right), \tag{21}$$

where the subscript of δ denote the discrete-time index, in essence, $\delta_k = \delta(kT_v)$. The expansion of the summation results in:

$$\begin{aligned} p = &\left(u_a - \frac{\delta_k - \delta_{k-1}}{T_v}\right) + \left(u_a - \frac{\delta_{k-1} - \delta_{k-2}}{T_v}\right) + \cdots \\ &+ \left(u_a - \frac{\delta_{k-\mu+2} - \delta_{k-\mu+1}}{T_v}\right) + \left(u_a - \frac{\delta_{k-\mu+1} - \delta_{k-\mu}}{T_v}\right) = \frac{T_p u_a - (\delta_k - \delta_{k-\mu})}{T_v}. \end{aligned} \tag{22}$$

The last form of p implies that the additive term should be $u_a = (\delta_k - \delta_{k-\mu})/T_p$. This means that internal sensor faults can be countered with information available at the time-steps of the external loop, no matter the values in between (at the cost of an approximation error). This is useful if the external loop receives little information about the internal one, which is likely the case when working with off-the-shelf drones. In reality, δ cannot be measured directly. For this reason, we propose the following expression:

$$u_a(t) = k_a r_2(t). \tag{23}$$

If both loops were to run at the same frequency, we would have nearly total fault cancellation with $k_a = 1$ since $r_2 \approx \dot{\delta}$. However, it is important to remember that the inner loop executes faster than the outer loop, so u_a remains constant until a new outer loop command is computed. In this case, choosing unitary k_a will result in an overcompensation of the fault. Moreover, the value of r_2 could be contaminated with noise. Therefore, we propose the range of this parameter to be $0 < k_a < 1$. Furthermore, this parameter could take different values for positive and negative disturbances, $k_a = \{k_a^+, k_a^-\}$. In the end, the first control sub-law is given by following equation:

$$u_p^f(t) = k_{p,1}[z_{ref}(t) - y_v(t)] + k_{d,1}\frac{d}{dt}[z_{ref} - y_v(t)] + \dot{z}_{ref}(t) + u_a(t). \tag{24}$$

Some literature has examined converting sensor faults into actuator faults through a transformation [41,42]. For comparison purposes, we consider sensor faults as actuator faults or disturbances and propose a second control sub-law based on Sliding Mode Control (SMC). Before going into detail, some remarks are in order. The time response of the internal loop is much smaller than that of the external loop, which means that the control of velocity is faster than that of the position. Therefore, we can neglect the response time of the velocity control and consider the model from u_v to $\ddot{z}$ as a proportion. Thence, the model from u_p to $\ddot{z}$ can be regarded as a double integrator. Also, we propose to activate the SMC only when a fault is detected, so $u_p \rightarrow u_p^f$. Considering disturbances or the actuator fault, the model of the z axis results in:

$$\ddot{z}(t) = u_p^f(t) + c(t), \tag{25}$$

where $c(t)$ is a composite disturbance term. We assume that $|c(t)| \leq C$ for some known $C > 0$. Define the sliding surface s with:

$$s(t) = \dot{e}_z(t) + k_e e_z(t),\ k_e > 0. \tag{26}$$

The second control sub-law becomes:

$$u_p^f(t) = \ddot{z}_{ref}(t) + k_e\dot{e}_z(t) + k_s \operatorname{sign}[s(t)],\ k_s > 0. \tag{27}$$

Proposition 3. *The control (27) drives the position error to the sliding surface (26) and keeps the error on the surface thereafter in the presence of the bounded disturbance $c(t)$.*

Proof. Recalling the position error e_z, it follows that:

$$\ddot{e}_z = \ddot{z}_{ref} - \ddot{z} = \ddot{z}_{ref} - u_p^f - c. \tag{28}$$

Let us select a candidate Lyapunov function as:

$$V = \frac{1}{2}s^2. \tag{29}$$

Now, let us take the derivative of V as in the usual Lyapunov method:

$$\dot{V} = s\dot{s} = s(\ddot{z}_{ref} - u_p^f - c + k_e\dot{e}_z) = -s[k_s \operatorname{sign}(s) + c] = -k_s|s| - sc. \tag{30}$$

Taking the absolute value of the second term and considering the disturbance bound, it results in:

$$\dot{V} \leq -k_s|s| + |s|C \leq -|s|(k_s - C). \tag{31}$$

Suppose k_s is chosen appropriately ($k_s > C$), it yields that $\dot{V} < 0$ for $s \neq 0$. Therefore, the region $s = 0$ must be invariant. □

Throughout the next section, we will refer as (Fault-Tolerant Controller) FTC 1 to the strategy that adopts (24), and as FTC 2 to the one that uses (27).

3.5. Equipment

For safety, we first tested the proposed control laws in simulation. The whole system was implemented in Simulink. The block diagram approach makes it easy to combine continuous models with discrete systems, such as control loops running on computers. In this way, we can select the sampling period of the external and internal loops individually. Sensor measurements are obtained by simply passing the signals of the plant through zero-order-hold blocks. Also, we emulate the range sensor faults by adding the output of the uniform random number block to the position variable at specified times and intervals.

In the real-world experiments, we used a Parrot Bebop 2 quadcopter for which a software driver is available, allowing the usage of simple commands for takeoff, landing and piloting. The Parrot Bebop 2 weighs 0.5 kg and has a 29 cm frame. It has three-blade propellers measuring 6 inches. This drone sends and receives data via WiFi. Among the data that the drone sends, we can find the video stream from its monocular frontal camera and the altitude gained by fusing data from its ultrasonic sensor and barometer. Also, the camera angle can be controlled via the Software Development Kit (SDK) (https://developer.parrot.com/docs/SDK3/). Lastly, we obtain the ground truth of the position with a Vicon Vantage motion capture system (https://www.vicon.com/hardware/cameras/vantage/). We placed reflective markers on the drone and configured the capture system to deliver measurements at 100 Hz. This system provides measurements with millimeter accuracy.

4. Results and Discussion

This section presents results comparing the two control sub-laws described in the previous section. We begin by describing the general behavior of sensor faults induced by the ground effect. Then, the proposed strategies will be evaluated in simulation. The strategies will also be evaluated in real-world experiments. For supplementary video see: https://youtu.be/uszilXBFKP4. The project's code is available at: https://github.com/AMatusV/mrotor-sfaults-control.

4.1. Altitude Sensor Faults

For low-level flights, the altitude of UAVs is measured with a combination of barometers and range sensors. The first type of sensor is affected by the ground effect because of the increment in the air pressure around the vehicle [3]. In general, the second type of sensor has limitations when there are abrupt changes in the measurement surface. Particularly, ultrasonic sensors, which are perhaps the most widely used for multirotors, are subjected to other problems such as acoustic noise and air turbulence (https://www.maxbotix.com/ultrasonic-sensor-operation-uav.htm). For the latter, the best results are obtained by mounting the sensor as far away from the propellers as possible. Nevertheless, the presence of the ground induces an upwash encountering the central part of the rotorcraft body [17,43]. Thus, flying close to the ground may cause sensor faults.

Technical manuals have reported the tendency of rotorcrafts to climb back into the air when close to the ground (https://docs.px4.io/v1.9.0/en/advanced_config/tuning_the_ecl_ekf.html, http://ardupilot.org/copter/docs/ground-effect-compensation.html). This is caused when the high-pressure zone below the drone affects the barometer. The result is a lower reading or sensor fault in pressure altitude, leading to the inner loop commanding an unwanted climb.

Applying ultrasonic wave propagation as an airflow velocity sensor is not a new concept. Ultrasonic sensors are used in several applications such as in gas, hydraulic and airflow meters [44,45]. This reinforces the idea that the ground effect could cause ultrasonic range sensor faults. Moreover, the authors' experience has shown the following strange behavior when flying quadrotors close to the ground—the rotorcraft suddenly descends without being commanded to do so. This behavior is not in line with the literature about the ground effect.

To demonstrate the general behavior of the altitude faults induced by the ground effect, we collected data from the quadcopter described in Section 3.5. With a proportional outer loop controller, we commanded the drone to hover close to the ground. The feedback for this controller is obtained from the vision algorithm described in Section 3.3. Figure 5 presents the results of three tests in which sensor faults actually occur. A sensor fault occurs when the altitude reported by the drone, y_r, deviates from the ground truth (Vicon). For example, a sensor fault appeared in Test 1 around the 40 s mark which last 3 s approximately. In Test 2, a sensor fault appeared between 27 s and 30 s. Remaining deviations in this test did not affect the vehicle significantly and could be ignored by a suitable detection threshold. Test 3 shows a fault between 25 s and 30 s, the deviation was in the opposite direction than the faults in the other tests. Another fault happened around 42 s which can be detected by an appropriate weighting of the residual. Deviations detected as sensor faults cause the inner loop to react, producing a disturbance for the outer controller, see Figure 6. As expected, the position disturbance is in the opposite direction of the sensor fault.

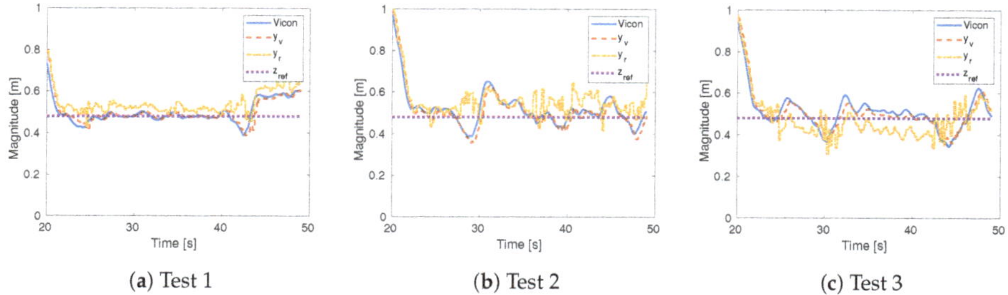

(a) Test 1 (b) Test 2 (c) Test 3

Figure 5. Position response while hovering close to the ground.

It is important to note that the vision-based measurement agrees well with the ground truth, even though it relies on the range sensor to estimate the metric pose. The vision algorithm, assuming planar ground and knowing the camera angle and distance to the ground, constructs a synthetic depth image, which is coupled to an RGB (Red-Green-Blue) frame. Then, the RGB-D (Red-Green-Blue-Depth) version of ORB-SLAM2 consumes this data to generate a pose estimate. Our results show that the SLAM system is robust against range sensor faults when the camera moves mainly along the principal axis. In this situation, the ability to perform relocalization and reuse the map yield robustness to the system. For this to happen, the initial synthetic depth image should be generated with faultless altitude readings, which can be easily obtained while hovering far from horizontal and vertical surfaces.

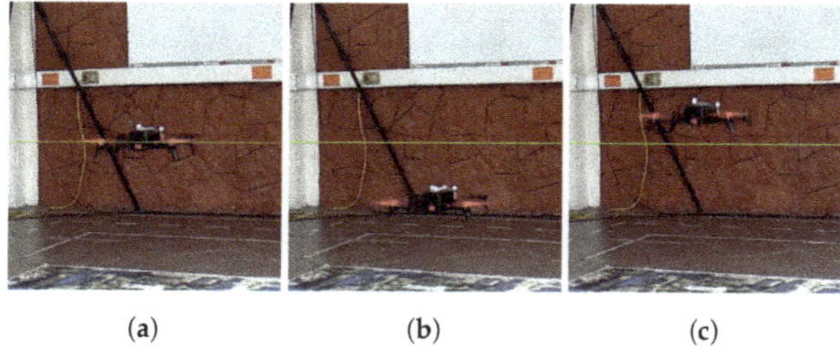

(a) (b) (c)

Figure 6. Photographs showing the quadrotor in three conditions: (**a**) faultless, (**b**) descending due to an upward sensor fault, and (**c**) ascending due to a downward sensor fault; we indicate the reference with a green line; for supplementary video check https://youtu.be/uszilXBFKP4.

Similar to this subsection, in the next ones, we will restrict to examining tests in hover conditions. This restriction is reasonable since it represents the situation in which the vehicle is most vulnerable to disturbances. Moreover, this situation extends to missions where the drone is tracking a trajectory in the x-y plane while maintaining a constant altitude.

4.2. Simulation

In the simulation, besides the double integrator model (4), we considered the following expression for the rotor thrust increment due to the ground effect:

$$\frac{T_{IGE}}{T_{OGE}} = \frac{1}{1 - \frac{R^2}{16z^2}}, \tag{32}$$

where T_{OGE} is the thrust generated by the rotorcraft flying out of ground effect, T_{IGE} is the thrust when the rotorcraft is in ground effect, R is the radius of the rotor, and z is the vertical distance of the rotor to the ground. Specifically, we set $R = 4r$, where $r = 0.0762$ m is the radius of one propeller of the Bebop 2.

The control objective is to track a reference of 0.5 m from initial conditions of 1 m and 0 m/s. The mass of the vehicle is 0.5 kg and the gravity acceleration is 9.81 m/s^2. The values of fault-free control parameters are $k_{p,1} = 1$, $k_{d,1} = 0.1$, and $k_{p,2} = k_{i,2} = 10$. The residual weighting matrix is $W = \text{diag}(2, 0.5)$ and the fault detection threshold is $J_{th} = 0.2$. For the FTC strategies, the parameters are $k_a^+ = 0.3$, $k_a^- = 0.1$, $k_e = 1.7$ and $k_s = 0.01$. The total simulation time is 30 s. To simulate sensor faults, we considered a uniform random generator with range $[0.01, 0.2]$, which is added to the internal loop measurement at times 5 s to 6 s, 15 s to 17, and 25 s to 28 s. We set the sampling times of the external and internal loops to 0.05 s and 0.01 s, respectively. Also, we assume that the internal loop shares its measurement with the external loop every 0.2 s.

Figure 7 shows the outputs of the sensors, the reference and the fault detection scheme. It can be seen that the sensor faults were properly detected at the time intervals at which the output of the random generator was added to the internal loop measurement. These graphical results demonstrate that the fault detector indirectly depends on the control strategy. Nevertheless, all significant sensor faults were recognized for each control strategy. Also, it can be noted that the fault can be tolerated after the control sub-laws are adopted.

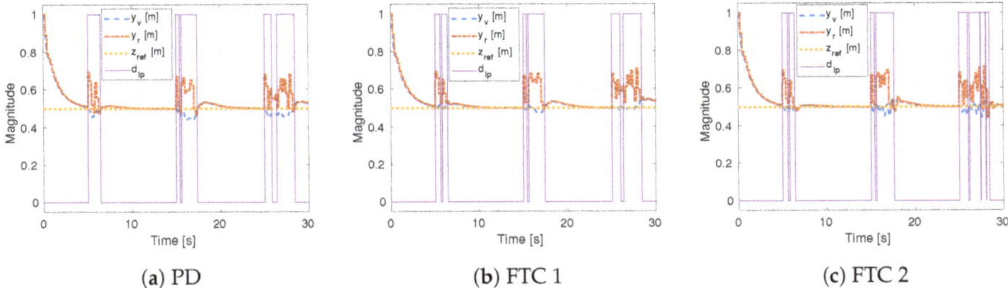

(a) PD (b) FTC 1 (c) FTC 2

Figure 7. Simulation response of three control strategies.

Figure 8 illustrates the control actions of each strategy. As anticipated, the PD output reached lower magnitudes than the FTC strategies. On the other hand, the FTC 2 scheme attained higher command magnitudes. This might indicate problems in real implementation. The rapid changes in commands could cause fast battery depletion. Also, the vision algorithm that estimates the position could fail due to the fast movements of the camera.

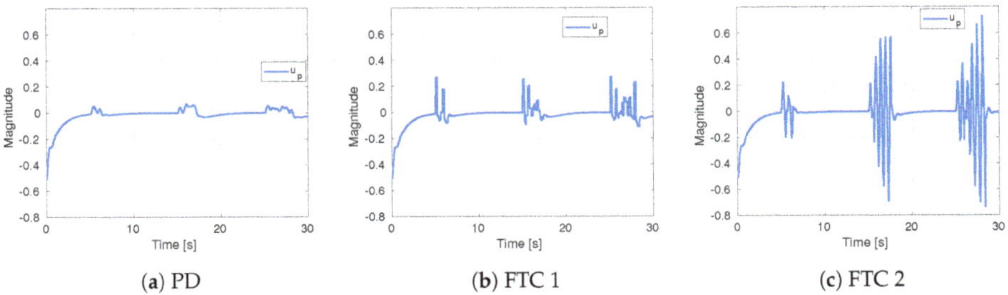

(a) PD (b) FTC 1 (c) FTC 2

Figure 8. Control actions of three strategies in simulation.

Qualitatively, the FTC 2 scheme yielded the best response. For a quantitative comparison, we computed the Root Mean Square Error (RMSE) and summarized the results in Table 1. Compared to the PD controller, the proposed strategies improved the error measure in 24.41% and 38.06%, respectively.

Table 1. Error measures of the control strategies in simulation.

	PD	FTC 1	FTC 2
RMSE	0.0214	0.0172	0.0155

An important take away from the simulation is the advantage of adopting asymmetrical values for k_a^+ and k_a^-. The simulation clarified that the integral action of the inner loop provides a negative offset to compensate for the ground effect. In consequence, negative commands of the external loop have more impact than positive commands until the integral reaches to zero.

4.3. Experiments

In the experiments, we exploit the features of the Bebop 2 to run the metric monocular SLAM algorithm to estimate the pose of the camera (refer to Section 3.3), which will be the feedback of the external controller. For all tests, the camera angle was set to -83° with respect to the horizon.

In the tests with the drone, we restrict the x-y position and the yaw angle with PID controllers: $u_x(t) = \mathrm{PID}(x;t)$, $u_y(t) = \mathrm{PID}(y;t)$, and $u_\psi(t) = \mathrm{PID}(\psi;t)$. To preserve the direction in global coordinates of the x-y projection of the vector generated by the position controllers, we apply the transformation in (33).

$$\begin{aligned} F &= \sqrt{u_x^2 + u_y^2} \\ u_{x,b} &= F\cos\left[\arctan\left(\frac{u_y}{u_x}\right) - \psi\right]. \\ u_{y,b} &= F\sin\left[\arctan\left(\frac{u_y}{u_x}\right) - \psi\right] \end{aligned} \tag{33}$$

The control objective is to track a reference of 0.48 m from an initial condition of 1 m. The values of fault-free control parameters are $k_{p,1} = 0.6$, and $k_{d,1} = 0.2$; the drone's inner loop parameters are kept as default. The residual weighting matrix is $W = \mathrm{diag}(1, 0.5)$ and the fault detection threshold is $J_{th} = 0.2$. For the FTC strategies, the parameters are $k_a^+ = 0.5$, $k_a^- = 0.2$, $k_e = 1.25$, and $k_s = 0.01$. In practice, we observed variable frequency in the vision feedback (external loop) with an average of 15 Hz (≈ 0.0667 s). Like in the simulation, the Bebop 2 publishes altitude measurements every 0.2 s.

Instead of emulating the sensor faults, we opted for evaluating flights with real faults, in essence, sensor faults actually occur. Given the apparent stochastic behavior of sensor faults induced by the ground effect, we collected results from five tests for each control strategy. Only tests with detected sensor faults and without vision tracking losses were examined. For comparison, we have plotted the error signals in Figure 9. Overall, the figure reveals that the best qualitative behavior was obtained with FTC 1. This observation is confirmed with Table 2, where the FTC 1 strategy has the lowest average RMSE. Taking as reference the PD strategy, the FTC techniques improved the average error measure in 85.55% and 8.43%, respectively.

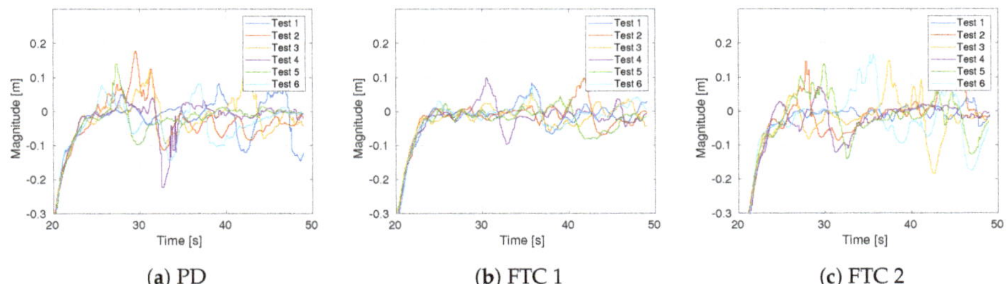

(**a**) PD (**b**) FTC 1 (**c**) FTC 2

Figure 9. Error response of the control strategies for five tests with the real drone.

Table 2. Error measures of the control strategies with the real drone.

RMSE	Test 1	Test 2	Test 3	Test 4	Test 5	Average
PD	0.0492	0.0591	0.0479	0.0460	0.0549	0.0514
FTC 1	0.0266	0.0304	0.0249	0.0317	0.0247	0.0277
FTC 2	0.0215	0.0576	0.0579	0.0417	0.0582	0.0474

For further comparison, we picked one test of each strategy and plotted their signals in Figure 10. Different from the simulations, in the experiments, a sensor fault may occur gradually (see the first detection in Figure 10b). In this case, a fault can still be detected (after a delay) since our detector depends on the position component of the residual. With J_{th}, we sacrifice the speed at which we can identify a gradual fault for robustness against noise. On the other hand, it can be observed that the worst-case scenario from FTC 1 dominates the median response of the other strategies. Unlike in the simulations, FTC 2 shows error peaks similar to the standard PD strategy.

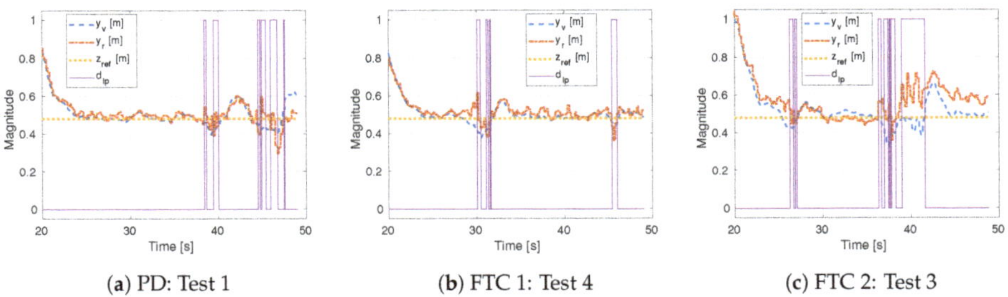

(**a**) PD: Test 1 (**b**) FTC 1: Test 4 (**c**) FTC 2: Test 3

Figure 10. Sensors, reference and fault detection signals for individual tests.

The control commands for the tests in the previous plots are shown in Figure 11. The same trend as in simulations can be seen in this figure—the commands of FTC 2 exhibits the highest peaks, followed by FTC 1. The noise in the measurement and the computation of the unsmoothed first derivative explain the peaks in the PD commands.

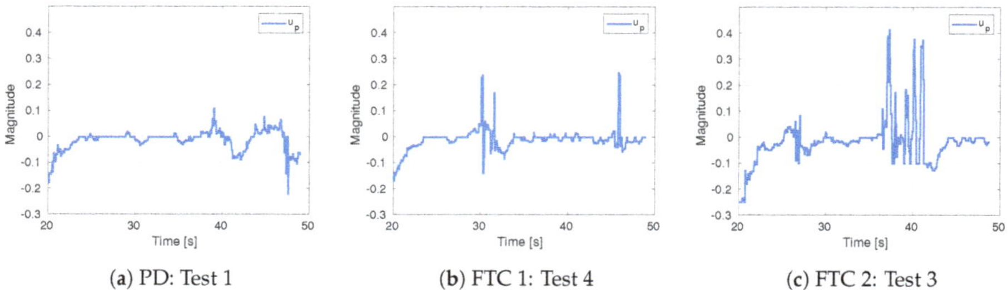

(a) PD: Test 1 (b) FTC 1: Test 4 (c) FTC 2: Test 3

Figure 11. Control signals for individual tests.

While collecting results for each strategy, we had complications with the vision algorithm. On one hand, the vision tracking was being lost when the camera approached the ground. On the other, the vision tracking was being lost when the camera moved too fast. For the PD controller, we were losing the vision feedback because the commands allowed the drone to move close to the ground. Whereas for FTC 2, we were losing the feedback because the commands moved the drone rapidly. FTC 1 reduced the occurrence of this problem by keeping the drone close to the reference while using moderate energy.

The performance of FTC 2 was significantly different in the experiments and the simulation. One evident reason is the reduction of parameter k_e in the experiments. We lowered this parameter because we were not obtaining tests without vision feedback losses with higher values of k_e. Another evident reason is the lower variable frequency of the real external loop. Together, these differences explain the performance degradation of FTC 2 in the experiments.

Focusing on the improvements of the error measure for FTC 1, the improvement was greater in the experiments than in the simulation. Different from the simulation, sensor faults in real tests worsen as the vehicle moves closer to the ground. This indicates that the PD controller is expected to have inferior performance in reality. Moreover, the opportune and moderate commands of FTC 1 somewhat preserved the performance of this strategy. The combined effect is the increase in error measure improvement for FTC 1 in the experiments.

5. Conclusions

In this article, we have presented a procedure for fault diagnosis and two control sub-laws for a multirotor with altitude sensor faults. In particular, we have considered sensor faults induced by the ground effect. In the fault-free case, a hierarchical control composed by an external PD and an internal PI controller has been developed for trajectory tracking. Exploiting the trend of onboard cameras, we have considered a vision-based feedback for the external loop based on a well known technique in robotics known as Monocular SLAM. This is the first time such a technique has been used to address any issue related to ground effect in multirotors

The fault diagnosis has been achieved using a weighted residual, which is obtained by comparing estimations from the metric monocular SLAM system against faulty internal readings. The first control sub-law has been proposed as a combination of the external PD controller and a function of the residual. The second control sub-law has been based on SMC. The performance of the strategies has been illustrated in simulations and experiments. It is important to remark that we have adopted onboard sensors only.

Regarding the results, we have discovered that both control sub-laws overcame the performance of the standard PD controller in simulations. However, the first control sub-law offered better behavior in experiments. The performance of the second control sub-law degraded due to a combination of its rapid switching nature and limitations of the vision algorithm. In the experiments, we have found that the first control sub-law improved the RMSE in 85% compared to the PD controller.

Results reported in this work indicate that our fault detection scheme is feasible for altitude faults induced by the ground effect. To our knowledge, this is the first time that the problem of detecting internal sensor faults is addressed by using a metric monocular SLAM system. Besides, we have shown that our first control sub-law enhances the flying performance when hovering close to the ground. Therefore, this controller can be used to improve missions such as take-off, landing, hovering and operating near the ground in general.

Future work will focus on considering model uncertainties and sensor noise. These terms could be dealt with either with a state observer or with a Kalman filter. To compensate for delays in the measurements, a mathematical model could be used for prediction. Also, rejection of altitude sensor faults caused by abrupt changes in the surface below the drone should be explored.

Author Contributions: Conceptualization, A.M.-V., G.R.-G., and J.M.-C.; methodology, A.M.V., G.R.-G., and J.M.-C.; software, A.M.-V., validation, A.M.-V.

Funding: This research received no external funding.

Acknowledgments: The first author is thankful to Consejo Nacional de Ciencia y Tecnología (CONACYT) for the scholarship (No. 540945).

Conflicts of Interest: The authors declare no conflict of interest.

List of Symbols

x, y, z	Position in the world-fixed frame
ϕ, θ, ψ	Roll, pitch, and yaw angles
v_z	Velocity of the z axis
m	Mass of the multirotor
g	Gravitational acceleration
J_x, J_y, J_z	Diagonal elements of the inertia matrix
u_z	Thrust force
u_ϕ, u_θ, u_ψ	Input torques
u_v, u_p	Control inputs of acceleration and velocity
z_{ref}, v_{ref}	Position and velocity references of the z axis
e_z, e_v	Position and velocity errors of the z axis
$k_{p,1}, k_{d,1}$	Parameters of the external controller
$k_{p,2}, k_{i,2}, k_{d,2}$	Parameter of the internal controller
h	Height above ground of the camera
α	Angle of the camera
f	Focal length of the camera
l	Vector from the camera's optical center through a pixel to the ground
n	Vector perpendicular to the ground from the camera's optical center to the ground
d	Distance at which l intersects the ground
δ	Sensor fault
y_r, y_v	Range-based and vision-based measurements of the altitude
y_i	Inertial-based measurement
r	Residual vector
J_{eval}	Residual evaluation function
J_{th}, d_{lp}	Fault detection threshold and logical variable
T_p, T_v	Sampling periods of the external and internal loops
u_p^n, u_p^f	Control input of velocity in normal case and fault case
u_a	Additive control term
$k_a = \{k_a^+, k_a^-\}$	Control gain (for positive and negative disturbances) of FTC 1
c, C	Composite disturbance term and its bound
s, k_e	Sliding surface and its parameter
k_s	Sliding mode control gain of FTC 2

References

1. Hesch, J.A.; Kottas, D.G.; Bowman, S.L.; Roumeliotis, S.I. Camera-IMU-based localization: Observability analysis and consistency improvement. *Int. J. Rob. Res.* **2013**, *33*, 182–201. doi:10.1177/0278364913509675. [CrossRef]
2. Moon, H.; Martinez-Carranza, J.; Cieslewski, T.; Faessler, M.; Falanga, D.; Simovic, A.; Scaramuzza, D.; Li, S.; Ozo, M.; Wagter, C.D.; et al. Challenges and implemented technologies used in autonomous drone racing. *Intell. Serv. Rob.* **2019**, *12*, 137–148. doi:10.1007/s11370-018-00271-6. [CrossRef]
3. Nakanishi, H.; Kanata, S.; Sawaragi, T. Measurement model of barometer in ground effect of unmanned helicopter and its application to estimate terrain clearance. In Proceedings of the 2011 IEEE International Symposium on Safety, Security, and Rescue Robotics, Kyoto, Japan, 1–5 November 2011; pp. 232–237.
4. Cheeseman, I.; Bennet, W. The Effect of the Ground on a Helicopter Rotor in Forward Flight. *Aeronaut. Rese. Counc. Rep. Memo* **1957**, *3021*, 1–10.
5. Hayden, J.S. The Effect of the Ground on Helicopter Hovering Power Required. In Proceedings of the 32nd Annual Forum of the American Helicopter Society, Washington, WA, USA, 10–12 May 1976. Available online: https://vtol.org/store/product/the-effect-of-the-ground-on-helicopter-hovering-power-required-9877.cfm (accessed on 11 November 2019).
6. Lee, T.E.; Leishman, J.G.; Ramasamy, M. Fluid Dynamics of Interacting Blade Tip Vortices with a Ground Plane. *J. Am. Helicopter Soc.* **2010**, *55*, 22005. doi:10.4050/jahs.55.022005. [CrossRef]
7. Nonaka, K.; Sugizaki, H. Integral sliding mode altitude control for a small model helicopter with ground effect compensation. In Proceedings of the 2011 American Control Conference, San Francisco, CA, USA, 29 June–1 July 2011; pp. 202–207.
8. Ryan, T.; Kim, H.J. Modelling of Quadrotor Ground Effect Forces via Simple Visual Feedback and Support Vector Regression. In Proceedings of the AIAA Guidance, Navigation, and Control Conference, Minneapolis, MN, USA, 13–16 August 2012; pp. 1–12.
9. Lee, D.; Kim, H.J.; Sastry, S. Feedback linearization vs. adaptive sliding mode control for a quadrotor helicopter. *Int. J. Control Autom. Syst.* **2009**, *7*, 419–428. doi:10.1007/s12555-009-0311-8. [CrossRef]
10. Mirzaei, M.; Nia, F.S.; Mohammadi, H. Applying adaptive fuzzy sliding mode control to an underactuated system. In Proceedings of the 2nd International Conference on Control, Instrumentation and Automation, Shiraz, Iran, 27–29 December 2011; pp. 654–659.
11. Matus-Vargas, A.; Rodriguez-Gomez, G.; Martinez-Carranza, J. Aerodynamic Disturbance Rejection Acting on a Quadcopter Near Ground. In Proceedings of the 2019 6th International Conference on Control, Decision and Information Technologies (CoDIT), Paris, France, 23–26 April 2019; pp. 1516–1521.
12. Robinson, D.C. Modelling and Estimation of Aerodynamic Disturbances Acting on a Hovering Micro Helicopter in Close Proximity to Planar Surfaces. Ph.D. Thesis, Monash University, Melbourne, Australia, 2016.
13. Lakshminarayan, V.K.; Kalra, T.S.; Baeder, J.D. Detailed Computational Investigation of a Hovering Microscale Rotor in Ground Effect. *AIAA J.* **2013**, *51*, 893–909. doi:10.2514/1.j051789. [CrossRef]
14. Kalra, T.S. CFD Modeling and Analysis of Rotor Wake in Hover Interacting with a Ground Plane. Ph.D. Thesis, University of Maryland, College Park, MD, USA, 2014.
15. Guenard, N.; Hamel, T.; Eck, L. Control Laws For The Tele Operation of An Unmanned Aerial Vehicle Known as An X4-flyer. In Proceedings of the 2006 IEEE/RSJ International Conference on Intelligent Robots and Systems, Beijing, China, 9–15 October 2006; pp. 3249–3254.
16. Powers, C.; Mellinger, D.; Kushleyev, A.; Kothmann, B.; Kumar, V. Influence of Aerodynamics and Proximity Effects in Quadrotor Flight. In *Experimental Robotics*; Springer: Heidelberg, Germany, 2013; pp. 289–302. doi:10.1007/978-3-319-00065-7_21. [CrossRef]
17. Sharf, I.; Nahon, M.; Harmat, A.; Khan, W.; Michini, M.; Speal, N.; Trentini, M.; Tsadok, T.; Wang, T. Ground effect experiments and model validation with Draganflyer X8 rotorcraft. In Proceedings of the 2014 International Conference on Unmanned Aircraft Systems (ICUAS), Orlando, FL, USA, 27–30 May 2014; pp. 1158–1166.
18. Sanchez-Cuevas, P.; Heredia, G.; Ollero, A. Characterization of the Aerodynamic Ground Effect and Its Influence in Multirotor Control. *Int. J. Aerosp. Eng.* **2017**, *2017*, 1–17. doi:10.1155/2017/1823056. [CrossRef]

19. Nobahari, H.; Sharifi, A. Continuous ant colony filter applied to online estimation and compensation of ground effect in automatic landing of quadrotor. *Eng. Appl. Artif. Intell.* **2014**, *32*, 100–111. doi:10.1016/j.engappai.2014.03.004. [CrossRef]
20. Hu, B.; Lu, L.; Mishra, S. Fast, safe and precise landing of a quadrotor on an oscillating platform. In Proceedings of the 2015 American Control Conference (ACC), Chicago, IL, USA, 1–3 July 2015. pp. 3836–3841.
21. Danjun, L.; Yan, Z.; Zongying, S.; Geng, L. Autonomous landing of quadrotor based on ground effect modelling. In Proceedings of the 2015 34th Chinese Control Conference (CCC), Hangzhou, China, 28–30 July 2015; pp. 5647–5652.
22. Bartholomew, J.; Calway, A.; Mayol-Cuevas, W. Learning to predict obstacle aerodynamics from depth images for Micro Air Vehicles. In Proceedings of the 2014 IEEE International Conference on Robotics and Automation (ICRA), Hong Kong, China, 31 May–7 June 2014; pp. 4967–4973.
23. Bartholomew, J.; Calway, A.; Mayol-Cuevas, W. Improving MAV control by predicting aerodynamic effects of obstacles. In Proceedings of the 2015 IEEE/RSJ International Conference on Intelligent Robots and Systems (IROS), Hamburg, Germany, 28 September–2 October 2015; pp. 4826–4873.
24. Tomic, T.; Haddadin, S. A unified framework for external wrench estimation, interaction control and collision reflexes for flying robots. In Proceedings of the 2014 IEEE/RSJ International Conference on Intelligent Robots and Systems, Chicago, IL, USA, 14–18 September 2014; pp. 4197–4204.
25. Du, H.; Pu, Z.; Yi, J.; Qian, H. Advanced quadrotor takeoff control based on incremental nonlinear dynamic inversion and integral extended state observer. In Proceedings of the 2016 IEEE Chinese Guidance, Navigation and Control Conference (CGNCC), Nanjing, China, 12–14 August 2016; pp. 1881–1886.
26. Robinson, D.C.; Ryan, K.; Chung, H. Helicopter hovering attitude control using a direct feedthrough simultaneous state and disturbance observer. In Proceedings of the 2015 IEEE Conference on Control Applications (CCA), Sydney, NSW, Australia, 21–23 September 2015; pp. 633–638. doi:10.1109/cca.2015.7320700. [CrossRef]
27. Xiao, B.; Yin, S. A New Disturbance Attenuation Control Scheme for Quadrotor Unmanned Aerial Vehicles. *IEEE Trans. Ind. Inf.* **2017**, *13*, 2922–2932. doi:10.1109/tii.2017.2682900. [CrossRef]
28. McKinnon, C.D.; Schoellig, A.P. Unscented external force and torque estimation for quadrotors. In Proceedings of the 2016 IEEE/RSJ International Conference on Intelligent Robots and Systems (IROS), Daejeon, South Korea, 9–14 October 2016; pp. 5651–5657.
29. Zhang, Y.; Chamseddine, A.; Rabbath, C.; Gordon, B.; Su, C.Y.; Rakheja, S.; Fulford, C.; Apkarian, J.; Gosselin, P. Development of advanced FDD and FTC techniques with application to an unmanned quadrotor helicopter testbed. *J. Franklin Inst.* **2013**, *350*, 2396–2422. [CrossRef]
30. Berbra, C.; Lesecq, S.; Martinez, J. A Multi-observer Switching Strategy for Fault-Tolerant Control of a Quadrotor Helicopter. In Proceedings of the 2016 16th Mediterranean Conference on Control and Automation, Ajaccio, France, 25–27 June 2008; pp. 1094–1099.
31. Rafaralahy, H.; Richard, E.; Boutayeb, M.; Zasadzinski, M. Simultaneous observer based sensor diagnosis and speed estimation of Unmanned Aerial Vehicle. In Proceedings of the 2008 47th IEEE Conference on Decision and Control, Cancun, Mexico, 9–11 December 2008; pp. 2938–2943.
32. Nguyen, H.; Berbra, C.; Lesecq, S.; Gentil, S.; Barraud, A.; Godin, C. Diagnosis of an Inertial Measurement Unit based on set membership estimation. In Proceedings of the 2009 17th Mediterranean Conference on Control and Automation, Thessaloniki, Greece, 24–26 June 2009; pp. 211–216.
33. Qin, L.; He, X.; Yan, R.; Zhou, D. Active Fault-Tolerant Control for a Quadrotor with Sensor Faults. *J. Intell. Rob. Syst.* **2017**, *88*, 449–467. doi:10.1007/s10846-017-0474-0. [CrossRef]
34. Cook, M.V. Systems of Axes and Notation. In *Flight Dynamics Principles*; Elsevier: Oxford, UK, 2013; pp. 13–32.
35. Bouabdallah, S.; Siegwart, R. Full control of a quadrotor. In Proceedings of the 2007 IEEE/RSJ International Conference on Intelligent Robots and Systems, San Diego, CA, USA, 29 October–2 November 2007; pp. 1–6. doi:10.1109/iros.2007.4399042. [CrossRef]
36. Bresciani, T. Modelling, Identification and Control of a Quadrotor Helicopter. Master's Thesis, Lund University, Lund, Sweden, 2008.
37. Xuan-Mung, N.; Hong, S.K. Improved Altitude Control Algorithm for Quadcopter Unmanned Aerial Vehicles. *Appl. Sci.* **2019**, *9*, 2122. doi:10.3390/app9102122. [CrossRef]

38. Khan, H.S.; Kadri, M.B. Attitude and altitude control of quadrotor by discrete PID control and non-linear model predictive control. In Proceedings of the 2015 International Conference on Information and Communication Technologies (ICICT), Karachi, Pakistan, 12–13 December 2015; pp. 1–11.
39. Mur-Artal, R.; Tardos, J.D. ORB-SLAM2: An Open-Source SLAM System for Monocular, Stereo, and RGB-D Cameras. *IEEE Trans. Rob.* **2017**, *33*, 1255–1262. doi:10.1109/tro.2017.2705103. [CrossRef]
40. Noura, H.; Theilliol, D.; Ponsart, J.C.; Chamseddine, A. *Fault-tolerant Control Systems*; Springer: Berlin/Heidelberg, Germany, 2009. doi:10.1007/978-1-84882-653-3. [CrossRef]
41. Tan, C.P.; Edwards, C. Sliding mode observers for robust detection and reconstruction of actuator and sensor faults. *Int. J. Robust Nonlinear Control* **2003**, *13*, 443–463. doi:10.1002/rnc.723. [CrossRef]
42. Zhang, K.; Jiang, B.; Cocquempot, V. Adaptive Observer-based Fast Fault Estimation. *Int. J. Control Autom. Syst.* **2008**, *6*, 320–326.
43. Griffiths, D.A.; Ananthan, S.; Leishman, J.G. Predictions of Rotor Performance in Ground Effect Using a Free-Vortex Wake Model. *J. Am. Helicopter Soc.* **2005**, *50*, 302–314. doi:10.4050/1.3092867. [CrossRef]
44. Raine, A.; Aslam, N.; Underwood, C.; Danaher, S. Development of an Ultrasonic Airflow Measurement Device for Ducted Air. *Sensors* **2015**, *15*, 10705–10722. doi:10.3390/s150510705. [CrossRef] [PubMed]
45. Davies, D.G.; Bolam, R.C.; Vagapov, Y.; Excell, P. Ultrasonic sensor for UAV flight navigation. In Proceedings of the 2018 25th International Workshop on Electric Drives: Optimization in Control of Electric Drives (IWED), Moscow, Russia, 31 January–2 February 2018; pp. 1–7.

Article

Honeycomb Map: A Bioinspired Topological Map for Indoor Search and Rescue Unmanned Aerial Vehicles

Ricardo da Rosa [1,2,*,†], Marco Aurelio Wehrmeister [2,†], Thadeu Brito [3,†], José Luís Lima [3,4,†] and Ana Isabel Pinheiro Nunes Pereira [3,†]

1 Federal Institute of Education, Science and Technology—Parana (IFPR), 85814-800 Campus Cascavel, Brazil
2 Campus Curitiba, Federal University of Technology—Parana (UTFPR), 80230-901 Curitiba, Brazil; wehrmeister@utfpr.edu.br
3 Campus de Santa Apolónia, Instituto Politécnico de Bragança (IPB), Research Centre in Digitalization and Intelligent Robotics (CeDRI), 5300-253 Bragança, Portugal; brito@ipb.pt (T.B.); jllima@ipb.pt (J.L.L.); apereira@ipb.pt (A.I.P.N.P.)
4 INESC TEC - INESC Technology and Science, 4200-465 Porto, Portugal
* Correspondence: ricardo.rosa@ifpr.edu.br; Tel.: +55-45-99141-8255
† These authors contributed equally to this work.

Received: 16 January 2020; Accepted: 4 February 2020; Published: 8 February 2020

Abstract: The use of robots to map disaster-stricken environments can prevent rescuers from being harmed when exploring an unknown space. In addition, mapping a multi-robot environment can help these teams plan their actions with prior knowledge. The present work proposes the use of multiple unmanned aerial vehicles (UAVs) in the construction of a topological map inspired by the way that bees build their hives. A UAV can map a honeycomb only if it is adjacent to a known one. Different metrics to choose the honeycomb to be explored were applied. At the same time, as UAVs scan honeycomb adjacencies, RGB-D and thermal sensors capture other data types, and then generate a 3D view of the space and images of spaces where there may be fire spots, respectively. Simulations in different environments showed that the choice of metric and variation in the number of UAVs influence the number of performed displacements in the environment, consequently affecting exploration time and energy use.

Keywords: multi-robot; UAV; bioinspired map; topologic mapping; map exploration

1. Introduction

Mobile robotics is being applied more often to not only solve problems found in industrial environments, but also applied to services and home uses. For example, robots can be used in the process of warehouse automation, space monitoring, and house cleaning. These new applications show that a mobile robot can perform complex tasks while navigating unknown environments and avoiding unexpected obstacles by reacting to environmental stimuli [1]. Another application of mobile robotics is in the support of rescue teams in natural-disaster or catastrophe situations. Exploration might put the life of rescue-team professionals in danger. The use of Unmanned Aerial Vehicles (UAV) may assist rescue activities, especially in indoor areas where the arrival or movement of a ground robot is sometimes impossible. Access to unknown indoor areas requires techniques for defining the space where a robot is positioned, generating environmental mappings in order to aid teams in the reconnaissance of these areas where the use of global positioning systems (GPS) is unavailable. Thus, an autonomous robot must deal with two critical problems to survive and navigate in its environment: mapping the environment, and searching for its own location in the map [2].

For rescue environments, the time for space recognition becomes critical. Thus, the use of multiple robots can reduce environment exploration time. The collective construction of a map that is used to displace both multiple robots and the rescue team must represent spaces where it is possible to move and points that need more attention, such as human-temperature recognition, toxic elements, fires, and other factors that could be life-threatening.

This work proposes a mapping approach that was bioinspired by honeycomb construction. Honeybees use hexagonal-pattern cylinders to progressively build a complex structure by adding wax produced and manipulated by several bees [3]. This hexagonal structure allows the construction of combs with less wax (material saving), with the capacity for more storage. The construction of a honeycomb structure starts from a cell floor. Then, the structure is progressively extended in depth by adding more materials around the cell walls. The hive combs are the result of the collective work of hundreds of bees. There is no central commander/master for the building process. The individuals follow simple rules related to environmental construction (e.g., only one bee at a time can build a particular comb, and a new cell must be adjacent to an existing cell), so that this environment influences behavior, which, in turn, transforms the environment, it being a mechanism of synergy [3].

The scope of this work is in the application of simulated models of UAVs with similar configuration, and in addition, it will make use of simulation environments to validate the developed method. In this way, details and restrictions of communication technologies are abstracted.

2. Related Works

2.1. Map Generation

Building an environment map is necessary for both robot exploration and in simultaneous localization and mapping (SLAM) tasks. In [4], map generation was partitioned into three parts: metric, topological, and hybrid maps. Cartographic maps are able to make use of Vector map ([5–7]); however, they are not the focus of this work.

2.1.1. Metric Maps

Metric maps try to extract the features and geometric properties of the environment, and they are represented as a grid, geometric, or feature map [8] . Often, metric maps are probabilistic [4], and establish methods for modeling noise and its effects on environmental modeling. The approaches are based on a Bayesian filter, graph-based SLAM, and submap-joining SLAM.

2.1.2. Topological Maps

Topological maps represent the environment in graphs, where nodes represent places and objects of interest, and edges represent the spatial relationship or path between nodes [4]. In addition to providing a more compact representation of the environment than metric maps, topological maps provide a higher-level symbolic understanding for planning and navigation tasks. While metric maps are achieved with odometry-error accumulation, topological maps are built without the worry of metric aspects. Odometry errors that are accumulated between graph nodes do not necessarily accumulate through the global map.

2.1.3. Hybrid Maps

Hybrid maps combine the advantages of metric and topological mapping. Topological mapping is applied for a global view of the environment, while metric mapping is applied to smaller areas, which reduces computational complexity during metric-information processing. A hybrid-map form is the use of each topological-map node to represent a small metric map, and edges between nodes represent the path from the center point of one metric map to the center point of the next metric map [4].

2.2. Multiple Robots in Environment Mapping

Solutions that use multiple robots are characterized by the application of homogeneous and heterogeneous robots. Many related works make use of SLAM algorithms, but the focus of this work is environment exploration. Thus, works that make use of SLAM were considered for understanding the way they build the maps.

In [9], the authors performed collaborative space mapping with UAV and Unmanned Ground Vehicle (UGV) modeling through complementary maps. While the UGV does 2D area mapping, the UAV does 3D mapping of orthogonal objects in the environment. In [10], the authors presented a practical application, which is the mapping of areas struck by earthquakes. This being an implementation that uses a UAV and UGV, operation is semiautonomous. That happens because the UGV is remotely controlled, but when it faces obstacles it cannot overcome, the UAV autonomously does the mapping of the area. The execution of a 3D SLAM is done by the UAV via an RGB-D sensor, and by the UGV with a laser scanner. In [11], the UAV implements a Parallel Tracking and Mapping (PTAM)on the basis of sonar readings, while the UGV executes a Visual SLAM (VSLAM) fed by RGB-D and laser sensors. The work's goal was heterogeneous exploration using integer programming. The UGV has its own VSLAM and, for places that it cannot explore, the UAV is put in action using PTAM. UAV data via PTAM are then sent to the UGV and integrated in a VSLAM.

Some works that only use UAVs are presented: [12] uses a swarm to distribute areas to be explored by the UAVs. The focus is the use of UAVs for both hunting and cleaning. Here, in a group of many UAVs, one is defined as a sentinel and partitions the area for exploration. The work of [13] modified the PTAM algorithm for multiple agents using monocular cameras. Environment exploration is done cooperatively with recognition of points of interest. The definition of exploration is done via auction, where each bid is the linear distance of each UAV to the point being explored. The shortest distance wins the auction. In [14], an adaptation of PTAM (Parallel Tracking and Multiple Mapping—PTAMM) with the use of RGB-D, inertial measurement unit (IMU), and infrared (IR) sensors was presented. The work did localization and mapping using RGB-D sensors. A characteristic of this work is that it decomposed a 3D SLAM problem in a monocular SLAM with sparse representation.

There are solutions that implemented cooperative indoor mapping by using only UGVs [15–19]. In [15], heterogeneous robots were used in 2D and 3D area mapping using laser scanners, performing 3D and 2D cooperative mapping via autonomous agent navigation. Here, each robot builds a local map and sends the relevant data to a central server, where the data are joined with existing data using join-compatibility branch and bound (JCBB) implementation. In [16], the authors adapted the FastSLAM algorithm for multiple agents by also using laser scanners. Presenting a version of FastSLAM adapted to multiple UGV robots, it could perform cooperative mapping with the stigmergic potential field (SPF) technique, which represents behavioral influences of gathered data from the operational environment of one of the agents. In [17], the UGVs executed a VSLAM via a monocular camera. The creation of cooperative SLAM was based on salient landmarks to represent prominent characteristics. For that, each robot performs its own monocular SLAM with Extended Kalman Filter (EKF). The merge algorithm uses duplicated landmarks to increase the accuracy of the centralized map. In [18], a laser and webcam were used to model an area. By employing multiple autonomous UGVs, this work performs exploration with teams of robots for learning. Each robot creates a partial 3D map that it shares with other robots in its communication range. A global map is created on the basis of matching poses and mutual characteristics found in individual maps. The authors in [19] presented an implementation of multiple GraphSLAM using a stereo camera. Here, autonomous UGVs perform 6D mapping of an area using graph topology to separate uncertainty estimates of the local filters of multiple robots in a SLAM graph.

3. Methodology—Bioinspired Mapping Method

For [20], an exploration task is the combination of both mapping and robot motion-control activity.

This work proposes an environment exploration method with multiple UAVs inspired by how bees build hives. The authors in [3] discussed how bees perform hive construction. Following the behavior of bees in the construction of each honeycomb, UAVs perform the build and exploration map in a similar way, where combs are represented as hexagons. Each honeycomb can have only one bee occupying its space, so each hexagon can hold a maximum of one UAV. The built map is a collection of hexagons.

The construction of a beehive begins with the work of the first bee, which begins construction of the first honeycomb using wax to build its walls. Similarly, in the proposed method, a first UAV, identified as the sentinel, generates the first map hexagon, checking whether there are adjacencies for each of the six sides (honeycomb walls). In this case, the term adjacency means the possibility for a UAV to move from one hexagon to another. Thus, a hexagon exists on the map if and only if it is possible for a UAV to fully access it from another hexagon on at least one of its six sides, so obstacles cannot exist between the center of one hexagon and the center of the other hexagon. Figure 1 shows a UAV exploring a hex that should rotate at six angles: $\pi/2, \pi/6, -\pi/6, -\pi/2, -5\pi/6$, and $5\pi/6$. Each evaluated hexagon with possible adjacency is marked with an identifier.

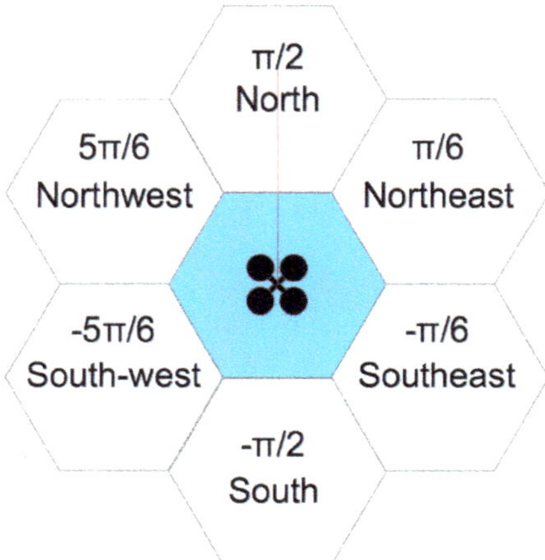

Figure 1. Unmanned Aerial Vehicle (UAV) in hexagon exploration.

Briefly, the UAV explores the hexagon in each of its six angles, sets a new hexagon to explore and moves to this, starts a new exploration. The Figure 2 shows this action.

Once the sentinel UAV finishes the first scan, all UAVs can start searching for spaces to explore. To control the hexagons identified in the reading process from each of the six angles, some structures are used. To record the identifiers (ids) of the explored hexagons, a list called "visited hexagon list" is used. When the UAV rotates and finds adjacency for a new hexagon, a new id is generated and added into a structure called a "not visited hexagon list". Thus, a UAV searching for a hexagon to explore should perform this search in the "not visited hexagon list".

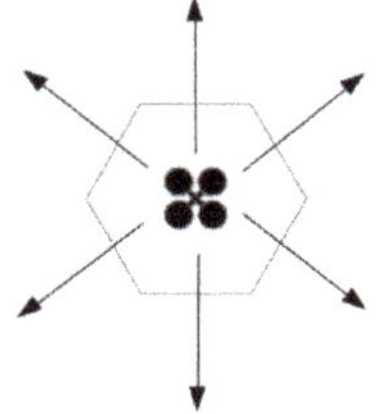

(a) Action 1: explore the first hexagon.

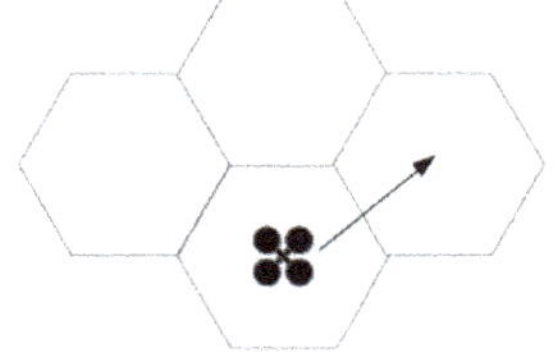

(b) Action 2: get a new hexagon to explore and scroll to it.

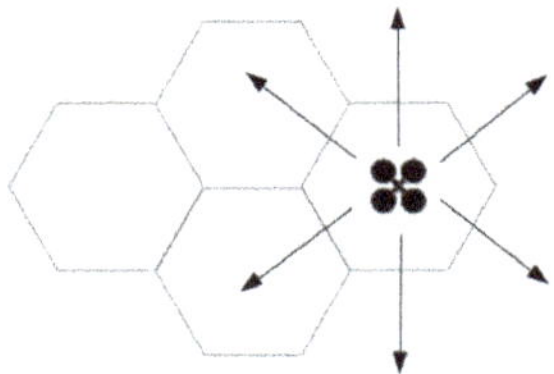

(c) Action 3: explore the new hexagon

Figure 2. Exploration stages.

At the end of hexagon exploitation, id is removed from the latter list. Figure 3 presents two UAVs exploring a given space. In this case, exploration started with hexagon 1, which was already explored. For illustration purposes, hexagon 1 is green, indicating that it was already fully explored. In its exploration process, adjacencies were identified with hexagons 2–4, which were inserted into the "not visited hexagon list". When a UAV began exploring hexagon 2, hexagons 5–7 were identified. Blue hexagons represent spaces in exploration, while yellow ones are those that were identified but not yet explored. The exploration process ends when the "not visited hexagon list" is empty.

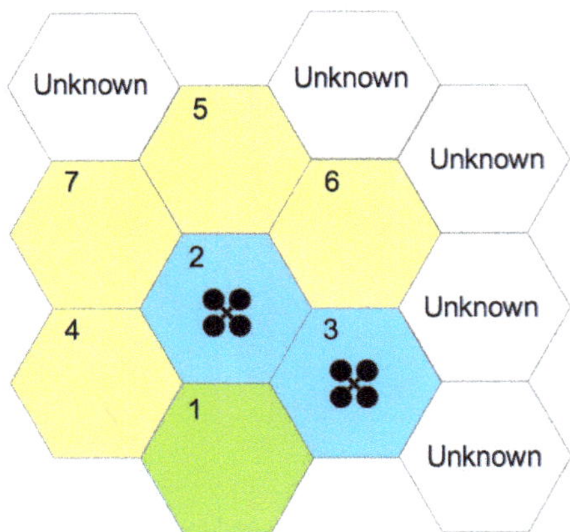

Figure 3. Multiple UAV exploration: green hexagon, explored place; yellow hexagons, places that have adjacency, but not yet explored; blue hexagons, places that are explored by UAV; white hexagons, unknown places that are mapped in future steps.

3.1. Environment Exploration

Figures 4 and 5 present state diagrams of the scanning activity of both sentinel and other UAVs. The sentinel UAV only behaves differently in the first exploration (where it generates the first *id* from point *xyz* from its placement); in the others, it has the default behavior of the other UAVs.

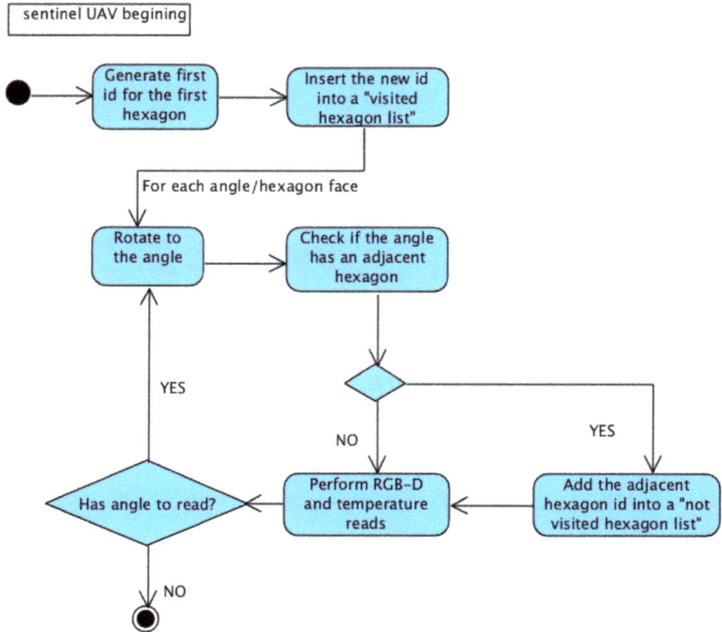

Figure 4. First sentinel UAV exploration.

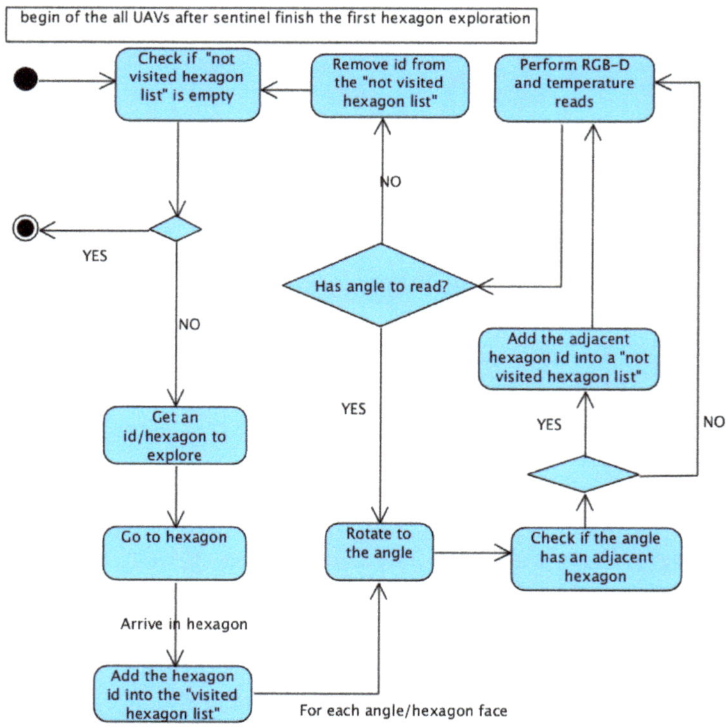

Figure 5. All UAVs after first sentinel exploration.

3.1.1. Checking If *"Not Visited Hexagon List"* Is Empty

This is the stopping criterion of the exploring algorithm. Each uncovered discovered hexagon is inserted into the "not visited hexagon list". When a UAV receives a id to explore, it remains in the "not visited hexagon list" to the end of the exploration, but the UAVs that it exploits are registered. This ensures that no UAV stops the exploration process without actually having any new spaces to explore. For example, at one point in the exploration, one UAV may have finished its exploration, while another is working. If no unvisited hexagons are currently available, the first UAV waits for possible discoveries of the second UAV, which is still in exploration activity. If no new hexagon is discovered, then exploration is finished, or new explorations process are done again.

3.1.2. Getting Id/Hexagon to Explore

When a UAV is free (no hexagons in the exploration process), it seeks a new place to explore, which is done in the "not visited hexagon list". To define which is assigned to the UAV, two metrics were defined for different simulations: First-In–First-Out (FIFO) and Euclidean distance. The FIFO metric assigns to the UAV that unvisited hexagon than has been awaiting exploration for the longest, so the first discoveries are the first to be explored. The second metric defines that the hexagon to be explored by the UAV is the one with the smallest Euclidean distance from the initial hexagon (id 1).

3.1.3. Go to Hexagon

Once the UAV gets a hexagon to explore, it must travel there. The UAV only transitions through familiar and accessible spaces. Thus, given the hexagon where the UAV is located and the target, one path is defined to go. This path is built from Dijkstra's algorithm [21], with an adapted version from [22]. With this path, the UAV travels the map until it reaches its target.

3.1.4. Add Hexagon Id into "Visited Hexagon List"

When the UAV finishes moving along the path defined by Dijkstra's algorithm, it is in the hexagon to explore. At the beginning of the exploration activity, hexagon id is inserted into the "visited hexagon list" . This ensures that a UAV identifies a hexagon already found and identified by another UAV, not creating a new id for the same space.

3.1.5. Rotate to Angle and Check If Angle Has Adjacent Hexagon

For each six sides of the hexagon (six angles), the UAV should rotate and check for adjacency: a sensor checks if it is possible for the UAV to access the center of the neighboring hexagon; in other words, if there are no obstacles between the two hexagons. If so, adjacency is added to an adjacency matrix. Each angle of the explored hexagon is identified in the honeycomb map with dashed lines if there is adjacency at that angle, or with continuous lines if there is none, as shown in Figure 6.

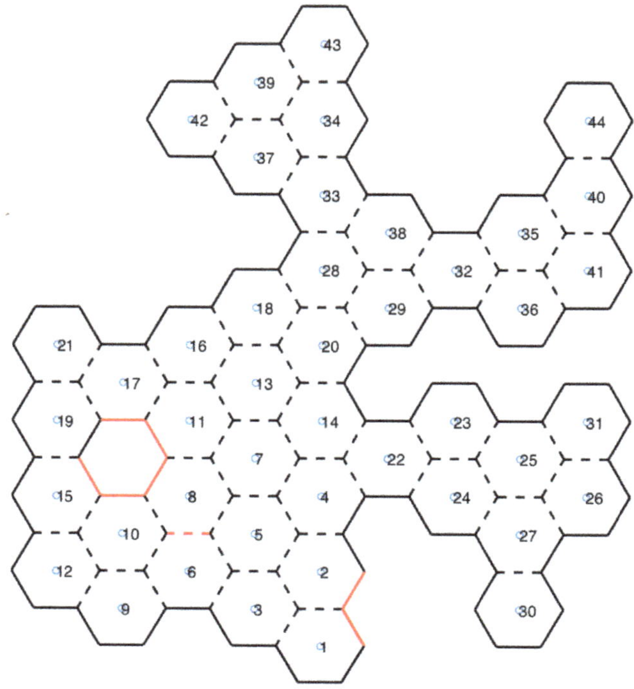

Figure 6. Honeycomb map.

3.1.6. Add Adjacent Hexagon Id into "Not Visited Hexagon List"

When the sensor reading discovers an adjacent hexagon for each of the six angles, and it is not in the "visited hexagon list", it sends it to the "not visited hexagon list", if it is not already there (this is a newly discovered hexagon).

3.1.7. Perform RFB-D and Temperature Reads

Map information is available for both UAVs to control their movements in the environment, and for rescue teams to know the space that can be navigated. In addition to obstacle sensors, RGB-D and temperature sensors are used. RGB-D sensors read the 3D angle of the UAV, and a cube view is then built to aid rescue teams in space recognition. At the same time, a thermal sensor reads the temperature from the same angle. If a temperature higher than a reference value is found, it is identified and a photo of the location is taken. In honeycomb map, this scenario is represented with red lines, as shown in Figure 6.

The RGB-D reading returns a matrix structure. The matrix size (nxm) and the range of RGB-D sensor are set in a V-REP simulator in the sensor settings. Matrix values are between 0 and 1, where 0 is very close to and 1 very far from the sensor. When the UAV reads RGB-D, the 3D data, and the angle and position of the UAV during the reading are recorded. These data are transformed from a perspective to a global point. For instance, let R be the cube size, $amplitudeRGBD$ the extent of the RGB-D sensor, $buffer$ the RGB-D matrix read, xn and yn the $buffer$ dimension, $angUAV$ the UAV Euler angle, and $posUAV$ the xyz UAV position. Algorithm 1 brings the data transformation.

Algorithm 1 RGB-D transformation algorithm.

```
function [xc,yc,zc] = TransformRGBD(R, amplitudeRGBD,
                                    buffer, xn, yn, angUAV, posUAV)
xc = 0;
yc = 0;
zc = 0;
deltaAngleRGBD = double(amplitudeRGBD)/double(xn);
for i=1:yn
    for j=1:xn
        if (double(buffer(i,j))<0.99)
            angUAVz=rad2deg(angUAV(3));
            ---nz is the distance in meter.
            ---RGB-D is set to 2m
            nz=double(buffer(i,j))*2;
            if (j==1)
                dtAng=0;
            elseif (j==xn)
                dtAng=amplitudeRGBD;
            else
                dtAng=deltaAngleRGBD*double(j);
            end
            if (dtAng < amplitudeRGBD/2)
                alfa=double(angUAVz)+((amplitudeRGBD/2)-dtAng);
            else
                alfa=double(angUAVz)-(dtAng-(amplitudeRGBD/2));
            end
            alfarad=deg2rad(alfa);
            ---Sine's Law
            dy = double(nz*sin(double(alfarad))/sin(deg2rad(90)));
            dx = nz*sin(double(deg2rad(180-90-alfa)))/sin(deg2rad(90));
            ---calculate dz
            angUAVx = rad2deg(angUAV(1));
            if (i==1)
                dtAngz=0;
            elseif (i==xn)
                dtAngz=amplitudeRGBD;
            else
                dtAngz = deltaAngleRGBD*double(i);
            end
            if (dtAngz < amplitudeRGBD/2)
                alfaz=double(angUAVx)+((amplitudeRGBD/2)-dtAngz);
            else
                alfaz=double(angUAVx)-(dtAngz-(amplitudeRGBD/2));
            end
            dz = nz*sin(double(deg2rad(alfaz)))/sin(deg2rad(90));
            xp = posUAV(1) + dx;
            yp = posUAV(2) + dy;
            zp = posUAV(3) + dz;
            ---Discretizing values.
            xc = round(xp/R)*R;
            yc = round(yp/R)*R;
            zc = round(zp/R)*R;
        end
    end
end
end
```

3.1.8. Remove Id from "Not Visited Hexagon List"

At the end of the reading of the six sides of the hexagon, id is removed from the "not visited hexagon list". Then, the UAV can begin the search for a hexagon to explore again if the "not visited hexagon list" is not empty; otherwise, the UAV's exploration activity is finished.

3.2. Lock Path Resolution

Throughout the exploration process, the various UAVs will be moving towards their targets, and consequently, their paths may cross. Avoiding collisions is a critical point for an environment with multiple robots. Several approaches have been presented, where means of prevention are proposed by optimized programming [23–25], potential fields [26], sampling-based methods [27], and others. In general, two concepts are applied [28]: one where robots are free and can change their paths, and another where robots have a fixed path with no possibility of changes. Thus, in the first concept, the focus is on changing paths, while in the second, the focus is on controlling movement and time. In [28] a method for treating deadlock for multiple robots where the path is fixed is discussed. To improve performance, some stopping policies are proposed. With these policies, each robot makes the decision to change or wait for another one. A correct-by-construction synthesis approach to multi-robot mission planning that guarantees collision avoidance with respect to moving obstacles are approach in [29], where has done an integration of a high-level mission planner with a local planner that guarantees collision-free motion in three-dimensional workspaces, when faced with both static and dynamic obstacles.

To avoid collisions, the proposed architecture defines that only one UAV can occupy one hexagon (honeycomb) at a time. Thus, it is necessary to have a record of the hexagon that the UAV currently occupies. To do this, each time a UAV moves from one hexagon to another, it records both which one it is in and what is the next move. Collision is avoided in this way; however, deadlock states can happen.

In the proposed approach, it is assumed that a UAV can be found in three possible states: "in exploration", "in displacement" or "stopped". The state "in exploration" means that the UAV is reading the six angles of the hexagon (honeycomb), and generating the mapping data. "In displacement" means that the UAV is moving to a hexagon and make their exploitation. The "stopped" state means that the UAV has no allocated exploration, and is not moving to any honeycomb.

A key element for resolving path blocks in this approach is the Adjacent Degree (AD), which is the number of adjacent hexagons, directly or indirectly, to which the UAV can travel, in order to free the paths. To obtain the AD, each UAV checks how many hexagons are directly adjacent to it, excluding those in which they are occupied by other UAVs. If the AD value is greater than 1, it means that there is space to perform a maneuver to release the passage. If the AD value is 1, the AD value for this adjacent single is searched. The AD value found for the adjacent one will be its value as well.

A comparison between the AD values of each UAV is made, and if there is a conflict between UAVs, the one with the highest AD must give way to the one with the smallest, moving to one of its adjacent hexagons to resolve the deadlock . When he finishes moving, he retraces his trajectory for his hexagon to explore and returns to his tasks. The Figure 7 presents a scenario with two UAVs, where the UAV in the blue hexagon has three adjacent hexagons directly and its AD value is 3, while the one in yellow has a single adjacent one; however, this, in turn, it has two adjacent hexagons, making the UAV AD in the yellow hexagon to be 2.

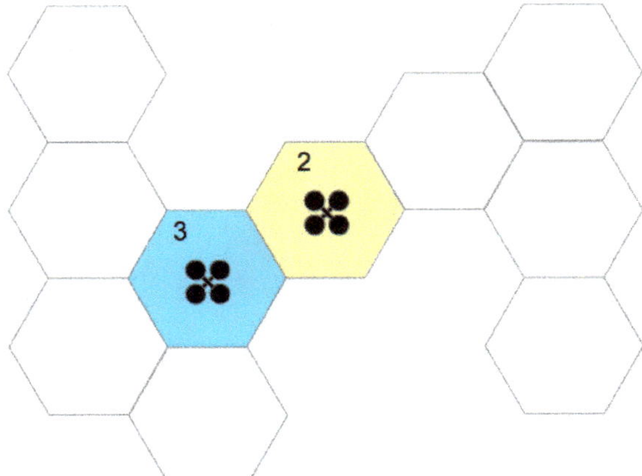

Figure 7. Adjacent Degree: blue hexagon has AD = 3, while yellow hexagon AD = 2.

Considering the existence of several UAVs in the environment, each one identified as $A, B, C, ..., Z$, and $A \rightarrow B$ representing the UAV who wants to move to the hexagon which is the UAV B. Some cases of path blocking can happen:

- **Case 1—$A \rightarrow B$ and $B \rightarrow A$:**

  In this scenario, UAV A wants to move to the hexagon of UAV B, and at the same time, UAV B wants to move to the hexagon where UAV A. Here, each UAV calculates its adjacent degree (AD). The UAV that has the largest AD will open the way to the other UAV.
- **Case 2—$A \rightarrow B$:**

  In this scenario, only UAV A shows that it wants to move to the hexagon of UAV B; however, B will not go to the hexagon of A is. In this case, UAV B may be in an "in exploration" or "stopped" state. If it is in an "in exploration" state, UAV A will recalculate a new path trying to deflect the hexagon occupied by B. If there is only one path, UAV A waits for UAV B to complete its exploration. On the other hand, if UAV B is in a "stopped" state, UAV B itself will identify that UAV A wants to go to the hexagon it occupies. That way, it will calculate your AD and compare it with the UAV A. If your AD is greater, it will move to a free adjacent hexagon, and otherwise, it will try to move to a hexagon adjacent to the UAV A, which causes them to find themselves in Case 1.
- **Case 3—$A \rightarrow B$, $B \rightarrow C$ and $C \rightarrow A$:**

  In this case, two UAVs are unable to mutually identify a deadlock. So, it is necessary to check if there is a cyclically blocking. Thus, from the hexagon to which you want to move, UAV A checks if there are any others that want to move to where it is. If this block is detected, the UAV calculates its AD, and if it is greater than 0, it will give space for the resolution of the deadlock. After that, the path to the defined hexagon will be recalculated, and then continue your task.

3.3. Simulation

To validate the proposed method, simulations were performed with different scenarios and UAV numbers. Through the simulation, it was possible to verify the proposed approach, i.e., to plan the UAV tasks to map a catastrophic environment. Simulations were created with the following setup: CPU, Intel Xeon with 3.33 GHz 6 Core, 6 GB 1333 MHz DDR3 memory, and GPU ATI Radeon HD 5770 1024 MB.

There are several robot simulation environments, such as Open HRP [30], Gazebo [31], Webots [32] and Virtual Robot Experimentation Platform (V-REP) [33]. In this work, we chose V-REP, which has application programming interfaces (API) that allow communication with many programming languages. The proposed approach was implemented in MATLAB [34].

Figure 8 shows the simulation scenarios. Both were 10 × 10 m locations. Scenario 1 (Figure 8a) presents a place characterized by rooms with furniture that were knocked down, like an earthquake scene, while Scenario 2 (Figure 8b) is a place with passages; red dots represent fire spots.

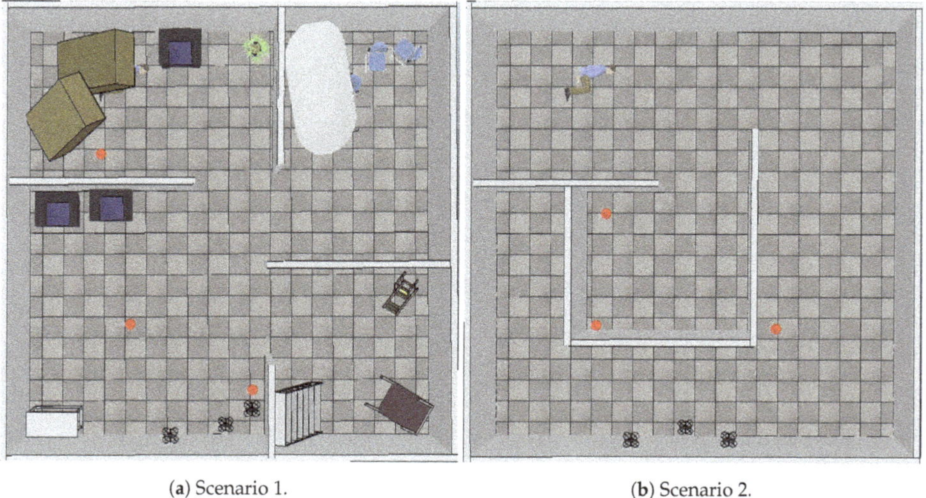

(**a**) Scenario 1. (**b**) Scenario 2.

Figure 8. V-REP simulation scenarios.

To perform exploration, the simulations made use of two and three similar UAVs. Figure 9 shows a used UAV. The UAV was equipped with an RGB-D camera, a thermal sensor, and a laser sensor. The laser sensor took a 0.5 cm radio to the honeycomb, so distance from a hexagon center to another was 1 m. For each scenario and each configuration (two or three UAVs), simulations were performed with the FIFO and Euclidean Distance algorithms.

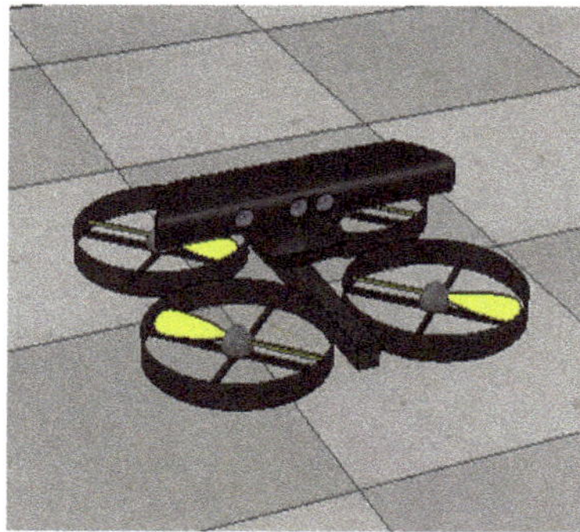

Figure 9. UAV in simulation.

4. Results

Figure 10 shows scenarios merged with the honeycomb-map build. After the simulations were performed, it was possible to verify the displacements of each UAV within the generated map, as well as the order of honeycomb exploration by each UAV.

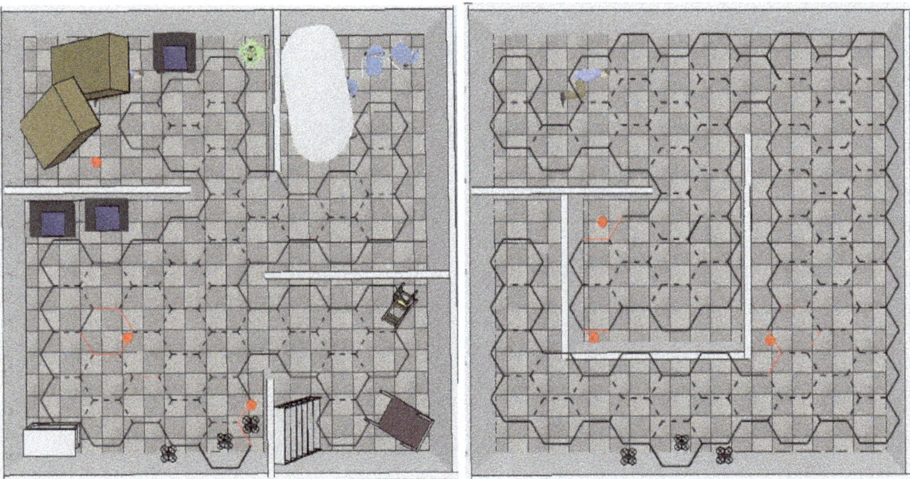

(a) Scenario 1 with honeycomb map. (b) Scenario 2 with honeycomb map.

Figure 10. Scenarios merged with honeycomb map.

4.1. Scenarios and Honeycomb-Map Generation

Figures 11 and 12 show the movements made by the UAVs in the simulations of Scenarios 1 and 2, respectively. Figures 13 and 14 present the exploration order of each UAV with the respective hexagon-definition algorithm to be explored, FIFO and Euclidean distance. The blue line correspond to UAV 1, the red is UAV 2, and the green is UAV 3 (when the simulation had three UAVs).

The yellow circle identifies the highest-traffic hexagon. Hexagon traffic means how many times a UAV went through the hexagon. Table 1 shows the max traffic number in the simulations. Considering the *ids* of Table 1, and relating them in Figures 11 and 12, these most accessed hexagons were located in places characterized as doors or passageways.

In the simulations, the movements of each UAV were recorded. Displacement means that a UAV moved from a hexagon to an adjacent one. Table 2 shows the displacement number and average per UAV in each simulation in Scenario 1. Table 3 shows the same for Scenario 2. Table 4 bring the exploration time. Tables 5 and 6 details data from both exploration order and displacement.

Table 1. Hexagon traffic.

Simulation	Scenario 1		Scenario 2	
	Id Hexagon	**Traffic Number**	**Id Hexagon**	**Traffic Number**
Two UAVs–FIFO	8	17	3	19
Two UAVs–Euclidean distance	4 and 12	8	3	13
Three UAVs–FIFO	5 and 10	15	2	21
Three UAVs–Euclidean distance	10 and 13	9	12	14

Table 2. Displacement number—Scenario 1.

Simulation	FIFO	Average/UAV	Euclidean Distance	Average/UAV
Two UAVs	196	98	143	71.5
Three UAVs	203	67	169	56.33
Variation	-	31.63%	-	21.21%

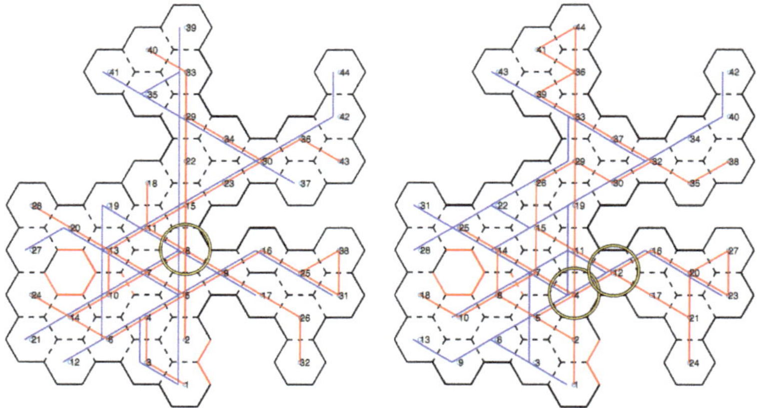

(**a**) Scenario 1 - 2 UAVs—First-In–First-Out (FIFO).

(**b**) Scenario 1. Two UAVs–Euclidean distance (ED).

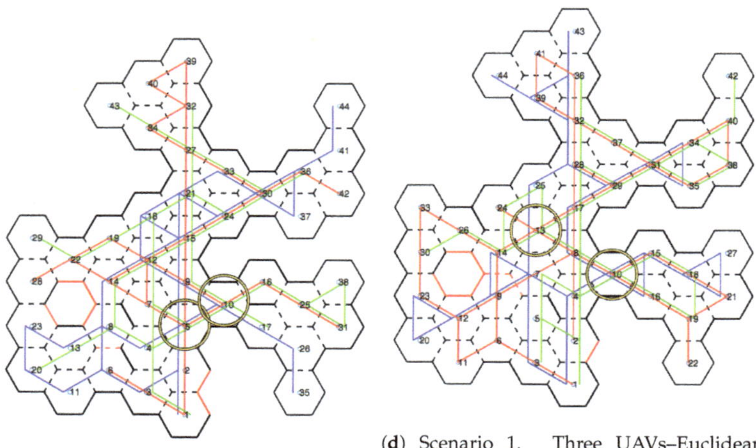

(**c**) Scenario 1. Three UAVs–FIFO.

(**d**) Scenario 1. Three UAVs–Euclidean distance.

Figure 11. Displacement—Scenario 1.

Table 3. Displacement number—Scenario 2.

Simulation	FIFO	Average/UAV	Euclidean Distance	Average/UAV
Two UAVs	290	145	206	103
Three UAVs	324	108	271	90.33
Variation	-	25.51%	-	12.29%

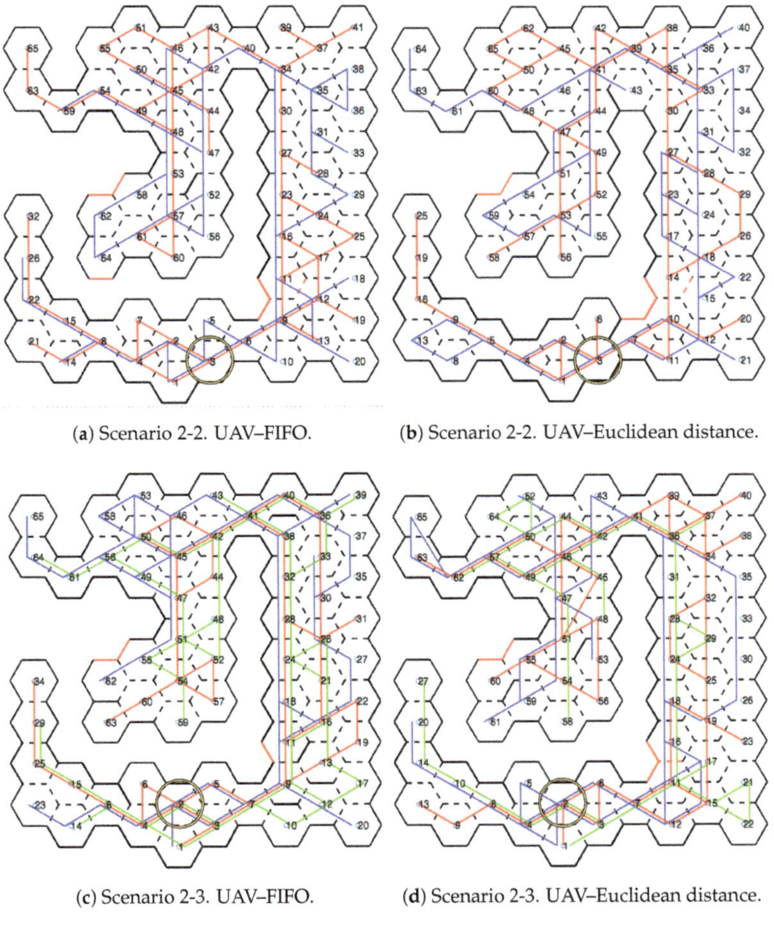

(**a**) Scenario 2-2. UAV–FIFO. (**b**) Scenario 2-2. UAV–Euclidean distance.

(**c**) Scenario 2-3. UAV–FIFO. (**d**) Scenario 2-3. UAV–Euclidean distance.

Figure 12. Displacement—Scenario 2.

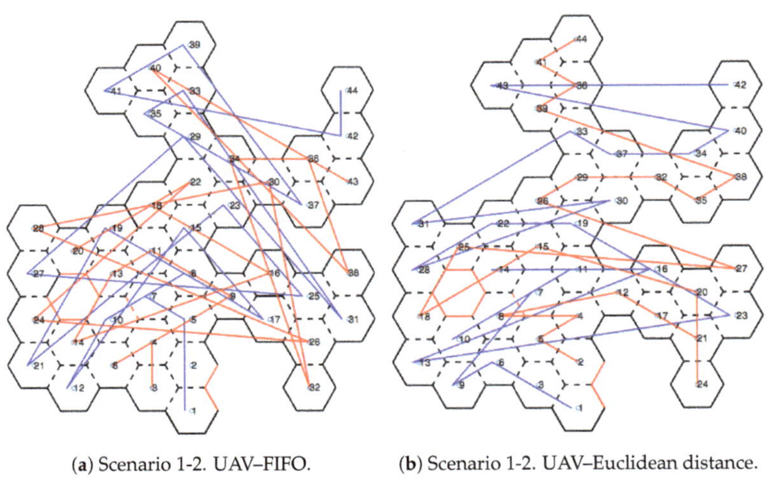

(**a**) Scenario 1-2. UAV–FIFO. (**b**) Scenario 1-2. UAV–Euclidean distance.

Figure 13. *Cont.*

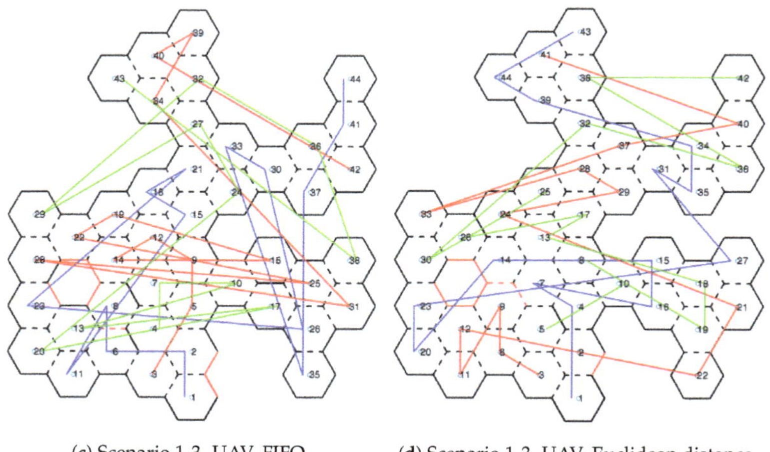

(**c**) Scenario 1-3. UAV–FIFO. (**d**) Scenario 1-3. UAV–Euclidean distance.

Figure 13. Exploration order—Scenario 1.

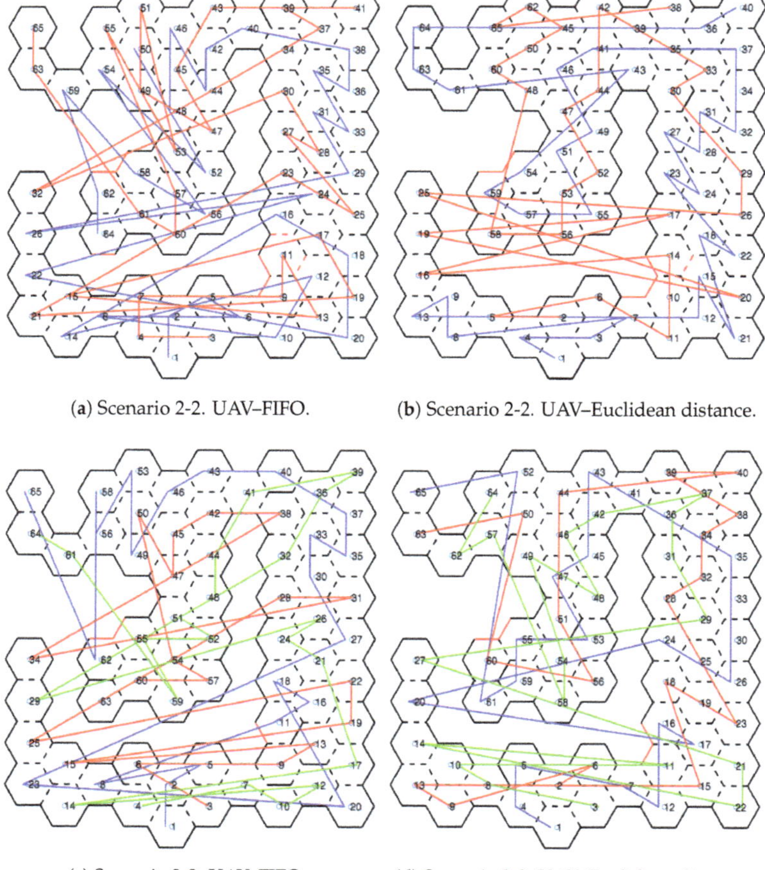

(**a**) Scenario 2-2. UAV–FIFO. (**b**) Scenario 2-2. UAV–Euclidean distance.

(**c**) Scenario 2-3. UAV–FIFO. (**d**) Scenario 2-3. UAV–Euclidean distance.

Figure 14. Exploration order—Scenario 2.

Table 4. Exploration time.

Simulation	Scenario 1	Scenario 2
Two UAVs–FIFO	2:30:32	3:00:17
Two UAVs–Euclidean distance	2:24:56	2:27:58
Three UAVs–FIFO	3:44:25	2:08:09
Three UAVs–Euclidean distance	3:54:17	1:56:08

Table 5. Exploration order.

Scenarios	UAV Number	Exploration Order
		Three UAV - Euclidean Distance
Scenario 1	UAV 1	1 2 4 7 16 15 14 20 23 27 31 35 34 39 44 43
	UAV 2	3 6 9 11 12 22 21 24 29 28 33 37 40 41
	UAV 3	5 10 8 19 18 13 17 26 25 30 32 38 36 42
		Two UAV - Euclidean Distance
	UAV 1	1 3 6 9 7 10 11 14 16 13 23 19 22 28 30 31 33 37 34 40 43 42
	UAV 2	2 5 4 8 12 17 21 24 20 15 18 25 27 26 29 32 35 38 39 36 41 44
		Three UAV - FIFO
	UAV 1	1 2 6 8 11 15 18 21 23 26 30 33 35 37 41 44
	UAV 2	3 5 9 12 14 16 19 22 25 28 31 34 39 40 42
	UAV 3	4 7 10 13 17 20 24 27 29 32 36 38 43
		Two UAV - FIFO
	UAV 1	1 2 5 7 8 10 12 15 17 19 21 23 25 27 29 31 33 35 37 39 41 42 44
	UAV 2	3 4 6 9 11 13 14 16 18 20 22 24 26 28 30 32 34 36 38 40 43
		Three UAV - Euclidean Distance
Scenario 2	UAV 1	1 4 5 7 12 16 17 20 24 26 30 33 35 41 43 45 47 53 55 59 61 52 65
	UAV 2	2 6 9 13 15 18 19 23 25 28 32 34 38 39 40 44 51 56 60 50 63
	UAV 3	3 8 10 11 14 22 21 27 29 31 36 37 42 46 48 49 54 58 57 62 64
		Two UAV - Euclidean Distance
	UAV 1	1 4 3 7 8 9 13 12 15 21 18 22 23 24 27 28 31 32 34 37 41 46 49 51 55 57 59 54 43 61 63 64 36 40
	UAV 2	2 5 6 11 10 14 16 17 19 20 25 26 29 30 33 35 39 42 44 47 52 53 56 58 48 60 50 45 62 65 38
		Three UAV - FIFO
	UAV 1	1 2 5 8 11 16 18 20 23 27 30 33 35 37 40 43 46 49 53 56 58 62 65
	UAV 2	3 6 9 13 15 19 22 25 28 31 34 38 42 45 47 50 54 57 60 63
	UAV 3	4 7 10 12 14 17 21 24 26 29 32 36 39 41 44 48 51 52 55 59 61 64
		Two UAV - FIFO
	UAV 1	1 2 5 6 8 10 12 14 16 18 20 22 24 26 29 31 33 35 36 38 40 42 44 46 48 50 52 54 56 58 59 62 64
	UAV 2	3 4 7 9 11 13 15 17 19 21 23 25 27 28 30 32 34 37 39 41 43 45 47 49 51 53 55 57 60 61 63 65

Table 6. Displacement order.

Scenarios	UAV Number	Exploration Order
		Three UAV - Euclidean Distance
Scenario 1	UAV 1	1 2 4 7 4 10 16 15 10 4 7 14 9 12 20 23 12 9 7 4 10 15 18 21 27 18 15 10 8 17 29 31 35 34 31 29 28 32 39 44 39 36 43 36 32 28 25 13 7 5 3 1
	UAV 2	1 3 6 9 6 11 12 9 7 8 10 16 19 22 19 21 18 15 10 8 13 24 13 17 29 28 17 13 14 9 12 23 30 33 26 14 13 17 28 32 37 31 35 38 40 34 31 29 28 32 36 41 39 32 28 17 8 4 2
	UAV 3	1 2 5 2 4 10 8 10 16 19 18 15 10 8 13 17 13 14 26 14 24 25 13 14 26 30 26 14 13 17 28 32 28 29 31 35 38 34 31 37 32 36 32 37 31 34 40 42 40 34 31 29 17 8 4 7 5 3
		Two UAV - Euclidean Distance
	UAV 1	1 3 6 9 6 5 7 8 10 8 7 11 7 14 7 4 12 16 12 4 5 6 9 13 9 6 5 4 12 16 20 23 20 16 12 11 19 15 22 25 28 25 14 15 19 30 19 15 14 25 31 25 22 26 29 33 37 32 34 40 34 32 37 33 39 43 39 33 37 32 34 40 42 40 34 32 30 19 11 4 5 3 1
	UAV 2	1 2 5 4 5 8 5 4 12 17 21 24 21 20 16 12 11 15 7 8 10 18 10 8 14 25 14 7 4 12 16 20 23 27 20 16 12 11 15 26 29 30 32 35 38 35 32 37 33 39 36 41 44 36 33 29 19 11 4 2
		Three UAV - FIFO
	UAV 1	1 2 5 2 3 6 8 6 11 6 8 4 5 9 15 18 21 15 12 14 8 6 11 20 23 13 8 4 5 10 17 26 17 10 9 15 24 30 33 21 15 9 10 17 26 35 26 17 10 9 12 18 21 24 30 37 36 41 44 41 36 30 24 15 9 5 4
	UAV 2	1 3 6 3 1 2 5 9 12 14 7 5 10 16 10 9 12 19 22 14 7 5 10 16 25 16 10 5 7 14 22 28 22 14 7 5 10 16 25 31 25 16 10 5 7 12 15 21 27 34 32 39 40 32 27 33 30 36 42 36 30 24 15 9 5 2
	UAV 3	1 3 4 7 5 10 5 4 8 13 8 4 5 10 17 10 5 4 8 13 20 13 8 14 12 15 24 21 27 21 18 19 22 29 22 19 18 21 27 32 27 33 30 36 30 24 15 9 10 16 25 31 38 25 16 10 9 15 21 27 34 43 34 27 21 18 12 7 4 3
		Two UAV - FIFO
	UAV 1	1 2 5 7 8 7 10 6 12 6 4 5 8 15 8 9 17 9 8 11 19 13 10 14 21 14 10 7 8 15 23 15 8 9 16 25 16 9 5 7 13 20 27 20 13 11 15 22 29 22 15 8 9 16 25 31 25 16 9 8 15 22 29 33 35 29 34 30 37 30 34 29 33 39 33 35 41 35 29 34 30 36 42 44 42 36 30 23 15 8 5 4 3 1
	UAV 2	1 3 4 6 4 5 9 8 11 13 10 14 6 4 5 9 16 9 8 11 18 11 13 20 13 11 15 22 15 8 7 10 14 24 14 6 4 5 9 17 26 17 9 5 7 13 20 28 20 13 11 15 23 30 23 15 8 9 17 26 32 26 17 9 8 15 22 29 34 30 36 30 23 15 8 9 16 25 31 38 25 16 9 8 15 22 29 33 40 33 29 34 30 36 43 36 30 23 15 8 5 2

Table 6. *Cont.*

Scenarios	UAV Number	Exploration Order
		Three UAV - Euclidean Distance
Scenario 2	UAV 1	1 4 5 2 3 7 12 11 16 17 11 7 3 2 4 8 10 14 20 14 10 8 4 2 3 7 12 15 17 16 18 24 18 19 26 30 33 35 34 36 41 43 42 45 47 46 47 48 53 48 47 51 55 59 61 59 54 51 47 46 44 52 44 50 57 62 65 63 62 57 49 46 42 41 36 31 28 24 18 16 11 7 6 2 1
	UAV 2	1 2 3 6 2 4 8 9 13 9 8 4 2 6 7 11 15 11 16 18 19 23 19 25 24 28 32 34 38 34 36 39 37 40 37 36 41 42 44 46 45 51 48 51 55 51 48 53 56 54 55 60 55 51 47 46 50 57 49 57 50 57 50 57 49 57 62 63 62 57 49 46 42 41 39 37 34 32 29 25 19 17 15 12 7 6 2
	UAV 3	1 3 7 3 2 4 8 10 8 4 2 3 7 11 7 3 2 4 8 10 14 10 8 4 2 3 7 11 15 22 21 15 11 17 11 7 3 2 4 8 10 14 20 27 20 14 10 8 4 2 3 7 11 16 18 24 29 28 31 36 37 36 41 42 46 45 48 45 46 49 47 51 54 58 54 51 47 49 57 62 57 50 57 50 52 64 50 44 42 41 36 31 28 24 18 16 11 7 3
		Two UAV - Euclidean Distance
	UAV 1	1 4 1 3 7 3 2 4 5 8 5 9 13 8 5 4 2 3 7 10 12 15 12 21 12 15 18 15 22 18 17 23 24 23 27 28 31 32 34 37 33 35 39 41 46 41 44 49 51 49 52 55 53 57 59 54 51 47 44 41 43 41 46 48 60 61 63 64 63 61 60 48 46 41 39 35 36 40 36 33 31 28 24 18 15 12 11 7 3 1
	UAV 2	1 2 4 5 4 2 3 6 3 7 11 10 14 10 7 3 1 4 5 9 16 9 5 4 2 3 7 10 14 17 14 10 7 3 2 4 5 9 16 19 16 9 5 4 2 3 7 10 12 20 12 10 7 3 2 4 5 9 16 19 25 19 16 9 5 4 2 3 7 10 14 18 26 29 28 27 30 33 35 39 42 41 44 47 49 52 53 56 53 57 58 57 53 51 47 48 60 50 45 62 65 50 45 41 39 38 35 30 27 23 17 14 10 7 3 2
		Three UAV - FIFO
	UAV 1	1 2 5 2 4 8 4 2 3 7 9 11 16 18 11 9 12 20 12 9 7 3 2 4 8 14 23 14 8 4 2 3 7 9 11 16 22 27 26 30 33 30 35 37 36 40 36 39 36 38 41 40 41 38 41 40 41 40 41 38 43 46 45 49 45 46 53 50 56 50 53 58 50 45 47 51 55 62 55 51 47 49 56 61 56 49 56 61 64 65 64 61 56 49 45 42 41 38 32 28 24 18 11 9 7 5 2 1
	UAV 2	1 3 2 3 2 3 2 4 6 2 3 7 9 13 9 7 3 2 4 8 15 8 4 2 3 7 9 13 19 22 16 11 9 7 3 2 4 8 15 25 15 8 4 2 3 7 9 11 18 24 28 26 31 26 21 16 11 9 7 3 2 4 8 15 25 29 34 29 25 15 8 4 2 3 7 9 11 18 24 28 32 38 41 42 45 47 44 42 46 50 45 47 51 54 52 57 54 60 63 60 54 51 47 44 42 41 40 36 33 30 26 21 16 11 9 7 5 2
	UAV 3	1 3 7 3 2 4 8 10 8 4 2 3 7 11 7 3 2 4 8 10 14 10 8 4 2 3 7 11 15 22 21 15 11 17 11 7 3 2 4 8 10 14 20 27 20 14 10 8 4 2 3 7 11 16 18 24 29 28 31 36 37 36 41 42 46 45 48 45 46 49 47 51 54 58 54 51 47 49 57 62 57 50 57 50 52 64 50 44 42 41 36 31 28 24 18 16 11 7 3
		Two UAV - FIFO
	UAV 1	1 2 3 5 6 3 2 4 8 4 2 3 6 10 9 12 9 6 3 2 4 8 14 8 4 2 3 6 9 11 16 11 12 18 12 13 20 13 9 6 3 2 4 8 15 22 15 8 4 2 3 6 9 11 16 24 16 11 9 6 3 2 4 8 15 22 26 22 15 8 4 2 3 6 9 11 16 24 29 28 31 33 31 35 36 38 35 34 40 42 44 42 46 45 48 45 50 45 44 47 52 47 48 49 54 49 48 47 52 56 57 53 58 53 48 49 54 59 54 49 48 53 58 62 64 61 57 52 47 44 42 40 34 30 27 23 16 11 9 6 3 1
	UAV 2	1 3 1 4 7 2 3 6 9 11 9 13 9 6 3 2 4 8 15 8 4 2 3 6 9 11 17 12 19 12 9 6 3 2 4 8 14 21 14 8 4 2 3 6 9 11 16 23 16 17 25 24 23 27 28 27 30 27 23 16 11 9 6 3 2 4 8 15 22 26 32 26 22 15 8 4 2 3 6 9 11 16 23 27 30 34 37 39 37 41 37 34 40 43 42 45 44 47 44 45 49 45 46 51 46 45 48 53 48 45 46 51 55 50 45 48 53 48 53 57 60 61 57 53 48 49 54 59 63 65 63 59 54 49 45 46 43 40 34 30 27 23 16 11 9 6 3 2

4.2. Cube View and Temperature Caption

In addition to performing the mapping honeycomb in a hexagonal shape, further information can be generated for rescue teams. With the RGB-D sensor, a cube view can be generated. Figure 15 shows the Scenario 1 cube projection (Figure 8a), while Figure 16 shows the honeycomb map.

A cutout of 41 and 44 hexagons from the generated map of Figure 16, and the location of these hexagons in the simulator, are shown in Figure 17. In Section 3.1.7, the TransformRGB-D Algorithm 1 shows how RGB-D points are converted into 3D cubes.

In Figure 16, red lines (continuous or dashed) indicate the temperature reading above a reference value. Then, a fire-spot photo was recorded. Figure 18 shows the caption of hexagon 14 in Scenario 1.

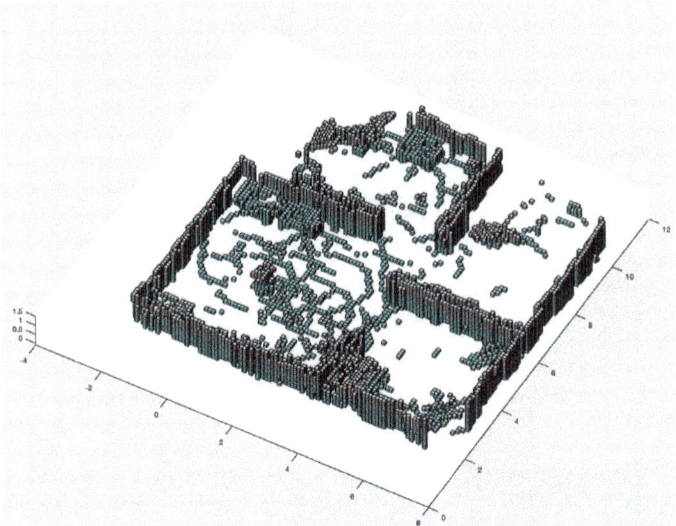

Figure 15. Three-dimensional cube view.

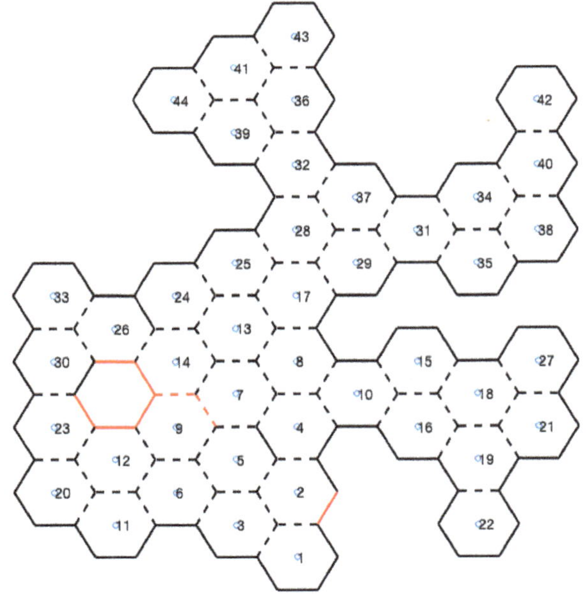

Figure 16. Honeycomb-map simulation—Scenario 1, three UAVs, Euclidean distance algorithm.

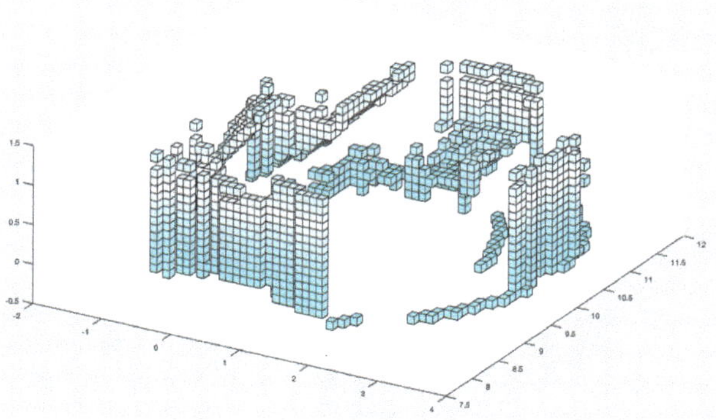

(a) Three-dimensional cube view—hexagons 41 and 44.

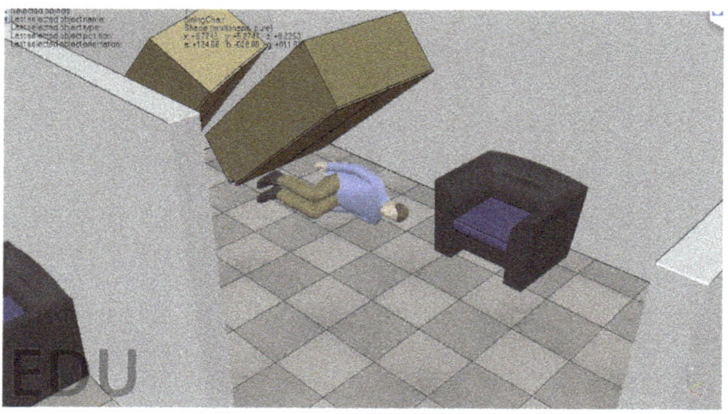

(b) V-REP scenario 1—hexagons 41 and 44.

Figure 17. Clipping of hexagons 41 and 44 of Figure 16.

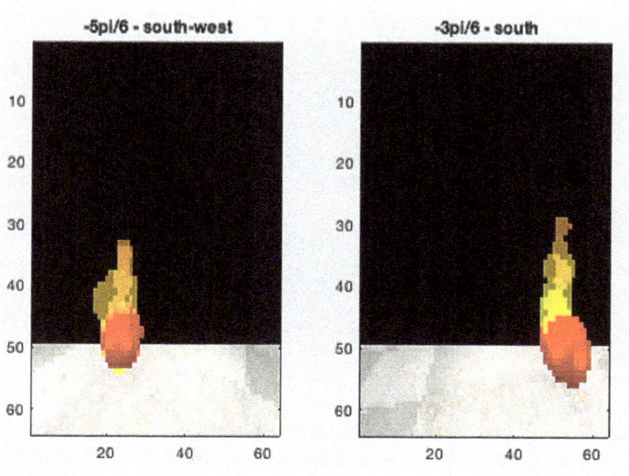

Figure 18. Fire-spot—hexagon 14 of Scenario 1.

5. Discussion

Topological mapping reduces information processing compared to metric mappings. The graph structure allows the execution of generic algorithms, such as the Dijkstra algorithm, used in trajectory planning. The data presented in Section 4 show the behavioral differences in the simulations considering number of UAVs, and algorithms in the definition of places to be explored, besides environment characteristics. By comparing the simulations, we verified that traffic in the hexagons was reduced when there was an algorithm change (FIFO for Euclidean Distance), as can be seen in Table 1. When comparing the change in UAV number, there was a slight increase in the maximum traffic value and exploration time, as can be seen in Table 4. When Scenario 2 is analyzed, the variation in the number of UAVs from two to three, in both FIFO and Euclidean Distance algorithms, reduces the exploration time. Already in Scenario 1 the opposite occurs. This happens due to the characteristics of the scenarios, where Scenario 2 has wide passages and more space for maneuvers, while Scenario 1 is composed of rooms and narrow doors, which influences the processing to avoid collisions in this points that were bottlenecks on the map. To decrease these values, an algorithm that considers not only Euclidean distance, but also the arrangement of UAVs and hexagons as a whole, should be evaluated.

On UAV displacement in the simulated scenario, Tables 2 and 3 exhibited a strong reduction in UAV movement when increasing the number of UAVs and changing the algorithm of choosing hexagons to explore. For Scenario 1, the change in the number of UAVs in the FIFO algorithm showed 31.63% reduction in average displacement per UAV. For the Euclidean distance algorithm, the reduction was 21.21%. By changing the simulation with two UAVs to three, and the FIFO algorithm for the Euclidean distance algorithm, reducing displacement in the scenario reached 42.52%; the same analysis for Scenario 2 showed a displacement decrease of 37.7%. This saves both energy and exploration time .

6. Conclusions

This work presented an environment mapping method inspired by how bees build their hives. Since only one bee constructs and occupies the space of a honeycomb, a topological map was constructed so that UAVs involved in the mapping process behaved similarly to bees. The definition of which honeycomb the UAV should map depends on a metric. The performed simulations considered two metrics to define which honeycomb should be mapped, FIFO and Euclidean Distance. In addition, simulations were performed by changing the number of UAVs. This demonstrated that setting the exploration order has direct impact on the number of offsets and a UAV in the environment, considering its position on the map. This can result in saving both energy and exploration time. Generating RGB-D and thermal-reading information enables rescuers to be prepared for obstacles and dropped objects, but also life-threatening elements such as high temperatures.

Future Work

Improvements in the definition of the spaces to be explored can be made, with metrics that consider not only distance from the initial hexagon (Euclidean distance), but also UAV location and environmental characteristics. In addition, in identifying points that may endanger the life of the rescue team, the use of gas or other toxic-element sensors may be applied. There is still the challenge of gathering this information and processing it with the use of game theory and machine learning. So far, each UAV works independently; however, it is not identified when a failure occurs with another one. A way of detecting failures and generating contingency plans needs to be implemented in future work. In this work, the representation of the hexagons is made in a projection of the x and y axes. In future work, the z axis will be added, so that this representation has several layers.

Author Contributions: R.d.R. developed the software and contributed to methodology, investigation, data curation, formal analysis, resourses, validation and writing (original draft). M.A.W. contributed to resourses, project administration, supervision, validation and writing (review and editing). T.B., J.L.L. and A.I.P.N.P. contributed to conceptualization, resourses and supervision. All authors have read and agreed to the published version of the manuscript.

Funding: This research was funded by the Federal University of Technology (UTFPR), Federal Institute of Education, Science and Technology (IFPR) and Polytechnic of Bragança (IPB).

Acknowledgments: We would like to thank UTFPR and IFPR for their support in providing the equipment to run the simulations.

Conflicts of Interest: The authors declare no conflict of interest.

Abbreviations

The following abbreviations are used in this manuscript:

AD	Adjacent Degree
ED	Euclidean Distance
EKF	Extended Kalman Filter
FIFO	First-In–First-Out
GPS	Global Positioning Systems
IMU	Inertial Measurement Unit
IR	Infrared
JCBB	Join-compatibility Branch and Bound
PTAM	Parallel Tracking and Mapping
PTAMM	Parallel Tracking and Multiple Mapping
RGB-D	Red, Green, Blue and depth
SLAM	Simultaneous Localization and Mapping
SPF	Stigmergic Potential Field
UAV	Unmanned Aerial Vehicle
UGV	Unmanned Ground Vehicle
VSLAM	Visual SLAM

References

1. Thrun, S. *Robotic Mapping: A Survey*; Research paper; School of Computer Science, Carnegie Mellon University: Pittsburgh, PA, USA, 2002.
2. Saeedi, S.; Trentini, M.; Seto, M.; Li, H. Multiple-Robot Simultaneous Localization and Mapping: A Review. *J. Field Robot.* **2016**, *33*, 3–46, doi:10.1002/rob.21620. [CrossRef]
3. Nazzi, F. The hexagonal shape of the honeycomb cells depends on the construction behavior of bees. *Sci. Rep.* **2016**, *6*, 28341, doi:10.1038/srep28341. [CrossRef] [PubMed]
4. Dhiman, N.K.; Deodhare, D.; Khemani, D. Where am I? Creating spatial awareness in unmanned ground robots using SLAM: A survey. *Sadhana* **2015**, *40*, 1385–1433. [CrossRef]
5. Douglas, D.H.; Peucker, T.K. Algorithms for the Reduction of the Number of Points Required to Represent a Diditized Line or its Caricature. *Cartographica* **1973**, *10*, 112–122. [CrossRef]
6. Jelinek, A. Vector Maps in Mobile Robotics. *Acta Polytech. CTU Proc.* **2015**, *2*, 22–28. [CrossRef]
7. Reimer, A. Cartographic Modelling for Automated Map Generation. Ph.D. Thesis, Technische Universiteit Eindhoven., Eindhoven, The Netherlands, 2015.
8. Stachniss, C. Coordinated multi-robot exploration. In *Robotic Mapping and Exploration*; Springer: Berlin, Germany, 2009; pp. 43–71.
9. Mahendran, A.; Dewan, A.; Soni, N.; Krishna, K.M. UGV-MAV Collaboration for Augmented 2D Maps. In *AIR '13, Proceedings of the Conference on Advances in Robotics*; ACM: New York, NY, USA, 2013; pp. 1–6, doi:10.1145/2506095.2506116. [CrossRef]
10. Michael, N.; Shen, S.; Mohta, K.; Mulgaonkar, Y.; Kumar, V.; Nagatani, K.; Okada, Y.; Kiribayashi, S.; Otake, K.; Yoshida, K.; et al. Collaborative mapping of an earthquake-damaged building via ground and aerial robots. *J. Field Robot.* **2012**, *29*, 832–841. [CrossRef]
11. Dewan, A.; Mahendran, A.; Soni, N.; Krishna, K.M. Heterogeneous UGV-MAV exploration using integer programming. In Proceedings of the 2013 IEEE/RSJ International Conference on Intelligent Robots and Systems, Tokyo, Japan, 3–7 November 2013; pp. 5742–5749.

12. McCune, R.R.; Madey, G.R. Agent-based Simulation of Cooperative Hunting with UAVs. In Proceedings of the Agent-Directed Simulation Symposium, Society for Computer Simulation International, San Diego, CA, USA, 7–10 April 2013; p. 8.
13. Williams, R.; Konev, B.; Coenen, F., Multi-agent Environment Exploration with AR.Drones. In *Advances in Autonomous Robotics Systems, Proceedings of the 15th Annual Conference, TAROS 2014, Birmingham, UK, 1–3 September 2014*; Mistry, M., Leonardis, A., Witkowski, M., Melhuish, C., Eds.; Springer International Publishing: Cham, Switzerland, 2014; pp. 60–71. doi:10.1007/978-3-319-10401-0_6. [CrossRef]
14. Loianno, G.; Thomas, J.; Kumar, V. Cooperative localization and mapping of MAVs using RGB-D sensors. In Proceedings of the 2015 IEEE International Conference on Robotics and Automation (ICRA), Washington, DC, USA, 26–30 May 2015; pp. 4021–4028.
15. Rogers, J.G.; Baran, D.; Stump, E.; Young, S.; Christensen, H.I. Cooperative 3D and 2D mapping with heterogenous ground robots. In *SPIE Defense, Security, and Sensing*; International Society for Optics and Photonics: Bellingham, WA, USA, 2012; p. 838708.
16. Stipes, J.; Hawthorne, R.; Scheidt, D.; Pacifico, D. Cooperative localization and mapping. In Proceedings of the 2006 IEEE International Conference on Networking, Sensing and Control, Ft. Lauderdale, FL, USA, 23–25 April 2006; pp. 596–601.
17. Wu, M.; Huang, F.; Wang, L.; Sun, J. Cooperative Multi-Robot Monocular-SLAM Using Salient Landmarks. In Proceedings of the 2009 International Asia Conference on Informatics in Control, Automation and Robotics, Bangkok, Thailand, 1–2 February 2009; pp. 151–155, doi:10.1109/CAR.2009.22. [CrossRef]
18. Chellali, R. A distributed multi robot SLAM system for environment learning. In Proceedings of the 2013 IEEE Workshop on Robotic Intelligence in Informationally Structured Space (RiiSS), Singapore, 16–19 April 2013; pp. 82–88, doi:10.1109/RiiSS.2013.6607933. [CrossRef]
19. Schuster, M.J.; Brand, C.; Hirschmuller, H.; Suppa, M.; Beetz, M. Multi-robot 6D graph SLAM connecting decoupled local reference filters. In Proceedings of the 2015 IEEE/RSJ International Conference on Intelligent Robots and Systems (IROS), Hamburg, Germany, 28 September–2 October 2015; pp. 5093–5100, doi:10.1109/IROS.2015.7354094. [CrossRef]
20. Makarenko, A.A.; Williams, S.B.; Bourgault, F.; Durrant-Whyte, H.F. An experiment in integrated exploration. In Proceedings of the IEEE/RSJ International Conference on Intelligent Robots and Systems, Lausanne, Switzerland, 30 September–4 October 2002; Volume 1, pp. 534–539, doi:10.1109/IRDS.2002.1041445. [CrossRef]
21. Dijkstra, E.W. A note on two problems in connexion with graphs. *Numer. Math.* **1959**, *1*, 269–271. [CrossRef]
22. Aryo, D. Dijkstra Algorithm. MATLAB Central File Exchange. 2020. Available online: https://www.mathworks.com/matlabcentral/fileexchange/36140-dijkstra-algorithm (accessed on 2 January 2020)
23. Fukushima, H.; Kon, K; Matsuno, F. Model Predictive Formation Control Using Branch-and-Bound Compatible With Collision Avoidance Problems. *IEEE Trans. Robot.* **2013**, *29*, 1308–1317, doi:10.1109/TRO.2013.2262751. [CrossRef]
24. Gan, S. K.; Fitch, R.; Sukkarieh, s. Real-time decentralized search with inter-agent collision avoidance. In Proceedings of the 2012 IEEE International Conference on Robotics and Automation, St Paul, MN, USA, 14–19 May 2012; pp. 504–510, doi:10.1109/ICRA.2012.6224975. [CrossRef]
25. Morgan, D.; Chung, S.; Hadaegh, F. Model Predictive Control of Swarms of Spacecraft Using Sequential Convex Programming. *J. Guid. Control Dyn.* **2014**, *37*, 1–16, doi:10.2514/1.G000218. [CrossRef]
26. Pamosoaji, A. K.; Hong, K. A Path-Planning Algorithm Using Vector Potential Functions in Triangular Regions. *IEEE Trans. Syst. Man Cybern. Syst.* **2013**, *43*, 832–842, doi:10.1109/TSMCA.2012.2221457. [CrossRef]
27. Bekris, K.E.; Grandy, D.K.; Moll, M.; Kavraki, L.E. Safe distributed motion coordination for second-order systems with different planning cycles. *Int. J. Robot. Res.* **2012**, *31*, 129–150, doi:10.1177/0278364911430420. [CrossRef]
28. Zhou, Y.; Hu, H.; Liu, Y.; Ding, Z. Collision and Deadlock Avoidance in Multirobot Systems: A Distributed Approach. *IEEE Trans. Syst. Man Cybern. Syst.* **2017**, *47*, 1712–1726, doi:10.1109/TSMC.2017.2670643. [CrossRef]
29. Alonso-Mora, J.; DeCastro, J.A.; Raman, V.; Rus, D.; Kress-Gazit, H. Reactive mission and motion planning with deadlock resolution avoiding dynamic obstacles. *Auton. Robot.* **2018**, *42*, 801–824, doi:10.1007/s10514-017-9665-6. [CrossRef]

30. Kanehiro, F.; Hirukawa, H.; Kajita, S. OpenHRP: Open Architecture Humanoid Robotics Platform. *Int. J. Robot. Res.* **2004**, *23*, 155–165, doi:10.1177/0278364904041324. [CrossRef]
31. Koenig, N.P.; Howard, A. Design and use paradigms for Gazebo, an open-source multi-robot simulator. In Proceedings of the 2004 IEEE/RSJ International Conference on Intelligent Robots and Systems (IROS) (IEEE Cat. No.04CH37566), Sendai, Japan, 28 September–2 October 2004; Volume 3, pp. 2149–2154.
32. Michel, O. WebotsTM: Professional Mobile Robot Simulation. *Int. J. Adv. Robot. Syst.* **2004**, *1*, doi:10.5772/5618. [CrossRef]
33. Rohmer, E.; Singh, S.P.N.; Freese, M. V-REP: A versatile and scalable robot simulation framework. In Proceedings of the 2013 IEEE/RSJ International Conference on Intelligent Robots and Systems, Tokyo, Japan, 3–7 November 2013; pp. 1321–1326, doi:10.1109/IROS.2013.6696520. [CrossRef]
34. The Mathworks, Inc. MATLAB Version 9.6.0.1114505 (R2019a). Available online: https://www.mathworks.com/ (accessed on 2 January 2020)

Article

Assessment of DSMs Using Backpack-Mounted Systems and Drone Techniques to Characterise Ancient Underground Cellars in the Duero Basin (Spain)

Serafín López-Cuervo Medina [1,*], Enrique Pérez-Martín [2], Tomás R. Herrero Tejedor [2], Juan F. Prieto [1], Jesús Velasco [1], Miguel Ángel Conejo Martín [2], Alejandra Ezquerra-Canalejo [3] and Julián Aguirre de Mata [1]

1 Departamento de Ingeniería Topográfica y Cartográfica, Escuela Técnica Superior de Ingenieros en Topografía, Geodesia y Cartografía, Universidad Politécnica de Madrid, Campus Sur, A-3, Km 7, 28031 Madrid, Spain; juanf.prieto@upm.es (J.F.P.); jesus.velasco@upm.es (J.V.); julian.aguirre@upm.es (J.A.d.M.)

2 Departamento de Ingeniería Agroforestal, Escuela Técnica Superior de Ingeniería Agronómica, Alimentaria y de Biosistemas, Universidad Politécnica de Madrid, Av. Puerta de Hierro, 2, 28040 Madrid, Spain; enrique.perez@upm.es (E.P.-M.); tomas.herrero.tejedor@upm.es (T.R.H.T.); miguelangel.conejo@upm.es (M.Á.C.M.)

3 Departamento de Ingeniería y Gestión Forestal y Ambiental, Escuela Técnica Superior de Ingeniería de Montes y del Medio Natural, Universidad Politécnica de Madrid, Ciudad Universitaria, 28040 Madrid, Spain; alejandra.ezquerra@upm.es

* Correspondence: s.lopezc@upm.es; Tel.: +34-910673986

Received: 30 October 2019; Accepted: 2 December 2019; Published: 4 December 2019

Abstract: In this study, a backpack-mounted 3D mobile scanning system and a fixed-wing drone (UAV) have been used to register terrain data on the same space. The study area is part of the ancient underground cellars in the Duero Basin. The aim of this work is to characterise the state of the roofs of these wine cellars by obtaining digital surface models (DSM) using the previously mentioned systems to detect any possible cases of collapse, using four geomatic products obtained with these systems. The results obtained from the process offer sufficient quality to generate valid DSMs in the study area or in a similar area. One limitation of the DSMs generated by backpack MMS is that the outcome depends on the distance of the points to the axis of the track and on the irregularities in the terrain. Specific parameters have been studied, such as the measuring distance from the scanning point in the laser scanner, the angle of incidence with regard to the ground, the surface vegetation, and any irregularities in the terrain. The registration speed and the high definition of the terrain offered by these systems produce a model that can be used to select the correct conservation priorities for this unique space.

Keywords: DSM assessment; backpack mobile mapping; UAV; underground cellars

---

1. Introduction

Instruments and techniques for the massive capture of data are increasingly being used to document all types of cultural landscapes and heritages. There are recommendations and criteria for adequate procedures to ensure the defence and preservation of these types of heritages and landscapes in the European context [1]. Many research projects on heritage management use geospatial technologies [2–5] to generate various products such as digital surface models (DSM) [6–13]. A large number of studies focus on optimising and improving actions with unmanned aircraft vehicles (UAV) [14].

Against this background, our study zone, the underground cellars of El Plantío in Atauta (Soria), declared an Asset of Cultural Interest (Bien de Interés Cultural, BIC) in March 2017, which represents a unique testimony of the life associated with work on the land and the traditional wine production system. It is important to ensure that they are not ultimately forgotten and destroyed, as they are a manifestation of the cultural identity of a large region in the Duero River basin (Figure 1).

In recent years, fixed terrestrial laser scanning (TLS) systems have been used to map and monitor various areas of interest from the point of view of their cultural heritage [15–18], in archaeological studies [19,20], underground studies [21,22], and civil engineering [23], among others [24]. 3D technologies include both the TLS system and sensors installed on UAV for the documentation, visualisation, and preservation of heritage [25–27]. The precision of a digital surface model (DSM) obtained by UAV photogrammetry for documenting surface structures [28] and forms in 3D models [29–31] and for identifying constructions in urban planning [32,33] has also been analysed. Chiabrando et al. [19] described a technique for preparing digital surface models in their archaeological studies. Norhafizi validated the use of UAV for creating DSMs of tide data [34]. Villanueva [35] studied DSMs and their application in zones at risk of flooding.

Komarek et al. [36] carried out studies to assess the precision of DSMs obtained by UAV on rural plots. In other cases, DSMs have been assessed with or without ground control points (GCP) taken by means of a real-time kinematic Global Positioning System (RTK-GPS) techniques [37–39]. Other studies have evaluated the use of software for generating DSM models from UAV with GCP taken in the field by means of other geomatic methods such as TLS, GPS [40], or UAV photogrammetry techniques [41].

Figure 1. Location of the study area. The underground cellars are located in Atauta (Soria), a region in the Duero Basin in the north-central part of Spain.

There are examples of studies on the precision of 3D models in indoor spaces and areas through backpack mobile mapping [42–44] and assessments of different TLS [45–47]. Other works assess the performance of a mobile mapping system (MMS) and backpack mobile mapping [48,49] compared to the use of a TLS [47], and highlight the superior performance of the mobile system, even though its overall precision is lower than with TLS. Campos [50] applied the backpack mobile mapping system in forest mapping where there is limited accessibility. Przemyslaw [51] also used backpack mobile mapping for the geolocation of tree trunks and their comparison with UAV records.

The DSMs obtained are used to compare and complement the information registered with ground penetrating geo-radar technology (GPR) and terrestrial laser scanning (TLS) [52]. In previous studies in the same area [53], DSMs enabled the definition and comparison of wall and ceiling thicknesses in underground cellars to ensure their stability.

The aim of this work is to assess the condition of the roofs of underground cellars in their natural state by obtaining an accurate DSM to characterise, detect, and prevent their collapse. This is accomplished using a backpack-mounted 3D mobile scanning system (backpack MMS) and a fixed-wing drone (UAV).

The study examines the use of backpack MMS in areas of irregular terrain such as the Atauta underground cellar, using data registration parameters and DSM generation including working widths and angles of incidence in the measurement and obstacles. The study also compared it with UAV equipment that enables analysis of the benefits of both systems for generating DSMs of key importance in assessing the current risk status of underground wine cellars.

2. Materials and Methods

2.1. Case Study Description

The study area is part of the series of underground wine cellars of El Plantío in Atauta, Soria (41°31′ N, 3°12′ W), shown in Figure 1, which were declared an Asset of Cultural Interest (BIC) as an "ethnographic collection" on 16 March 2017. Located at the foot of the village of Atauta on an area of 1.9 ha, they are a testimony of the life associated with work on the land and the artisanal wine production system. These underground constructions were used to store and preserve wine.

2.2. Equipment Used to Take the Images

The topographic and photogrammetric survey using UAV techniques was carried out with a Mavinci fixed-wing drone with a Lumix-GX1 camera with a focal distance of 14 mm and a resolution of 4592 × 3448 pixels. The flight was made at a height of 90 m, obtaining a ground sample distance (GSD) of 2.16 cm and taking 344 images by following a flight plan with a regular grid pattern defined in a northwest-southeast direction. The longitudinal overlap was 70%, and the transversal overlap was 50%. Precise coordinates were obtained throughout the flight due to an RTK-GPS receiver located on the UAV, which received corrections from a base on the ground. Also using RTK-GPS techniques, high-precision control points were observed in order to improve the geo-referencing of the photogrammetric model with points on the ground and for a quality check assessment. The deformation in the images caused by the camera [54] is compensated with an autocalibration performed during the aerotriangulation process (Figure 2).

2.3. System Used for Mass Registration with a Laser

A Leica Pegasus data registration system was used as a backpack mobile mapping model (Figure 3). This system incorporates five cameras and two LiDAR (light detection and ranging) scanners for the registration of the 3D point cloud and images. It also has two Velodyne sensors (VLP16) that rotate at 10 Hz and acquire 600,000 points per second at a distance of up to 100 m, even though this may be influenced by several factors, as indicated in its specifications [48,55]. It weighs approximately 13 kg and has a scanning autonomy of three hours. The system includes a SLAM (simultaneous localisation and mapping) algorithm and an IMU (inertial measurement unit) as an aid for generating the 3D model.

Figure 2. (**a**) Mavinci UAV used during the registration process. (**b**) Image orientation process task in Agisoft Photoscan® professional software (http://www.agisoft.com/) showing flight paths and photocenters. (**c**) 3D view diagram of the recorded images.

Figure 3. (**a**) GPS Leica GX1230 GG equipment, acting as a reference base. (**b**) Pegasus backpack system for taking the point cloud in the study area before obtaining the DSM model to be assessed.

Leica Pegasus Backpack allows the acquisition of LiDAR and image data with precise positioning of outdoor and indoor data [18] in a system that is easily transportable by one person. This makes this type of equipment very useful in environments with limited space, underground environments, [20] and in areas with dense vegetation, as well as for managing data on disasters [56] and documenting industrial facilities. It offers the option of including external sensors such as GPR equipment, thermal cameras, noise and pollution sensor, etc., assisted by a flash module for working in conditions of low light. The work software has tools for extracting LiDAR and photogrammetric data and detecting changes, and is compatible with workflows by means of AutoCAD and ArcGIS. The only limitation to its use is its autonomy of four hours due to its batteries.

It took 30 min to configure and calibrate the backpack MMS data registration system, and one hour to survey the study area. The main features of the UAV and the Pegasus backpack are shown in Table 1.

Table 1. UAV and MMS system specifications.

Main Features	Mavinci UAV	Pegasus Backpack
Technology	Lumix-G1 16 Mpx camera	Velodyne VLP16 laser scanner
Measurement technology	Computation from images	Polar distance measurement
Distance measurement	90 m	5–100 m
System resolution	GSD 2.16 cm	Dist. acc. 3 cm at 100 m
DSM resolution	600 points/m^2	36,000 points/m^2

2.4. Control and Assessment System Obtained by GPS

High-precision coordinates were computed on a fixed base using GPS techniques in post-processing to obtain both Ground Evaluation Points (GEP) and Ground Control Points (GCP). The points are distributed to cover the entire area of characterisation and possible collapses on the surface of the underground cellars, which are the same as the UAV flight and the tracks taken by the backpack, as shown in Figure 4.

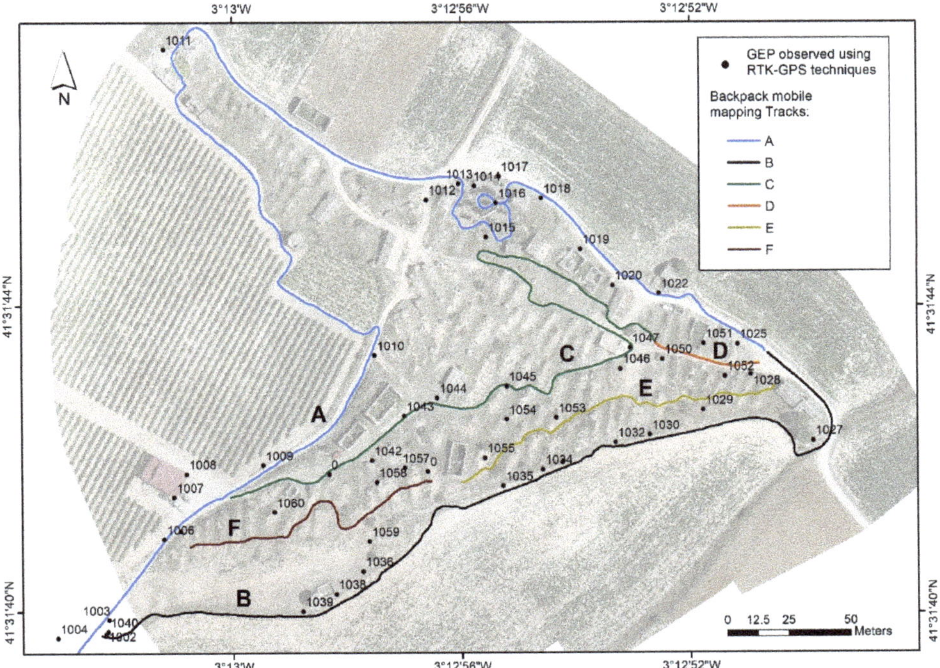

Figure 4. GEP observed using RTK-GPS techniques for the characterisation surveys of collapse zones. Lines depict the backpack mobile mapping tracks (Tracks A to F) used to assess the 3D model.

This GPS base served as a reference for all the topographic and photogrammetric surveys (see Figure 5, Section 1, centre). The equipment used was a Leica Geosystems GX1230 GG, and the data were processed with the Leica Geo Office software, with a relative average precision of more than 0.02 m in all the points. The base GPS receiver was situated in the centre of the study area and all the GCPs and GEPs observed were located at a radius of less than 200 m and measured with another two GPS rover receivers. From this base, 12 GCPs were computed in post-processing to support the UAV. The GCPs were positioned on the periphery and inside the study zone. This same base served as a support for the backpack MMS, and all its tracks were within a range of less than 250 m from the base.

Furthermore, 59 GEP coordinates were also computed by RTK-GPS techniques from the same base, and, subsequently, used in the assessment of the DSMs generated.

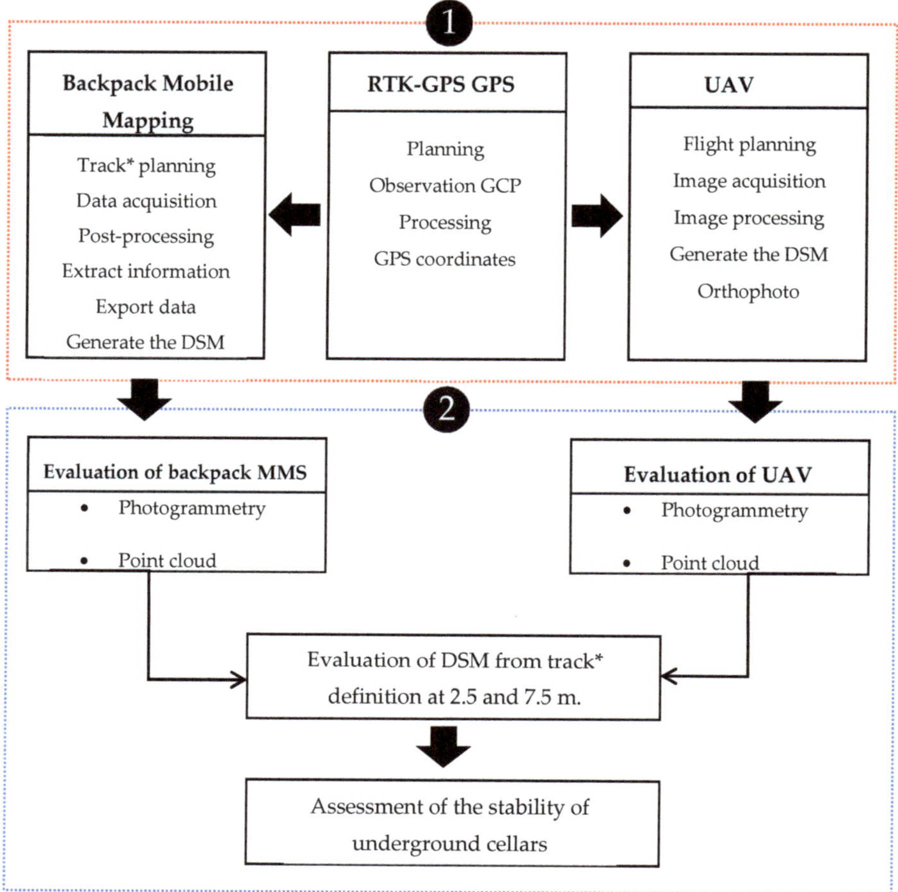

Figure 5. Workflow of the methodology for data acquisition where (**1**) refers to data acquisition and post-processing for backpack mobile mapping, UAV, and RTK-GPS, and (**2**) represents DSM processing and stability evaluation. Note the interactions between the benchmark survey (GPS) and backpack mobile mapping and UAV surveys.

2.5. Methods

As stated, the aim of this research is to obtain digital surface models (DSM) on the terrain corresponding to the natural cover of the underground cellars. A methodology was defined in several phases for this purpose, as shown in the diagram in Figure 5. In the first phase, the 45 GEP points were analysed after obtaining the data with methods based on backpack MMS and UAV technologies. These points were identified using Agisoft Photoscan® Professional software and included in the aerotriangulation process, and then identified and extracted from the DSM generated from the dense point cloud in the UAV photogrammetric project. Each GEP point was located in the data from the photographs registered by backpack MMS and measured by stereoscopy with ArcExplorer (Esri, USA) software. The GEP were also identified in the LiDAR point cloud from the backpack MMS. These four measures were compared with the GEP coordinates obtained by the RTK-GPS method.

The results of the comparisons led to the selection of the point cloud-based methods and the rejection of the photogrammetric methods, since they were insufficiently dense to generate DSMs that could be guaranteed to detect possible collapses. However, the photogrammetric backpack MMS was the most accurate method (see results section).

The data obtained with the backpack MMS define the tracks from which to obtain the point cloud covering the terrain. These tracks were used as a common element to establish the zones that cover the total study area and allow the evaluation of the DSM.

In a second phase, the results from the two massive data recording techniques are statistically analysed to compare the numerical and graphic products obtained and to define the DSM more clearly.

As has been mentioned, a supported network of control points was defined with RTK-GPS to serve as the basis for the UAV flight and the MMS backpack. This network established the real precision of both models and allowed the study of parameters such as the distance to the track in the case of backpack MMS, the point density according to the method used, and the veracity of the model with regard to walls, roofs, steeply sloping areas, and other elements. These points acted as a geometric control of the parameters to be assessed. The methodology applied in each system is shown in the diagram in Figure 5 and described below in more detail.

2.6. Processing the UAV Point Cloud

The previously mentioned software was used to generate the point cloud with the images recorded with UAV, and also included their corresponding metadata and the RTK-GPS coordinates of the photocentres of the images referenced on the ground. This type of software uses sfM-MVS algorithms for the orientation and calculation of point clouds and is widely used in UAV work processes due to the large number of images, type of cameras, and auxiliary data taken into account [57–60].

The sfM-MVS processing workflow used was as follows: initially a feature detection is performed by identifying a large number of key points in each image. With these points, an image matching process is carried out to identify and match these features in all the images in which they are registered. Subsequently, a blunder adjustment is performed with the camera's self-calibration, photocenter coordinates, and other parameters. Thus, the photogrammetric solution of the external orientation parameters of the images is obtained together with the 3D sparse model composed of the feature points detected in these images, as described in Reference [61].

This study has specifically included the use of GCPs as field control, GEPs as manual photogrammetric tie points, and feature points as automatic tie points or sparse points (see Figure 5, right) in the same photogrammetric adjustment process. Lastly, the digital terrain model (DTM) was created to produce the orthophotograph on which to identify the assessment points and verify the quality of the LiDAR DSM.

2.7. Processing the Backpack MMS Point Cloud

The point clouds from the LiDAR data registered with the backpack MMS platform were computed using the Leica Pegasus Manager software, which also allows the management of the data captured by means of MMS. It is composed of several modules in a workflow that ranges from the prior planning of the work to be done, the acquisition of the data, the subsequent processing and refinement with other sensors and algorithms, and the automatic and manual extraction of the characteristics of interest within the point clouds.

A total of six tracks were obtained (labelled A to F, Figure 4). The points were measured by photogrammetry with the Leica ArcExplorer application. The software includes a tool that allows the stereoscopic measurement of any point in the images from the perspective centres recorded and their rotations, and, thereby, obtains the coordinates of the GEPs. The last step was to identify the GEPs on the orthophotograph obtained by UAV and to measure them in the LiDAR MMS point cloud (see Figure 5, left).

2.8. Comparison of the UAV-Backpack MMS Point Clouds

After comparing the coordinates for each GEP with the DSMs obtained from the UAV and backpack MMS data, the information from each registration system was analysed. This comparison uses the tracks defined with backpack MMS to identify the different sections for analysis, while taking into account the following factors.

- The surface of the terrain in its natural state is covered with herbaceous and shrubby vegetation, which may pose an obstacle for measurement in both methods. Two types of zone were identified to assess the DSMs: one with exterior corridors with less influence of vegetation, and another on the interior with tracks with a greater presence of grassland vegetation. Some tracks were made to pass over the roofs of the cellars in order to obtain more detailed information in critical areas.
- Taking the axis of the tracks as a reference for the LiDAR scanning distances, the point cloud obtained on both sides of these tracks was computed and analysed to a limit of between 2.5 and 7.5 m from their axis.

The backpack MMS tracks A-F were used as identifiers for the assessment, and grouped according to the similarity of their features. Tracks A and B are perimetral, defined in areas of broad corridors with little influence of vegetation at a distance of 2.5 m each side of the axis. Tracks C, D, E, and F are located on the interior and, therefore, have a presence of vegetation and irregularities in the terrain, low walls, etc. These six tracks were divided into sections based on the similarity of the type of terrain in order to enable a better comparison between UAV and MMS data. Hence, Track A was divided into five sections, Track B was divided into four sections, Tracks C and E were divided into two sections, and, lastly, Tracks D and F each have only one section. Obstacles caused by constructions were eliminated during the digital processing of the point cloud to avoid errors due to vertical and/or horizontal measuring in the two systems.

For the assessment, a DSM was generated with a resolution of 0.05 m from the point clouds obtained with backpack MMS, and a grid with a point-to-point resolution of 0.10 m was projected on the DSM. The same procedure was followed for the UAV point cloud. Therefore, the resolution and geolocation of the DSMs coincide and allow their subsequent comparison.

Lastly, for the assessment of the DSM comparison at 7.5 m from the axis of the track taken by backpack MMS, the sections with no errors due to obstacles were maintained, and sections A4, A5, B1, B2, and C2 were removed.

3. Results

Before comparing the results of the different methods, it should be noted that a mean square error of 1.9 cm in height was obtained in the calculation of the GEP points.

Ten of the 59 initial GEPs were eliminated since they could not be measured in all four methods including some that were impossible to identify and others had insufficient resolution for providing real coordinates with regard to other methods, such as the eaves of a warehouse in the testing area. Four more points were also eliminated due to problems in identifying them, such as the corner of a bench which—although it could be adequately measured in photogrammetry—was difficult to discern from the LiDAR point clouds.

The following results were obtained from the various methods, according to the distance criteria 0 to 2 m, 2 to 3.5 m, and 3.5 to 10 m.

3.1. Point Cloud Processed with UAV

- The assessment of coordinates was obtained by triangulation with the UAV images. The GEP points included as linkage points in the aerotriangulation process gave the following results. The mean square error for the three distances is 4.7 cm in distance and 8.9 cm in height. The points were measured on an average of 16 photographs and their internal precision was 1.2 pixels. The standard deviation was up to 6.2 cm in height. This is a "low density" point cloud (2 points/square metre) (Table 2 and Figure 6). Five minutes were required to measure each point in photogrammetry, since the points must be measured in all the photographs. Sixty points were recorded.

- This section discusses the assessment of points in the UAV dense point cloud. The GEPs identified and measured in the dense point cloud gave the following results. The mean square error for the three distances was 5.0 cm in distance and 10.7 cm in height. The standard deviation of the points measured in the dense point cloud was 7.5 cm (Table 2). The points in the dense clouds were measured. Ten hours were required to make the dense model of 2,400,000 points extracted from this work.

Table 2. Precision comparison by method and distance to the backpack MMS track (in metres).

Product	Coordinate	0–2		2–3.5		3.5–10	
		Mean	Dev	Mean	Dev	Mean	Dev
Backpack MMS photogrammetry	Horizontal	0.089	0.051	0.049	0.035	0.161	0.109
	Vertical	0.031	0.032	0.025	0.024	0.051	0.033
Backpack MMS point cloud	Horizontal	0.066	0.057	0.032	0.020	0.054	0.027
	Vertical	0.068	0.114	0.090	0.107	0.056	0.049
UAV photogrammetry	Horizontal	0.053	0.034	0.040	0.018	0.047	0.023
	Vertical	0.073	0.053	0.102	0.079	0.096	0.057
UAV point cloud	Horizontal	0.067	0.051	0.033	0.020	0.047	0.022
	Vertical	0.089	0.077	0.118	0.088	0.116	0.063

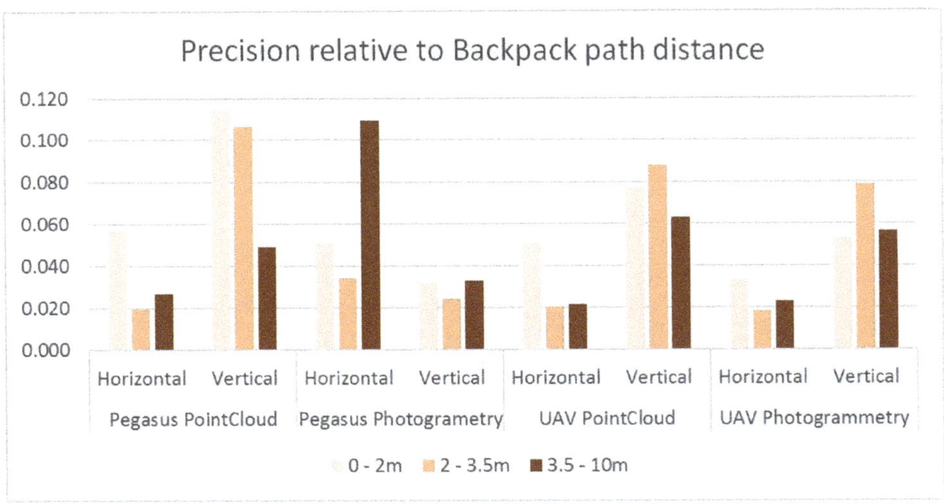

Figure 6. Comparison of precision based on the method used and the distance from the axis of backpack MMS.

3.2. Point Cloud Processed with Backpack MMS

- Assessment of LiDAR points recorded by backpack MMS. The mean square error for the three distances was 5.2 cm in distance and 7.0 cm in height. The standard deviation in height was up to 9.2 cm (Table 2 and Figure 6). The point measurement with MMS is an automatic and dynamic process, and the points are obtained while covering the work zone. Two hours were required to record the 115 million points included in the trajectories in the project.
- The assessment of coordinates was obtained by photogrammetry with ArcExplorer software. The mean square error for the three distances was 10.3 cm in distance and 3.6 cm in height.

The standard deviation in height was up to 3.2 cm (Table 2). This is the least productive measuring process, with one point measured every 10 min. The points are identified manually, and the parallax must be manually cancelled in each point.

Table 2 shows the precision and densities obtained with each method. They are then compared with the RTK-GPS points network measured in the area, and the data are grouped by distance to the track with the laser scanner located in the backpack: 0 to 2 m (measurement angle of the laser scanner on the terrain from 90° to 5°), 2 to 3.5 m (measurement angle of 45° to 22°) and over 3.5 m (angle of less than 22°). The greatest precisions, particularly in terms of height, are obtained in the photogrammetric measurement on the backpack images, even though this is the least efficient method in terms of density and production. When the point is clearly defined, no significant loss of precision was observed with regard to distance.

3.3. Comparison of Backpack MMS vs. UAV DSMs

The result of the precision analysis for the DSM at a distance of 2.5 m from the axis of the track is shown in Table 3 and Figure 7. The two DSMs are evaluated with a square grid with a resolution of 0.1 m based on the point clouds obtained with UAV and backpack MMS. The height precision for the tracks on the exterior is 3.8 cm, whereas the precision in interior areas is 5.7 cm in height. Tracks A4, A5, B1, B4, D1, and E2 have a vertical displacement between 0.10 and 0.15 m. All these areas have an influence due to construction walls or other obstacles.

Table 3. Results of the comparison between UAV and backpack MMS DSMs with points measured at a distance of at least 2.5 m from the backpack MMS track.

Track	Perimeter (m)	Area (m^2)	Distance (m)	UAV Resolution	Backpack MMS Resolution	Grid Resolution	Precision (m)	RMS (m)
A1	370	564	184	357,165	13,694,022	55,578	−0.02	0.04
A2	189	420	90	255,611	11,128,095	41,606	−0.1	0.04
A3	135	196	63	116,012	4,222,443	19,171	−0.1	0.01
A4	95	117	44	69,940	5,290,734	11,562	−0.09	0.03
A5	59	90	26	52,414	7,885,791	8869	−0.16	0.02
B1	98	138	45	80,114	2,921,953	13,628	−0.16	0.02
B2	157	182	77	111,226	5,390,511	17,678	−0.1	0.05
B3	106	159	50	94,737	4,219,886	15,508	−0.08	0.06
B4	236	286	116	180,494	13,471,066	27,982	−0.12	0.07
C1	339	348	168	226,162	9,852,318	33,758	−0.04	0.04
C2	262	339	126	243,346	16,488,277	33,363	−0.09	0.08
D1	69	83	31	52,109	1,691,763	8062	−0.15	0.02
E1	169	232	83	167,189	10,230,845	22,552	−0.07	0.07
E2	109	153	53	105,739	4,740,500	14,961	−0.16	0.07
F1	236	464	116	302,089	14,207,240	45,617	−0.06	0.06

If the distance from the backpack MMS track increases from 2.5 m to 7.5 m on each side of the axis from which the point clouds defining the DSM are obtained, the maximum angle of measurement of backpack MMS on the terrain would decline from 45° to 20°, which increases the error due to the noise produced in the measurement. There are also a higher number of errors due to obstacles. The number of zones in the study was, therefore, reduced in order to avoid errors caused by built elements or obstacles. The study for these distances at 7.5 m is limited to zones A1, A2, A3, B3, and B4 in the exterior part and to zones C1, D1, E1, E2, and F1 in the interior. In overall terms, the mean square error in height ranges from 0.04 m to 0.21 m (Table 3 and Figure 8).

Table 4 shows a comparison between the data obtained with points located at a distance of up to 2.5 and 7.5 m from the backpack MMS track. The correlation coefficient R^2 allows the incidence of isolated errors to be analysed when comparing both DSMs.

Figure 7. Distribution of the precise comparison of the point clouds based on the backpack MMS track. Letters denote the different tracks while numbers depict the different sections in the tracks.

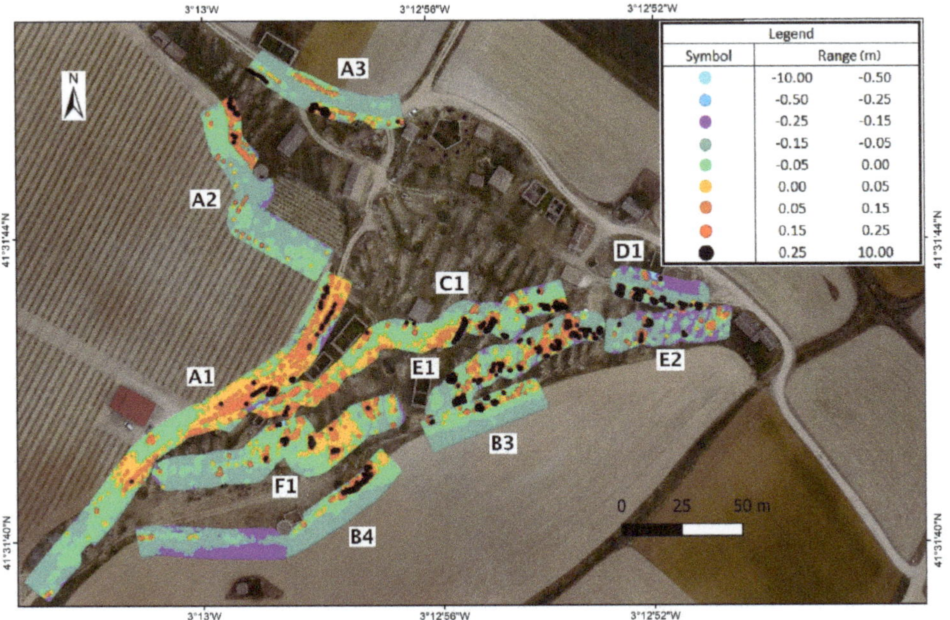

Figure 8. Precision distribution with the influence of 7.5 m with regard to the backpack MMS track. Letters denote the different tracks while numbers depict the different sections in the tracks.

Table 4. Comparison of precision when the width is increased from 2.5 to 7.5 m.

	Grid Points		Precision		RMS		R^2	
Track	5 m	15 m	5 m	15 m	5 m	15 m	5 m	15 m
A1	55,578	233,444	−0.02	−0.07	0.04	0.14	0.9998	0.9959
A2	41,606	135,901	−0.10	−0.19	0.04	0.28	0.9985	0.8108
A3	19,171	88,832	−0.10	−0.1	0.01	0.15	0.9928	0.9806
B3	15,508	71,411	−0.08	−0.1	0.06	0.11	0.9969	0.9792
B4	27,982	144,769	−0.12	−0.12	0.07	0.24	0.9911	0.9968
C1	33,758	197,137	−0.04	−0.09	0.04	0.73	0.9998	0.9598
D1	8062	46,599	−0.15	−0.16	0.02	0.33	0.9979	0.8295
E1	22,552	103,978	−0.07	−0.14	0.07	0.36	0.9991	0.9848
E2	14,961	72,501	−0.16	−0.2	0.07	0.17	0.9967	0.9567
F1	45,617	158,948	−0.06	−0.09	0.06	0.11	0.9975	0.9908

4. Discussion

The methods for acquiring and processing the data provided by UAV and backpack MMS are of sufficient quality to generate valid DSMs in the study area or similar.

The first part of the work develops the methods of data capture using photogrammetry and a laser scanner with UAV and backpack MMS. The most challenging task was to select the points that could be identified using the different methods. Additionally, 20% of the control points were eliminated due to of the fact that identifying these points in all the methods is not possible.

Except in the case of measurement by photogrammetry, where backpack MMS provides the best precision results with a little under 5 cm in height, the rest of the methods had a precision of around 5 and 10 cm. Better overall results are obtained with UAV methods than with backpack MMS, but this depends on the distance of the points from the axis of the backpack MMS track. Backpack MMS is also better over a short distance under 3 m, with 6.8 cm. At higher distances, the UAV methods obtain a lower precision of over 8 or 9 cm.

In addition to precision, performance and production were also analysed. Manual photogrammetric methods were discarded since they required longer execution times, which represent a higher production cost. The registration of the dense cloud points with UAV proved to be the fastest and most economical method, whereas the registration via MMS gave the best performance, but had a higher production cost.

Given these results, the comprehensive comparison of UAV and backpack MMS point clouds in different zones that are more or less devoid of vegetation and small obstacles allows the analysis of their precision and the influence of each method on the definition of the DSM to be characterised. The precision in clear zones or zones near the track trajectory is 5 cm (Figure 9).

More substantial differences are found at greater distances, and measurement problems also arise due to the lack of information or measurements in an orography such as the one in the study area, with holes/gaps or even obstacles, as can be seen at the most extreme points of the model (Figure 10). Although they are few and more distant from the backpack trajectory, they present a proportionally higher error, which can be seen in Table 3 and Figures 7 and 8.

The clearest example of these differences can be seen in the correlation graphs in Figure 11 (external tracks) and Figure 12 (internal tracks), where the different backpack MMS trajectories and their comparison with the UAV points not only show the R^2 coefficients and the correlation equations indicated above, but also highlight the difference in the number and height of the points outside the trend line for the analysis of each track.

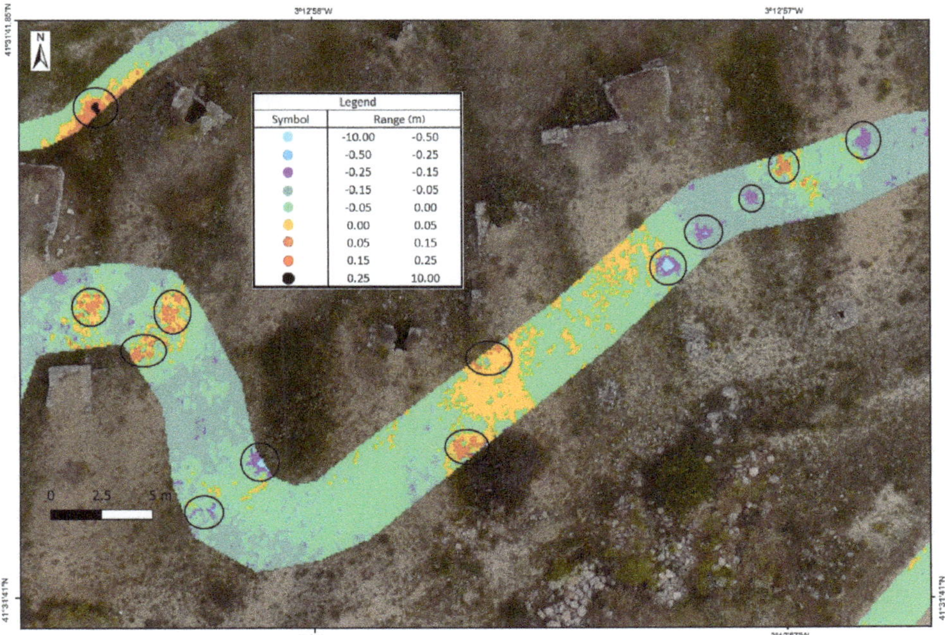

Figure 9. Identification of zones with significant differences between UAV and MMS DSMs.

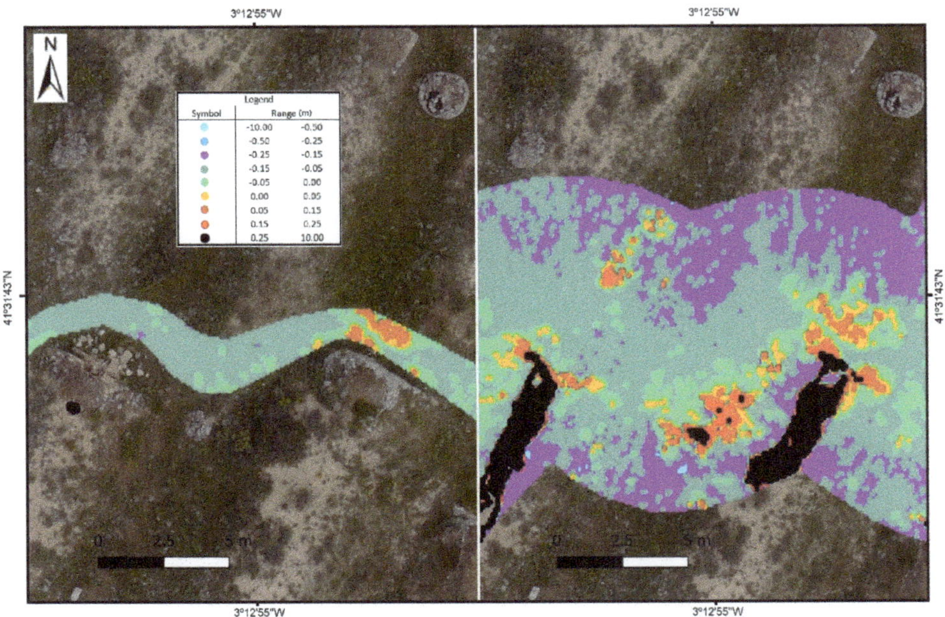

Figure 10. Differences in height according to the type of system, the width of the scan, and the obstacles or concealed elements. (**Left**): example of a zone with a 2.5 m track width. (**Right**): same area with a 15 m track width. Black areas depict concealed zones not registered in the backpack MMS.

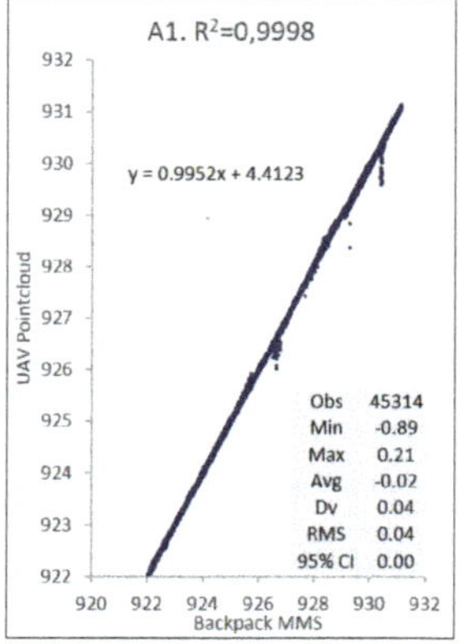

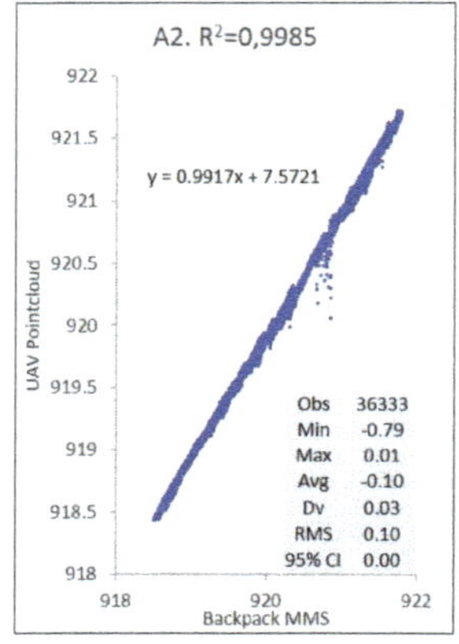

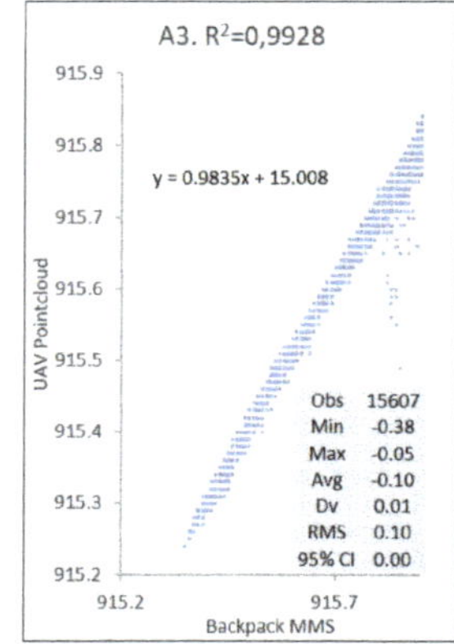

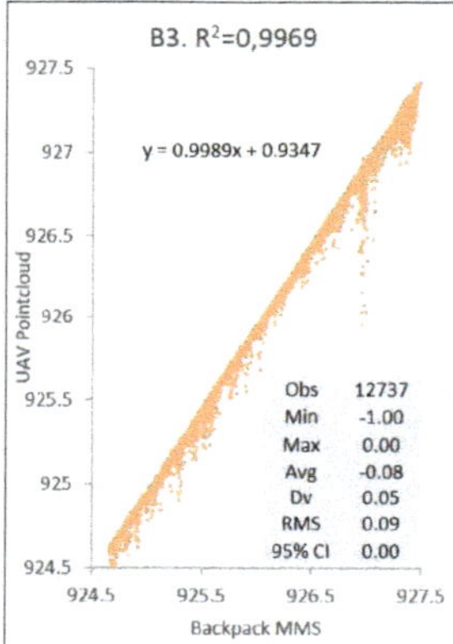

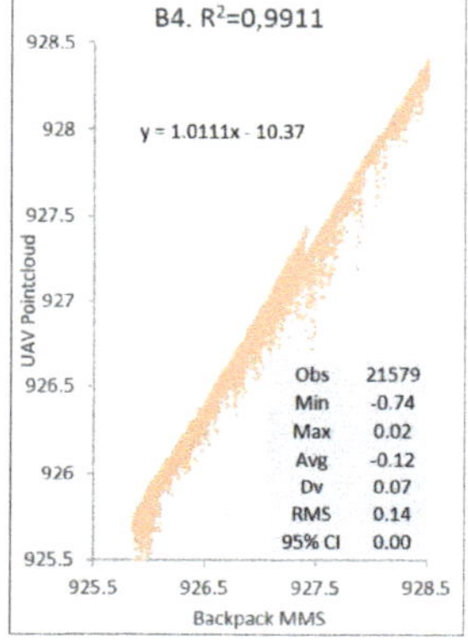

Figure 11. Comparison of UAV vs. backpack MMS by external tracks (**A**,**B**).

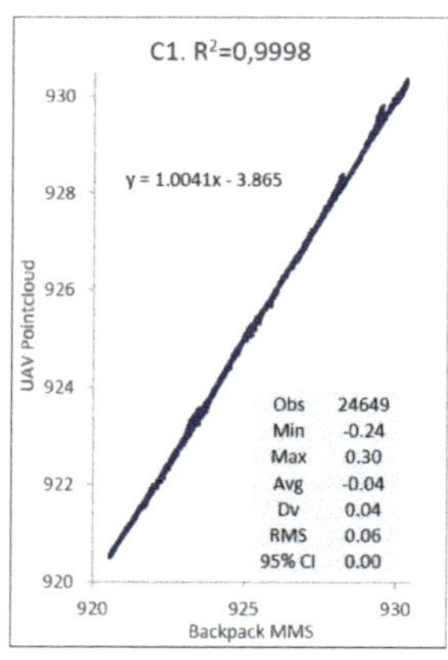

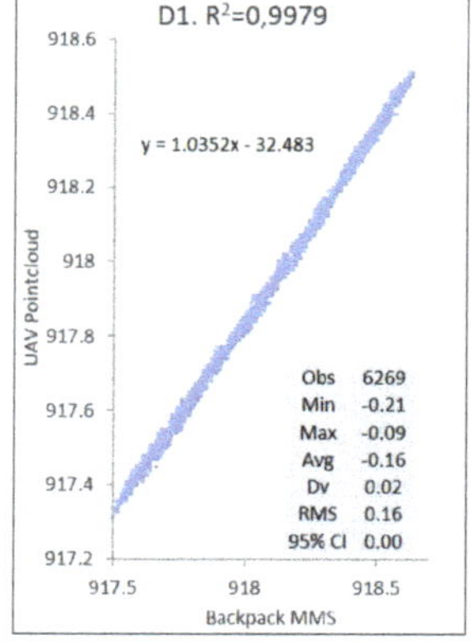

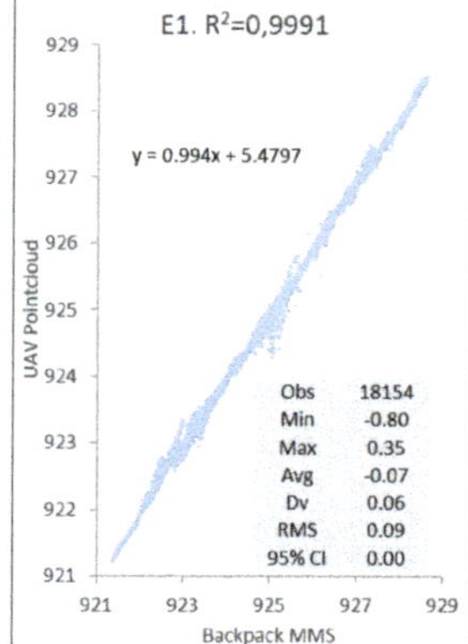

Figure 12. *Cont.*

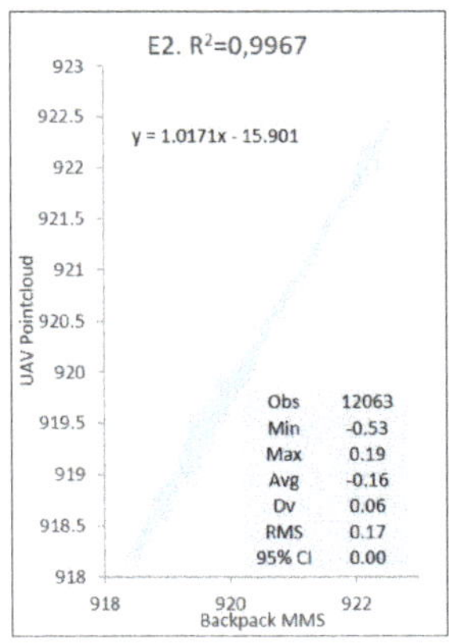

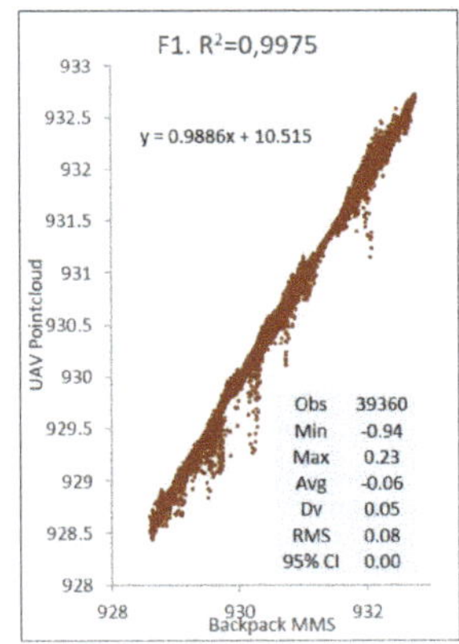

Figure 12. Comparison of UAV vs. backpack MMS by internal tracks (**C–F**).

5. Conclusions

When comparing the data extracted individually, a similar precision is obtained with an average of around 5 cm in height with both methods, and an average of 10 cm with distance or identification problems. This precision is perfectly acceptable for defining DSMs in this work environment.

Since the precision of both systems is known to be similar in small environments or at short distances from backpack MMS, denser trajectories must be used to reduce errors due to obstacles. A mixed method such as the proposed MMS–UAV technique represents a useful tool for identifying areas that are difficult to determine in the DSM and which lead to the most significant differences in height in the comparison. Given the very high number of points obtained with these procedures, the detection and elimination of these points would make it possible to obtain a DSM with greater precision and a greater level of detail.

The assessment of the DSMs reveals that the tracks followed by backpack mobile mapping present a scarcity of information in some spaces. There are various obstacles and hidden areas in this terrain. In summary, the UAV provides a more homogeneous or stable DSM even though the DSM obtained with backpack mobile mapping is more accurate. This is due to the optimum distance range and the spaces where there are no obstacles or strong rupture lines. Both techniques can, therefore, be considered complementary and reliable for obtaining DSMs for the area where these underground cellars are located. This type of application can help detect deformations in the ground a posteriori.

The use of methods of mass capture offers an excellent opportunity in such a complex area as the exterior surface of the underground cellars of El Plantío in Atauta. These methods have different limitations, such as the irregularity of the terrain, difficult-to-isolate low-growing vegetation, and mobile or fixed obstacles. However, the general precision is high and in line with the data results necessary for their study and preservation.

A novel development is the inclusion of parameters such as the distance to the scanning point, the angles of incidence with regard to the ground, and the study of irregularities in the terrain. A clear comparison of both technologies that conclusively reveals the pros and cons of their use would be impossible without considering these aspects. Their combined use has also been proposed, which may be a source of further improvements in future studies.

The results obtained point to the conclusion that both techniques—albeit not without difficulty—provide DSMs that are capable of defining terrain stability. Due to the registration speed and the precision achieved, both systems allow the assessment of the underground wine cellars. Their use over time will make it possible to establish the necessary priorities to guarantee the conservation of such unique and important spaces such as this site of El Plantío in Atauta.

Author Contributions: Conceptualisation, S.L.-C.M. and E.P.-M., Methodology, S.L.-C.M., E.P.-M., J.F.P., J.V., and T.R.H.T. Investigation, J.F.P., J.V., and T.R.H.T. Data backpack MMS registration M.Á.C.M., A.E.-C., and E.P., Data UAV registration S.L.-C.M. and J.A.d.M. GPS field observation, J.F.P. and J.V. Backpack MMS data processing E.P.-M., M.Á.C.M., and A.E.-C. UAV data processing, S.L.-C.M. and J.A.d.M. GPS data processing, J.F.P., J.V., and T.R.H.T. Visualisation, E.P.-M., M.Á.C.M., and J.A.d.M. Supervision, T.R.H.T. and A.E.-C. All the authors contributed to writing and original draft preparation.

Funding: This work is part of the results of the "Red de Investigación en Paisajes Culturales" research project in the national R+D plan with reference number RED2018-102558-T. It is also supported by the programme for collaboration among the GIAPSI (ETSIINF), GESYP (ETSIAAB) and GIPC (ETSAM) research groups promoted by the Universidad Politécnica de Madrid (UPM) with reference VJOVUPM17IMS.

Acknowledgments: We would like to thank the Atauta Town Hall (San Esteban de Gormaz) and the companies TOPCON and LEICA for their help with the cultural landscape, including administrative and technical support (materials used for experiments). We also want to acknowledge Pru Brooke-Turner (M.A. Cantab.) for her English language and style review.

Conflicts of Interest: The authors declare that there are no conflicts of interest.

References

1. Council of Europe. *Recommendation CM/Rec (2008) 3 of the Committee of Ministers to Member States on the Guidelines for the Implementation of the European Landscape Convention*; Council of Europe: Strasbourg, France, 2008.
2. Brown, G.; Brabyn, L. An analysis of the relationships between multiple values and physical landscapes at a regional scale using public participation GIS and landscape character classification. *Landsc. Urban Plan.* **2012**, *107*, 317–331. [CrossRef]
3. Skalos, J.; Molnarova, K.; Kottova, P. Land reforms reflected in the farming landscape in East Bohemia and in Southern Sweden—Two faces of modernisation. *Appl. Geogr.* **2012**, *35*, 114–123. [CrossRef]
4. Mills, J.W.; Curtis, A.; Kennedy, B.; Kennedy, S.W.; Edwards, J.D. Geospatial video for field data collection. *Appl. Geogr.* **2010**, *30*, 533–547. [CrossRef]
5. Koch, A.; Heipke, C. Quality assessment of digital surface models derived from the Shuttle Radar Topography Mission (SRTM). In Proceedings of the Geoscience and Remote Sensing Symposium (IGARSS 2001), Sydney, Australia, 9–13 July 2001; pp. 2863–2865.
6. Levoy, M.; Pulli, K.; Curless, B.; Rusinkiewicz, S.; Koller, D.; Pereira, L.; Ginzton, M.; Anderson, S.; Davis, J.; Ginsberg, J.; et al. The digital Michelangelo project: 3D scanning of large statues. In Proceedings of the 27th Annual Conference on Computer Graphics and Interactive Techniques (SIGGRAPH 2000), New Orleans, LA, USA, 23–28 July 2000; pp. 131–144.
7. Beraldin, J.-A.; Picard, M.; El-Hakim, S.F.; Godin, G.; Valzano, V.; Bandiera, A.; Latouche, C. Virtualizing a Byzantine crypt by combining high-resolution textures with laser scanner 3D data. In Proceedings of the 8th International Conference on Virtual Systems and MultiMedia (VSMM 2002), Oral Session 1: Virtual Heritage 1 (VH1), Gyeongju, Korea, 25–27 September 2002.
8. Avrami, E.; Mason, R.; De la Torre, M. *Values and Heritage Conservation: Research Report*; Getty Conservation Institute: Los Angeles, CA, USA, 2000.
9. Stumpfel, J.; Tchou, C.; Yun, N.; Martinez, P.; Hawkins, T.; Jones, A.; Emerson, B.; Debevec, P.E. Digital Reunification of the Parthenon and its Sculptures. In Proceedings of the 4th International conference on Virtual Reality, Archaeology and Intelligent Cultural Heritage (VAST 2003), Brighton, UK, 5–7 November 2003; pp. 41–50.
10. Stovel, H. Effective use of authenticity and integrity as world heritage qualifying conditions. *City Time* **2007**, *2*, 3.
11. Ikeuchi, K.; Miyazaki, D. *Digitally Archiving Cultural Objects*; Springer Science & Business Media: Berlin, Germany, 2008; pp. 1–504. ISBN 978-0-387-75807-7.
12. Remondino, F.; El-Hakim, S.; Girardi, S.; Rizzi, A.; Benedetti, S.; Gonzo, L. 3D Virtual Reconstruction and Visualization of Complex Architectures—The 3D-ARCH Project. *Int. Arch. Photogramm. Remote Sens. Spat. Inf. Sci.* **2009**, 38.
13. Aber, J.S.; Marzolff, I.; Ries, J.B. *Small-Format Aerial Photography: Principles, Techniques and Geoscience Applications*; Elsevier: Amsterdam, The Netherlands, 2010; pp. 1–266. ISBN 978-0-444-53260-2.

14. Hassanalian, M.; Abdelkefi, A. Classifications, applications, and design challenges of drones: A review. *Prog. Aerosp. Sci.* **2017**, *91*, 99–131. [CrossRef]
15. Liang, H.; Li, W.; Lai, S.; Zhu, L.; Jiang, W.; Zhang, Q. The integration of terrestrial laser scanning and terrestrial and unmanned aerial vehicle digital photogrammetry for the documentation of Chinese classical gardens—A case study of Huanxiu Shanzhuang, Suzhou, China. *Journal of Cultural Heritage.* **2018**, *33*, 222–230. [CrossRef]
16. Anton, D.; Medjdoub, B.; Shrahily, R.; Moyano, J. Accuracy evaluation of the semi-automatic 3D modeling for historical building information models. *Int. J. Archit. Herit.* **2018**, *12*, 790–805. [CrossRef]
17. Sanchez-Aparicio, L.J.; Del Pozo, S.; Ramos, L.F.; Arce, A.; Fernandes, F.M. Heritage site preservation with combined radiometric and geometric analysis of TLS data. *Autom. Constr.* **2018**, *85*, 24–39. [CrossRef]
18. Jo, Y.H.; Hong, S. Three-dimensional digital documentation of cultural heritage site based on the convergence of terrestrial laser scanning and unmanned aerial vehicle photogrammetry. *ISPRS Int. J. Geo Inf.* **2019**, *8*, 53. [CrossRef]
19. Chiabrando, F.; Nex, F.; Piatti, D.; Rinaudo, F. UAV and RPV systems for photogrammetric surveys in archaeological areas: Two tests in the Piedmont region (Italy). *J. Archaeol. Sci.* **2011**, *38*, 697–710. [CrossRef]
20. Masiero, A.; Fissore, F.; Guarnieri, A.; Piragnolo, M.; Vettore, A. Comparison of low cost photogrammetric survey with TLS and Leica pegasus backpack 3D models. *Int. Arch. Photogramm. Remote Sens. Spat. Inf. Sci.* **2017**, *42*, 147. [CrossRef]
21. Pukanska, K.; Bartos, K.; Bella, P.; Rakay, S.; Sabova, J. Comparison of non-contact surveying technologies for modelling underground morphological structures. *Acta Montan. Slovaca* **2017**, *22*, 246–256.
22. Aziz, A.S.; Stewart, R.R.; Green, S.L.; Flores, J.B. Locating and characterizing burials using 3D ground-penetrating radar (GPR) and terrestrial laser scanning (TLS) at the historic Mueschke Cemetery, Houston, Texas. *J. Archaeol. Sci. Rep.* **2016**, *8*, 392–405. [CrossRef]
23. Kukko, A.; Kaartinen, H.; Hyyppa, J.; Chen, Y. Multiplatform mobile laser scanning: Usability and performance. *Sensors* **2012**, *12*, 11712–11733. [CrossRef]
24. Ouedraogo, M.M.; Degre, A.; Debouche, C.; Lisein, J. The evaluation of unmanned aerial system-based photogrammetry and terrestrial laser scanning to generate DEMs of agricultural watersheds. *Geomorphoogy* **2014**, *214*, 339–355. [CrossRef]
25. Sun, Z.; Zhang, Y.Y. Using drones and 3D modeling to survey Tibetan architectural heritage: A case study with the multi-door stupa. *Sustainability* **2018**, *10*, 2259. [CrossRef]
26. Herrero, T.; Pérez-Martín, E.; Conejo-Martín, M.A.; de Herrera, J.L.; Ezquerra-Canalejo, A.; Velasco-Gómez, J. Assessment of underground wine cellars using geographic information technologies. *Surv. Rev.* **2014**, *47*, 202–210. [CrossRef]
27. Bolognesi, M.; Furini, A.; Russo, V.; Pellegrinelli, A.; Russo, P. Testing the low-cost rpas potential in 3D cultural heritage reconstruction. *Int. Arch. Photogramm. Remote Sens. Spatial Inf. Sci.* **2015**, *40–45*, 229–235. [CrossRef]
28. Kršák, B.; Blišťan, P.; Pauliková, A.; Puškárová, P.; Kovanič, Ľ.; Palková, J.; Zelizňaková, V. Use of low-cost UAV photogrammetry to analyze the accuracy of a digital elevation model in a case study. *Measurement* **2016**, *91*, 276–287. [CrossRef]
29. Tilly, N.; Kelterbaum, D.; Zeese, R. Geomorphological mapping with terrestrial laser scanning and uav-based imaging. *Int. Arch. Photogramm. Remote Sens. Spat. Inf. Sci.* **2016**, *41*, 591–597. [CrossRef]
30. Raeva, P.; Filipova, S.; Filipov, D. Volume computation of a stockpile-a study case comparing GPS and UAV measurements in an open pit quarry. In Proceedings of the 23rd ISPRS Congress, Prague, Czech Republic, 12–19 July 2016.
31. Pichon, L.; Ducanchez, A.; Fonta, H.; Tisseyre, B. Quality of digital elevation models obtained from unmanned aerial vehicles for precision viticulture. *OENO ONE* **2016**, *50*, 101–111. [CrossRef]
32. Yan, Y.; Tan, Z.; Su, N.; Zhao, C. Building extraction based on an optimized stacked sparse autoencoder of structure and training samples using LIDAR DSM and optical images. *Sensors* **2017**, *17*, 1957. [CrossRef]
33. Yan, Y.; Gao, F.; Deng, S.; Su, N. A hierarchical building segmentation in digital surface models for 3D reconstruction. *Sensors* **2017**, *17*, 222. [CrossRef]
34. Mohamad, N.; Khanan, M.F.A.; Ahmad, A.; Din, A.H.M.; Shahabi, H. Evaluating water level changes at different tidal phases using UAV photogrammetry and GNSS vertical data. *Sensors* **2019**, *19*, 3778. [CrossRef]

35. Escobar Villanueva, J.R.; Iglesias Martinez, L.; Perez Montiel, J.I. DEM generation from fixed-wing UAV imaging and LiDAR-derived ground control points for flood estimations. *Sensors* **2019**, *19*, 3205. [CrossRef]
36. Komarek, J.; Kumhalova, J.; Kroulik, M. Surface modelling based on unmanned aerial vehicle photogrammetry and its accuracy assessment. In Proceedings of the International Scientific Conference Engineering for Rural Development, Jeglava, Latvia, 25–27 May 2016.
37. Yeh, M.; Chou, Y.; Yang, L. The evaluation of GPS techniques for UAV-based photogrammetry in urban area. In Proceedings of the 23rd ISPRS Congress, Prague, Czech Republic, 12–19 July 2016.
38. Chang, H.C.; Ge, L.L.; Rizos, C.; Milne, T. Validation of DEMs derived from radar interferometry, airborne laser scanning and photogrammetry by using GPS-RTK. In Proceedings of the International Geoscience and Remote Sensing Symposium Science for Society (IGARSS 2004): Exploring and Managing a Changing Planet, Anchorage, AK, USA, 20–24 September 2004; pp. 2815–2818.
39. Skarlatos, D.; Procopiou, E.; Stavrou, G.; Gregoriou, M. Accuracy assessment of minimum control points for UAV photography and georeferencing. In Proceedings of the 1st International Conference on Remote Sensing and Geoinformation of the Environment (RSCy), Paphos, Cyprus, 8–10 April 2013.
40. Tong, X.; Liu, X.; Chen, P.; Liu, S.; Luan, K.; Li, L.; Liu, S.; Liu, X.; Xie, H.; Jin, Y. Integration of UAV-based photogrammetry and terrestrial laser scanning for the three-dimensional mapping and monitoring of open-pit mine areas. *Remote Sens.* **2015**, *7*, 6635–6662. [CrossRef]
41. Tokunaga, M. Accuracy verification of DSM obtained from UAV using commercial software. In Proceedings of the 2015 IEEE International Geoscience and Remote Sensing Symposium (IGARSS), Milan, Italy, 26–31 July 2015; pp. 3022–3024.
42. Kurian, A.; Morin, K.W. A fast and flexible method for meta-map building for ICP based slam. In Proceedings of the 23rd ISPRS Congress, Prague, Czech Republic, 12–19 July 2016; pp. 273–278. [CrossRef]
43. Velas, M.; Spanel, M.; Sleziak, T.; Habrovec, J.; Herout, A. Indoor and outdoor backpack mapping with calibrated pair of velodyne LiDARs. *Sensors* **2019**, *19*, 3944. [CrossRef]
44. Chen, H.; Yu, P. Application of laser SLAM technology in backpack indoor mobile measurement system. In Proceedings of the 4th International Conference on Advances in Energy Resources and Environment Engineering (ICAESEE), Chengdu, China, 29 March 2018.
45. Lehtola, V.V.; Kaartinen, H.; Nuchter, A.; Kaijaluoto, R.; Kukko, A.; Litkey, P.; Honkavaara, E.; Rosnell, T.; Vaaja, M.T.; Virtanen, J.P.; et al. Comparison of the selected state-of-the-art 3D indoor scanning and point cloud generation methods. *Remote Sens.* **2017**, *9*, 796. [CrossRef]
46. Nocerino, E.; Menna, F.; Remondino, F.; Toschi, I.; Rodriguez-Gonzalvez, P. Investigation of indoor and outdoor performance of two portable mobile mapping systems. In Proceedings of the Conference on Videometrics, Range Imaging, and Applications XIV, Munich, Germany, 26–27 June 2017.
47. Masiero, A.; Fissore, F.; Guarnieri, A.; Pirotti, F.; Visintini, D.; Vettore, A. Performance evaluation of two indoor mapping systems: Low-cost UWB-aided photogrammetry and backpack laser scanning. *Appl. Sci.* **2018**, *8*, 416. [CrossRef]
48. Leica Geosystems. Leica Pegasus: Backpack Wearable Mobile Mapping Solution. Available online: https://leica-geosystems.com/products/mobile-sensor-platforms/capture-platforms/leica-pegasus-backpack (accessed on 1 November 2019).
49. Lauterbach, H.A.; Borrmann, D.; Hess, R.; Eck, D.; Schilling, K.; Nuechter, A. Evaluation of a backpack-mounted 3D mobile scanning system. *Remote Sens.* **2015**, *7*, 13753–13781. [CrossRef]
50. Campos, M.B.; Garcia Tommaselli, A.M.; Honkavaara, E.; Prol, F.d.S.; Kaartinen, H.; El Issaoui, A.; Hakala, T. A backpack-mounted omnidirectional camera with off-the-shelf navigation sensors for mobile terrestrial mapping: Development and forest application. *Sensors* **2018**, *18*, 827. [CrossRef]
51. Polewski, P.; Yao, W.; Cao, L.; Gao, S. Marker-free coregistration of UAV and backpack LiDAR point clouds in forested areas. *ISPRS J. Photogramm. Remote Sens.* **2019**, *147*, 307–318. [CrossRef]
52. Angel Conejo-Martin, M.; Ramon Herrero-Tejedor, T.; Perez-Martin, E.; Lapazaran-Izargain, J.; Otero-Garcia, J.; Francisco Prieto-Morin, J.; Velasco-Gomez, J. Characterization of underground cellars in the duero basin by GNSS, LIDAR and GPR techniques. *Math. Planet Earth* **2014**, 277–280. [CrossRef]
53. Conejo Martin, M.A. Propuesta Metodológica para el Estudio de Sistemas Topográficos Aplicados a la Reresentación Gráfica de Bodegas Subterráneas Tradicionales. Ph.D. Thesis, E.T.S.I. Agronomos, Madrid, Spain, July 2014.

54. Gruen, A.; Beyer, H.A. *System Calibration Through Self-Calibration*; Springer: Berlin/Heidelberg, Germany, 2001; pp. 163–193.
55. Leica Geosystems. Leica Pegasus: Backpack. Mobile reality capture. Backpack specifications. Available online: https://leica-geosystems.com/-/media/files/leicageosystems/products/datasheets/leica_pegasusbackpack_ds.ashx?la=en-gb (accessed on 1 November 2019).
56. Sayama, T.; Matsumoto, K.; Kuwano, Y.; Takara, K. Application of backpack-mounted mobile mapping system and rainfall-runoff-inundation model for flash flood analysis. *Water* **2019**, *11*, 963. [CrossRef]
57. Han, X.; Thomasson, J.A.; Bagnall, G.C.; Pugh, N.A.; Horne, D.W.; Rooney, W.L.; Jung, J.; Chang, A.; Malambo, L.; Popescu, S.C.; et al. Measurement and calibration of plant-height from fixed-wing UAV images. *Sensors* **2018**, *18*, 4092. [CrossRef]
58. Smith, M.W.; Carrivick, J.L.; Quincey, D.J. Structure from motion photogrammetry in physical geography. *Prog. Phys. Geogr. Earth Environ.* **2016**, *40*, 247–275. [CrossRef]
59. Fleming, Z.; Pavlis, T. An orientation based correction method for SfM-MVS point clouds—Implications for field geology. *J. Struct. Geol.* **2018**, *113*, 76–89. [CrossRef]
60. Sanz-Ablanedo, E.; Chandler, J.H.; Rodríguez-Pérez, J.R.; Ordóñez, C. Accuracy of unmanned aerial vehicle (UAV) and SfM photogrammetry survey as a function of the number and location of ground control points used. *Remote Sens.* **2018**, *10*, 1606. [CrossRef]
61. Eltner, A.; Kaiser, A.; Castillo, C.; Rock, G.; Neugirg, F.; Abellán, A. Image-based surface reconstruction in geomorphometry—Merits, limits and developments. *Earth Surf. Dyn.* **2016**, *4*, 359–389. [CrossRef]

Article

Towards Automatic UAS-Based Snow-Field Monitoring for Microclimate Research

Petr Gabrlik [1,*], Premysl Janata [2], Ludek Zalud [1] and Josef Harcarik [2]

1 Central European Institute of Technology, Brno University of Technology, Purkynova 123, 612 00 Brno, Czech Republic; ludek.zalud@ceitec.vutbr.cz
2 The Krkonose Mountains National Park Administration, Dobrovskeho 3, 543 01 Vrchlabi, Czech Republic; pjanata@krnap.cz (P.J.); jharcarik@krnap.cz (J.H.)
* Correspondence: petr.gabrlik@ceitec.vutbr.cz

Received: 11 March 2019; Accepted: 23 April 2019; Published: 25 April 2019

Abstract: This article presents unmanned aerial system (UAS)-based photogrammetry as an efficient method for the estimation of snow-field parameters, including snow depth, volume, and snow-covered area. Unlike similar studies employing UASs, this method benefits from the rapid development of compact, high-accuracy global navigation satellite system (GNSS) receivers. Our custom-built, multi-sensor system for UAS photogrammetry facilitates attaining centimeter- to decimeter-level object accuracy without deploying ground control points; this technique is generally known as direct georeferencing. The method was demonstrated at Mapa Republiky, a snow field located in the Krkonose, a mountain range in the Czech Republic. The location has attracted the interest of scientists due to its specific characteristics; multiple approaches to snow-field parameter estimation have thus been employed in that area to date. According to the results achieved within this study, the proposed method can be considered the optimum solution since it not only attains superior density and spatial object accuracy (approximately one decimeter) but also significantly reduces the data collection time and, above all, eliminates field work to markedly reduce the health risks associated with avalanches.

Keywords: snow mapping; UAS; photogrammetry; remote sensing; direct georeferencing; snow field; snow-covered area; snow depth

---

1. Introduction

Environmental mapping embodies a relevant target field for unmanned aerial system (UAS)-based photogrammetry. The low cost, safety, and flexibility of operation allow us to employ aerial mapping in domains where manned aircraft cannot be used profitably. One of the possible applications consists of snow-cover mapping, which is beneficial within, for example, avalanche and flood forecasting, local and regional climate research, and hydropower energy situation analysis. Dependable information about snow conditions is especially important for northern countries, where the snow cover is present for a significant part of the year.

There are various methods to estimate certain snow-cover parameters, and each of these techniques is beneficial at a different scale and in diverse applications. The basic parameter rests in determining the presence of snow, namely the snow-covered area (SCA). Such information is of interest mainly for the investigation of trends in large areas (or, more concretely, at the level of regions and countries) and finds use in climate and hydrological research. In general, terms, two determining approaches are typically employed: optical and microwave-based survey. Snow coverage is recognizable from aerial imagery (collected from manned aircraft or satellites) only with a clear, cloudless view [1,2]. Conversely, radar-based observation is independent of the weather conditions, although the resolution and accuracy are typically lower [2–4]. In any case, the information value of the SCA is limited because it does not describe the amount of snow or the relevant content of water.

The snow depth (SD) indicator provides us with a better insight into the amount of the accumulated snow. The SD in a certain area can be estimated via either interpolating data from a network of observation stations [5] or as a result of a fusion of point measurements and satellite-based observations [6]. In certain conditions, SD is also estimable from radar-based measurements or by using the light detection and ranging (LiDAR) technology; terrestrial and airborne laser scanning (TLS, ALS) have been recently used for this purpose [7–9]. The most meaningful indicator in this context is the snow water equivalent (SWE), describing the amount of water contained in the snow cover. This type of information is crucial for hydrologists, especially as regards flood prediction during snow melting periods. As with SD estimation, the SWE is frequently obtained from in situ measurements. The spatial resolution is increased either via fusing space-born radiometric measurements [10] and ground observations [11] or by using snow models [4].

Manned aircraft and satellite-based estimation of the snow-cover parameters are not suitable for small areas and applications that require high accuracy, the reasons being the low resolution, low estimation accuracy, and substantial cost. Considering other relevant approaches, the resolution of the point measurements is not sufficient due to the sparse network of observation stations. In such cases, unmanned aircraft can be employed effectively. Micro and light UASs (below 5 kg and 50 kg [12], respectively), the categories addressed within this study, are profitably employed in many fields of aerial mapping, such as agriculture [13], forestry [14], geodesy [15], archaeology [16], or environmental mapping [17]. However, due to their limited endurance, sensitivity to weather conditions, and relevant legal restrictions, the vehicles are not suitable for global area mapping, including flights at the regional level. Conversely, the concept ensures fast, safe, low-cost, and flexible operation, and it enables us to reach superior accuracy and resolution.

Most of the recent UAS-based snow mapping projects use aerial photogrammetry for SD estimation. Studies [7,8,18], for example, compare photogrammetry and indirect georeferencing-based SD estimation with terrestrial laser scanning in mountainous regions. The results indicate that both methods are suitable for the given purpose and achieve comparable accuracies. Indirect georeferencing (IG), a technique relying on ground targets, can be successfully replaced with visual landmark-based co-registration; however, as pointed out within sources [19–21], common visible points must exist in both snow-covered and snow-free scenes. According to [7,18], photogrammetric accuracy in snow mapping can be enhanced using near-infrared (NIR) cameras instead of visible-spectrum ones since the former perform well even in weak lighting conditions. In terms of the UAS type, all the referenced projects (except for [7]) use multi-copters to carry out the discussed photogrammetric task, mainly because of their greater wind resistance and higher payload capacity. The articles further indicate that UAS-based snow mapping is appropriate for areas up to tens of thousands of square meters. The idea of the photogrammetry-based depth estimation lies in that the actual surface model of the snow cover is compared with a reference snow-free model. Relevant reference data can be obtained in multiple ways. The common approach is to create the snow-free surface model via the same technique, namely UAS photogrammetry, in a period when there is no snow cover [7,8,18–21]. Another option rests in using national terrain models (typically produced through LiDAR measurements), which are generally available in many countries.

The SD map is obtained from the height difference between the snow-covered and the snow-free surface models. To acquire such a difference, both models must be accurately georeferenced. All the above-mentioned research projects, except for [21], use IG or visual landmark-based co-registration. IG typically achieves an object accuracy slightly better than that of direct georeferencing (DG) (centimeter to decimeter level); however, a serious health risk may arise during the ground survey due to avalanches, especially in mountainous regions. Despite benefits of the DG, the approach is not widely employed presently. Difficulties addressed in the study [21] and other articles using DG in UAS photogrammetry [22–24] relate to the complexity of the hardware equipment onboard UAS. This fact often causes reliability issues. Furthermore, such systems require calibration and precise time synchronization, higher payload capacity, and finally, the overall cost is higher.

As in the case of manned aircraft and satellite-based snow-cover mapping, passive microwave (radar) devices can be carried by UASs as well. Such an approach, discussed within [25], is nevertheless rather uncommon.

This article embodies a case study on UAS-based SD mapping performed with a multi-sensor system specially designed for DG in aerial photogrammetry. The study area, namely the Mapa Republiky snow field, has attracted the interest of scientists because of its particular characteristics; thus, various techniques for estimating SD and SCA have been tested there to this day. Our approach has the potential to reduce health risks and man-performed field work without significant accuracy limitations. The benefits and drawbacks of the presented method are discussed and compared with the previously employed techniques.

2. Materials and Methods

2.1. Study Area and the Monitoring History

Our method for estimating snow-field parameters was examined at the location called Mapa Republiky (Map of the Republic), a snow field with a shape similar to the outlines of former Czechoslovakia; the area is situated in the Krkonose (Giant Mountains, Figures 1 and 2a). This snow field, at an altitude of approximately 1430 m above mean sea level (AMSL), is rather specific as the snow often persists until the summer season, several months longer compared to other locations in the Krkonose or the Czech Republic in general. This is somewhat unusual, considering the fact that the area is located on a south-facing slope exposed to sunlight during most of the day; the reason rests in the amount of snow contained within the field, where the SD significantly exceeds the average value in the given location (the maximum SD of 15.7 m was established in 2000, Figure 3). The accumulation is caused by northern and northwestern winds depositing snow in the lee. The phenomenon is further intensified by the terrain depression formed by the snow itself [26,27].

Depending on the weather conditions during the melting season, together with the precipitation as well as the wind speed and direction in the winter, the snow persists until June to August or, occasionally, does not melt at all (see Figure 3 to read the annual statistics). For this reason, the place is of substantial interest to scientists; the examined problems include, for example, the impact of the SD on the vegetation pattern [26], the relationship between the geo- and biodiversity in the given area [28], and the water balance of the drainage basin. Such analysis, however, requires relevant input information about the amount of snow.

The first attempts to estimate the SD were conducted more than 100 years ago; throughout the 20th century, the research was nevertheless hampered by a lack of appropriate technical equipment [29,30]. Wire probes (collapsible avalanche probes) are not applicable if the snowpack is deeper than 3–4 m, and fixed steel poles may fail due to the shear stress of the snow mass [26]. Systematic investigation was, in fact, made possible only after the introduction of the accurate global positioning system (GPS) technology. Since 1999, the Krkonose Mountains National Park Administration have been monitoring the location via kinematic carrier phase-enabled GPS receivers. The surface of the snow cover is reconstructed by interpolation of GPS position data collected during walking or slow horizontal skiing. These data collection techniques were further supported by the "stop and go" method (walking with stops to facilitate static collection of point data, Figure 2b), which ensured accuracy verification; during periods characterized by a serious avalanche risk, a viable alternative consisted of using a rope-driven sledge with a GPS receiver. Employing the same equipment, data for the construction of a snow-free surface model were collected after the melting season. A SD map was then computed as the difference between these two models. The described approach enabled us to estimate the depth of thick snow layers for the first time [26,31], although the spatial resolution was rather low with respect to the amount of time needed for the data collection.

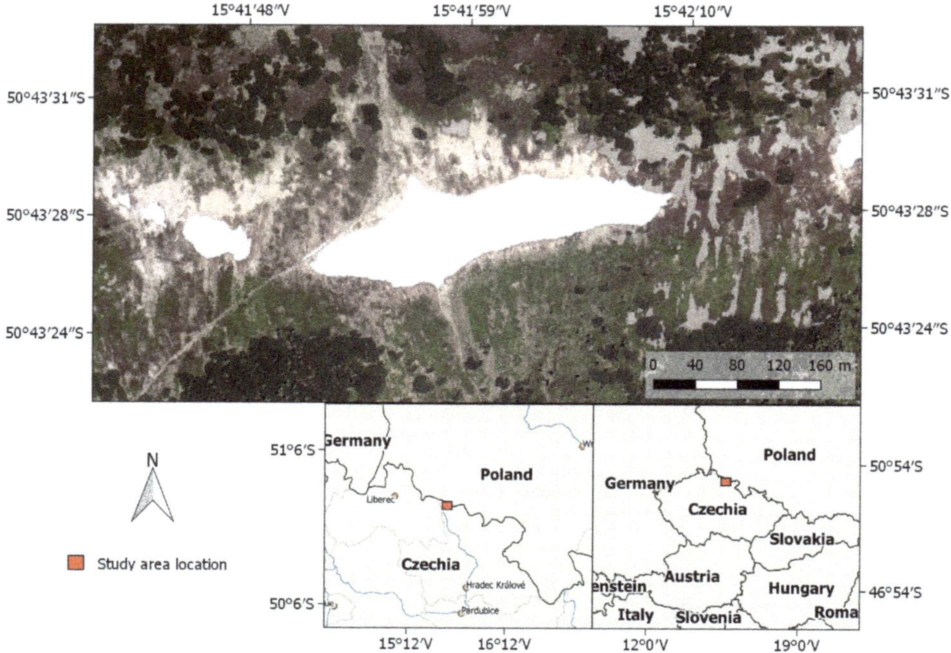

Figure 1. The study site location (the orthophoto obtained during the melting season 2012).

(a) (b)

Figure 2. The isolated Mapa Republiky snow field during the spring season (**a**), and ground survey-based point determination of the snow field shape (**b**).

Another turning point came with the use of UASs. Since 2016, a camera-equipped UAS has been used to acquire relevant aerial image data to be further employed for the photogrammetry-based surface reconstruction. This method significantly reduced the data collection time and, moreover, increased the spatial resolution by two orders of magnitude. Except for the SD, computed in the same manner as within the aforementioned method, this approach enabled us to estimate the SCA employing the actual orthophoto. A major drawback of this method is the necessity of ground targets used for the georeferencing. This task still comprises certain risk due to the avalanche hazard.

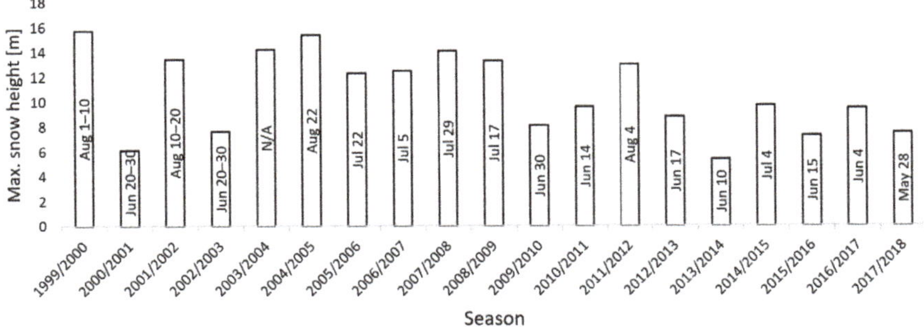

Figure 3. The data on the maximum snow height and melting days since 1999 (the year when the global positioning system (GPS)-based monitoring was performed for the first time).

2.2. Overall Concept

This paper discusses a vision-based snow-field monitoring method that uses ground control point (GCP)-free UAS photogrammetry (Figure 4). The approach eliminates a substantial portion of the common field work and is thus usable primarily in inaccessible or dangerous areas. Thanks to its capabilities, the technique offers major potential for automated data processing and, importantly, reduces the data acquisition and processing time. Compared to the other relevant methods, however, complex hardware equipment and more specialized operator skills are needed.

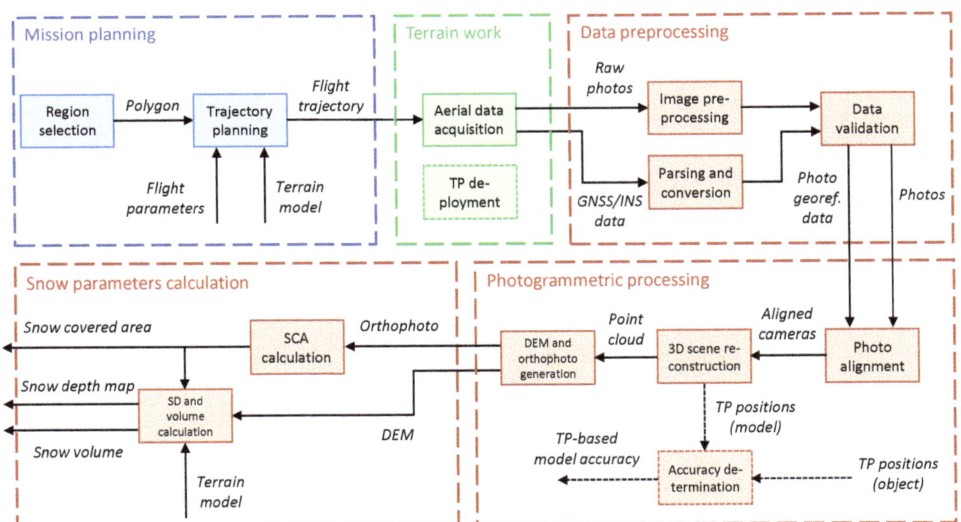

Figure 4. The workflow illustrating the entire snow-field monitoring procedure. TP—test point; GNSS—global navigation satellite system; INS—inertial navigation system; DEM—digital elevation model; SCA—snow-covered area; SD—snow depth.

Our technique employs aerial photography-based products for SCA calculations; SD and volume calculations nevertheless require another component, namely a snow-free terrain model. Since all aerial data are acquired using a UAS operated in an automatic mode, special attention must be paid to mission planning. This is essential especially in regions with rugged terrain. The field work comprises the aerial data acquisition, or the UAS flight itself, and—if required—test point (TP) deployment. Our UAS is equipped with sensors for DG of the imagery, and thus no ground targets (GCPs) are necessary for the processing. In the proposed experiment, however, we employ several ground targets to enable the previously used IG technique, which will be later compared with our method. By extension, the targets can be used for the DG quality assessment.

The processing part of the workflow is further divisible into three portions. First, the aerial photographs and the data from the global navigation satellite and inertial navigation systems (GNSS and INS, respectively) must be preprocessed and converted into an appropriate format. An important procedure rests in validation, which determines whether the acquired data are consistent and suitable for further processing. The second portion subsumes the photogrammetric processing, comprising tasks well known from UAS photogrammetry, such as the structure-from-motion (SfM)-based 3D scene reconstruction. In our case study, the quality of this process is assessed thanks to the ground targets, or test points (TPs). The orthophoto and digital elevation model (DEM) as photogrammetric products are then used in the final part of the processing to estimate the snow-field parameters. The SCA is computed from the orthophoto, and the SD and volume are estimated primarily from the height differences between the snow-covered and the snow-free terrain models.

Important aspects of the workflow in Figure 4 are addressed in more detail within the following sections. First, the employed UAS and sensors are described in Section 2.3. The ground measurement, a process involving test point deployment (green segment), is then outlined in Section 2.4. Section 2.5 characterizes the mission planning and aerial data acquisition, strongly related topics highlighted in blue and green, respectively. The remaining (red) portions of the Figure display the actual data processing. The first of these segments, data preprocessing, comprises operations necessary for subsequent processing; such steps are not considered in depth, because their relevance to the presented application is limited. The workflow portions comprising photogrammetric processing and snow parameter calculation are examined within Sections 2.6 and 2.7, respectively.

2.3. UAS and Onboard Sensors

To obtain the aerial data, we used a Mikrokopter Ookto XL UAS, a 95 cm span commercial octocopter capable of flying for approximately 10 min with the payload of 2–3 kg. The UAS supports automatic flight based on selected waypoints; the device therefore carries a low-accuracy GPS receiver. The hoverability and high payload capacity are the central reasons why multi-rotor UASs are often employed in photogrammetry missions similar to that outlined herein [24,32,33].

All necessary equipment used for the remote sensing is comprised in the custom-built multi-sensor system mounted on the UAS (Figure 5b). The device was developed previously at CEITEC laboratories and embodies a part of the ATEROS (Autonomous Telepresence Robotic System) robotic system [34,35]. The multi-sensor system consists of a Sony Alpha A7 digital camera, a Trimble BD982 GNSS receiver, an SBG Ellipse-E INS, and a single board computer Banana Pi R1. The GNSS receiver measures the position with centimeter-level accuracy when real-time kinematic (RTK) correction data are transmitted, and as it is equipped with two antennas for vector measurement, the device also measures the orientation around two axes. The position and orientation data are used as an auxiliary input for the INS, which provides data output at a frequency of up to 200 Hz. All the sensors are precisely synchronized; thus, once an image has been captured, the position and orientation data are saved into the onboard solid-state drive (SSD) data storage (more parameters are contained in Table 1).

The aforementioned system is relatively uncommon since it is completely independent of the UAS while integrating all equipment necessary for GCP-free aerial photogrammetry. The relevant testing cycles were performed on various unmanned platforms (such as that visible in Figure 5b) during several missions, with the resulting accuracy assessed in our recent study [36]. Similar setups designed for the DG of aerial imagery but not allowing portability had been published previously [24,32].

Table 1. The parameters of the custom-built multi-sensor system for UASs. The position and attitude accuracy according to the INS manufacturer's specifications (the RTK mode, airborne applications).

Parameter	Value
Position accuracy	hor.: 20 mm, ver.: 40 mm
Attitude accuracy	roll/pitch: 0.1°, heading: 0.4°
Camera sensor resolution	6000 × 4000 px
Camera sensor size	36 × 24 mm
Camera principal distance	21 mm
Camera aperture	f/4.5
Operational time	120 mins
Max. distance from base	1000 m
Dimensions	1.5 × 0.2 × 0.2 m
Weight	2.6 kg

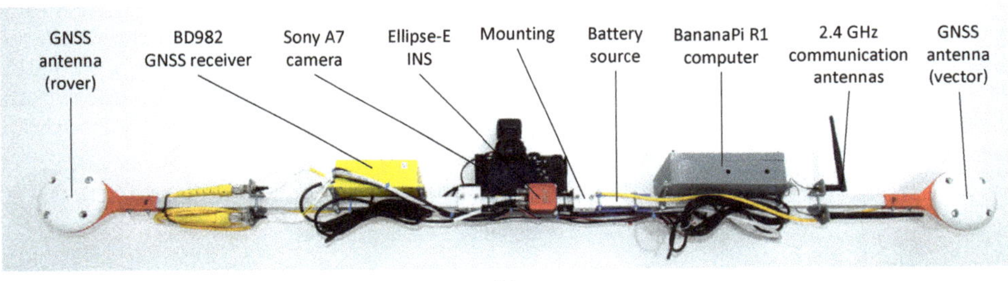

(**a**)

(**b**)

Figure 5. The custom-built multi-sensor system for direct georeferencing (**a**), and a UAS fitted with the system during a previous mapping mission (**b**).

The performance of the applied system may satisfy the requirements of the discussed application. With respect to the used SD determination method (Section 2.7), the height accuracy of the photogrammetry-based DEM should be comparable to or better than the snow-free model height accuracy (0.15 m RMSE in our case). Similarly, the horizontal accuracy should reach this level too to prevent height errors caused by inaccurate alignment of the elevation models, an effect apparent especially in steep slopes. The applied multi-sensor system meets such requirements: the object error of the model is typically below a decimeter root mean square (RMS) for both the horizontal and the vertical coordinates when flying at 50 m above ground level (AGL) [36]. In addition to the elevation model, our mapping method uses the actual orthophoto for the SCA calculation (Section 2.7). With respect to the parameters of the used camera and lens (Table 1), the ground sample distances are approximately 14 mm and 29 mm for the AGL altitudes of 50 m and 100 m, respectively. These parameters are far beyond the value necessary for an accurate area estimation in a snow field with a size in the order of hundreds of meters.

2.4. Ground Measurements

As already mentioned in the previous section, the proposed concept does not require ground targets for the georeferencing and data processing stages; in fact, the UAS flight is the only field work activity in our concept. Despite this, several ground targets were employed to ensure backward compatibility with the previously used IG technique. Moreover, the targets enable us to determine the accuracy of the directly georeferenced photogrammetric products.

During late April 2018, the time originally selected for our investigation, the Mapa Republiky snow field became completely isolated from the remaining snow cover in the vicinity. This fact allowed the positioning of six ground targets just around the snow field on the peripheral part of the area to be mapped (the distribution is visible in Figure 6). Such a solution was commonly used in the past, enabling us to compare both georeferencing techniques. The horizontal separation of the peripheral GCPs is 140 m on average, or approximately one ground base. The spatial distribution and number of GCPs play a substantial role in the quality of IG in photogrammetry. This aspect has already been addressed within numerous publications [37,38]. Our peripheral location of the targets meets the basic GCP distribution requirement; however, a higher density and extra GCPs located in the central part would help us to increase the georeferencing accuracy.

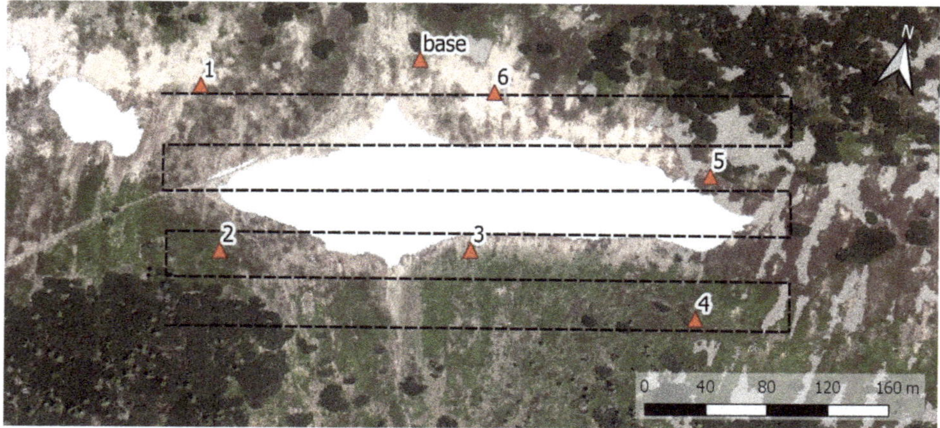

Figure 6. The locations of the ground targets and the base station, together with the flight trajectory for the UAS (the orthophoto obtained during the melting season 2012).

We used 20 cm squared, black-and-white-patterned paper targets with clearly defined centers. All these targets were glued onto a solid support and fixed to the ground by using iron nails. The position of every single target was acquired using a survey-grade Topcon HiPer HR RTK GNSS receiver obtaining correction data from the TopNet (Czech provider of correction data). With the same equipment, the position of the GNSS base station (Figure 7) was determined. The base station, also indicated within Figure 6, provides the correction data for the GNSS receiver aboard the UAS, and its accurate position is crucial as regards the quality of the snow-field parameters estimation.

Figure 7. The base station (located close to the snow field) performing real-time GNSS corrections for the sensors aboard the UAS.

2.5. Mission Planning and Aerial Data Acquisition

Mission planning comprises standard tasks known from UAS-based photogrammetry. First, a mapping region must be selected; in our experiment, this was a rectangular area around the isolated snow field. Both horizontal coordinates of the region are enlarged by approximately 50% because the exact size and location of the snow field during the melting period is difficult to estimate. Subsequently, we design the trajectory (the waypoints for the automatic flight) within the region, satisfying the following photogrammetric requirements: the desired forward and side image overlaps; ground resolution; and altitude restrictions. The trajectory planning thus also depends on the applied photographic equipment, its intrinsic parameters in particular; these include, for example, the principal distance and sensor size. We use the common parallel strips flight pattern known from both manned and unmanned aerial photogrammetry [39,40], as shown in Figure 6. The parameters of the flight trajectory and image data acquisition for the presented case study are summarized in Table 2.

The mission planning process is, in our case, performed in the field, immediately after assessing the actual situation. For this purpose, a laptop with the MikroKopter-Tool software for automatic waypoint-based mission planning is used. Since the location is a steep slope (the height difference between the highest and lowest spots is approximately 100 m), a constant flight altitude AMSL would cause high variation in altitude AGL and thus also ground resolution. For this reason, we adjusted the AMSL altitude of the individual survey lines to minimize variation in ground resolution.

Except for the take-off and landing, the UAS operates in the automatic mode based on the aforementioned settings, uploaded to the device's memory just before the flight. With the employed photogrammetric parameters, the mapping of the area takes approximately 10 min. The image and GNSS/INS data, which are hardware-synchronized, are recorded throughout the flight on the SD card and SSD storage, respectively.

Table 2. The parameters of the flight trajectory and image acquisition; the values that depend on the flight altitude (AGL) are stated for the average altitude.

Parameter	Value
Distance between strips	30 m
Strip length	400 m
Number of strips	6
Base (distance between consecutive images)	10 m
Flying altitude AGL	90–130 m
Flying speed	5 m s^{-1}
Flying time	10 min.
Time between images	2 s
Photo scale	1:5200
Forward overlap	92%
Side overlap	84%
Image footprint	190 × 125 m
Ground resolution	3.1 cm px^{-1}
Shutter speed	1000^{-1} s
Aperture	5.6
ISO	Auto (100–400)

2.6. Photogrammetric Processing

The method proposed in this paper uses photogrammetry to reconstruct the given 3D scene, namely snow-covered terrain. For this purpose, we used Agisoft Photoscan Professional (version 1.4.2), a complex software package to execute all photogrammetric processing stages, from the image and georeferencing-related data to the orthophoto, DEM, and other products. The workflow starts with the align phase, where the exterior and interior camera orientation [41] is estimated based on the feature points detected in the overlapping images. Furthermore, the locations of the feature points are determined via the structure-from-motion procedure, resulting in a sparse point cloud [42]. Within the following step, a dense point cloud can be generated, using multi-view stereo (MSV) reconstruction. As the operation is performed at the level of pixels, even small details are reconstructed to yield a point cloud containing millions of points.

The camera poses determined in the previous step being relative, Photoscan offers two methods of georeferencing, namely transformation into geographic coordinates (WGS-84 in our case): one employing the image position data measured by the onboard sensors (Section 2.3), and the other performed via the GCPs positions obtained during a ground measurement (Section 2.4). Both procedures rely on similarity transformation comprising translation, rotation, and scaling [39,43]. The techniques are also known as direct (DG) and indirect georeferencing (IG), respectively. Although we used a multi-sensor system to carry out the GCP-free photogrammetry, both georeferencing options were tested. For this reason, the georeferencing phase of the workflow was executed twice to evaluate the two methods separately (employing the same image dataset and identical photogrammetric processing settings). In practice, most of today's photogrammetry missions use GCPs to reach a centimeter-level spatial accuracy even with consumer-grade equipment onboard a UAS; this scenario is presented in several sources, such as [22,44,45]. DG, commonly supported by computer vision (CV) algorithms, typically leads to a slightly lower accuracy; this approach, however, brings certain benefits, as outlined within the above chapters and elsewhere [22–24].

Based on the georeferenced dense point cloud, a DEM was generated; the model can be considered a digital surface model (DSM). This product is essential for the formation of the orthophoto. The quality of these outputs was determined by using test points, or ground targets with accurately known spatial positions (Section 2.4). We used the targets in two manners, namely as the TPs in DG and the GCPs in IG. The majority of the referenced papers employed Photoscan, whose workflow and algorithms are described in more detail within [46].

Photogrammetry-based 3D scene reconstruction fully depends on the features visible in the images, resulting in certain restrictions as regards the presented study. Snow-covered terrain typically lacks a texture because a snow surface normally contains white color that does not vary substantially. Such a property then causes the reconstruction process to fail; however, relevant studies show that snow often contains a texture sufficient for photogrammetric processing [18,20,21]. Finally, low visibility caused by weather effects, including fog or clouds, will also produce a negative impact on the quality of the results.

2.7. Snow-Field Parameter Estimation

For many UAS-based mapping applications, the final output consists of photogrammetry products, such as a georeferenced orthophoto and DEM. However, in our case, these products form the input data for the final processing phase, namely the calculation of the snow-field parameters (see the process flow diagram in Figure 4). As described in the Introduction, the three main parameters related to snow mapping are SCA; SD, or snow volume; and SWE. Our method considers merely the first two of these as the last one typically requires in situ measurements.

The method uses optical-based SCA estimation employing the orthophoto obtained within the previous step. This approach relies on visual separability of the snow cover from the rest of the snow-free terrain, typically covered with vegetation, rocks, water, or artificial objects. The desired effect is achieved through segmentation, an elementary CV task that consists of partitioning an image into multiple homogeneous and meaningful regions [47]. A common segmentation technique is thresholding [48], which in its basic form, separates an object and the background in a grayscale image. The key assumption is that the object class includes levels of gray different from those exhibited by the background class, enabling the user to find the thresholding level (Figure 8a).

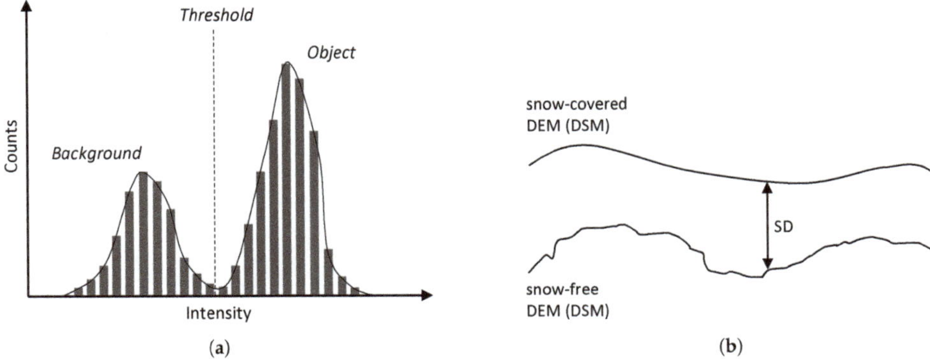

Figure 8. Segmentation by thresholding: a grayscale image histogram and the threshold separating an object and the background (**a**); the snow depth determination principle (**b**).

The snow field (representing the object, with a color approaching white) in our study area is clearly separable from the snow-free terrain (or the background, covered with vegetation) via thresholding since the two classes are characterized by different levels in both red-green-blue (RGB) channels and the grayscale interpretation. Within our research, the threshold level was established primarily manually to facilitate precise SCA determination; we nevertheless also employed the well-known

Otsu method [49] (implemented within the `graythresh` function in MATLAB) to test the automatic threshold level determination. Once the snow has been separated from the background, the snow field, Mapa Republiky, must be separated from the snow cover in its vicinity. This step was performed using the `bwlabel` function, which labels connected components in the binary image. The largest component then corresponds to the snow field, and its area is determinable thanks to the known pixel size of the orthophoto.

The aforementioned semi-automatic and automatic SCA estimation methods were also compared with the manual technique exploiting an area measurement tool contained in geographic information systems (GISs); this solution allows the operator to measure any area manually. All the approaches are discussed in the final sections of this article.

Snow depth, expressible with a positive real number, describes the height of the snow cover above the terrain at a certain location; a snow depth map is then the 2D SD representation of the given area georeferenced into geographic coordinates. SD, unlike the above-described SCA, cannot be estimated using an orthophoto only. Our method interprets SD as the height difference between a snow-covered and a snow-free DEM, as illustrated in Figure 8b. The former model was obtained during a melting season via photogrammetry (Section 2.6); the latter, snowless elevation model was acquired from CUZK (the Administration of Land Surveying and Cadastre of the Czech Republic [50]). The Administration provides an airborne laser scanning (ALS)-based digital terrain model (DTM) covering the entire Czech Republic; its height accuracies are 0.15 m and 0.25 m for uncovered and wooded terrain, respectively, and the average density corresponds to 5 points m^{-2} in the very least. This model was resampled to obtain a DTM with the resolution of 0.05 m px^{-1} (the shaded DTM of the study area is illustrated in Figure 9). Unlike the DSM produced by photogrammetry, the DTM does not involve artificial objects and vegetation. However, this property should not lead to major difficulties, because our study area is mostly covered with grass, and no objects are present. The calculation was performed in QGIS (version 3.0.2) GIS to allow calculations with raster layers and to support various coordinate reference systems. The input raster layers, or DEMs, were obtained using different methods; thus, their pixel resolutions do not coincide. To deal with this issue, the nearest neighbor resampling method was applied.

Figure 9. The fifth generation, shaded DTM of the Krkonose (obtained from CUZK). The study area is highlighted by the dashed line.

The drawback of the SD calculation technique is that the height difference between the DEMs does not have to be caused by the snow only. In mountainous regions, a rapid surface change can occur due to, for example, vegetation growth or a rock slide. To minimize the impact of such effects, we consider only areas where the presence of the snow was detected using an orthophoto, as described within the SCA determination earlier in this section. The masked (snow-free) area is thus excluded from the SD calculation.

The snow-cover volume is another meaningful indicator. As with the SCA, it informs us about the quantity of snow via a single number. Once the SD map has been obtained, the volume can be computed as the sum of the individual heights multiplied by the map resolution (pixel size). From this point, there remains only a small step to determining the SWE, namely the amount of water contained in the snow field. Since the snow density oscillates between 50 kg m^{-3} (new snow) and 900 kg m^{-3} (ice, [51]), the parameter must necessarily be obtained. Such a task, however, is not feasible using the sensors onboard our UAS; to obtain relevant data, we would therefore have to rely on in situ observation.

3. Results

3.1. Photogrammetry

The aerial data acquisition lasted 14 min; during this time, 403 images were collected. In addition to the automatic waypoint-based flight, the operational period included the take-off and landing, namely manually controlled phases. The data acquired within these latter steps are not relevant; thus, 222 images only were used for the processing. This image set was preprocessed using Adobe Photoshop Lightroom (version 6.0) to compensate lens vignetting and to convert the RAW image files into JPG at minimum compression. One of the images is displayed within Figure 10a.

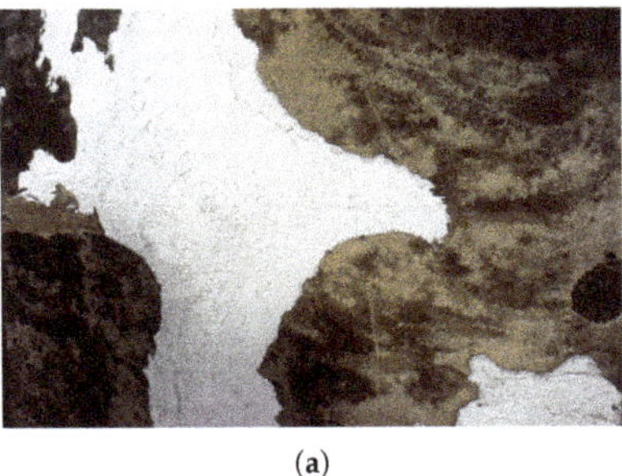

(a)

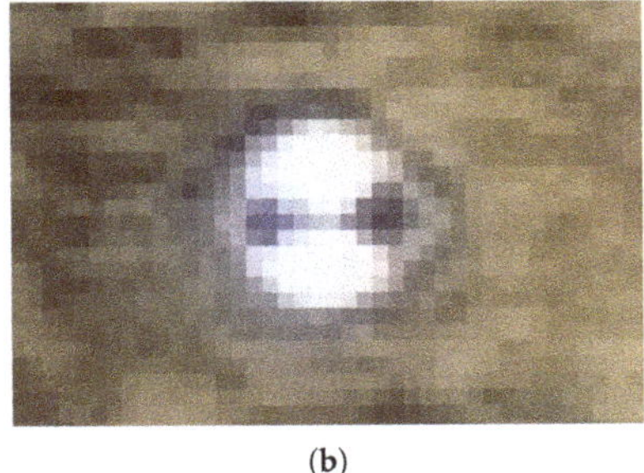

(b)

Figure 10. A photo captured from the height of 100 m AGL (**a**); a zoomed ground target (**b**).

The GNSS/INS logging was triggered with the camera shutter, and the number of records contained in the log file equaled that of images captured. The data were converted into an appropriate format and subsequently analyzed. Figure 11 shows the measured positions of the cameras in the local system as well as the solution quality indicator estimated by the onboard GNSS. The RTK fixed solution was available for 73% of the time; the RTK float for 5%; the differential GNSS (DGNSS) for 18%; and the autonomous fix for 4%. The RTK fixed solution outages were caused by interruption of the data link that facilitates the transmission corrections. The age of the corrections has a direct impact on the quality of the GNSS solution; the fixed solution was thus lost for several moments, especially in the southeastern part of the flight trajectory, as is evident from Figure 11. This fact will likely negatively affect the accuracy of the DG and, therefore, also that of the photogrammetric products. The aforementioned figure further comprises the actual locations of the GCPs employed for the IG; one of the targets (as captured in an aerial image) is displayed within Figure 10b.

The datasets from the onboard sensors, namely the images and the GNSS/INS derived data were imported into the Photoscan, together with the locations of the ground targets. The photogrammetric processing was performed once with DG and once with IG, using the same settings. These techniques, except for georeferencing itself, require identical processing phases. In the case of DG, the camera locations (and estimated errors) were simply imported from a text file; IG, conversely, involves manual placement of markers on the targets visible in the images. Despite this, the processing times do not differ much, as indicated in Table 3. However, if we also consider the data collection phase, IG is more time-intensive due to the deployment of ground targets (GCPs).

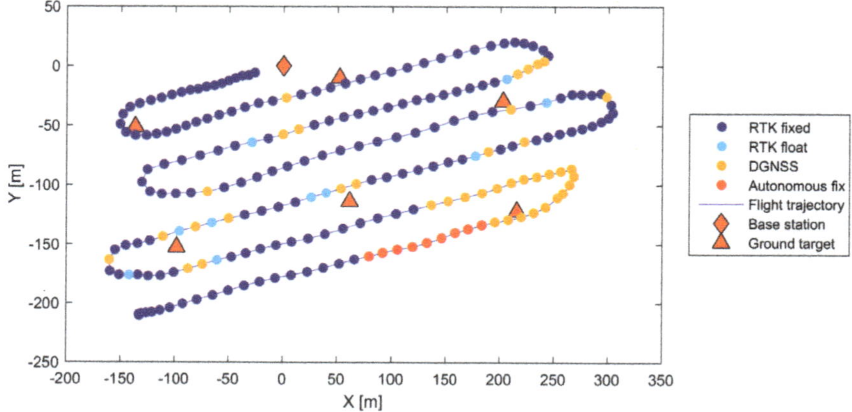

Figure 11. The camera locations and the GNSS quality indicator estimated by the onboard GNSS/INS.

The parameters of the photogrammetric products are also comparable to a certain extent: Both dense point clouds contain approximately 190 points m^{-2}, the resolutions of the DEMs and the orthophotos are 10.9 cm px^{-1} and 2.7 cm px^{-1}, respectively (Figure 12). As expected, the difference shows itself in the georeferencing accuracy. According to the 6 TPs, DG reached the spatial RMS error of 11.4 cm; the RMSE of IG is slightly better, equaling 7.6 cm (Table 4). It should be noted that although using such a low number of TPs to express accuracy is not sufficiently credible in terms of statistics, we can still benefit from the actual information of whether a problem occurred during the data collection and processing. A detailed analysis of DG accuracy was proposed within our previous study [36]. Furthermore, IG accuracy stated using GCPs (with the targets employed for georeferencing) is also of mainly informative character.

Table 3. The data collection and processing times of the applied techniques; the photogrammetric processing was executed on a personal computer with an Intel Core i7-6700 CPU, 32 GB RAM, and an NVIDIA GeForce GTX 1051 Ti GPU.

Processing Phase	Automatic/Manual	DG [h:mm:ss]	IG [h:mm:ss]
Aerial data acquisition	A	0:13:36	0:13:36
Target deployment	M	—	~0:30:00
Data collection in total	—	0:13:36	~0:44:00
Photo alignment	A	0:18:33	0:21:00
Marker placement	M	—	0:09:55
Camera calibration	A	0:00:06	0:00:06
Dense point cloud generation	A	1:23:59	1:25:33
DEM generation	A	0:00:36	0:00:26
Orthophoto generation	A	0:09:36	0:08:04
Processing in total	—	1:52:50	2:05:04
Entire process	—	2:06:26	2:49:04

Table 4. The accuracy of direct and indirect georeferencing, determined by using 6 ground targets. In the IG procedure, the accuracy was assessed via the GCPs.

	DG [cm]			IG [cm]		
Target	*XY*	*Z*	*XYZ*	*XY*	*Z*	*XYZ*
1	1.3	−1.3	1.8	4.1	−7.7	8.8
2	2.7	2.8	3.9	1.8	6.6	6.9
3	2.2	−0.3	2.2	3.2	3.3	4.6
4	7.5	−6.4	9.9	5.2	−6.6	8.4
5	21.8	10.7	24.3	10.0	1.5	10.1
6	3.5	7.7	8.5	5.6	2.2	5.6
Mean	6.5	2.2	8.4	4.9	−0.1	7.4
RMSE	9.7	6.1	11.4	5.5	5.3	7.6

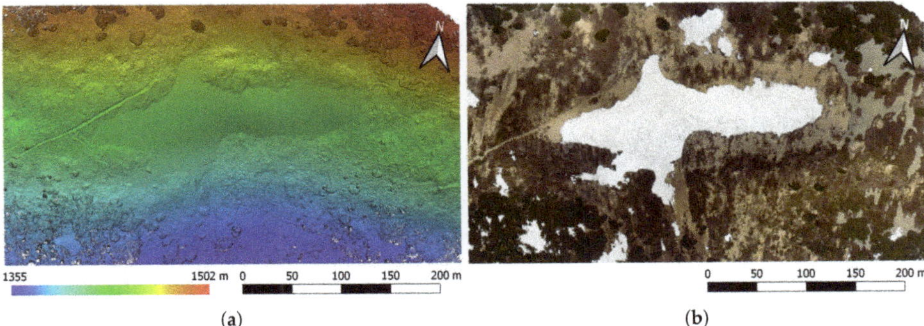

(**a**) (**b**)

Figure 12. The directly georeferenced, shaded DEM (**a**) and the relevant orthophoto (**b**), both representing the output of the photogrammetric processing phase.

The analysis indicates that in our study, the georeferencing techniques do not differ significantly as regards the time intensity and the final accuracy (IG needed 34% more time, Table 3; DG was slightly less accurate, Table 4). The main benefit of DG rests in the elimination of human involvement, especially during the data collection procedure, which is potentially hazardous to human health. Moreover, such a scenario enables automation of all major data collection and processing phases.

3.2. Snow-Field Parameters

The calculation of the snow-field parameters started with the orthophoto-based SCA estimation, as discussed in Section 2.7. First, the orthophoto was converted into an 8-bit grayscale image (Figure 13b) for segmentation purposes. We determined two threshold levels in the respective histogram: one level, 206, was established manually, while the other, 144, was computed using the Otsu method (Figure 13a). The Otsu technique was found to produce unsatisfactory results in this application as it does not separate the object from the background precisely; the threshold should lie in the local minimum between the high valued counts (whites) and the darker background. To automate this procedure, a reliable segmentation algorithm would have to be designed; such a step, however, would require an adequate test dataset. For this reason, we applied the manual threshold level. In the segmented image, connected components were found; the component comprising the largest number of pixels corresponded to the snow field (Figure 13c). Based on the known pixel size, the SCA of 15,802 m^2 was estimated. This result was then manually verified using the QGIS area measurement tool, yielding 15,493 m^2, a value 2% lower than the automatically delivered one.

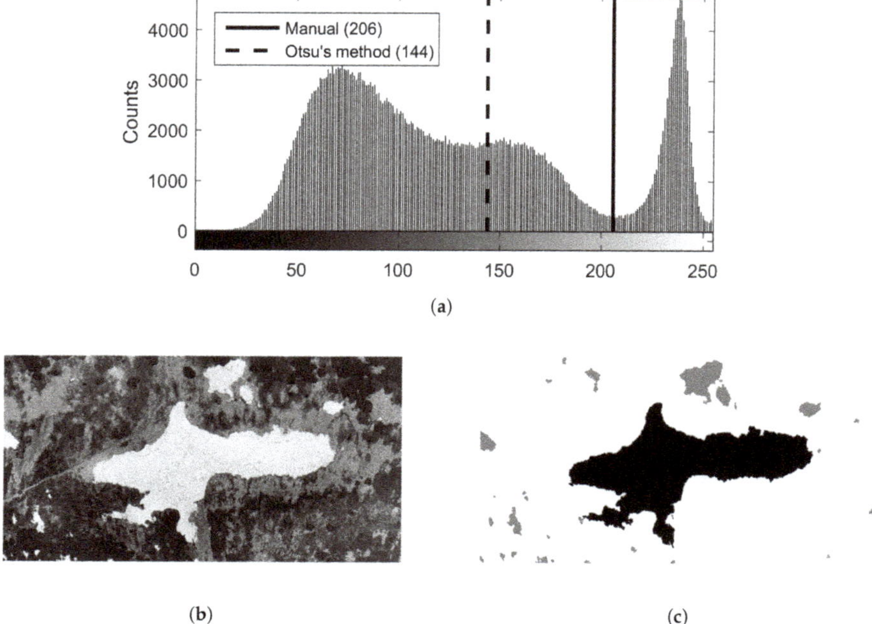

Figure 13. The histogram (**a**) of the gray-scaled orthophoto (**b**), with the snow-covered area extracted (the snow field is highlighted in black) (**c**).

The SD map was derived from the elevation models. Since both the snow-free ALS-based DTM and the snow-covered photogrammetry-based DSM are georeferenced in the same coordinate system, they were used for direct calculation. At the initial stage, we employed the QGIS raster calculator to compute the height difference between the DSM and the DTM. Since the former, unlike the latter, generally also includes vegetation and artificial objects, the difference map does not contain the snow only. This effect is represented in Figure 14a, where, especially in the upper right corner, the vegetation (scrub mountain pine) causes considerable height differences. To exclude such objects from the SD calculation, we applied the SCA mask computed within the previous step. After that, the difference map contained the snow cover only, as displayed in Figure 14b. The maximum height of 5.45 m corresponds to the maximum depth of the Mapa Republiky snow field, whose volume, computed as the sum of the heights in the individual pixels, equals 26,367 m^3.

As is evident, the study area contains various types of vegetation. Since we used the DTM as the snow-free reference, it cannot be excluded that some of the vegetation is under the snow cover. Such a scenario would negatively affect the accuracy of the SD determination, meaning that the estimated SD would be higher than it really is. To deal with this issue, a DSM created via, for example, UAS-based photogrammetry, should be employed as the snow-free reference model.

The high-resolution models introduced within our research allow us to analyze cross-sections in any direction, leading to better understanding of how the snow is accumulated in relation to the terrain shape. An example of two cross-sections through the SD maximum in the east-west and north-south directions is provided in Figure 15. The east-west direction, for instance, illustrates that the maximum is located exactly in the terrain depression present within the area of interest.

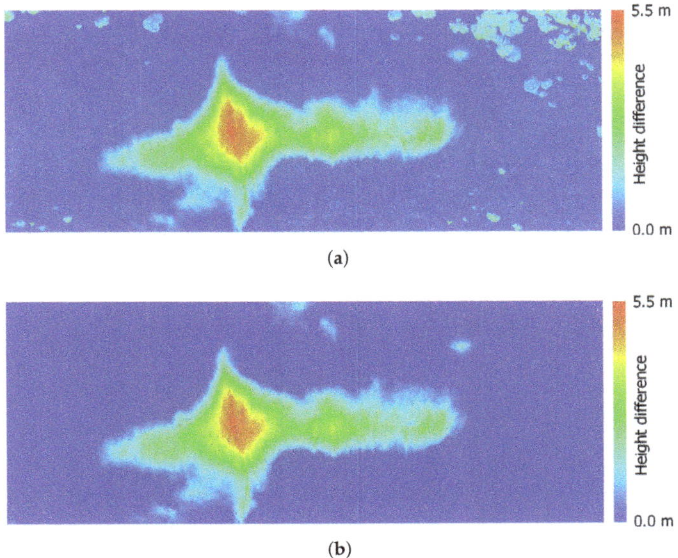

Figure 14. The height difference between the snow-covered, photogrammetry-based DSM and the snow-free, ALS-based DTM (**a**). This result involves differences caused by not only the snow itself but also other aspects, including, for example, vegetation (upper right corner). After the SCA mask has been applied, the difference map contains the SD only (**b**).

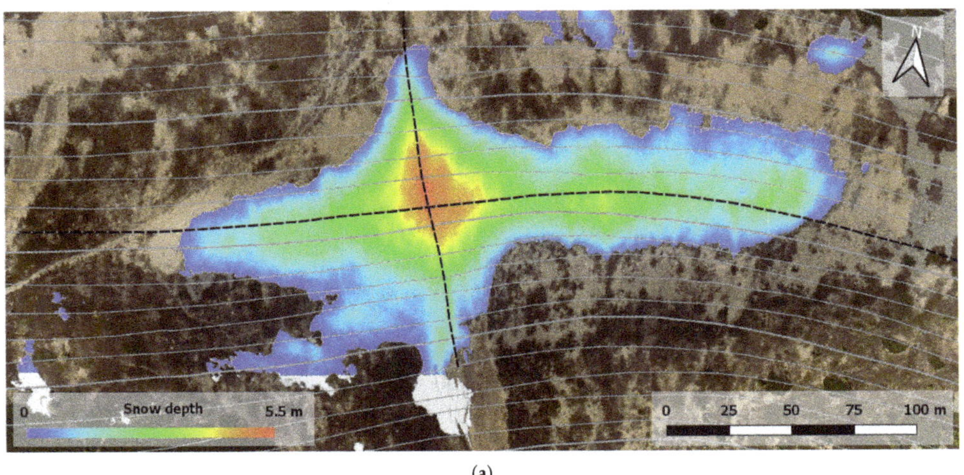

(a)

Figure 15. *Cont.*

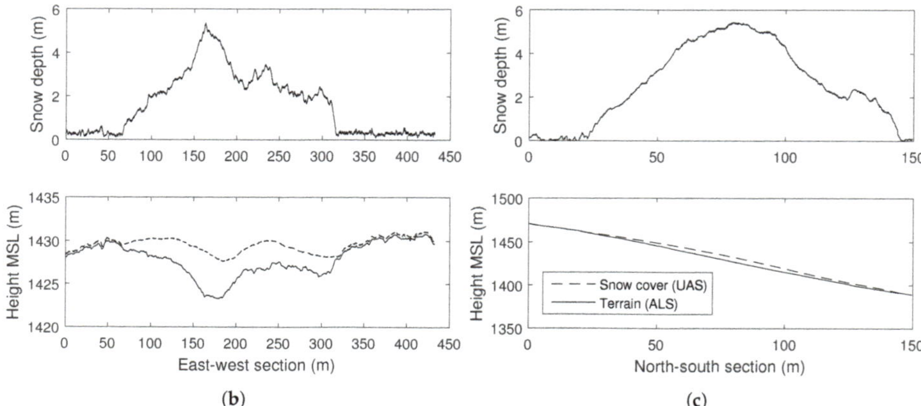

Figure 15. An interpretation of the mapping results (**a**). The UAS photogrammetry-based orthophoto is supplemented with a layer illustrating the snow depth; both items were created with the multi-sensor system for direct georeferencing. The map includes contour (solid gray) and section (dashed black) lines to facilitate the cross-section analysis. The cross-sections of the SD map, the UAS-based surface model, and the ALS-based terrain model, all passing through the SD maximum in the east-west direction, are displayed in (**b**); those that run in the north-south direction are then indicated in (**c**).

4. Discussion

Within this study, we introduced an innovative, state-of-the-art technique for snow-field monitoring and tested its performance at Mapa Republiky, a location of major scientific interest within the discussed research problem. The method exploits various advanced technologies, such as micro UASs to perform automatic aerial data collection; an embedded RTK GNSS receiver allowing centimeter-level positioning; and photogrammetric software providing powerful tools to create digital orthophotos and elevation models. Our solution thus effectively eliminates most drawbacks inherent in previous approaches to the estimation of SD, volume, and covered area.

In the context of the mapping, we can specify several categories to evaluate the quality of each method, and these are as follows: spatial density (resolution) of the collected data; SD estimation accuracy; data acquisition time; safety risks; and automation potential. We comprehensively evaluated each of the methods employed previously on Mapa Republiky as well as the procedure proposed herein; the results are summarized in Table 5. As regards the earliest technique, which relied on wire probes, the main drawbacks consisted of the impossibility to penetrate thick snow layers (above 3–4 m); very long measurement time; and low spatial density. The SD estimation accuracy is sufficient (less than 10 cm, as established upon comparison with the GNSS-based method [26]) if the probe reaches the ground. As with all methods requiring manual data collection, wire probing is a low safety procedure due to the avalanche risk. The approach shares most of its parameters with the GNSS-based ground procedure, introduced within [26,31]; significantly, the latter method is nevertheless capable of measuring thick snow layers without accuracy reduction. Since the data collection is realized through continuous walking or skiing, higher space density data can be obtained in less time. Furthermore, a rope-driven sledge can be used instead of the walking and skiing to decrease the safety risk during the avalanche season.

Table 5. A comparison of the snow-field mapping methods.

Type	Method	Resolution (Meter-Order)	Accuracy (Meter-Order)	Time Consumption	Safety	Automation Potential
Ground	wire probe survey	10^0–10^1	10^{-1}–10^0	hrs	avalanche risk	no
Ground	GNSS survey	10^0–10^1	10^{-2}	hrs	avalanche risk	no
Ground + aerial	UAS photo. (IG)	10^{-2}	10^{-2}	mins–hrs	avalanche risk	partial
Aerial	UAS photo. (DG)	10^{-2}	10^{-2}–10^{-1}	mins	no	high

The UAS-based aerial data acquisition markedly reduced the data collection time without decreasing the accuracy and, simultaneously, ensured superior density (at the centimeter level). In the remaining categories, the quality varies with the applied georeferencing technique. As mentioned in Section 2.6, the IG in UAS-based photogrammetry outperforms the direct technique in terms of the object accuracy. Both approaches, however, allow us to reach a centimeter-level object accuracy [22,32,52]. This fact was confirmed during our previous research [36] and within this study as well. We achieved the spatial RMS error of 11.4 cm in the DG, securing sufficient performance for the relevant application, given that the height accuracy of the applied snow-free model was 15 cm RMS. Yet, the SD estimation accuracy was not assessed in this study; this topic is discussed, for example, within articles [7,8]. IG, compared to DG, does not require very accurate and expensive positioning equipment onboard the UAS; despite this, a survey-grade GNSS receiver is still needed to measure the positions of the GCPs, an operation that consumes substantial extra time (Table 3 and [7]). In the remaining respects, DG nevertheless brings a major advantage in that the field work comprises only the automatic UAS flight, enabling the operator to control the entire mission from a safe distance. Such a scenario then considerably reduces the health risks and time intensity. Additionally, DG enables us to significantly increase the portion of automatic processing. In terms of the aforementioned aspects, the method that uses UAS-based photogrammetry and DG can be considered the most appropriate solution for snow mapping at present.

In the context of Mapa Republiky, our approach brings undoubtable benefits, the most prominent one being that the snow monitoring can begin already in early spring, namely at a time when the avalanche risks are usually too high to employ other techniques, terrestrial-based procedures in particular. This advantage enables us to collect complex data over a longer time period, leading to a better insight into the snow melting process and its impact on the local microclimate and hydrological situation. Moreover, the lower time intensity and higher automation potential may facilitate more frequent observations in the future. The method's usability, however, is not limited to the discussed site only: Many other research-relevant, interesting locations in the Krkonose cannot be easily monitored by UASs today, mainly due to the avalanche risk, inaccessibility, or dense vegetation disallowing GCP deployment.

The only partially unsuccessful portion of the experiment was the correction transmission during the aerial data acquisition, where the datalink outages probably slightly affected the object accuracy. As regards the data processing, the segmentation phase will have to be refined to eliminate operator interventions; we nevertheless believe that the entire data processing cycle can be executed automatically in the future.

It should be noted that the proposed approach exhibits certain limitations. The operation of the UASs, the small ones in particular, depends on the weather conditions: Flying in rain or snowfall is typically prohibited, and the wind speed cannot exceed the pre-defined level. The discussed conditions may embody a markedly limiting factor, especially in mountainous regions and during the winter season, namely parameters specific to our research. Moreover, poor visibility (due to fog, for example) can affect the quality of the photogrammetric processing or even make this phase impossible.

Further development of the technology will most likely increase the already high effectivity of mapping missions. Thus, our solution relying on a custom-built multi-sensor system and a UAS, with the total weight of the components being approximately 8 kg, may be possibly replaced with a more compact design. RTK GNSS-equipped UASs whose weight ranges below 2 kg have been recently available on the market [53]; using such equipment, the field work could be easily carried out by one person. The quality of this solution, however, is yet to be assessed.

5. Conclusions

This article proposed a UAS-based snow-field mapping technique and evaluated its performance at Mapa Republiky, a mountainous location of substantial scientific interest. The main benefit of the method rests in using ground control point-free UAS photogrammetry enabled by a custom-built multi-sensor system; this approach allowed us to perform an entire mapping mission from a safe distance, eliminating a major portion of the safety risks associated with avalanches. The parameters observed within the experiment, namely the georeferencing accuracy (approximately one decimeter RMSE) and spatial resolution of the photogrammetric products (centimeter-level), fully satisfied the requirements of the discussed application. Regrettably, the thick snow cover at the study site did not enable us to evaluate the SD mapping results achievable with a direct method, such as wire probing. In terms of the time consumption, our approach proved to be more efficient than the previously employed techniques, facilitating automatic execution of most processing steps. Despite this, there remains space for improvements, especially in view of the fact that some processing procedures currently depend on user intervention. Exploiting the outcomes presented herein, those related to safety in particular, the technique generally allows the user to collect data at Mapa Republiky very early in spring; such an operation was not feasible previously, meaning that the associated microclimate research at the given location had to face substantial objective limitations. This impediment can now be effectively overcome by using the novel approach.

Author Contributions: Conceptualization, P.G. and P.J.; Data curation, P.G.; Formal analysis, P.G. and P.J.; Funding acquisition, L.Z.; Investigation, P.G. and P.J.; Methodology, P.G. and P.J.; Project administration, L.Z.; Resources, P.G. and P.J.; Software, P.G.; Supervision, L.Z.; Validation, P.G., P.J. and J.H.; Visualization, P.G.; Writing—original draft, P.G.; Writing—review & editing, P.J., L.Z. and J.H.

Funding: This work was supported by the European Regional Development Fund under the project Robotics for Industry 4.0 (reg. No. CZ.02.1.01/0.0/0.0/15_003/0000470).

Conflicts of Interest: The authors declare no conflict of interest.

References

1. Anttila, S.; Metsamaki, S.; Pulliainen, J.; Luojus, K. From EO data to snow covered area (SCA) end products using automated processing system. In Proceedings of the 2005 IEEE International Geoscience and Remote Sensing Symposium, Seoul, Korea, 29 July 2005; Volume 3, pp. 1947–1950. [CrossRef]
2. Shi, J.; Dozier, J.; Rott, H. Snow mapping in alpine regions with synthetic aperture radar. *IEEE Trans. Geosci. Remote Sens.* **1994**, *32*, 152–158. [CrossRef]
3. Luojus, K.; Karna, J.P.; Hallikainen, M.; Pulliainen, J. Development of Techniques to Retrieve Snow Covered Area (SCA) in Boreal Forests from Space-borne Microwave Observations. In Proceedings of the 2006 IEEE International Symposium on Geoscience and Remote Sensing, Denver, CO, USA, 31 July–4 August 2006; pp. 2180–2182. [CrossRef]
4. Saloranta, T.M. Operational snow mapping with simplified data assimilation using the seNorge snow model. *J. Hydrol.* **2016**, *538*, 314–325. [CrossRef]
5. Tveito, O.E.; Bjørdal, I.; Skjelvåg, A.O.; Aune, B. A GIS-based agro-ecological decision system based on gridded climatology. *Meteorol. Appl.* **2005**, *12*, 57–68. [CrossRef]
6. Foppa, N.; Stoffel, A.; Meister, R. Synergy of in situ and space borne observation for snow depth mapping in the Swiss Alps. *Int. J. Appl. Earth Observ. Geoinf.* **2007**, *9*, 294–310. [CrossRef]
7. Adams, M.S.; Bühler, Y.; Fromm, R. Multitemporal Accuracy and Precision Assessment of Unmanned Aerial System Photogrammetry for Slope-Scale Snow Depth Maps in Alpine Terrain. *Pure Appl. Geophys.* **2018**, *175*, 3303–3324. [CrossRef]

8. Avanzi, F.; Bianchi, A.; Cina, A.; De Michele, C.; Maschio, P.; Pagliari, D.; Passoni, D.; Pinto, L.; Piras, M.; Rossi, L. Centimetric Accuracy in Snow Depth Using Unmanned Aerial System Photogrammetry and a MultiStation. *Remote Sens.* **2018**, *10*, 765. [CrossRef]
9. Deems, J.S.; Painter, T.H.; Finnegan, D.C. Lidar measurement of snow depth: A review. *J. Glaciol.* **2013**, *59*, 467–479. [CrossRef]
10. Snow Mapping, Natural Resources Canada. 2015. Available online: https://www.nrcan.gc.ca/node/9561 (accessed on 8 March 2019).
11. Takala, M.; Luojus, K.; Pulliainen, J.; Derksen, C.; Lemmetyinen, J.; Kärnä, J.P.; Koskinen, J.; Bojkov, B. Estimating northern hemisphere snow water equivalent for climate research through assimilation of space-borne radiometer data and ground-based measurements. *Remote Sens. Environ.* **2011**, *115*, 3517–3529. [CrossRef]
12. Arjomandi, M.; Agostino, S.; Mammone, M.; Nelson, M.; Zhou, T. *Classification of Unmanned Aerial Vehicles*; Report for Mechanical Engineering Class; University of Adelaide: Adelaide, Australia, 2006.
13. Torres-Sánchez, J.; López-Granados, F.; Borra-Serrano, I.; Peña, J.M. Assessing UAV-collected image overlap influence on computation time and digital surface model accuracy in olive orchards. *Precis. Agric.* **2017**, 1–19. [CrossRef]
14. Mikita, T.; Janata, P.; Surový, P. Forest Stand Inventory Based on Combined Aerial and Terrestrial Close-Range Photogrammetry. *Forests* **2016**, *7*, 165. [CrossRef]
15. James, M.R.; Robson, S.; Smith, M.W. 3-D uncertainty-based topographic change detection with structure-from-motion photogrammetry: Precision maps for ground control and directly georeferenced surveys. *Earth Surf. Process. Landf.* **2017**. [CrossRef]
16. Saleri, R.; Cappellini, V.; Nony, N.; Luca, L.D.; Pierrot-Deseilligny, M.; Bardiere, E.; Campi, M. UAV photogrammetry for archaeological survey: The Theaters area of Pompeii. In Proceedings of the Digital Heritage International Congress (DigitalHeritage), Marseille, France, 28 October–1 November 2013; Volume 2, pp. 497–502. [CrossRef]
17. Lazna, T.; Gabrlik, P.; Jilek, T.; Zalud, L. Cooperation between an unmanned aerial vehicle and an unmanned ground vehicle in highly accurate localization of gamma radiation hotspots. *Int. J. Adv. Robot. Syst.* **2018**, *15*, 1–16. [CrossRef]
18. Bühler, Y.; Adams, M.S.; Stoffel, A.; Boesch, R. Photogrammetric reconstruction of homogenous snow surfaces in alpine terrain applying near-infrared UAS imagery. *Int. J. Remote Sens.* **2017**, *38*, 3135–3158. [CrossRef]
19. Bühler, Y.; Adams, M.S.; Bösch, R.; Stoffel, A. Mapping snow depth in alpine terrain with unmanned aerial systems (UASs): Potential and limitations. *Cryosphere* **2016**, *10*, 1075–1088. [CrossRef]
20. Cimoli, E.; Marcer, M.; Vandecrux, B.; Bøggild, C.E.; Williams, G.; Simonsen, S.B. Application of Low-Cost UASs and Digital Photogrammetry for High-Resolution Snow Depth Mapping in the Arctic. *Remote Sens.* **2017**, *9*, 1144. [CrossRef]
21. Vander Jagt, B.; Lucieer, A.; Wallace, L.; Turner, D.; Durand, M. Snow Depth Retrieval with UAS Using Photogrammetric Techniques. *Geosciences* **2015**, *5*, 264–285. [CrossRef]
22. Fazeli, H.; Samadzadegan, F.; Dadrasjavan, F. Evaluating the Potential of RTK-UAV for Automatic Point Cloud Generation in 3D Rapid Mapping. *Int. Arch. Photogramm. Remote Sens. Spat. Inf. Sci.* **2016**, *XLI-B6*, 221–226. [CrossRef]
23. Lo, C.F.; Tsai, M.L.; Chiang, K.W.; Chu, C.H.; Tsai, G.J.; Cheng, C.K.; El-Sheimy, N.; Ayman, H. The Direct Georeferencing Application and Performance Analysis of Uav Helicopter in Gcp-Free Area. *Int. Arch. Photogramm. Remote Sens. Spat. Inf. Sci.* **2015**, *XL*, 151–157. [CrossRef]
24. Turner, D.; Lucieer, A.; Wallace, L. Direct Georeferencing of Ultrahigh-Resolution UAV Imagery. *IEEE Trans. Geosci. Remote Sens.* **2014**, *52*, 2738–2745. [CrossRef]
25. Tan, A.; Eccleston, K.; Platt, I.; Woodhead, I.; Rack, W.; McCulloch, J. The design of a UAV mounted snow depth radar: Results of measurements on Antarctic sea ice. In Proceedings of the 2017 IEEE Conference on Antenna Measurements Applications (CAMA), Tsukuba, Japan, 4–6 December 2017; pp. 316–319. [CrossRef]
26. Hejcman, M.; Dvorak, I.J.; Kocianova, M.; Pavlu, V.; Nezerkova, P.; Vitek, O.; Rauch, O.; Jenik, J. Snow Depth and Vegetation Pattern in a Late-melting Snowbed Analyzed by GPS and GIS in the Giant Mountains, Czech Republic. *Arct. Antarct. Alp. Res.* **2006**, *38*, 90–98. [CrossRef]
27. Dvořák, J. Odhalené tajemství Mapy republiky. *Krkonoše—Jizerské hory* **2005**, *2005*, 4–8.

28. Dvořák, I.J.; Kociánová, M.; Hejcman, M.; Treml, V.; Vaněk, J. Linkage between Geo- and Biodiversity on Example of Snow-patch "Map of Republic" (Modrý důl Valley). In Proceedings of the International Conference Geoecological Problems of the Giant Mountains, Szklarska Poręba, Poland, 5–11 November 2003; Volume 41, pp. 100–110.
29. Jeník, J. Geobotanická studie lavinového pole v Modrém dole v Krkonoších. *Acta Univ. Carol. Biol.* **1958**, *5*, 49–95.
30. Vrba, M. Snow Accumulation in the Avalanche Region of the Modry Dul – Valley in Krkonose Mountains. *Opera Corcon.* **1964**, *1*, 55–69.
31. Dvořák, I.J.; Kociánová, M.; Pírková, L. Example of utilization GPS and GIS technologies by study of cryoplanation terraces on the Luční Mt. and Studniční Mt. In Proceedings of the International Conference Geoecological Problems of the Giant Mountains, Szklarska Poręba, Poland, 5–11 November 2003; Volume 41, pp. 18–24.
32. Eling, C.; Wieland, M.; Hess, C.; Klingbeil, L.; Kuhlmann, H. Development and Evaluation of a UAV Based Mapping System for Remote Sensing and Surveying Applications. *Int. Arch. Photogramm. Remote Sens. Spat. Inf. Sci.* **2015**, *XL-1/W4*, 233–239. [CrossRef]
33. Rieke, M.; Foerster, T.; Geipel, J.; Prinz, T. High-precision Positioning and Real-time Data Processing of UAV-Systems. *Int. Arch. Photogramm. Remote Sens. Spat. Inf. Sci.* **2012**, *XXXVIII-1/C22*, 119–124. [CrossRef]
34. Kocmanova, P.; Zalud, L. Multispectral Stereoscopic Robotic Head Calibration and Evaluation. In *Modelling and Simulation for Autonomous Systems*; Lecture Notes in Computer Science; Springer: Cham, Swizerlands, 2015; pp. 173–184.
35. Zalud, L.; Kocmanova, P.; Burian, F.; Jilek, T.; Kalvoda, P.; Kopecny, L. Calibration and Evaluation of Parameters in A 3D Proximity Rotating Scanner. *Elektronika Ir Elektrotechnika* **2015**, *21*, 3–12. [CrossRef]
36. Gabrlik, P.; Cour-Harbo, A.l.; Kalvodova, P.; Zalud, L.; Janata, P. Calibration and accuracy assessment in a direct georeferencing system for UAS photogrammetry. *Int. J. Remote Sens.* **2018**, *39*, 4931–4959. [CrossRef]
37. Rangel, J.M.G.; Gonçalves, G.R.; Pérez, J.A. The impact of number and spatial distribution of GCPs on the positional accuracy of geospatial products derived from low-cost UASs. *Int. J. Remote Sens.* **2018**, *39*, 7154–7171. [CrossRef]
38. Vericat, D.; Brasington, J.; Wheaton, J.; Cowie, M. Accuracy assessment of aerial photographs acquired using lighter-than-air blimps: Low-cost tools for mapping river corridors. *River Res. Appl.* **2009**, *25*, 985–1000. [CrossRef]
39. Kraus, K. *Photogrammetry: Geometry from Images and Laser Scans*; Walter de Gruyter: Berlin, Germany, 2007.
40. Cabreira, T.M.; Brisolara, L.B.; Ferreira, P.R., Jr. Survey on Coverage Path Planning with Unmanned Aerial Vehicles. *Drones* **2019**, *3*, 4. [CrossRef]
41. Hartley, R.; Zisserman, A. *Multiple View Geometry in Computer Vision*, 2nd ed.; Cambridge University Press: Cambridge, UK, 2004.
42. Szeliski, R. *Computer Vision: Algorithms and Applications*; Texts in Computer Science; Springer: London, UK, 2011.
43. Cramer, M. Performance of GPS/Inertial Solutions in Photogrammetry. In *Photogrammetric Week 01*; Wichmann: Heidelberg, Germany, 2001; pp. 49–62.
44. Barry, P.; Coakley, R. Field Accuracy Test of RPAS Photogrammetry. *Int. Arch. Photogramm. Remote Sens. Spat. Inf. Sci.* **2013**, *XL-1/W2*, 27–31. [CrossRef]
45. Panayotov, A. (University of Colorado Denver, Denver, CO, USA). Photogrammetric Accuracy of Real Time Kinematic Enabled Unmanned Aerial Vehicle Systems. Unpublished work, 2015.
46. Verhoeven, G.; Doneus, M.; Briese, C.; Vermeulen, F. Mapping by matching: A computer vision-based approach to fast and accurate georeferencing of archaeological aerial photographs. *J. Archaeol. Sci.* **2012**, *39*, 2060–2070. [CrossRef]
47. Acharya, T.; Ray, A.K. *Image Processing: Principles and Applications*; John Wiley & Sons: Hoboken, NJ, USA, 2005.
48. Chang, J.H.; Fan, K.C.; Chang, Y.L. Multi-modal gray-level histogram modeling and decomposition. *Image Vis. Comput.* **2002**, *20*, 203–216. [CrossRef]
49. Otsu, N. A Threshold Selection Method from Gray-Level Histograms. *IEEE Trans. Syst. Man Cybern.* **1979**, *9*, 62–66. [CrossRef]
50. CUZK, Geoportal. Available online: https://geoportal.cuzk.cz/ (accessed on 8 March 2019).

51. Paterson, W.S.B. *Physics of Glaciers*, 3rd ed.; Butterworth-Heinemann: Oxford, UK, 1994.
52. Pfeifer, N.; Glira, P.; Briese, C. Direct Georeferencing with on Board Navigation Components of Light Weight Uav Platforms. *Int. Arch. Photogramm. Remote Sens. Spat. Inf. Sci.* **2012**, *39B7*, 487–492. [CrossRef]
53. Phantom 4 RTK—Next Gen Mapping Solution—DJI. Available online: https://www.dji.com/cz/phantom-4-rtk (accessed on 8 March 2019).

 *sensors*

Article

UAVs for Structure-From-Motion Coastal Monitoring: A Case Study to Assess the Evolution of Embryo Dunes over a Two-Year Time Frame in the Po River Delta, Italy

Yuri Taddia [1,*], Corinne Corbau [2], Elena Zambello [2] and Alberto Pellegrinelli [1]

1 Engineering Department, University of Ferrara, Saragat 1, 44122 Ferrara, Italy; alberto.pellegrinelli@unife.it
2 Physics and Earth Science Department, University of Ferrara, Saragat 1, 44122 Ferrara, Italy; corinne.corbau@unife.it (C.C.); elena.zambello@unife.it (E.Z.)
* Correspondence: yuri.taddia@unife.it; Tel.: +39-0532-974918

† This paper is an extended version of our conference paper published in Taddia, Y.; Corbau, C.; Zambello, E.; Russo, V.; Simeoni, U.; Russo, P.; Pellegrinelli, A. UAVs to assess the evolution of embryo dunes. In Proceedings of the International Conference on Unmanned Aerial Vehicles in Geomatics, 4–7 September 2017, Bonn, Germany.

Received: 4 March 2019; Accepted: 8 April 2019; Published: 10 April 2019

Abstract: Coastal environments are usually characterized by a brittle balance, especially in terms of sediment transportation. The formation of dunes, as well as their sudden destruction as a result of violent storms, affects this balance in a significant way. Moreover, the growth of vegetation on the top of the dunes strongly influences the consequent growth of the dunes themselves. This work presents the results obtained through a long-term monitoring of a complex dune system by the use of Unmanned Aerial Vehicles (UAVs). Six different surveys were carried out between November 2015 and December 2017 in the littoral of Rosolina Mare (Italy). Aerial photogrammetric data were acquired during flight repetitions by using a DJI Phantom 3 Professional with the camera in a nadiral arrangement. The processing of the captured images consisted of the reconstruction of a three-dimensional model using the Structure-from-Motion (SfM). Each model was framed in the European Terrestrial Reference System (ETRS) using GNSS geodetic receivers in Network Real Time Kinematic (NRTK). Specific data management was necessary due to the vegetation by filtering the dense cloud. This task was performed by both performing a slope detection and a removal of the residual outliers. The final products of this approach were thus represented by Digital Elevation Models (DEMs) of the sandy coastal section. In addition, DEMs of Difference (DoD) were also computed for the purpose of monitoring over time and detecting variations. The accuracy assessment of the DEMs was carried out by an elevation comparison through especially GNSS-surveyed points. Relevant cross sections were also extracted and compared. The use of the Structure-from-Motion approach by UAVs finally proved to be both reliable and time-saving thanks to quicker in situ operations for the data acquisition and an accurate reconstruction of high-resolution elevation models. The low cost of the system and its flexibility represent additional strengths, making this technique highly competitive with traditional ones.

Keywords: coastal mapping; coastal monitoring; Digital Elevation Models (DEMs); geomorphological evolution; photogrammetry; Structure-from-Motion (SfM); Unmanned Aerial Vehicles (UAVs)

---

1. Introduction

Beaches are usually characterized by vegetated dune systems performing a variety of functions valuable to the ecosystem, such as protection against the ingression of the sea water during storms and a sand reservoir that supplies sand to eroded beaches during storms and buffers windblown sand and salt spray. In addition, the dunes act as an ecological niche for both flora and fauna.

In a natural context, coastal dunes can be classified into three main types: incipient foredunes, established foredunes, and foredune plains [1,2]. The incipient foredunes, or embryo dunes [3], are low dunes formed by aeolian sand deposition within pioneer plant communities on the backshore of beaches [2,4]. Their formation is due to an increase in surface roughness due to the presence of some elements on the surface of the backshore responsible for a reduction of wind flow velocities, resulting in sediment deposition.

From a geomorphological point of view, the embryo dunes present the most interesting shapes to be assessed because they are particularly sensitive to even the smallest changes in any of the coastal environment factors. Indeed, they can quickly be destroyed as they are created. Due to this importance, the presence of embryo dunes hence represents a crucial parameter for the final selection of the case study location.

Consequently, the monitored site (Figure 1a) presents all the typical characteristics of a natural dune system previously mentioned. It is located in the northern Adriatic Sea close to the Po River delta (Italy), near the town of Rosolina Mare. The beach of the monitored area extends up to 300 m in the North-South direction. The back-dunes environment is characterized by the presence of several infra-dune depressions in which the rainwater is collected. That allows the growth of a particular hygrophilous vegetation. A patch of bushy plants covers the innermost established foredunes and spatially precedes the arboreal formations. The forest is made of both autochthonous holm oaks and pines of artificial origin.

Moreover, the inclusion of this coastal stretch within the "Coastal Botanical Gardens" of Porto Caleri ensures both the complete conservation and the natural evolution of the monitored coastal environment. No significant impact of human activities has been noticed during the last few years preceding the beginning of the surveys; hence the site was found to be particularly suitable for the assessment of the natural geomorphological evolution of dunes, focusing on the embryonic stage.

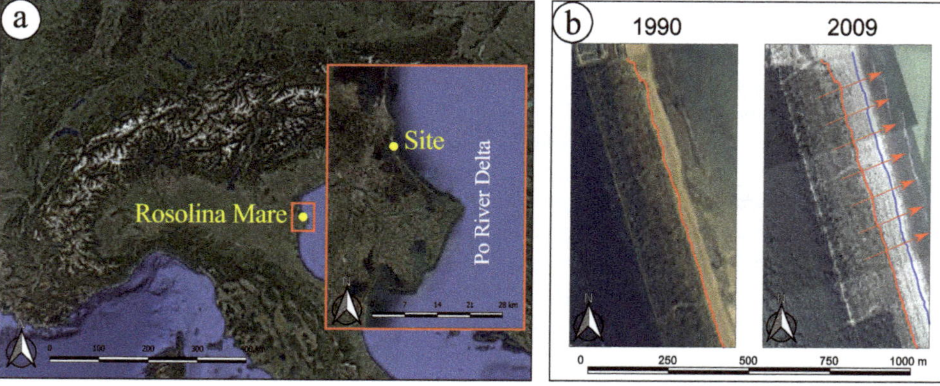

Figure 1. (**a**) The case study location, northern Italy (by Google Earth); (**b**) The progression (*progradation*) of the embryo dunes' limit over 19 years in the past.

During recent decades, all the southern part of the Rosolina Mare littoral experienced a significant progradation of the shoreline. Figure 1b highlights the progression of the embryo dunes' limit from 1990 to 2009 as it was observed by aerial orthophotos and satellite imageries. In fact, survey techniques based on remote sensing from satellite were successfully used in the past to assess these phenomena by some authors [5]. The overall progression detected at Rosolina Mare was of about 50 m over almost 20 years while a maximum rate of 5 m per year was noticed. The progradation is still active at the site selected for this work where the embryo dunes generally grow and join each other, due to the aeolian transport [6].

The reconstruction of Digital Elevation Models (DEMs) represents a primary tool to detect and characterize the morphology of complex systems. Their comparison enables an accurate assessment, especially when the shapes change rapidly in both time and space. Different techniques are currently available for generating DEMs at a variable spatial resolution. Using discrete approaches, a large amount of points may be directly surveyed through GNSS geodetic receivers. When carried out by means of a Network Real Time Kinematic (NRTK) technique, the survey is hence already framed in an official reference system. Alternatively, the same detection can be done in a local reference system by using a total station. Regardless of the instrumentation used, the manual mapping of the shapes is a very time-consuming task and cannot be applied to broad extensions.

Further approaches, with a dense (and almost continuous) data collection, such as Terrestrial Laser Scanners (TLSs), airborne LiDAR (Light Detection And Ranging) and digital photogrammetry by Unmanned Aerial Vehicles (UAVs), enable the remote sensing of wide areas, still with a good level of accuracy. LiDAR has been perhaps the most common technique used for the purpose of mapping long coastline sections. In fact, the raw data acquired by this technique have high spatial resolutions. However, the subsequent data processing is usually imprinted to provide a resampled model to the users, characterized by a Ground Sample Distance (GSD) of about 0.5–1 m. Moreover, neither the characteristics of the instruments used to acquire the raw data nor the processing procedures (filtering and georeferencing techniques) further applied to them to generate the final products (clouds, DEMs) are actually known for all the LiDAR datasets currently available in the Rosolina Mare area. Despite this particular issue, the LiDAR approach is certainly an effective method for monitoring broad extensions. However, the sandy coastal environment evolves too much rapidly to be assessed by airborne LiDAR recognitions and the alternative use of helicopters has prohibitive costs. For advanced studies in coastal geomorphology, a different method with a great flexibility and able to provide better spatial resolutions at a more reasonable cost is hence needed.

A solution is represented by the recent availability of lightweight and compact digital cameras along with the continuous development of newer and more powerful algorithms in computer vision. In addition, the use of small UAVs has revolutionized the classical aerial photogrammetry leading to a new digital photogrammetry based on a Structure-from-Motion (SfM) approach that provides excellent results in terms of spatial resolution at a low cost [7].

The SfM process (Figure 2) is an advanced method of the computer vision branch. This approach enables the three-dimensional geometry reconstruction (the so-called *structure*) starting from a set of two-dimensional images of a scene (the so-called *motion*). The first stages are the detection of features on each image, their characterization through descriptors and the final matching of the established correspondences. The resulting matches, i.e., the tie points, constitute a sparse point cloud: this is a first raw model of the detected object. It is important noticing that no *a priori* assumption, such as the use of Ground Control Points (GCPs), is needed to reconstruct the geometry, since the well-known collinearity equations [8] can be solved in any arbitrary scale. A further advantage of the SfM lies in the concurrent estimation of both the exterior and interior orientation of the camera, performing the so-called *self-calibration* [9]. This is particularly useful whenever an uncalibrated camera is used for the acquisition of aerial imageries. Both the symmetric and asymmetric radial distortions can be modelled through a self-calibration.

To perform an accurate georeferencing of the point cloud model, a set of GCPs was surveyed by GNSS geodetic receivers through NRTK. The accuracies of such a technique are completely comparable with the Root Mean Square Error (RMSE) of a 2 cm GSD imagery in a SfM approach [10]. To frame the model in the same reference system of GCPs, a final block bundle adjustment (BBA) procedure was adopted [11], performed within the same software used to reconstruct the geometry (Agisoft

PhotoScan Professional). BBA can account for all the non-linearities, hence it provides better results than adopting a simpler similarity transformation. Moreover, a wide distribution of GCPs can help to prevent any possible bowl-effect on the model due to nadiral imageries, since no oblique images that may have minimized this issue [12] were ever captured in this work.

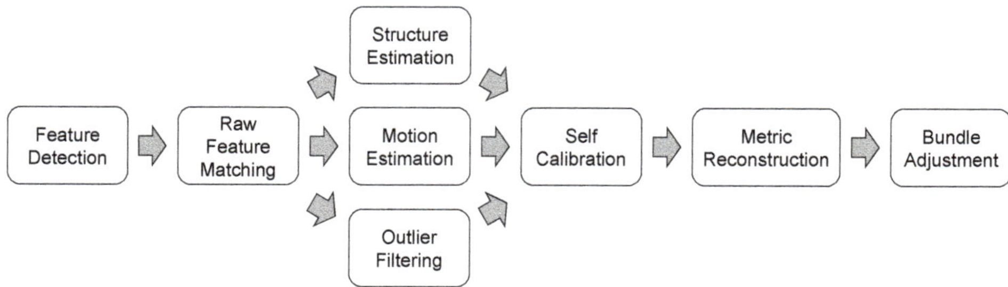

Figure 2. The Structure-from-Motion workflow.

Once the three-dimensional geometry has been reconstructed thanks to the tie points detection, a point cloud with higher density can be generated using specialized algorithms capable to reconstruct depth maps and searching for additional features to be matched. For this reason, the final dense cloud represented the starting-point for any further data processing. On average, each dense point cloud in this work contained up to ca. 80 million points.

There are increasingly reports in the literature regarding applications of UAV systems, also for the aim of a coastal monitoring [13]. Gonçalves and Henriques [14] studied a similar sandy coastal environment, while more recent applications regarding the Adriatic littoral have been shown by Scarelli et al. [15] in the Emilia-Romagna region. Taddia et al. [16] illustrated a very first approach about embryo dunes surveying techniques, but without showing the results of a long-term assessment. In addition, other authors also tested the SfM approach by UAV assessing the vertical accuracy of the products compared to TLS technology [17] or investigated the accuracy of georeferenced point clouds produced via multi-view stereopsis from UAV imagery flying at an altitude very close to ours [18]. All these studies confirm the reliability of the use of UAV-acquired imageries in coastal environments.

The SfM by UAVs therefore represents an innovative technique able to speed up the acquisition process and it finally provides accurate and reliable products. These considerations, along with its versatility, low cost, effectiveness [19] and significantly higher spatial resolutions than other methods, led us to adopt the SfM by UAV as remote sensing approach in the context of geomorphological coastal monitoring.

The innovative aspect of the present study lies in the use of the SfM technique not only for the simple mapping of a coastal section, but instead for the repetition of the survey with the purpose of monitoring the embryo dunes' morphological evolution. Indeed, we tested the applicability of these methodologies in the study of morphological changes of a dune system due to natural factors on a seasonal time scale. In particular, we will highlight the accuracy of the UAV-borne surveys and quantify incipient foredune morphodynamics. Such results may also contribute to the application of UAV surveys in coastal management studies.

2. Materials and Methods

The characterization of the dune system morphology was performed using UAVs [20], in particular for the detection of the embryo dunes with a small size. Indeed, LiDAR resolution, especially whenever resampled at a 0.5–1 m GSD, is definitely unsuitable for this latter purpose [21]. As previously mentioned, the decision to adopt SfM by UAVs was dictated by the great potential of the technique itself, including a fast image acquisition at a low cost and the ability to generate very high-resolution products (orthomosaics, DEMs) [22].

Six survey repetitions were carried out respectively on November 2015, March 2016, June 2016, November 2016, May 2017 and December 2017 (Figure 3). For each flight, a specific planned mission was set up. Several flights between two and three (Figure 4a) per repetition was necessary to map the full extent of the monitored area due to both the limited autonomy of the UAV (related to battery capacity) and the need to maintain an altitude of about 40 m. That was necessary to achieve the desired spatial resolution of 2 cm.

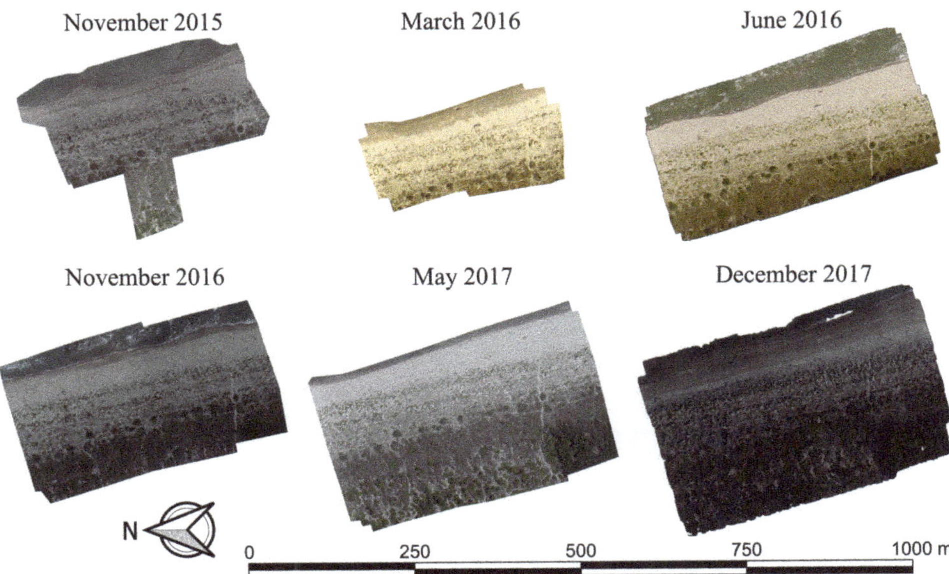

Figure 3. Orthomosaics of the six different survey repetitions from November 2015 to December 2017 showing the extent of the surveyed regions.

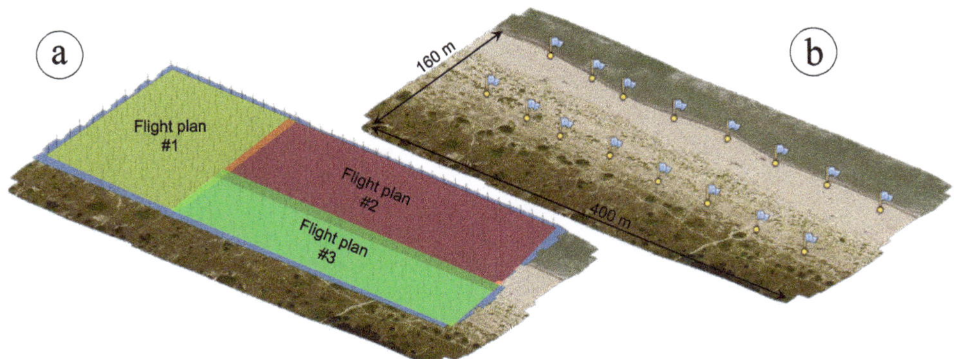

Figure 4. (**a**) Different flight plans' coverage with overlap for including all the study area; (**b**) Ground Control Points location and extent of the site.

Two aircraft were used: a DJI Phantom 2 Vision and a DJI Phantom 3 Professional, even though the DJI Phantom 2 was actually used for the first survey only (on November 2015), whereas the DJI Phantom 3 was used for all the subsequent survey repetitions. They were equipped respectively with a Panasonic Lumix DMC-GM1 camera (16-megapixel RGB sensor) and a DJI FC300X camera (12-megapixel RGB sensor). Specifications of aircraft and cameras are listed in Table 1. The main reason for using the DJI Phantom 3 was that it was stably equipped with a low-cost RGNIR sensor (Sentera Single) with the capability to acquire information in the near infrared (NIR) wavelengths in an independent way with respect to the DJI FC300X camera (Figure 5). The availability of an additional spectral information (potentially to be used for further analysis or processing) argued for using the DJI Phantom 3 aircraft.

Table 1. Aircraft and camera specifications.

Aircraft Specifications			
		DJI Phantom 2 Vision	*DJI Phantom 3 Professional*
Take-off weight		1160 g	1280 g
Max flight speed		15 m/s	16 m/s
Max flight time		16 min	18 min
Hovering accuracy	Horizontal	±2.5 m	±0.3 ÷ 1.5 m
	Vertical	±0.8 m	±0.1 ÷ 0.5 m
Cameras Specifications			
		Panasonic Lumix DMC–GM1	*DJI FC300X*
Sensor format		17.3 mm × 13.0 mm	6.5 mm × 4.9 mm
Focal length		12 mm	3.6 mm
35 mm equiv. focal length		24 mm	20 mm
Image resolution		4592 × 3448	4000 × 3000
Field of View		72°	84°
Pixel size		3.8 µm	1.6 µm
GSD at 40 m altitude		1.3 cm	1.8 cm

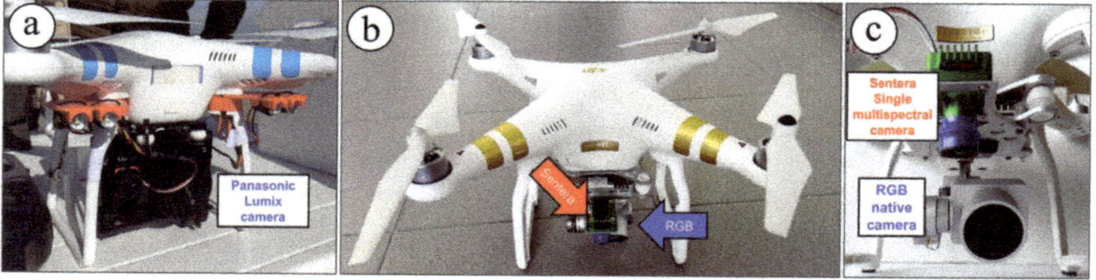

Figure 5. (**a**) The DJI Phantom 2 Vision equipped with the Panasonic Lumix camera; (**b**) The DJI Phantom 3 Professional equipped with both the RGB native camera (highlighted in blue) and a Sentera Single multispectral sensor (highlighted in red); (**c**) A detail on both the DJI Phantom 3 cameras, showing the nadiral fixed arrangement of the Sentera Single sensor.

Thanks to the ability of the combination of SfM and BBA to model any kind of uncalibrated camera through a self-calibration procedure with a very good estimation of all interior parameters and lens distortion, especially whenever a redundant set of images is available, the differences induced by using different cameras on different aircraft were negligible.

Generally, the flight plans parameters were imposed as shown in Table 2. The camera was always in a nadiral arrangement. Some camera settings (e.g., exposure time) were varied according to the real brightness of the scene at the time of each flight: in this case a range of values is given.

Each survey was framed within the European Terrestrial Reference System (ETRS) horizontal datum, in its ETRF 2000 (2008.0) realization. The National Dynamic Network (RDN) materializes this reference system in Italy, thanks to the same Continuously Operating Reference Stations (CORSs) used to stream corrections for the NRTK approach. Targets were distributed over the ground to almost completely enclose the surveyed area (Figure 4b) and were thus used as Ground Control Points. Generally, an amount between 14 and 18 was deployed. Each target, designed to be easily recognized on the images, with a clear and unmistakable position of its center, was surveyed with a GNSS geodetic receiver in NRTK mode. The accuracy achievable with a kinematic "stop and go" technique, with coordinates collected averaging 30 s of observations and corrections transmitted in real time from a network of CORS (the ItalPoS service was used for this purpose), is sufficient and comparable with the RMSE of a 2 cm GSD SfM-derived model. In addition, the need for any post-processing of the GNSS collected data for the final computation of the targets' coordinates is suppressed in this way. All the elevations were framed within the Italian vertical datum by applying the ITALGEO 05 geoidal separation model to convert ellipsoidal heights to orthometric elevations.

Table 2. Flight plan specifications.

Flight Plans Specifications	
Altitude	40 m
Longitudinal overlap	80%
Side overlap	60%
Speed of aircraft	ca. 3 m/s
Exposure time	1/400 ÷ 1/1000 s
ISO sensitivity	100 ÷ 400

The need to number each GCP (Figure 6) to be recognized on the images was overcome thanks to the GPS Exif metadata in the images acquired with the DJI Phantom 3 native camera. Indeed, relative distances from a target to the others were much higher than the absolute georeferencing error of a first raw alignment carried out based on the Exif camera locations. This implied that the closer target reprojected on an image was also the correct one to be specified on the image itself. The recognition of the targets on each image was thus performed manually: this operation was not particularly complex thanks to the mentioned reprojection of a marker once the target has been placed on two aligned images at least.

Figure 6. Targets used as Ground Control Points to be recognized on the aerial images placed (**a**) on the shoreline (initially numbered to be used for the processing of the Panasonic Lumix DMC–GM images) and (**b**) on the back-dunes (not numbered thanks to the Exif metadata in the DJI FC300X images).

A second alignment was performed once all the GCPs were identified on the images. In this latter alignment, all the camera locations were ignored due to their very poor accuracy. Thus, they were assigned with a very low weight or even simply deleted. A final optimization for reprojecting was performed after this second and final alignment in order to best fit the interior orientation parameters of the camera (focal length f, principal point's coordinates c_x, c_y, radial distortion coefficients k_1, k_2, k_3, tangential distortion coefficients p_1, p_2) and to achieve the maximum accuracy of the 3D model. A summary of residuals for each survey repetition is shown in Table 3.

The successful reconstruction of the three-dimensional geometry based on the matched features (tie points) was then followed by the creation of a dense cloud. This task has been performed using the same software used for the alignment of the images (Agisoft PhotoScan Professional), through the means of robust densification algorithms already implemented in it. However, the dense cloud represents a model of the outer surface of the detected object, therefore it contains points belonging to the vegetation on the top of the dunes instead of the sand at their bases. This means that it is not helpful to detect and assess any morphological change of the embryo dunes over time. To remove those points from any further data processing (including the creation of DEMs) a filter on a slope detection was used. The filter is a feature already available within Agisoft PhotoScan and it performs a ground points classification. Essentially, it [23] works through two steps: in the first one, the entire dense cloud is divided into several square cells of a user-specified size. The value adopted in this work 5 m × 5 m. The detected lowest point of each cell is hence assumed to belong to the ground and a first raw model of the ground itself is then created using the lowest points from all the cells. In the second step, all other points are classified by detecting both their distance and slope from the ground-classified points of step one. For this further task, threshold values for distances and slopes must be specified. In this work we adopted respectively 0.15 m and 15°. The final dense cloud of ground-classified points is shown in Figure 7, where all the points classified as effective ground are highlighted in red. Examples of botanical species found in the dune system and responsible for this issue are *Ammophila arenaria* (Figure 8a) and *Echinophora spinosa* (Figure 8b).

Table 3. Summary of the residuals.

Residuals		Survey Date					
		November 2015	March 2016	June 2016	November 2016	May 2017	December 2017
East	*RMSE* [m]	0.029	0.011	0.023	0.044	0.036	0.052
	Min [m]	−0.058	0.024	−0.096	−0.068	−0.102	−0.078
	Max [m]	0.050	0.020	0.043	0.050	0.054	0.107
North	*RMSE* [m]	0.018	0.016	0.026	0.02	0.037	0.032
	Min [m]	−0.037	−0.031	−0.102	−0.036	−0.069	−0.061
	Max [m]	0.03	0.027	0.034	0.045	0.090	0.057
Up	*RMSE* [m]	0.006	0.003	0.032	0.040	0.024	0.011
	Min [m]	−0.010	−0.006	−0.055	−0.041	−0.054	−0.027
	Max [m]	0.008	0.003	0.032	0.040	0.038	0.019
3D	*RMSE* [m]	0.035	0.019	0.036	0.053	0.057	0.062
	Min [m]	0.011	0.005	0.013	0.029	0.009	0.018
	Max [m]	0.063	0.032	0.142	0.073	0.104	0.121

Figure 7. Dense point cloud after the detection of the ground points by the use of the slope-based classification algorithm in Agisoft Photoscan Professional.

Since outliers were still present after the classification, a further statistical outlier removal (SOR) filter was applied to the dense cloud to remove residual points before proceeding with the creation of a DEM of the sandy terrain. This task was performed using the CloudCompare software. Figure 9 shows the effect of the SOR filter in detail: on the top (Figure 9a) a zoom on the raw dense cloud of ground-classified points, on the bottom (Figure 9b) the same region after the application of the filter, showing the effective outliers removal. The values for the parameters to set in such a filter were carefully assumed each time to maximize the filtering effect. Figure 9 was obtained assuming 50 points for the mean distance estimation and 1.5 as the standard deviation multiplier. After the removal of the residual vegetation outliers, a set of special NRTK surveyed points was then added to the filtered dense cloud to integrate all the regions where the ground data was missing.

Figure 8. Botanical species of vegetation on the dunes: (**a**) *Ammophila arenaria*; (**b**) *Echinophora spinosa*.

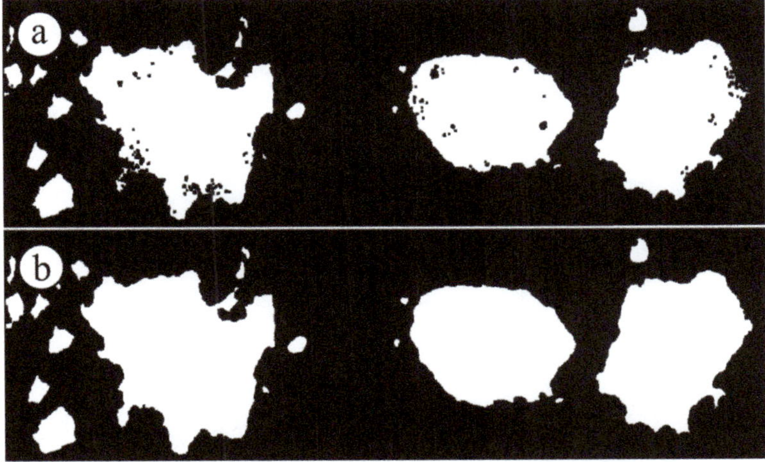

Figure 9. (**a**) Quality of the dense cloud after the detection of ground points on a detail. Some outliers are still present; (**b**) Quality of the dense cloud after the application of the statistical outlier removal (SOR) filter on the same detail. Outliers have been removed.

Such a process, from the raw dense cloud to the final integration with GNSS-acquired information, is shown in Figure 10 and it is summarized in the following. The raw dense cloud profile (Figure 10a) was eliminated by the application of the slope detection-based algorithm, thus obtaining a classification of the solely ground points (Figure 10b). Finally, an integration with specially surveyed GNSS points was performed to reconstruct a more faithful profile of the dense cloud itself (Figure 10c). The final GNSS-integrated dense point cloud was hence rasterized creating the DEM (Figure 11) Due to the very high density of points was so high (up to 1250 points per square meter), the simple average of the in-cell points' height was assumed as the elevation for each DEM's pixel, making it unnecessary to create a Triangulated Irregular Network (TIN) as intermediate step.

The DEM of each survey repetition was thus preventively checked [24] by comparing the model's elevations with another set of GNSS-surveyed points (Figure 12a). Their coordinates were collected independently from those points used for the integration process described above. Moreover, a particular care was adopted when surveying directly on the sand, e.g., by using a plate to prevent the pole tip from sinking, which could have been a remarkable source of errors. The possible presence of systematic errors was also accurately investigated and statistically assessed (Figure 12b). Indeed, this check was used to validate the real accuracy of the DEM with respect to the actual morphology of the dunes system. The validation results for the DEM on November 2016, which represents the survey in which the widest set of validation points was collected, are shown in Figure 12. Both the average and the standard deviation values reconfirm the full compatibility between the SfM approach and the NRTK method of surveying used to reference the model. Similar values, even if not explicitly reported, were obtained for the validation of the DEMs generated for the other repetitions.

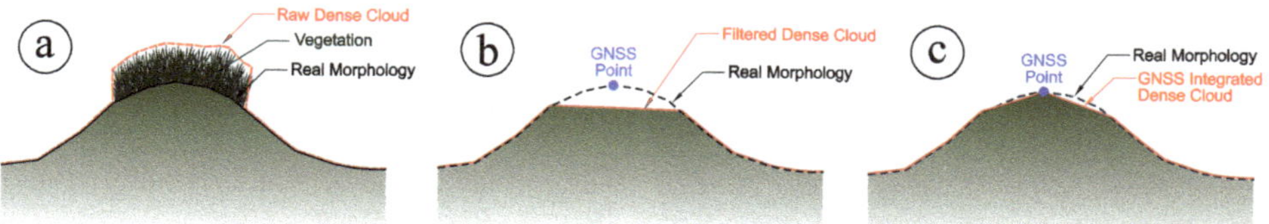

Figure 10. Schematization of the GNSS integration process by (**a**) filtering out the vegetation; (**b**) the survey of a point by GNSS and (**c**) the subsequent reconstruction of a more faithful shape.

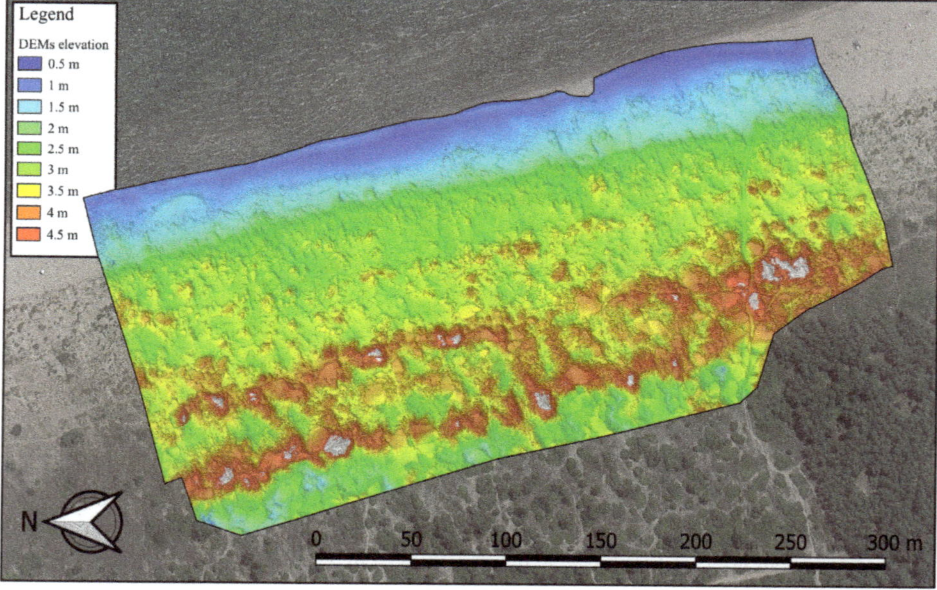

Figure 11. Digital Elevation Model (DEM) of the survey carried out on November 2016.

For a further and last validation, a cross-section profile was also reconstructed through GNSS NRTK during the solely survey on November 2016, to both assess the quality of the profiles reconstructed by a DEM extraction and the effectiveness of the plate under the pole. Figure 12c shows the results of such a comparison between this profile and its DEM derived one. Except for some outliers where the influence of vegetation was still slightly present, the computed average value of −1.2 cm for the difference between DEM (red line) and GNSS profiles (blue line) shows, once again, that the accuracy of the DEM is the same achievable by the use of the NRTK technique that was used for providing the vertical datum to the model.

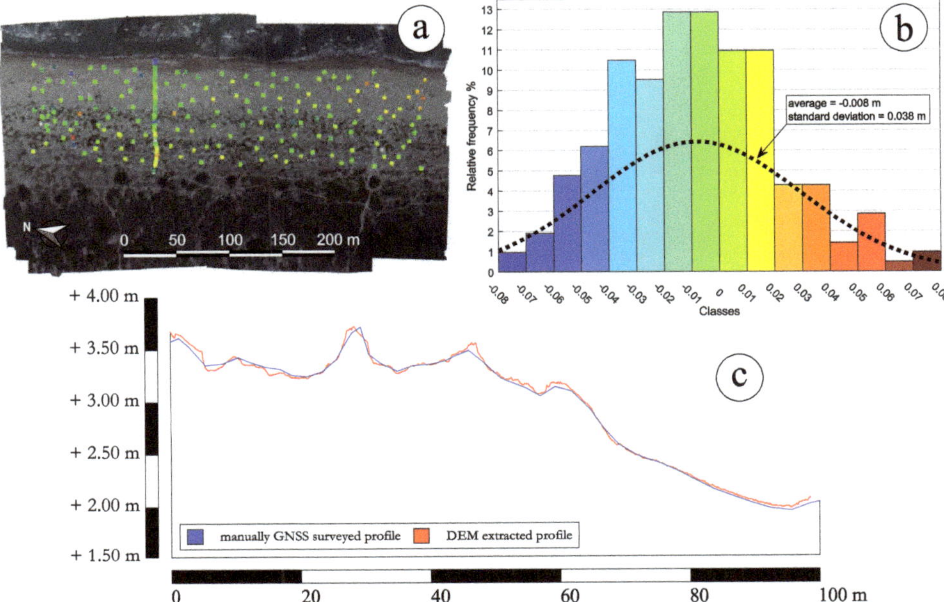

Figure 12. (**a**) Set of ground points used for the comparison between the DEM of November 2016 and the GNSS-surveyed elevations; (**b**) Validation results by histograms: the standard deviation value is comparable with the assumed accuracies of the SfM approach with a 2 cm GSD; (**c**) Comparison between GNSS and DEM on a profile. Discrepancies are still comparable with the SfM approach.

3. Results

Once successfully performed the validation of each computed DEM generated, with discrepancies similar to those mentioned above, a series of further information was finally extracted from the DEMs and then compared each other. In fact, the strength of the aerial imageries by UAVs, with respect to any traditional discrete and time-consuming approach based on the use of total stations or GNSS geodetic receivers in NRTK mode, lies in the possibility to make cross sections in any region of the model at any time, thanks to a practically continuous data.

In particular, to monitor the changes in the morphology of the studied coastal area, ten cross sections and three longitudinal sections (Figure 13a) were used to extract profiles.

Figure 13b illustrates a profile analysis conducted on the cross-section 2 to evaluate the effects of the first winter season. Cross-shore profile changes highlight the effects of the occurred winter sea-storms (November 2015 to March 2016). On this profile, five dunes may be observed. Dunes 5 and 4 appear stable, dunes 3 and 2 are characterized by a crest erosion while sediment deposition occurred in the interdune between 3 and 2. Finally, the seaward side of dune 1 and the backshore are in erosion.

Moreover, an elevation difference was computed generating a DEM of Difference (DoD) for the whole extent of the overlapping area between different surveys. This enabled to compute a sediment budget through the compared area extents in a simple pixel per pixel computation, exploiting the accurate georeferencing of every DEM. Figure 14 shows the elevation differences between the last and the first survey, respectively performed in December 2017 and November 2015. The expression of the computed quantity is given by $H^{ortho}_{Dec.2017} - H^{ortho}_{Nov.2015}$, since the orthometric height was used. Red zones represent accumulation regions (positive sediment budget), while blue zones represent erosion areas (negative budget).

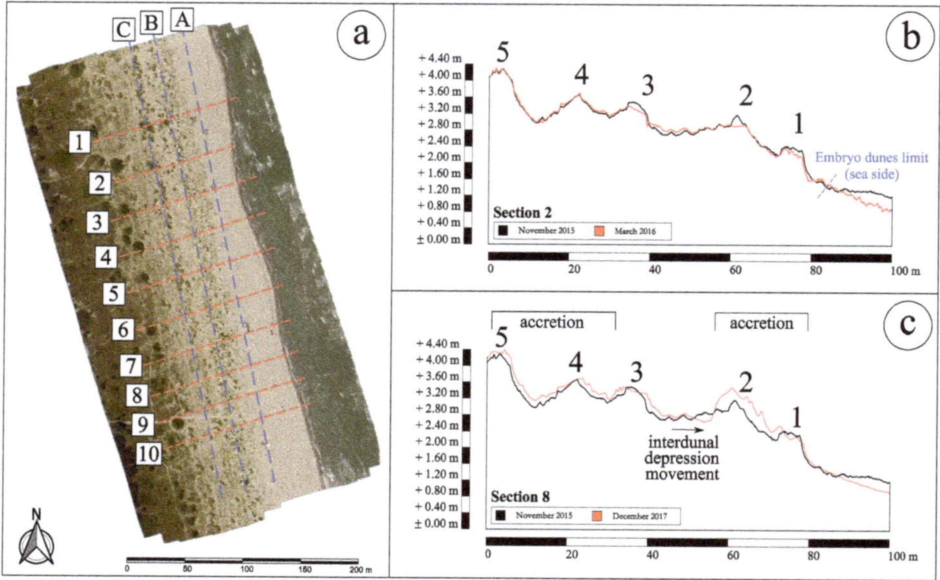

Figure 13. (**a**) Location of the cross and longitudinal sections; (**b**) Detection of the morphological evolution of dunes during first winter season by the comparison of the profiles of November 2015 (black) and March 2016 (red) on cross-section 2; (**c**) Detection of the morphological evolution of dunes over two years by the comparison of the profiles of November 2015 (black) and December 2017 (red) on cross-section 8.

Furthermore, Figure 13c shows the evolution of the cross-section 8 during all the two-year period (November 2015 to December 2017), which is still characterized by the presence of five foredunes. The evolution of the dune system is positive with height variations reaching +0.40 m. Locally, erosion is also observed and in particular on the interdunal depression and on the upper backshore. We may also observe that foredunes 1 and 2 merged resulting in a higher foredune (about +0.20 ÷ 0.40 m).

The analysis of the DoD in Figure 14c thus shows how the progradation phenomenon is clearly detectable through two years of monitoring. In particular, the back-zone is stable. Zone *A* reported in Figure 14 represents the latest developed embryo dunes, while zones *B* and *C* indicate a seaward displacement of the dune system. Indeed, zone *B* is lower than two years before while zone *C* represents the area where the interdunal depression itself has been filled. Also, the cross-section in Figure 13c helps to better understand both the ongoing progradation and the interdunal depression progression. The comparison between profiles over two years hence confirms what DoD shows: a general increase of elevations that is the proof of a progradation phenomenon that is still active in the Rosolina Mare region where the site was selected.

Even though a thorough geospatial analysis is certainly beyond the aim of this work, in order to further support the DoD results some basic statistics have been computed on the previously highlighted regions: stable dunes, zone *A*, zone *B* and zone *C*. For each of these regions, a polygon was selected to be the most representative as possible (shown in Figure 14c). Nonetheless, high standard deviations of the computed elevation differences within the polygons may be due not only to an actual noise on the input DEMs, but they also might be caused by a slight heterogeneity of the region itself. Table 4 summarizes pixel counts, mean values, standard deviations and min/max values extracted from the DoD on each region.

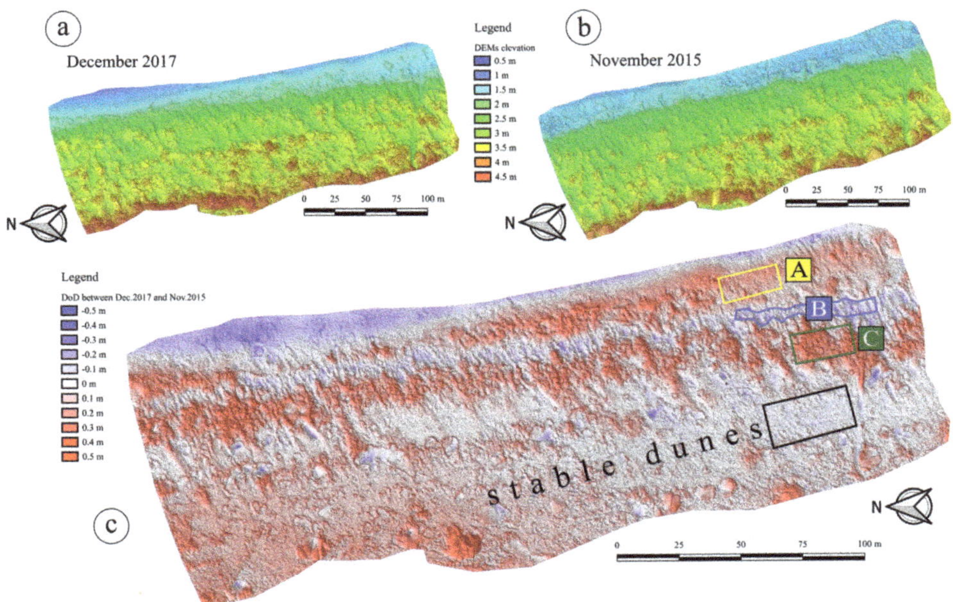

Figure 14. (**a**) DEM of the survey on December 2017; (**b**) DEM of the survey on November 2015; (**c**) Computation of the DEM of Difference (DoD) as $H^{ortho}_{Dec.2017} - H^{ortho}_{Nov.2015}$. Polygons describing stable dunes, zone *A*, zone *B* and zone *C* have been specified.

Table 4. Zonal statistics on the selected DoD regions.

Zone Classification	Pixel Count	Mean [m]	Std. Dev. [m]	Min [m]	Max [m]
stable dunes	92,662	−0.029	0.046	−0.190	0.269
zone *A*	37,350	0.229	0.055	0.000	0.433
zone *B*	43,610	−0.116	0.076	−0.412	0.126
zone *C*	43,851	0.359	0.131	−0.052	0.716

It is worth noting that the stable dunes' region shows a mean value close to zero, with a discrepancy (−0.029 m) that is comparable to the 0.046 m standard deviation. This fact is in accordance with the very low rate of evolution of stabilized back-dunes, especially if assessed over a two-year period only. Conversely, zones *A* and *C* are characterized by significantly positive mean values (respectively +0.229 m and +0.359 m) and thus they describe an actual ongoing deposition caused by the aeolian transport. Again, this fact confirms the high dynamics of the dune system seaward. Min/max values point out that few outliers are still present in the stable dunes' region despite the performed data filtering and that the polygons of the other zones not being perfectly homogeneous. Similarly, the negative mean value computed in zone *B* also shows a real interdunal depression progradation with a significant lowering of the beach in those intermediate areas.

From a geomorphological point of view, the definition of control regions such as the ones selected above may be useful for the purpose of monitoring the progradation of the dune system over time, as well as to assess the formation of new embryo dunes near the shoreline.

4. Discussion

DEMs represent one of best procedures to detect complex system's morphology, both in terms of high spatial resolution and accuracy. Presently, different techniques can collect data to be processed for finally obtaining DEMs. However, many are the aspects to be considered to adopt the most suitable approach, both in terms of expected results and costs.

The need for both high spatial and temporal resolutions to assess the evolution of critical environments subjected to rapid morphological changes, such as complex dune systems in coastal environments, can be certainly achieved with a SfM approach using UAVs as tools for the acquisition of the aerial imageries. This kind of technique represents an excellent compromise between the accuracy and the resolution typical of laser scanning systems with a very low cost, making it possible to repeat surveys also in shorter time intervals. Hence, the SfM approach can be successfully used to reconstruct any three-dimensional geometry. Moreover, it can model both the interior and exterior orientation parameters of the camera, allowing the use of different and even uncalibrated cameras. The BBA can take into account all the non-linearities and enables to achieve the best final accuracies.

The effects of the vegetation on any reconstructed surface can be mitigated by following a procedure based on the application of slope detection algorithms and SOR filters. This allows to minimize the influence of these factors on the subsequent DoD analysis. The use of a NRTK GNSS technique speeds up the survey of the GCPs needed for the georeferencing of dense clouds, DEMs, and orthomosaics, with accuracies comparable with the same precision achievable with the SfM approach. Although particular care should be taken for both on-field operations to prevent data biases, e.g., for the sinking of the pole's tip during the survey, and for the validation stage of the products, the entire process may be applied also from non-specialist geomatic staff, since its workflow is not particularly difficult.

The results of this work confirm both the reliability and the affordability of the SfM approach by UAVs and indicates that it is a reliable tool for the interpretation of the change over time of very small shapes by the geologists.

The commercial availability of both lightweight UAVs, often very easy to operate, and software implementing SfM algorithms will expand the use of these techniques, not only for coastal environments but wherever rapid acquisition is needed, and high spatial resolution must be achieved. Many types of applications can be already found in the scientific and technical literature, from the mapping of coastal areas [25] to the detection of river morphology [26], from the monitoring of melting glaciers dynamics [27] to the assessment of landslides [28].

Further developments of this research may be imprinted to exploit the multispectral data acquired by the Sentera Single sensor shown in Figure 5c and likely using more recent spectral sensors in the future to support the develop even more accurate filtering procedures. In fact, some small residual effects caused by vegetation are still recognizable in some regions of the models, especially approaching to the back-dunes. The use of an on-board NRTK may also certainly speed up the survey, avoiding the deploying of a high number of GCPs, while small LiDAR on-board instrumentations may also represent an interesting alternative technique. The application of Digital Image Correlation (DIC) techniques to the imagery datasets to quantitatively trace dune progradation may also represent an interesting technique for further geomorphological quantification of the progression phenomenon. Indeed, the DIC has been applied successfully for monitoring the migration rates of subarctic dune fields [29] and active landslides [30].

Despite a variety of many other alternative surveying methods, the digital photogrammetry represents, at present, an excellent technique able to achieve both high spatial and temporal resolution at a truly affordable cost.

Author Contributions: Conceptualization, Y.T., C.C. and A.P.; Methodology, Y.T. and A.P.; Software, Y.T. and E.Z.; Validation, Y.T. and A.P.; Formal Analysis, Y.T. and A.P.; Investigation, Y.T. and C.C.; Resources, C.C. and A.P.; Data Curation, Y.T., C.C., E.Z. and A.P.; Writing—Original Draft Preparation, Y.T.; Writing—Review & Editing, Y.T., C.C. and A.P.; Visualization, Y.T.; Supervision, A.P.; Project Administration, C.C. and A.P.; Funding Acquisition, C.C. and A.P.

Funding: This research was funded by the University of Ferrara.

Acknowledgments: The authors are grateful to Paolo Russo and Umberto Simeoni for their precious suggestions during this research. The authors would also like to thank the undergraduate students of the University of Ferrara who participated during both the fieldwork and the data processing stages.

Conflicts of Interest: The authors declare no conflict of interest. The founding sponsors had no role in the design of the study; in the collection, analyses, or interpretation of data; in the writing of the manuscript, and in the decision to publish the results.

Abbreviations

The following abbreviations are used in this manuscript:

BBA	Block Bundle Adjustment
CORS	Continuously Operating Reference Stations
DEM	Digital Elevation Model
DIC	Digital Image Correlation
DoD	DEM of Difference
ETRF	European Terrestrial Reference Frame
ETRS	European Terrestrial Reference System
Exif	Exchangeable image file format
GCP	Ground Control Point
GNSS	Global Navigation Satellite System
GPS	Global Positioning System
GSD	Ground Sample Distance
LiDAR	Light Detection And Ranging
NRTK	Network Real Time Kinematic
RDN	Rete Dinamica Nazionale (Dynamic National Network)
RGB	Red Green Blue
RMSE	Root Mean Square Error
SfM	Structure-from-Motion
SOR	Statistical Outlier Removal
TLS	Terrestrial Laser Scanner
UAV	Unmanned Aerial Vehicle

References

1. Short, A.; Hesp, P. Wave, beach and dune interactions in southeastern Australia. *Mar. Geol.* **1982**, *48*, 259–284. [CrossRef]
2. Hesp, P. Foredunes and blowouts: Initiation, geomorphology and dynamics. *Geomorphology* **2002**, *48*, 245–268. [CrossRef]
3. Davidson-Arnott, R. *Introduction to Coastal Processes and Geomorphology*; Cambridge University Press: Cambridge, UK, 2010.
4. Hesp, P.A.; Walker, I.J.; Chapman, C.; Davidson-Arnott, R.; Bauer, B.O. Aeolian dynamics over a coastal foredune, Prince Edward Island, Canada. *Earth Surf. Process. Landf.* **2013**, *38*, 1566–1575. [CrossRef]
5. Alesheikh, A.A.; Ghorbanali, A.; Nouri, N. Coastline change detection using remote sensing. *Int. J. Environ. Sci. Technol.* **2007**, *4*, 61–66. [CrossRef]
6. Lynch, K.; Jackson, D.W.; Cooper, J.A.G. Foredune accretion under offshore winds. *Geomorphology* **2009**, *105*, 139–146. [CrossRef]
7. Westoby, M.; Brasington, J.; Glasser, N.; Hambrey, M.; Reynolds, J. 'Structure-from-Motion' photogrammetry: A low-cost, effective tool for geoscience applications. *Geomorphology* **2012**, *179*, 300–314. [CrossRef]
8. Cramer, M.; Stallmann, D. System Calibration for Direct Georeferencing. *Int. Arch. Photogramm. Remote Sens. Spat. Inf. Sci.* **2002**, *34*, 79–84.

Article

Evaluating Water Level Changes at Different Tidal Phases Using UAV Photogrammetry and GNSS Vertical Data

Norhafizi Mohamad [1], Mohd Faisal Abdul Khanan [1,*], Anuar Ahmad [1], Ami Hassan Md Din [1,2] and Himan Shahabi [3]

1 Department of Geoinformation, Faculty of Built Environment and Surveying, Universiti Teknologi Malaysia (UTM), Johor Bahru 81310, Malaysia
2 Geomatics Innovation Research Group, Faculty of Built Environment and Surveying, Universiti Teknologi Malaysia (UTM), Johor Bahru 81310, Malaysia
3 Department of Geomorphology, Faculty of Natural Resources, University of Kurdistan, Sanandaj 66177-15175, Iran
* Correspondence: mdfaisal@utm.my; Tel.: +607-553-0858

Received: 2 July 2019; Accepted: 21 August 2019; Published: 31 August 2019

Abstract: Evaluating water level changes at intertidal zones is complicated because of dynamic tidal inundation. However, water level changes during different tidal phases could be evaluated using a digital surface model (DSM) captured by unmanned aerial vehicle (UAV) with higher vertical accuracy provided by a Global Navigation Satellite System (GNSS). Image acquisition using a multirotor UAV and vertical data collection from GNSS survey were conducted at Kilim River, Langkawi Island, Kedah, Malaysia during two different tidal phases, at high and low tides. Using the Structure from Motion (SFM) algorithm, a DSM and orthomosaics were produced as the main sources of data analysis. GNSS provided horizontal and vertical geo-referencing for both the DSM and orthomosaics during post-processing after field observation at the study area. The DSM vertical accuracy against the tidal data from a tide gauge was about 12.6 cm (0.126 m) for high tide and 34.5 cm (0.345 m) for low tide. Hence, the vertical accuracy of the DSM height is still within a tolerance of ±0.5 m (with GNSS positioning data). These results open new opportunities to explore more validation methods for water level changes using various aerial platforms besides Light Detection and Ranging (LiDAR) and tidal data in the future.

Keywords: water level changes; UAV photogrammetry; tidal phase; GNSS; Kilim River

---

1. Introduction

Exposure to tidal influence causes some rivers to have similar characteristics to coastal zones that is, they experience tidal inundations every 24 hours. One method used to expand our knowledge about tidal inundation in tidal rivers is field surveys using tide gauge instruments [1–5]. Tidal inundation evolves dynamically according to the alignment of the sun and moon, the pattern of tides in the deep ocean, and the shape of the coastline and near-shore bathymetry [6,7]. Hence, multi-source data with different epochs are required, and many solutions have been developed over the last decade. Measuring the water level using satellite images was discussed by [8–11] and was used to analyze water level changes at different tidal phases [12–16]. Instead of using Structure from Motion (SFM), digital surface models (DSMs) could be produced from high-resolution satellite images [17–23]. The other method to produce DSMs is through Light Detection and Ranging (LiDAR) and Terrestrial Laser Scanning (TLS), which allow the generation of accurate DSMs with comparable spatial resolution [24–30]. However, the availability of satellite images is limited because of uncertain weather, while the high cost of LiDAR and TLS acquisition limits the number of field measurements.

To measure the water level at different tidal phases, an unmanned aerial vehicle (UAV) combined with a Global Navigation Satellite System (GNSS) appears to be an efficient solution. Several studies showed the performance of this technique in tidal and intertidal areas; the vertical accuracy of the DSM was about ±10 cm [31–36]. Although a UAV is affected by weather and meteorological conditions such as rain and strong wind, it is still able to collect images that allow the production of a three-dimensional (3D) point cloud and DSM using the SFM process. Bad weather and meteorological conditions could be avoided by planning data acquisition on a fine day and in good weather conditions by referring to the weather forecast. The accuracy of the DSM was already proved to be similar to that with LiDAR data [37]. In [38], the authors summarized several advantages of UAVs: (1) a high level of automation of photographic survey; (2) very low operating cost; (3) high repeatability of the survey; and (4) the possibility to get aerial photography with centimetric resolution [39]. In addition, UAVs are also used for a wide range of hydrology and hydraulic applications such as fluvial monitoring, erosion detection, river bathymetry, and geomorphology using photogrammetric techniques [40–44].

In this study, we demonstrate the evaluation of water level changes at different tidal phases using a DSM from UAV devices supported by high-accuracy positioning using a GNSS receiver. The DSM from UAV data provides water level measurement with higher accuracy as measured by its similarity to the water level from tide gauge measurement. Low-orbital flight with a GNSS positioning system is the best combination to measure water level changes in a small river and an inaccessible environment, such as in a mangrove forest.

2. Study Area

The selected study area was at Kilim River, Langkawi Island, Kedah, Malaysia (Figure 1). This study area received recognition as the UNESCO Kilim Karst Geoforest Park (KKGP) in 2007. The Kilim River is located at 6°21.518′–6°26.093′ N and 99°51.159–99°51.159′ E, allowing this river to have a semi-diurnal tide, and this river also experiences two high and two low tides per day [45]. Although the Kilim River is approximately 3 kilometers from the coastal area, the tidal effect still exists and affects the river. During the tidal phenomenon, the water level increases as much as 12 hours and 25 minutes apart, and it takes 6 hours and 12.5 minutes to go from low to high tide and vice versa. The situation is similar to the coastal area but different in some aspects such as the narrow width of the channel, high slope of the riverbank, and thick mangrove forest along the river. During high tides, the impact of tidal inundation is obvious and visible because of the sinking of the riverbank, especially in the flat slope area.

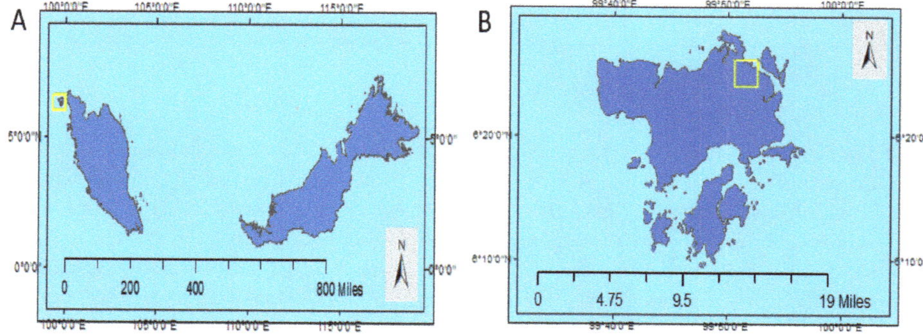

Figure 1. *Cont.*

Figure 1. Location of the study area; (A) Location of the study area on a map of Peninsular Malaysia; (B) Location of the study area on Langkawi Island; (C) Location of the study area on the Kilim River.

3. Materials and Methods

3.1. Specifications of the UAV

The UAV equipment used in this study was a Da-Jiang Innovations (DJI) Phantom 4 Advanced model, developed by SZ DJI Technology Corporation Limited (Shenzhen, China). This model is of medium size, with a net weight of 1388 g and a diagonal wheelbase (propeller size excluded) of only 350 mm (Figure 2A). Figure 2B shows an example of an aerial image captured by a DJI Phantom 4 UAV. The DJI Phantom 4 Advanced is able to fly with a maximum altitude of 6000 m and a flying range of 5000 m. Equipped with a 5870 mAH LiPo 4S battery, the DJI Phantom 4 Advanced has approximately 30 minutes of maximum flight time, and its maximum wind speed resistance ranges from 29 to 38 kph. It is equipped with a remote controller with a 2.4–2.483 GHz operating frequency with a maximum transmission distance of 5 km according to the Federal Communication Commission (FCC) and 3.5 km according to Conformité Européenne (CE). The DJI GO 4 was used with a mobile app that allowed us to plan the flight before the mission and to interact with the UAV during the flight using a 2.4 GHz ISM line view working frequency.

The UAV was mounted with a 1″ Complementary Metal Oxide Semiconductor (CMOS) sensor with 20 megapixels. The DJI Phantom 4 Advanced has focal length 84° 8.8 mm/24 mm (35 mm format equivalent), an f/2.8-11 aperture with 1 m to ∞ (auto focus) shooting range, and shutter speed around 8-1/8000 s (electronic) and 8-1/2000 s (mechanical). An FC 6310 was attached to this UAV with focal length 8.8 mm, pixel size 2.61 × 2.61 µm, and resolution 4864 × 3648. The photography modes comprise single shot, burst shooting: 3/5/7 frames, auto exposure bracketing system (AEB): 3/5 bracketed frames at 0.7, exposure bias, time lapse, and High Dynamic Ranges (HDR) [46]. This model has ±0.1 m (with vision positioning) or ±0.5 m (with GPS positioning) vertical hover accuracy, while the horizontal accuracy is ±0.3 m (with vision positioning) or ±1.5 m (with GNSS positioning) [46].

Figure 2. (**A**) The DJI Phantom 4 model used in this study; (**B**) An aerial view of study area captured by unmanned aerial vehicle (UAV).

3.2. Data Collection

3.2.1. UAV Image Acquisition

Two epochs of data collection were executed to identify the water level at different tidal phases of the Kilim River. Two flights were required to cover the whole study area for each epoch. Epoch 1 (20 December 2017) was collected during high tide, and Epoch 2 (20 December 2017) was collected during low tide. The average flying height for Epoch 1 was 184 m, while that for Epoch 2 was 228 m above ground level, which yielded images with a spatial resolution of 4.7 cm at Epoch 1 and 5.6 cm at Epoch 2. The flight plan was prepared using DJI GO 4 software, and the mission area was saved for each epoch. In total, 116 images were captured at Epoch 1 and 252 images were captured at Epoch 2 to cover the entire study area of about 0.688–0.689 km^2, shown in Figure 3A.

The meteorological conditions and the tidal level were regular for both epochs. For the meteorological conditions, there was no strong wind or rain affecting the flight parameters (yaw, pitch, roll). However, the tidal range between both epochs was different since the tide was within a transition phase from high to lower tide conditions. At Epoch 1, the river was at low tide conditions, while at Epoch 2, the river was at high tide conditions (Table 1).

Table 1. Tidal conditions with respect to marine chart data for both epochs (20 December 2017).

Epoch	Period	Hour	Tidal Reading per Hour (m)	Tidal Range (m)	Average Tidal Level (m)
1	10.56 a.m.–12.29 p.m.	1000 1100 1200 1300	1.499 1.301 1.337 1.593	0.292	1.433
2	13.58 p.m.–16.16 p.m.	1300 1400 1500 1600	1.593 2.056 2.624 3.136	1.872	2.575

3.2.2. GNSS Surveys

GNSS data are crucial to the image geo-referencing process and provide UAV products with high horizontal and vertical accuracy. Hence, ground control points (GCPs) were required for this study. Artificial targets (each a rubber mat with a highlighted "X" mark) were laid on the ground during the image acquisition process and were used later during image geo-referencing, as shown in Figure 3B. Eight GCPs were used since the study area was small, and two Topcon Trimble GR-5 models (Trimble Inc, Sunnyvale, CA, USA) were used as a GNSS receiver. The accuracy of the Topcon GR-5 model 3.0 mm + 0.5 ppm (horizontal) and 5.0 mm + 0.5 ppm (vertical), while that for Real-Time Kinematics (RTK) is 5 mm + 0.5 ppm (horizontal) and 10 mm + 0.8 ppm (vertical) [47]. Observation using a static technique was used within one hour for each station. The GPS observations used GDM 2000 as a local state coordinate system in the Kedah and Perak regions, and all stations were converted into the latitude and longitude format. As shown in Table 2, GPS stations were successfully measured at each location. The altitude of the ground surface is referred to as the ellipsoidal height and not the height above mean sea level (MSL).

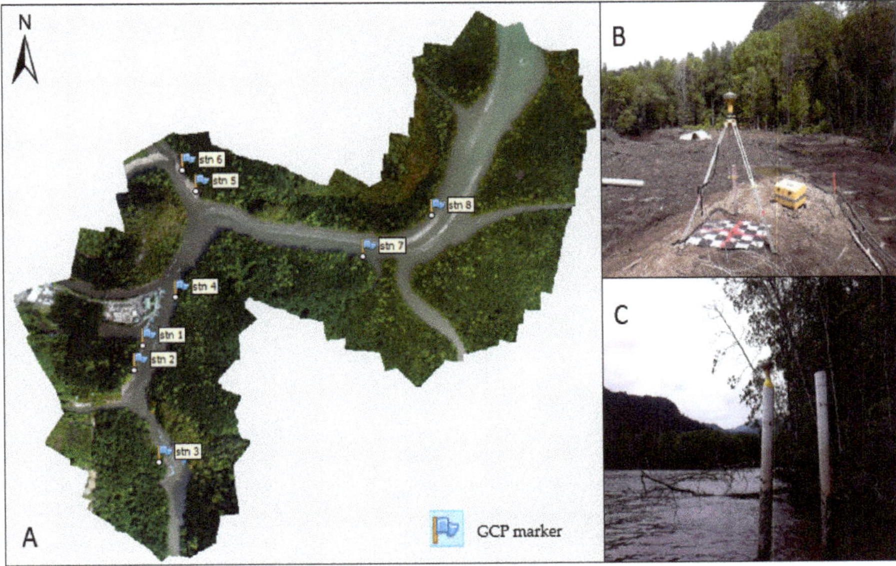

Figure 3. (**A**) Ground Control Points (GCP) distributions at the study area; (**B**) GCP 6 on the ground surface; (**C**) GCP 8 on the ground surface.

Table 2. Details of GCPs at the study area.

GCP	Longitude (°)	Latitude (°)	Ellipsoid Height h (m)	Geoid Height N (m)	Orthometric Height H (m)
1	99.8583050	6.404147	−13.542	−15.476	1.934
2	99.858142	6.403674	−13.506	−15.476	1.97
3	99.858608	6.401791	−12.892	−15.47	2.578
4	99.858971	6.405156	−13.395	−15.476	2.081
5	99.859376	6.40732	−13.495	−15.479	1.984
6	99.85912	6.4077478	−13.693	−15.481	1.788
7	99.862651	6.40594	−13.451	−15.464	2.013
8	99.86401	6.406795	−13.344	−15.46	2.116

To identify the height above MSL, three heights, which comprise the ellipsoidal, geoid, and orthometric heights, should be identified. The ellipsoidal height is the height derived from GNSS equipment, while the orthometric height is known as the height above mean sea level (Figure 4). The geoid model refers to gravity data which are collected through ground, airborne, or space gravity survey equipment. Once the geoid height is known, the orthometric height (H) can be measured based on the MyGeoid model, i.e., the Malaysian geoid model (Equation (1)). The geoid determination of Malaysia is based on gravimetry (airborne, surface, and satellite altimetry), which is located downward relative to the surface of the topography, after removal of a spherical harmonic reference field expansion [48,49]. The orthometric height is simplified as the difference between the geoid height and ellipsoidal height, as shown in Equation (1):

$$H = h - N, \tag{1}$$

where

H = orthometric height;
h = ellipsoidal height;
N = geoid height.

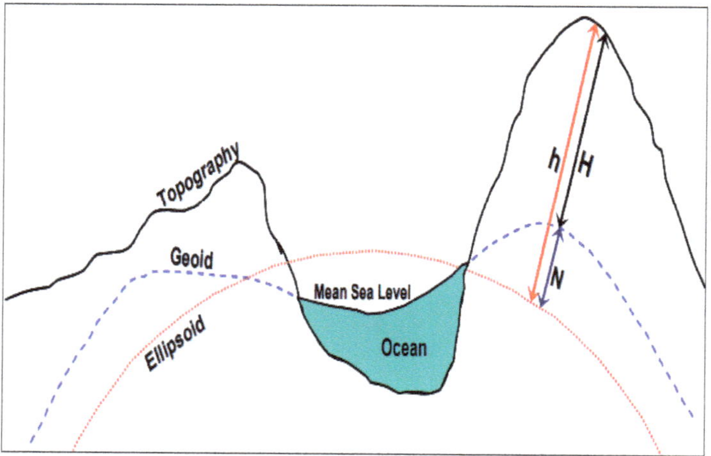

Figure 4. Relationship between the orthometric, ellipsoidal, and geoid heights [48].

3.3. Image Processing Using the SFM Algorithm

Measuring the water level at different tidal phases requires a riverbank 3D model. To generate a 3D model, running the images through a photogrammetric process called SFM is required. SFM reconstructs a relief from several stereoscopic images of the same object and reconstructs a 3D scene geometry from a set of images of a static scene by matching features on multiple images [50]. The SFM algorithm based on multi-views of the scene and the redundancy of the information allow the success of this process [51].

The SFM algorithm is available in several software products to generate DSMs and orthomosaics. Agisoft Photoscan®Professional Edition software (version 1.4.3, Agisoft LLC, St. Petersburg, Russia) was chosen as the processing software in this study. The workflow comprises several processes, including image alignment, camera calibration, camera optimization, point cloud and dense point cloud building, followed by the last step, which is mesh and model texture (Figure 5).

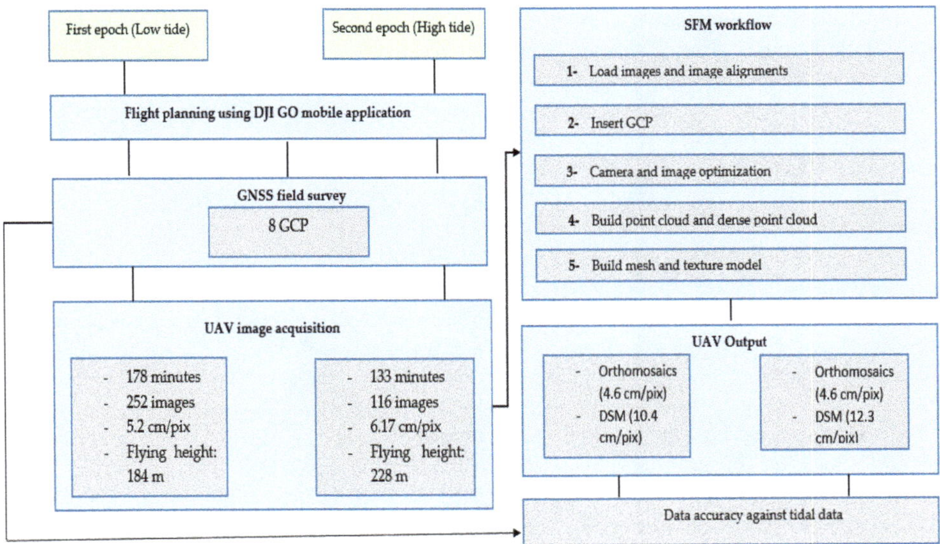

Figure 5. General overview of methods from flight planning to accuracy assessment of the digital surface model (DSM) with tidal data.

Generation of the DSM and orthomosaic image was started by aligning all aerial photos, followed by building a mesh (Figure 6).

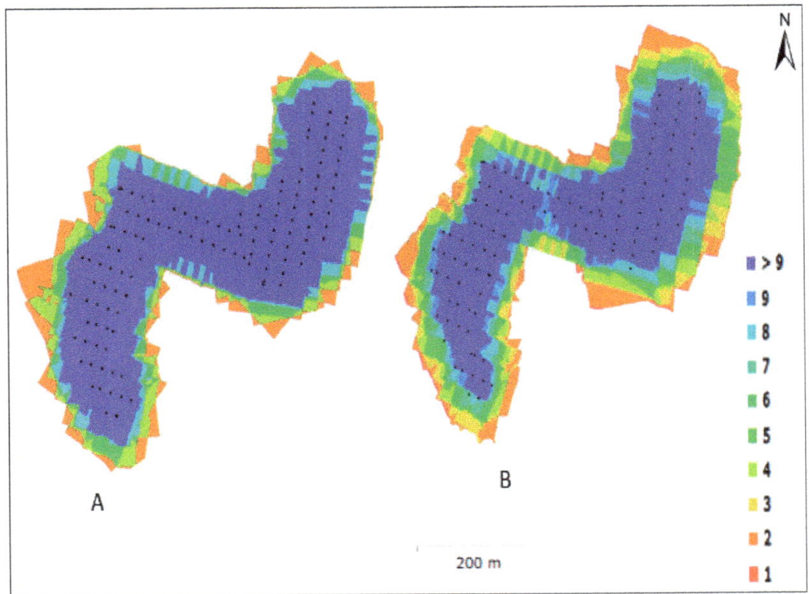

Figure 6. The camera locations and image overlap: (**A**) Epoch 1; (**B**) Epoch 2.

Later, the eight input marker coordinates containing GCP coordinates from GPS data were inserted manually (Table 2), and for altitude, the orthometric height was used instead of the ellipsoid or geoid height. Then, the process continued with optimizing the camera alignment so as to achieve higher accuracy in calculating the camera exterior orientation parameters and internal parameters and to correct the distortion. Subsequently, the aligned aerial photos were combined into a dense point cloud using the "build dense cloud" command. Afterward, the dense point clouds were generated using the "build mesh" command followed by "build texture" (with an optional polygonal model produced as a result). The final step in the Agisoft Photoscan software workflow included the "build DSM" process and the "build orthomosaic" command. The DSM and orthomosaic image were exported to an image file format such as JPEG, TIFF, or PNG for subsequent data processing.

3.4. DSM Generation

The dense point cloud was constructed at high quality and to an aggressive mode depth using Agisoft Photoscan software. These settings were time consuming since there were numerous tie points, especially in Epoch 2. Subsequently, the aligned aerial photos were combined into a dense point cloud using the "build dense cloud" command. Afterward, the dense point cloud was generated using the "build mesh" command followed by "build texture" (with an optional polygonal model produced as a result).

The UAV data product comprises DSM and orthomosaic data types (Figure 7). Figure 7A displays the DSM of the Kilim River during high tide (Epoch 1), while Figure 7B displays the Kilim River during low tide (Epoch 2). Both the DSMs in Figure 7A,B show that the terrain condition ranges from 100 m to 90 m, specifically referring to the red area as hill and dark blue as the river surface. One focus area was selected for further analysis regarding water level changes during high and low tide. Two cross section lines were established in Figure 7C,D to mark the strategic area used to determine the water level using the DSM orthometric height. Both cross section lines were compared with tidal data for water level verification.

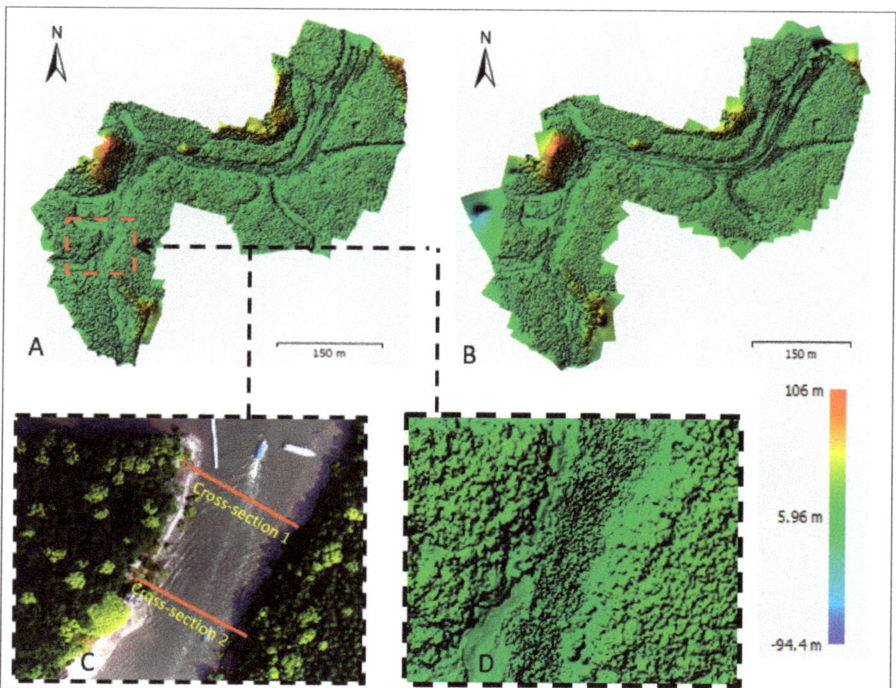

Figure 7. The DSM of the Kilim River; (**A**) At high tide (Epoch 1); (**B**) At low tide (Epoch 2); (**C**) Cross sections 1 and 2 in the orthomosaic image; (**D**) DSM at the area of cross sections 1 and 2.

4. Results and Discussion

The following section presents the output of image acquisition by UAV comprising an image geo-referencing assessment of the UAV photogrammetry and the vertical accuracy of the DSM. The vertical accuracy of the DSM was compared with the water level between high and low tides, and both tide levels were validated with tidal data. The last part of the discussion is about the relevance of UAV–GNSS measurement compared to LiDAR and Satellite Altimetry.

4.1. UAV Output

4.1.1. Results of SFM Image Processing

The results of SFM image processing include image alignments, generation of a point cloud, and dense cloud creation, as well as mesh and texture establishment. Figure 8 shows the results of each photogrammetric process at Epochs 1 and 2.

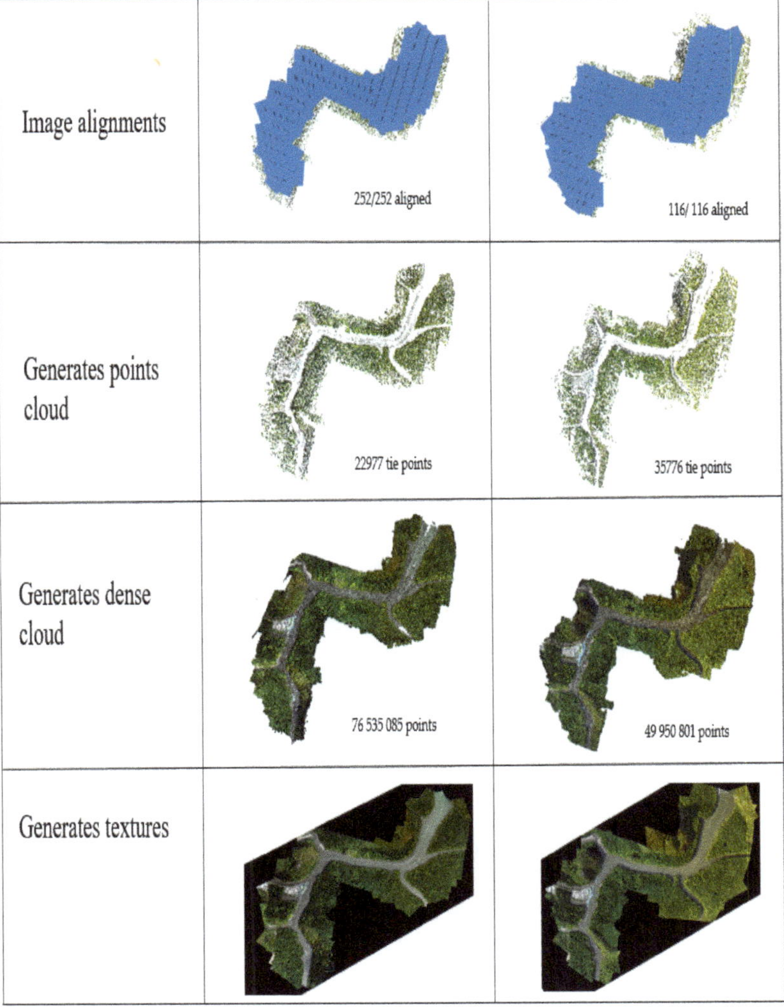

Figure 8. Image photogrammetric results at different stages for Epoch 1 (**left**) and Epoch 2 (**right**).

4.1.2. Image Geo-Referencing Assessment

Table 3 displays the assessment of camera parameters for Epochs 1 and 2 based on focal length (F), principal point coordinates (Cx and Cy), radial distortion polynomial coefficients (K1, K2, and k3), and tangential distortion coefficients (P1, P2, P3, and P4). Both epochs show different values. Epoch 1 shows lower values of all parameters compared to Epoch 2, which shows significant changes.

Table 3. Camera parameters for both epochs of flight.

Camera Parameters	Epoch 1	Epoch 2
F	3183.75	4002.38
Cx	−2.86094	−27.3149
Cy	25.4618	10.0872
B1	−37.2172	121.949
B2	−1.73538	82.3614
K1	0.00250922	0.0102225
K2	−0.0126191	−0.0258016
K3	0.00931268	0.037059
P1	-7.4061×10^{-5}	-9.27189×10^{-8}
P2	7.29483×10^{-5}	5.55727×10^{-8}
P3	22.0224	1724.83
P4	−19.0707	228.112

Table 4 contains the results for both epochs. In all cases, over ten thousand tie points were obtained. The table also shows over one million dense cloud points for both epochs. The final value of the root-mean-square error (RMSE) in pixels is shown. In all cases, the RMSE was below 1 pixel. Epoch 2 shows a higher value of RMSE than Epoch 1, as shown in Table 4. After the use of GCPs in image processing, the orientation procedure was completed, and the pixel error is shown in a histogram (Figure 9). Based on Figure 9, the pixel error was in the range 0.00 to 0.2 for Epoch 1, while for Epoch 2, the pixel error ranged from 0.2 to 1.4; this shows that the pixel error for Epoch 2 was higher than that for Epoch 1. Residuals for the GCPs for Epoch 1 were also calculated to be 0.196 (m), while the total error was 0.890 (pixels). For Epoch 2, the residual for GCPs was 0.010 (m), while the total error was 0.069 (pixels). This process also included the camera self-calibration that gave the results in Table 3. The focal length for both epochs was 8.8 mm, while the pixel size for both epoch was 2.61 × 2.61 μm.

Table 4. Assessment of the image geo-referencing error from GCPs for both epochs.

Epoch	GSD (cm)	No. of Photos Used	Tie Points	Dense Cloud Points	Error (pixel)
1	5.2	252	22,977	76,535,085	0.069
2	6.17	116	35,776	49,950,801	0.890

4.2. Vertical Accuracy of the DSM

4.2.1. Comparison of Water Level between High and Low Tide

To compare the water level between high and low tide, the DSM in Figure 7 was used. The line profile shown in Figure 10 visualizes the difference in water level during different tidal phases and also shows the condition of the water surface. The two cross section lines established in Figure 7C,D illustrate the noise or disturbance on the water surface affected by many factors such as the movement of boats or natural factors such as waves. Significant movement on the water surface only appeared in the middle of the line profile, which signified the middle of the river. It was assumed that boat or vessel movement disturbed the water surface, which caused the surface of the water to appear as in Figure 10.

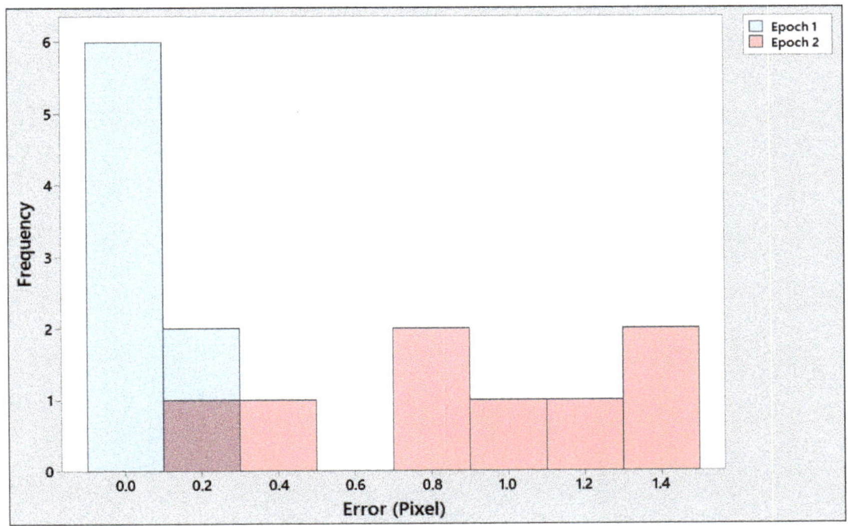

Figure 9. Histogram of the image geo-referencing error; (**A**) Error in pixels for Epoch 1; (**B**) Error in pixels for Epoch 2.

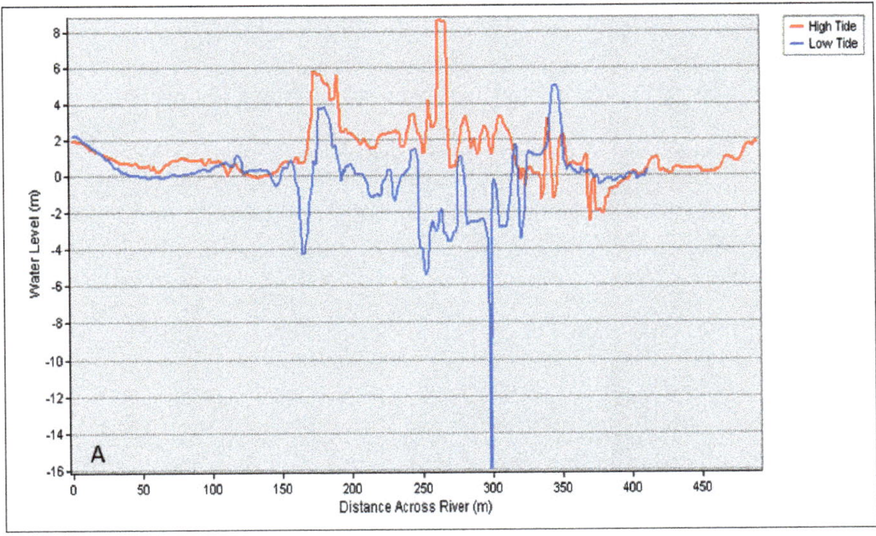

Figure 10. *Cont.*

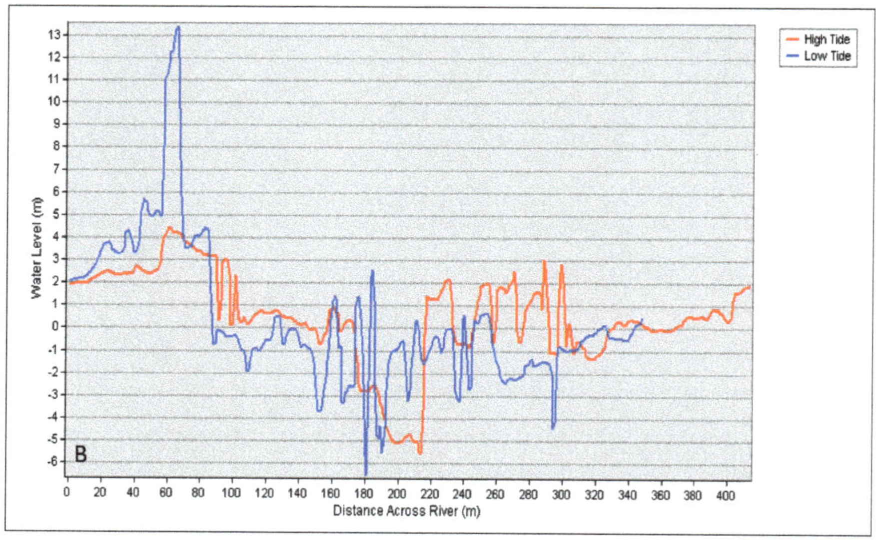

Figure 10. Water level profile (**A**) at cross Section 1; (**B**) at cross section 2.

According to Figure 10A, between 150 and 350 m across the river, the river profile showed noise or disturbance on the water surface because of interference by a vessel or boat. A similar situation happened in Figure 10B, which also displayed disturbance on the water surface. However, a significant difference between high and low tide still appeared at the edge of the line. Figure 11 shows a histogram of water level against frequency which displays the average of the water level at high and low tide. During high tide at cross section 1, the water level ranged from −0.514 to 3.395 m in Figure 11A,B, while at low tide, the water level ranged from −4.437 to 2.2886 m. Meanwhile, at cross section 2 in Figure 11C,D, the water ranged from −1.076 to 2.432 m during high tide, while at low tide, the water level ranged from −3.195 to 1.232 m.

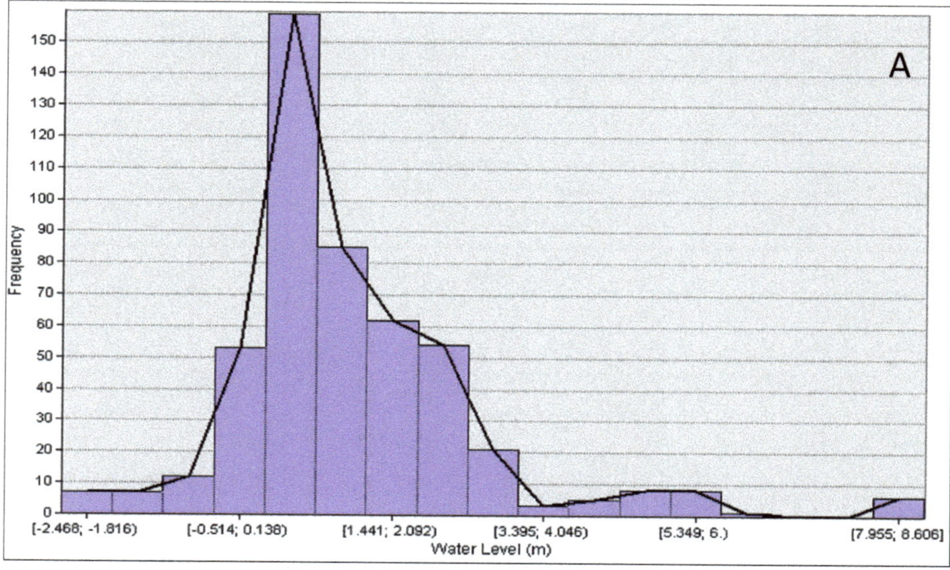

Figure 11. *Cont.*

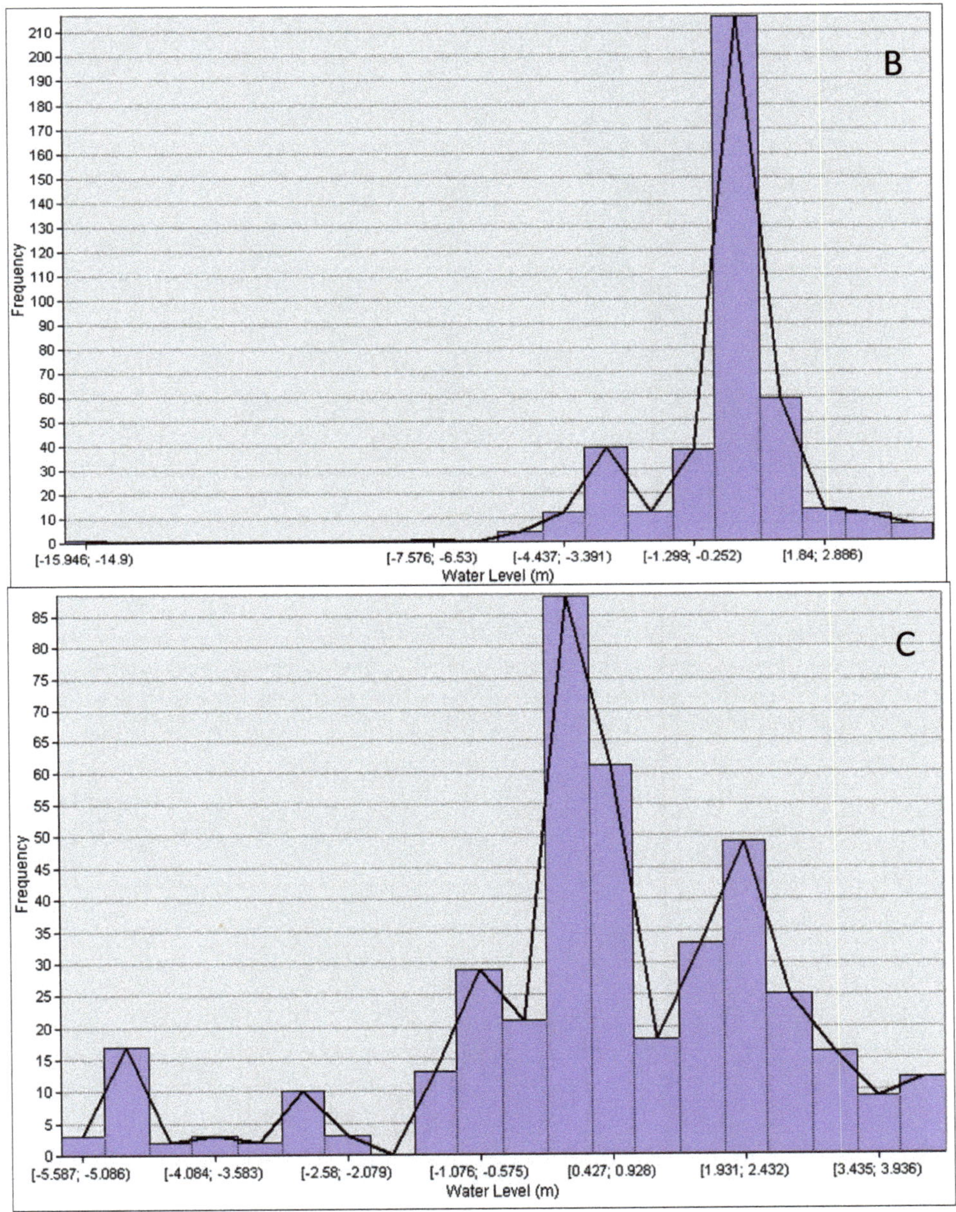

Figure 11. *Cont.*

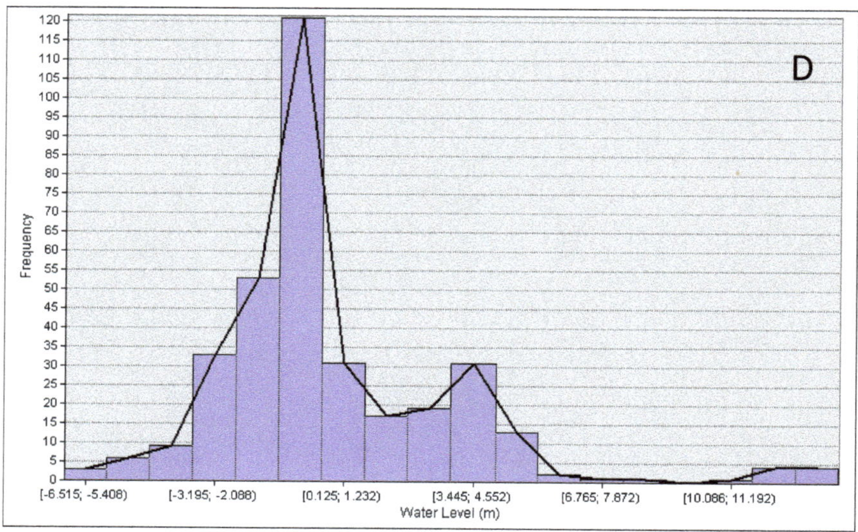

Figure 11. Histogram of water level against frequency; (**A**) Cross section 1 (high tide); (**B**) Cross section 1 (low tide); (**C**) Cross section 2 (high tide); (**D**) Cross section 2 (low tide).

4.2.2. Comparison between DSM Orthometrics and Tidal Data

To compare the water level extracted from the DSM orthometric height with tidal data, cross sections 1 and 2 were also used. The line profile in Figure 12 represents the water level surface during high tide, while tidal data were plotted in a curve fit line to represent the average tidal level.

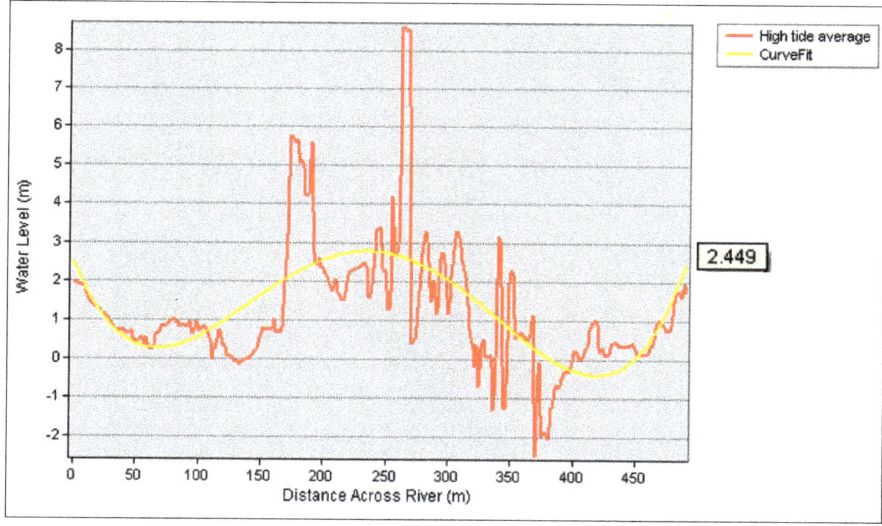

Figure 12. High tide average against tidal value at cross section 1.

Figure 12 represents the water level at cross section 1 from the DSM against tidal values at high tide (from 13.58 p.m. to 16.16 p.m.) which resulted in 2.449 m as the average tidal value. Meanwhile, Figure 13 illustrates the water level at cross section 1 from the DSM against tidal values at low tide (from 10.56 a.m. to 12.29 p.m.) which resulted in 1.088 m as the average tidal value. A curve fit was established to represent the average tidal data level (Table 1) in the tidal range in either high or low tide. The residual of the water level from the DSM orthometric height and tidal data for high tide was 0.126 m (12.6 cm), while that for low tide was 0.345 m (34.5 cm).

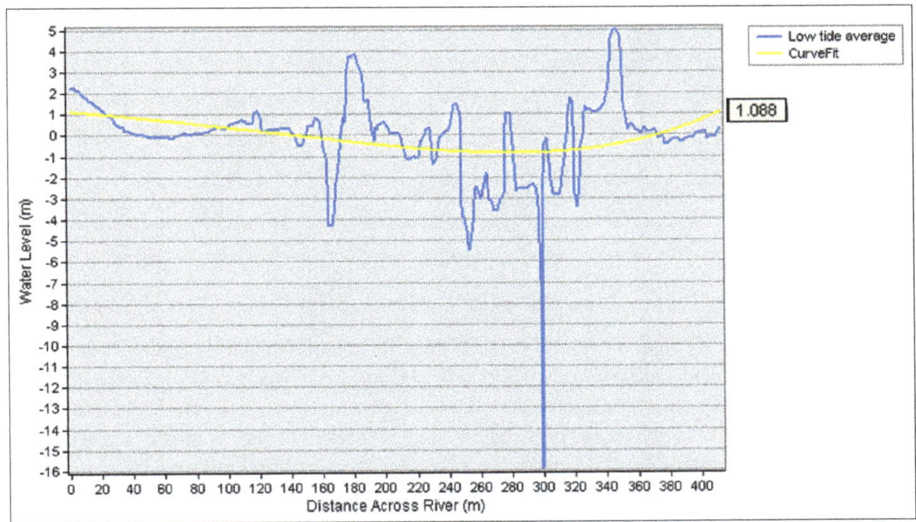

Figure 13. Low tide average against tidal value at cross section 1.

4.2.3. Relevance of UAV–GNSS Measurement Compared to LiDAR and Satellite Altimetry

The vertical accuracy of a UAV with GNSS positioning data shown in the previous section was 0.126 m (12.6 cm) for high tide, while for low tide it was 0.345 m (34.5 cm). According to DJI [32], the vertical accuracy for this model is ±0.5 m (with GNSS positioning), and the results show acceptable accuracy. This vertical accuracy was attained using the DSM orthometric height against tidal data, which were considered the reference data. This combination with GNSS positioning data provides UAV photogrammetry with high accuracy for vertical data, almost comparable with LiDAR and TLS.

In terms of data acquisition, UAV and GNSS combined measurement is flexible since the possibility to encounter bad weather can be avoided. During flight planning, the pilot is able to choose the time for data collection by considering weather problems such as rain intensity, wind, and other meteorological problems that could affect the flights. Data acquisition using a satellite platform has limitations since the user fully relies on satellite conditions to get good data, and the user cannot control the data according to the needs of their study. If the data do not satisfy the user, new data must be collected until the needs of their research are met. For LiDAR data, the cost per flight is expensive, especially when using rented instrumentation. The flight should be planned wisely to avoid any error because the cost for re-collection is expensive.

The Kilim River is surrounded by mangrove forest and experiences tidal phenomena, making it difficult to access for field measurement of water level changes. Using LiDAR is expensive for a 0.689 km^2 area as it would require more than two flights and the setup is very complicated. Its cheapness and similar accuracy when compared with LiDAR and TLS encourage the usage of UAV–GNSS for the measurement of water level changes at Kilim River. The SFM algorithm in Agisoft Photoscan software allows the generation of high-accuracy DSMs and orthomosaics which are able to identify water level changes at different tidal phases.

5. Conclusions

This study showed the potential of UAV photogrammetry and GNSS vertical data for the generation of DSMs to identify water level changes at different tidal phases. This method allowed for water level measurements over an intertidal area of 0.689 km^2 along the Kilim River with dense spatial resolution (5.2 cm at Epoch 1 and 6.17 cm at Epoch 2). On 20 December 2017, two epochs of image acquisition were acquired at the study area during low and high tide. In this study, eight GCPs were used for geo-referencing with total errors of 0.069 (pixels) for Epoch 1 and 0.890 (pixels) for Epoch 2. The SFM algorithm, implemented through software, was employed to generate a DSM and orthomosaics; this included several processes such as image alignment, camera optimization, building a point cloud and dense cloud, and building a mesh and texture model.

Using the DSM, cross-sectional lines were used to extract the profiles of water level surfaces for comparison between high and low tide, as well as for comparison with tidal data. Cross section lines were established across the river at a strategic location. During high tide at cross section 1, the water level ranged from −0.514 to 3.395 m, while at low tide, the water level ranged from −4.437 to 2.2886 m. Meanwhile, at cross section 2, the water ranged from −1.076 to 3.936 m during high tide, while at low tide, the water level ranged from −3.195 to 1.232 m. The results coincide with the conditions of the water level, which increases during high tide and decreases during low tide with such particular values. By comparison between the DSM and tidal data, the residual of the water level was 0.126 m (12.6 cm) for high tide, while for low tide, it was 0.345 m (34.5 cm). This value verifies that the UAV vertical accuracy is within ±0.50 m (50 cm).

In conclusion, the combination of UAV and GNSS vertical data is vital to identifying water level changes during different tidal phases at the Kilim River. Each type of data plays its role in identifying the water level in the 3D profile with high vertical accuracy supported by GNSS data. In addition, the availability of additional data, for example, MyGeoid data from DSMM, is highly appreciated since these data are valuable to calculating the orthometric height. Integrating aerial photogrammetry from a UAV platform with field measurement data could verify the accuracy of both approaches and the relationship between them. The results show only a slight bias for vertical accuracy (within ±0.5 m) between the DSM water level height and tidal data. Perhaps this study could benefit future research in exploring validation methods of water level surface measurements using UAV against tidal data from tide gauges.

Author Contributions: N.M. and M.F.A.K. were responsible for providing the idea of the whole structure of this study. Both persons started the study by identifying the issue and by conducting the literature review, data processing, result interpretation, and writing. M.F.A.K. provided understanding on theoretical aspects while N.M. focused on the aspects of the technical process and data analysis. Meanwhile, A.A., A.H.M.D., and H.S. provided data essential to accomplishing this study.

Funding: This study was supported by Ministry of Higher Education Malaysia (MOHE) with funding under Trans Disciplinary Research Grant Scheme (TRGS) Vote No: RJ130000.7827.4L855. Great appreciation is also due to the people who were directly or indirectly involved by sharing ideas and technical support throughout this study.

Conflicts of Interest: The authors declare no conflict of interest.

References

1. Dias, J.A.; Taborda, R. Tidal gauge data in deducing secular trends of relative sea level and crustal movements in Portugal. *J. Coast. Res.* **1992**, 655–659.
2. Mourre, B.; De Mey, P.; Ménard, Y.; Lyard, F.; Le Provost, C. Relative performance of future altimeter systems and tide gauges in constraining a model of North Sea high-frequency barotropic dynamics. *Ocean Dyn.* **2006**, *56*, 473–486. [CrossRef]
3. Moftakhari, H.R.; Jay, D.A.; Talke, S.A. Estimating river discharge using multiple-tide gauges distributed along a channel. *J. Geophys. Res. Ocean.* **2016**, *121*, 2078–2097. [CrossRef]
4. Withnell, A.J. Relative Sea-Level Variations Revealed by Tide-Gauge Records of Long Duration. Ph.D. Thesis, Massachusetts Institute of Technology and Woods Hole Oceanographic Institution, Woods Hole, MA, USA, 1990.
5. Hamid, A.I.A.; Din, A.H.M.; Hwang, C.; Khalid, N.F.; Tugi, A.; Omar, K.M. Contemporary sea level rise rates around Malaysia: Altimeter data optimization for assessing coastal impact. *J. Asian Earth Sci.* **2018**, *166*, 247–259. [CrossRef]

6. Consoli, S.; Recupero, D.R.; Zavarella, V. A survey on tidal analysis and forecasting methods for Tsunami detection. *arXiv* **2014**, arXiv:1403.0135.
7. Marfai, M.A.; King, L. Tidal inundation mapping under enhanced land subsidence in Semarang, Central Java Indonesia. *Nat. Hazards* **2008**, *44*, 93–109. [CrossRef]
8. Koblinsky, C.J.; Clarke, R.T.; Brenner, A.C.; Frey, H. Measurement of river level variations with satellite altimetry. *Water Resour. Res.* **1993**, *29*, 1839–1848. [CrossRef]
9. Birkett, C.M. Contribution of the TOPEX NASA radar altimeter to the global monitoring of large rivers and wetlands. *Water Resour. Res.* **1998**, *34*, 1223–1239. [CrossRef]
10. Morris, C.S.; Gill, S.K. Variation of Great Lakes water levels derived from Geosat altimetry. *Water Resour. Res.* **1994**, *30*, 1009–1017. [CrossRef]
11. Kouraev, A.V.; Zakharova, E.A.; Samain, O.; Mognard, N.M.; Cazenave, A. Ob'river discharge from TOPEX/Poseidon satellite altimetry (1992–2002). *Remote Sens. Environ.* **2004**, *93*, 238–245. [CrossRef]
12. Blanton, J.O.; Lin, G.; Elston, S.A. Tidal current asymmetry in shallow estuaries and tidal creeks. *Cont. Shelf Res.* **2002**, *22*, 1731–1743. [CrossRef]
13. Bandini, F.; Jakobsen, J.; Olesen, D.; Reyna-Gutierrez, J.A.; Bauer-Gottwein, P. Measuring water level in rivers and lakes from lightweight Unmanned Aerial Vehicles. *J. Hydrol.* **2017**, *548*, 237–250. [CrossRef]
14. Bandini, F.; Butts, M.; Jacobsen, T.V.; Bauer-Gottwein, P. Water level observations from unmanned aerial vehicles for improving estimates of surface water–groundwater interaction. *Hydrol. Process.* **2017**, *31*, 4371–4383. [CrossRef]
15. Domeneghetti, A.; Castellarin, A.; Tarpanelli, A.; Moramarco, T. Investigating the uncertainty of satellite altimetry products for hydrodynamic modelling. *Hydrol. Process.* **2015**, *29*, 4908–4918. [CrossRef]
16. Ridolfi, E.; Manciola, P. Water level measurements from drones: A pilot case study at a dam site. *Water* **2018**, *10*, 297. [CrossRef]
17. Li, Z.; Gruen, A. Automatic DSM generation from linear array imagery data. *Int. Arch. Photogramm. Remote Sens. Spat. Inf. Sci.* **2004**, *35*, 128–133.
18. Eisenbeiss, H.; Baltsavias, E.; Pateraki, M.; Zhang, L. Potential of Ikonos and Quickbird imagery for accurate 3D-Point positioning, orthoimage and DSM generation. *Int. Arch. Photogramm. Remote Sens. Spat. Inf. Sci.* **2004**, *35*, 522–528.
19. Poli, D.; Li, Z.; Gruen, A. Orientation and automated DSM generation from SPOT-5/HRS stereo images. In Proceedings of the 25th ACRS Conference, Chiang Mai, Thailand, 22–26 November 2004; Asian Association on Remote Sensing, 2004.
20. Zhang, L.; Gruen, A. Multi-image matching for DSM generation from IKONOS imagery. *ISPRS J. Photogramm. Remote Sens.* **2006**, *60*, 195–211. [CrossRef]
21. Zhang, L. *Automatic Digital Surface Model (DSM) Generation from Linear Array Images*; ETH Zurich: Zürich, Switzerland, 2005.
22. Baltsavias, E.; Li, Z.; Eisenbeiss, H. DSM generation and interior orientation determination of IKONOS images using a testfield in Switzerland. *Photogramm. Fernerkund. Geoinf.* **2006**, *2006*, 41.
23. Zhang, C.; Fraser, C. Generation of digital surface model from high resolution satellite imagery. *Int. Arch. Photogramm. Remote Sens. Spat. Inf. Sci.* **2008**, *37*, 785–790.
24. Poon, J.; Fraser, C.S.; Chunsun, Z.; Li, Z.; Gruen, A. Quality assessment of digital surface models generated from IKONOS imagery. *Photogramm. Rec.* **2005**, *20*, 162–171. [CrossRef]
25. Priestnall, G.; Jaafar, J.; Duncan, A. Extracting urban features from LiDAR digital surface models. *Comput. Environ. Urban. Syst.* **2000**, *24*, 65–78. [CrossRef]
26. Song, J.H.; Han, S.H.; Yu, K.Y.; Kim, Y.I. Assessing the possibility of land-cover classification using LiDAR intensity data. *Int. Arch. Photogramm. Remote Sens. Spat. Inf. Sci.* **2002**, *34*, 259–262.
27. Smith, M.J.; Asal, F.F.F.; Priestnall, G. The use of photogrammetry and LIDAR for landscape roughness estimation in hydrodynamic studies. In Proceedings of the International Society for Photogrammetry and Remote Sensing XXth Congress, Istanbul, Turkey, 12–23 July 2004; Orhan Altan, M., Ed.; ISPRS Archives: Hamburg, Germany, 2004; Volume XXXV. Part B2.
28. Kociuba, W.; Kubisz, W.; Zagórski, P. Use of terrestrial laser scanning (TLS) for monitoring and modelling of geomorphic processes and phenomena at a small and medium spatial scale in Polar environment (Scott River—Spitsbergen). *Geomorphology* **2014**, *212*, 84–96. [CrossRef]

29. Grun, A.; Zhang, L. Automatic DTM generation from three-line-scanner (TLS) images. *Int. Arch. Photogramm. Remote Sens. Spat. Inf. Sci.* **2002**, *34*, 131–137.
30. Alba, M.; Longoni, L.; Papini, M.; Roncoroni, F.; Scaioni, M. Feasibility and problems of TLS in modeling rock faces for hazard mapping. *ISPRS WG III/3 III/4* **2005**, *3*, 12–14.
31. Persad, R.A.; Armenakis, C. Alignment of point cloud DSMs from TLS and UAV platforms. *Int. Arch. Photogramm. Remote Sens. Spat. Inf. Sci.* **2015**, *40*, 369. [CrossRef]
32. Haarbrink, R.B.; Eisenbeiss, H. Accurate DSM production from unmanned helicopter systems. *Int. Arch. Photogramm. Remote Sens. Spat. Inf. Sci.* **2008**, *37*, 1259–1264.
33. Bhandari, B.; Oli, U.; Pudasaini, U.; Panta, N. Generation of high resolution DSM using UAV images. In Proceedings of the FIG Working Week 2015, Sofia, Bulgaria, 17–21 May 2015.
34. Nagai, M.; Chen, T.; Ahmed, A.; Shibasaki, R. UAV Borne mapping by multi sensor integration. *Int. Arch. Photogramm. Remote Sens. Spat. Inf. Sci.* **2008**, *37*, 1215–1221.
35. Mancini, F.; Dubbini, M.; Gattelli, M.; Stecchi, F.; Fabbri, S.; Gabbianelli, G. Using unmanned aerial vehicles (UAV) for high-resolution reconstruction of topography: The structure from motion approach on coastal environments. *Remote Sens.* **2013**, *5*, 6880–6898. [CrossRef]
36. Watanabe, Y.; Kawahara, Y. UAV photogrammetry for monitoring changes in river topography and vegetation. *Procedia Eng.* **2016**, *154*, 317–325. [CrossRef]
37. Haala, N.; Cramer, M.; Weimer, F.; Trittler, M. Performance test on UAV-based photogrammetric data collection. *Int. Arch. Photogramm. Remote Sens. Spat. Inf. Sci.* **2011**, *38*, 7–12. [CrossRef]
38. Gonçalves, J.A.; Henriques, R. UAV photogrammetry for topographic monitoring of coastal areas. *ISPRS J. Photogramm. Remote Sens.* **2015**, *104*, 101–111. [CrossRef]
39. Long, N.; Millescamps, B.; Guillot, B.; Pouget, F.; Bertin, X. Monitoring the topography of a dynamic tidal inlet using UAV imagery. *Remote Sens.* **2016**, *8*, 387. [CrossRef]
40. Muste, M.; Fujita, I.; Hauet, A. Large-scale particle image velocimetry for measurements in riverine environments. *Water Resour. Res.* **2008**, *44*, 1–14. [CrossRef]
41. Woodget, A.S.; Carbonneau, P.E.; Visser, F.; Maddock, I.P. Quantifying submerged fluvial topography using hyperspatial resolution UAS imagery and structure from motion photogrammetry. *Earth Surf. Process. Landf.* **2015**, *40*, 47–64. [CrossRef]
42. Jodeau, M.; Hauet, A.; Paquier, A.; Le Coz, J.; Dramais, G. Application and evaluation of LS-PIV technique for the monitoring of river surface velocities in high flow conditions. *Flow Meas. Instrum.* **2008**, *19*, 117–127. [CrossRef]
43. Fujita, I.; Muste, M.; Kruger, A. Large-scale particle image velocimetry for flow analysis in hydraulic engineering applications. *J. Hydraul. Res.* **1998**, *36*, 397–414. [CrossRef]
44. Nobi, E.P.; Umamaheswari, R.; Stella, C.; Thangaradjou, T. Land use and land cover assessment along Pondicherry and its surroundings using Indian remote sensing satellite and GIS. *Am. Eurasian J. Sci. Res.* **2009**, *4*, 54–58.
45. Mohamad, N.; Khanan, M.F.A.; Musliman, I.A.; Kadir, W.H.W.; Ahmad, A.; Rahman, M.Z.A.; Jamal, M.H.; Zabidi, M.; Suaib, N.M.; Zain, R.M. Spatio-temporal analysis of river morphological changes and erosion detection using very high resolution satellite image. In *IOP Conference Series: Earth and Environmental Science, Proceedings of the 9th IGRSM International Conference and Exhibition on Geospatial & Remote Sensing (IGRSM 2018), Kuala Lumpur, Malaysia, 24–25 April 2018*; IOP Publishing: Bristol, UK, 2018.
46. Da Jiang Innovations (DJI). DJI Phantom 4. Available online: https://www.dji.com/phantom-4/info (accessed on 26 August 2019).
47. Im, S.B.; Hurlebaus, S.; Kang, Y.J. Summary review of GPS technology for structural health monitoring. *J. Struct. Eng.* **2011**, *139*, 1653–1664. [CrossRef]
48. Department of Surveying and Mapping Malaysia (DSMM). *Malaysian Geoid Model (MyGEOID) Guidelines*; Department of Surveying and Mapping Malaysia (DSMM): Kuala Lumpur, Malaysia, 2005.
49. Ismail, M.K.; Din, A.H.; Uti, M.N.; Omar, A.H. Establishment of New Fitted Geoid Model in Universiti Teknologi Malaysia. In *the International Archives of the Photogrammetry, Remote Sensing and Spatial Information Sciences, Proceedings of the International Conference on Geomatics and Geospatial Technology (GGT 2018), Kuala Lumpur, Malaysia, 3–5 September 2018*; ISPRS Archieves: Hamburg, Germany, 2018; Volume XLII-4/W9.

50. Laksono, D. Open source stack for Structure from Motion 3D reconstruction: A geometric overview. In Proceedings of the 6th International Annual Engineering Seminar, Yogyakarta, Indonesia, 1–3 August 2016; IEEE: Piscataway, NJ, USA, 2016.
51. Bianco, S.; Ciocca, G.; Marelli, D. Evaluating the performance of structure from motion pipelines. *J. Imaging* **2018**, *4*, 98. [CrossRef]

Article

Automatic Change Detection System over Unmanned Aerial Vehicle Video Sequences Based on Convolutional Neural Networks

Víctor García Rubio [1,*], Juan Antonio Rodrigo Ferrán [2], Jose Manuel Menéndez García [2], Nuria Sánchez Almodóvar [1], José María Lalueza Mayordomo [1] and Federico Álvarez [2]

[1] Visiona Ingeniería de Proyectos S.L., 28020 Madrid, Spain; nsanchez@visiona-ip.es (N.S.A.); jlalueza@visiona-ip.es (J.M.L.M.)

[2] Grupo de Aplicación de Telecomunicaciones Visuales, Escuela Técnica Superior de Ingenieros de Telecomunicación, Universidad Politénica de Madrid, 28040 Madrid, Spain; jrf@gatv.ssr.upm.es (J.A.R.F.); jmm@gatv.ssr.upm.es (J.M.M.G.); fag@gatv.ssr.upm.es (F.Á.)

* Correspondence: vgarcia@visiona-ip.es

Received: 31 August 2019; Accepted: 11 October 2019; Published: 16 October 2019

Abstract: In recent years, the use of unmanned aerial vehicles (UAVs) for surveillance tasks has increased considerably. This technology provides a versatile and innovative approach to the field. However, the automation of tasks such as object recognition or change detection usually requires image processing techniques. In this paper we present a system for change detection in video sequences acquired by moving cameras. It is based on the combination of image alignment techniques with a deep learning model based on convolutional neural networks (CNNs). This approach covers two important topics. Firstly, the capability of our system to be adaptable to variations in the UAV flight. In particular, the difference of height between flights, and a slight modification of the camera's position or movement of the UAV because of natural conditions such as the effect of wind. These modifications can be produced by multiple factors, such as weather conditions, security requirements or human errors. Secondly, the precision of our model to detect changes in diverse environments, which has been compared with state-of-the-art methods in change detection. This has been measured using the Change Detection 2014 dataset, which provides a selection of labelled images from different scenarios for training change detection algorithms. We have used images from dynamic background, intermittent object motion and bad weather sections. These sections have been selected to test our algorithm's robustness to changes in the background, as in real flight conditions. Our system provides a precise solution for these scenarios, as the mean F-measure score from the image analysis surpasses 97%, and a significant precision in the intermittent object motion category, where the score is above 99%.

Keywords: change detection; convolutional neural networks; moving camera; image alignment; UAV

---

1. Introduction

The use of change detection algorithms is crucial in high precision surveillance systems. The methods that make use of those algorithms aim to detect the differences between information acquired at the same location, e.g., an image captured in different moments. Unmanned aerial vehicles (UAVs) became a revolution in the surveillance sector due to the lower cost and reduced human workload needed compared to previous systems. In addition, UAV operations can be automatized. This need of automation increases the importance of change detection methods. These methods are based on image sequences analysis, usually acquired by mobile vehicles. Image acquisition from the mentioned vehicles entails a considerable issue for change detection algorithms: The camera movement.

This is the fundamental challenge of the algorithms, as the movement produces a variable background, thus the flight's route will be moderately modified from one flight to another. Furthermore, the weather conditions and the precision of GPS positioning influence the relation between the acquired frames and the location of the UAV. Compared to moving cameras, stationary cameras significantly reduce the complexity of the change detection problem, as the background is common to every image [1].

Moving cameras introduce complexity to the problem because the reference image is continuously changing. As a result, the system needs to detect the background of each image to provide precise detection. Therefore, variable backgrounds introduce a considerable computational load. This generates a complex problem to solve for real-time video surveillance tasks. An interesting approach for change detection with moving cameras is studied in [2]. Our method is based on reconstruction techniques. However, the reconstruction process alone would not be precise enough for our task. This is a consequence of the limited precision of CNNs to generate detailed images. In addition, it conforms an additional process that could slow down the system. Nevertheless, it provides a distinct perspective from the stationary camera algorithm mentioned above. After reviewing the different state-of-the-art approaches such as [2–4], we have observed an increasing tendency to use CNNs on image processing systems for change detection. Furthermore, studied implementations do not consider a moving camera, as in [3,5]. Consequently, a supplementary component has been considered to overcome the problems introduced by a variable background. Therefore, we have developed our own complete approach based on image alignment and CNNs.

In this document, we have reviewed the state-of-the-art implementations for change detection and image feature extraction in Section 2. Section 3 includes a detailed explanation of the process followed to develop our implementation. Section 4 describes the metrics obtained by our algorithm and compares them with the state-of-the-art implementations analysed previously. Finally, in Section 6, the most relevant inferences obtained from the development are explained.

2. Related Work

In this section, a review of traditional change detection is performed in Section 2.1. In Section 2.2, an explanation of convolutional neural networks is presented. In Section 2.3, we provide an analysis of change detection implementations based on convolutional neural networks. Finally, in Section 2.4, different traditional image processing techniques to extract and find common features between images are reviewed.

2.1. Change Detection Using Traditional Image Processing

Traditional image processing methods for change detection are based on pixel intensity variation modelling. This is the prime feature of most background subtraction algorithms. Implementations such as the Gaussian mixture model [6] are exceptionally popular among them. This method uses multiple Gaussian distributions to model the intensity value of each pixel to differentiate between background and foreground elements. However, this method does not perform efficiently in complex environments where movement is permanent. Other relevant approaches are based on kernel density estimation (KDE) [7]. This method also employs probability density functions to model the intensity of background pixels. Based on the mentioned distributions, the foreground pixels are detected. These methods are founded on the intensity regions to estimate background probability. Most of them are unable to model frequent events. They also adapt poorly to dynamic background situations. Modern approaches in terms of change detection are SuBSENSE [8] and ViBe [9]. SuBSENSE relies on spatio-temporal binary features as well as colour information to detect changes. In case of ViBe, for each pixel, a set of values taken in the past at the same location is acquired. This set is compared to the current pixel value to determine if it belongs to the background. After that, the model is adapted by choosing randomly which values to substitute from the background model. Both methods are based on elemental features, such as colour analysis. The lack of high-level information brings a decrease in performance in complex environments, which represent our principal target.

2.2. Convolutional Neural Networks

CNNs applied to image analysis usually constrain their input to remain as a three-dimensional element with the following parameters: Width, height and colour channels (depth). The network is conformed by three types of layers: Convolutional, activation and pooling. The task of each layer is:

- Convolutional layers perform a dot product between each region of the input and a filter. These filters, also known as kernels, define the size of the mentioned regions. With this operation, the layer is capable of obtaining the most significant features of the input. The number of filters typically varies from each convolutional layer, in order to obtain higher or lower level features. The resultant structure depends on the padding employed on the convolution.
- Activation layers execute a predefined function to their inputs. The aim of these layers is to characterize the features obtained from the previous layers to obtain the most relevant ones. This process is performed using diverse functions, which may vary among the systems.
- Pooling layers down-sample their input, obtaining the most significant values from each region. Most CNNs employ a 2×2 kernel in their pooling layers. With this region size, the input is down-sampled by half in both dimensions, width and height.

The concatenation of these structures allows the system to learn different level features and generate a more precise prediction than traditional methods, based on low-level features such as colour. Therefore, the application of convolutional neural networks has supposed a revolution in image processing. A more in-depth explanation of these neural networks can be found in [10].

2.3. CNN-Based Change Detection

As mentioned in the introductory section, researchers have developed methods to detect changes in images based on CNN architectures. Two methods have been studied and analysed to obtain insights about the architectures and accuracy levels that form the state-of-the-art nowadays. In [3], both background and foreground images are sliced in patches with reduced dimensions. The two patches (back and foreground) are concatenated in the depth axis, obtaining a unique structure to feed the CNN as a result. Convolution applied to images stacked depth-wise produces faster and more appropriate results than in any other dimension as detailed in [11]. As the result of introducing the concatenated structure to the CNN-based architecture, this model returns a black and white binary patch, with the detected foreground zones in white and background zones in black. Nevertheless, the restriction of using a moving camera is unsolved as the system is indicated for static acquisition devices. In [12], a variation of the Visual Geometry Group Net (VGGNet) architecture is connected to a deconvolutional neural network. VGGnet is a state-of-the-art implementation developed by VGG (Visual Geometry Group) from the University of Oxford [13]. The resultant architecture is applied to obtain an image of the same dimensions as the input picture. To do so, the number of deconvolution layers is equivalent to the CNN layers in the VGGNet. As a result, the model returns a binary image with identical size to the input image. The stated method does not employ a background image. It employs only the foreground and, for training purposes, the ground-truth images. As a result, we can retrieve two conclusions: First, without a background reference, the system can be employed for moving cameras if the dataset is properly conformed; next, the algorithm has to learn a different background for each scene. Therefore, the system will have to learn on each scene, which slows processing time in real-world applications. From the results of both methods, it may be concluded that CNNs have improved considerably the accuracy of traditional change detection algorithms.

2.4. Image Matching Techniques

On stationary camera situations, acquiring a precise reference image represents an effortless task, while on moving camera scenarios it remains a challenging problem. One of the recent progresses of UAV technology consists on the automation of flights [14]. This system provides human–UAV interaction by allowing the programming of UAV's routes, specifying the coordinates at each point and other parameters, namely speed, height, mode of flight or behaviour, in case of any event possible. However, the mentioned innovation is subject to the real-world conditions that affect the behaviour of UAVs. As explained in Section 1, the image acquisition is assumed to vary from two flights on the same route. It can be caused by multiple reasons, in particular, modifications in UAV's height, position because of GPS precision variations, wind conditions or movement of the camera. Consequently, the reference pictures obtained on one of those flights are not circumscribed to exactly the same area. In this paper we have selected an approach based on ORB (oriented FAST and rotated BRIEF) algorithm [15] to resolve the alignment problem. ORB descriptor is based on BRIEF (binary robust independent elementary features) [16] and FAST (features from accelerated segment test) [17] algorithms. This descriptor extracts the key points of an image by implementing FAST method. It employs a Harris corner measure [18] to discover the optimal points. To obtain the orientation Equation (3), ORB uses a rotation matrix based on the computation of intensity centroid [19], Equation (2) from the moments Equation (1). Moments are obtained using:

$$m_{pq} = \sum_{x,y} x^p y^q I(x,y), \tag{1}$$

where $I(x,y)$ represents pixel intensities, x and y denote the co-ordinates of the pixel and finally, q and p indicate the order of moments. The centroid is obtained from the calculated moments as:

$$C = (\frac{m_{10}}{m_{00}} \frac{m_{01}}{m_{00}}) \tag{2}$$

Finally, a vector from corner's center to the centroid is constructed. The orientation is:

$$\theta = atan2(m_{01}, m_{10}) \tag{3}$$

where atan2 is the arctangent's quadrant-aware version given by Equation (4):

$$atan2(x,y) = \begin{cases} arctan(y \cdot x^{-1}) & x > 0, \\ arctan(y \cdot x^{-1} + \pi) & y \geq 0, x < 0, \\ arctan(y \cdot x^{-1} - \pi) & y < 0, x > 0, \\ \pi/2 & y > 0, x = 0, \\ -\pi/2 & y > 0, x = 0, \\ undefined & x = y = 0. \end{cases} \tag{4}$$

Other descriptors considered during the development have been SIFT (scale-invariant feature transform) [20] and SURF (Speeded-up robust features) [21] algorithms. SIFT algorithm transforms a picture into feature vectors. Each vector is invariant to image translation, scaling, and rotation. After that, it compares each vector from the new image to the ones obtained from the reference picture. Next, it provides a candidate by matching features based on Euclidean distance. SURF algorithm obtains the feature vectors based on the sum of the Haar wavelet response around the points of interest. These points have been detected previously by using an integer approximation of the determinant of the Hessian matrix. However, the ORB descriptor is able to perform two orders of magnitude faster than SIFT and several times faster than SURF. Because of this performance difference, we have selected ORB to process our videos and reduce the computational cost of the system.

3. Methodology

Image alignment techniques employed on the approach are explained in Section 3.1. After that, the architecture of the deep neural network model is explained in detail in Section 3.3. The dataset generation for the process has been described in Section 3.4. Subsequently, the training process is detailed in Section 3.5. Lastly, the post-processing part is described in Section 3.6 to introduce how we have obtained our desired image from the model output. In Figure 1, a block diagram of a system is shown to detail our pipeline.

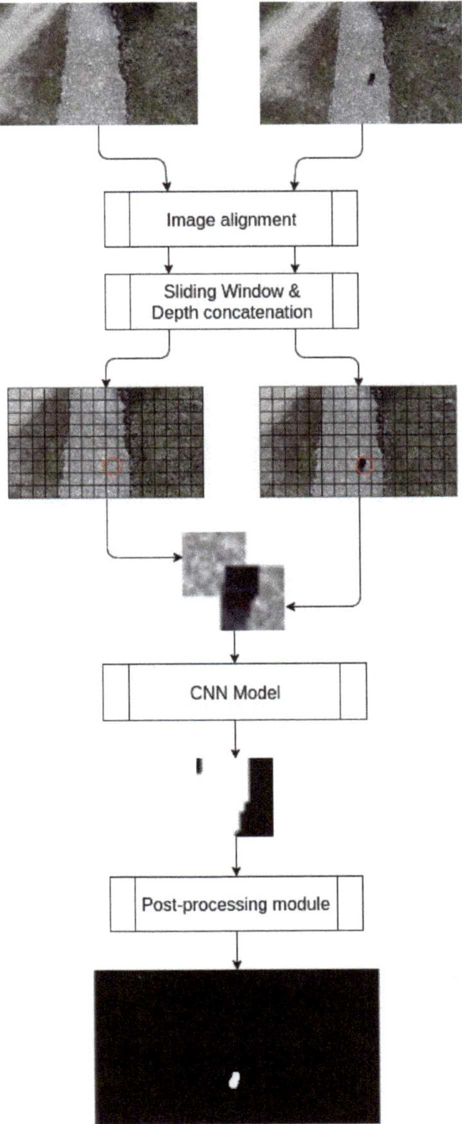

Figure 1. Block diagram of the system. Both reference and foreground images are introduced into our image alignment system. After that, sliding window algorithm is applied. The two resultant patches are concatenated along the depth axis. Ultimately, our CNN model predicts a grayscale image, which is post processed to obtain the final binary patch depicted.

3.1. Image Alignment

As stated on previous sections, our system is applied to moving cameras. Our objective is to compare two video sequences (background and foreground) to detect the changed regions between them. Because of implementation purposes, a reference video from the UAV's route is required, which will be considered as background. More recent videos from the same route are considered as our foreground scenarios. In real-world situations, the new images will not be completely aligned to the reference video. This can be caused by multiple reasons such as lack of GPS precision or weather variations. To solve the problem, an image alignment system has been developed using ORB [15]. The idea is similar to the feature alignment performed in [22]. Both the reference and the new route's images are compared. Feature extraction is performed with ORB algorithm to obtain the most significant zones of each picture. After that, a descriptor matcher algorithm from [23] is created. The descriptor performs an analysis of the obtained values and outputs the relation between them by distance difference. This output is filtered and sorted to obtain the most adjacent descriptive points of both images. Lastly, a geometric transformation is performed to generate a modified version of the acquired image, aligned to the reference. An example of the results obtained by this system is depicted in Figure 2. As can be observed, the resultant image contains a black zone that represents the pixels from the foreground image not included in the reference. To prevent the appearance of false positives because of these black zones, only the mid section of the image is selected automatically to perform the change detection. The alignment system entails an innovation to other implementations such as [3,12], which consider a scenario with a static camera.

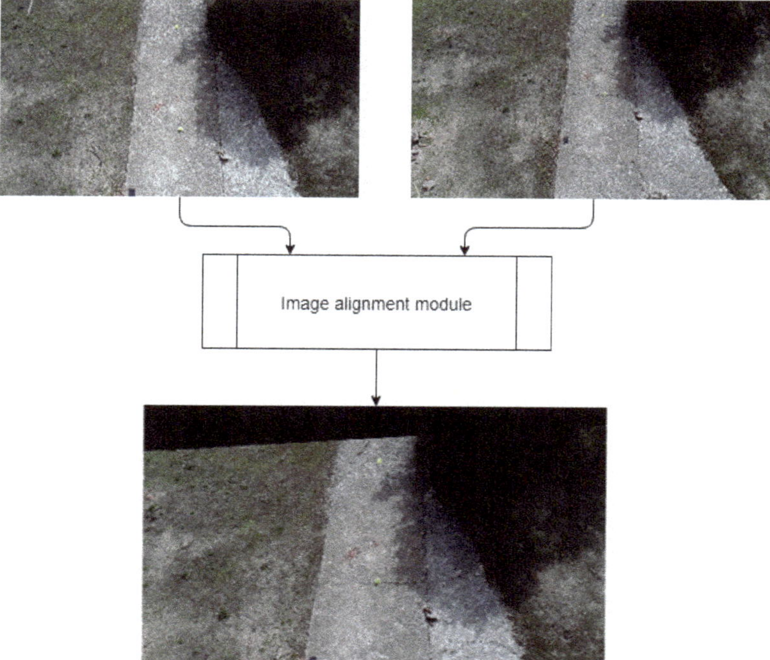

Figure 2. Example of our image alignment module's output with a foreground image with significant difference from the reference.

3.2. Sliding Window

The image provided to our system can vary in size depending primarily on the UAV's camera device or time processing requirements. For instance, the processing requirements of a real-time detection system varies from those of a post-flight analysis. Moreover, deep learning models usually struggle to work with very high-resolution images. Our approach to resolve these problems is described in this subsection. In the first place, the maximum input size is defined as a parameter of the system. If the input images overcome the defined dimensions, they are resized to the specified size. After that, we have to divide the pictures into small regions. This is a consequence our network's requirements to process images in reasonable time. Therefore, we can obtain pixel-level precision results.

To do so, we have employed an algorithm named sliding window. This algorithm iterates over the image's dimensions, retrieving a matrix of a specific size (window) which contains a region of the initial image. The dimensions of the window are identical in width and height to prevent any further complication of the segmentation process. Another parameter of the algorithm is defined as the step. The step of a sliding window algorithm represents the distance, in each dimension, from the starting point of a window in iteration i to the starting point of the region in iteration i + 1. Adaptive to each dimension, the step remains a crucial factor in terms of computational cost, just as the dimension of the window. With these two parameters we can control the overlapping of zones among adjacent windows. If overlapping occurs, the system will benefit from an increment of precision on the analysis. In this case, four predictions are generated for the most part of the image, except the limits of each dimension. As a conclusion, the variation of both parameters provides an extremely efficient tool that can modify the algorithm's performance to adapt it for multiple purposes: real-time segmentation, maximum precision prediction or sample generation. The application of this algorithm in both the reference image and background image is depicted in Figure 3.

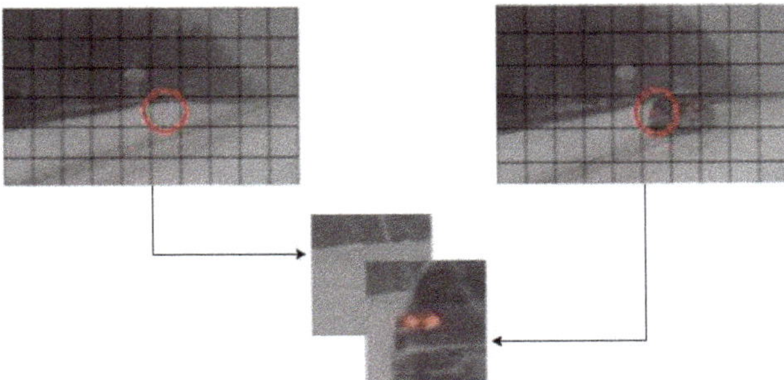

Figure 3. Representation of the sliding window and depth concatenation process on a reference image and a foreground image.

3.3. Deep Neural Network Architecture

The model is based on the concatenation of two input images: The reference background scene and the updated scene image which may contain some changes. Both images are merged in depth dimensions, as it is performed in other state-of-the-art methods such as [3,4,12]. This input form allows the model to learn hierarchical features. As a result, CNNs conform an effective tool to obtain relevant information from images. An exhaustive description of the CNN architecture is described in detail in Tables 1 and 2. Figure 4 illustrates the complete architecture. The deep neural network is composed by four CNNs with increasing number of filters and a kernel size of 3×3 pixels. As our inputs consist of images with reduced dimensions, we employ this kernel size to extract the features as detailed as possible to obtain a precise detection. For the activation layers of the CNN, we have used Rectified Linear Unit (ReLU) activation [24]. ReLU activation applies Equation (5) to its inputs. This function is widely implemented in CNNs as mentioned in Section 2.2 because of its reduced computational cost and the acceleration of the optimization process. Following this layer, we have employed a max-pooling layer with 2×2 kernel size.

$$R(z) = max(0, z) \tag{5}$$

After the last CNN structure, we have vectorized its output neurons. The first one-dimensional layer contains a dropout layer [25] to avoid overfitting [26]. Continuing the structure, we have employed a batch normalization [27], which subtracts the mean of its inputs and divides them by the standard deviation as in Equation (6):

$$\hat{x}_i = \frac{x_i - \mu_B}{\sqrt{\sigma_B^2 + \epsilon}} \quad (6)$$

denoting μ_B as the average value of the batch, σ_B as the standard deviation of the batch, and ϵ a constant added for numerical stability. The dropout layer deactivates part of the neurons from our densely connected layer during the training process. As a result, the model improves its generalization, as it forces the layer to predict the same output using different neurons. The aim of batch normalization is to increase the stability of the model by normalizing the output of the previous layer. As a consequence, the model will be adaptable to new scenarios, which is one of the essential features of the proposed system. As our output consists of pixels from changed regions or pixels from unchanged regions, we are in a binary classification problem. Therefore, the final output layer uses a sigmoid activation function as depicted in Equation (7):

$$\sigma(z) = \frac{1}{1 + e^{-z}} \quad (7)$$

The resulting normalized vector is the initial prediction of our system. It conforms the input of the post-processing methods described in Section 3.6. From this vector we construct a one-channel image that represents the changes between the reference image and the updated image.

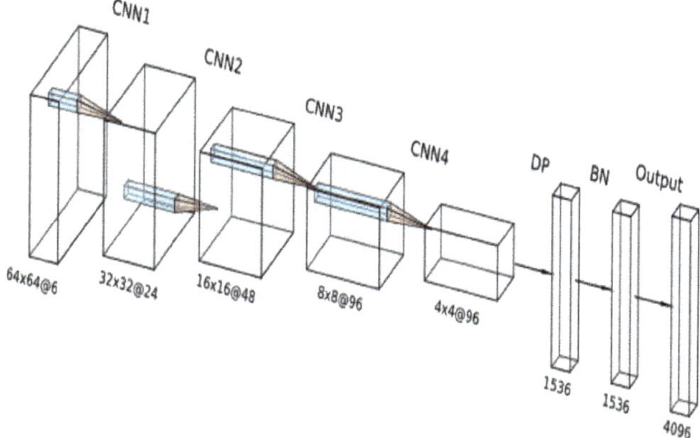

Figure 4. Diagram of the deep neural network architecture used in our system, containing layer types and dimensions.

Table 1. Layer description of the convolutional neural networks (CNNs) which conform the first part our model.

Name	Input Size	Kernel	Stride	Output Size
conv2d1	64 × 64 × 6	3 × 3	1	64 × 64 × 24
pool1	64 × 64 × 24	-	2	32 × 32 × 24
conv2d2	32 × 32 × 24	3 × 3	1	32 × 32 × 48
pool2	32 × 32 × 48	-	2	16 × 16 × 48
conv2d3	16 × 16 × 48	3 × 3	1	16 × 16 × 96
pool3	16 × 16 × 96	-	2	8 × 8 × 96
conv2d4	8 × 8 × 96	3 × 3	1	8 × 8 × 96
pool4	8 × 8 × 96	-	2	4 × 4 × 96

Table 2. Layer description of the fully connected layers which define the final part of our model.

Name	Input Size	Function	Output Size
flatten	$4 \times 4 \times 96$	-	1536
bn	1536	Batch Normalization	1536
dp	1536	Dropout	1536
final	1536	Sigmoid	4096

3.4. Dataset

To train our model, we have selected images from CD2014 dataset [1]. This dataset contains images categorized in: "Baseline", "Dynamic Background", "Camera Jitter", "Intermittent Object Motion", "Shadows", "Thermal", "Bad Weather", "Low Framerate", "PTZ", "Night" and "Air Turbulence". We have picked "Bad Weather", "Dynamic Background" and "Intermittent Object Motion" for our training process. This election has been based on the similarity of these images with real conditions where our model is to be applied. As stated before, the target of this system is to perform change detection on images obtained with variable weather and with dynamic scenarios. Therefore, the mentioned categories represent a stable base for our model to learn.

To build our dataset, we have selected one background image from each category to conform the background for each foreground picture. This has been done to provide a unique reference for the complete dataset and prepare the model for slight changes as mentioned in previous sections. To do so, each background image has been replicated to obtain a reference for each foreground picture. Because of computational limitations, the input size for our model represents a restrictive value. Two complete images could not be established as an input. Therefore, a division of each image must be performed to obtain slices from them with a reduced size. Image preprocessing has been applied to the CD2014 dataset using OpenCV [23]. This library allows us to resize the input images to dimensions multiple of our desired input size: 64×64. As a result, we generate blocks of 64×64 pixels from the three images: Reference, target and ground truth. This is achieved using the sliding window algorithm explained in Section 3.2. The resultant patches obtained from the previous process are automatically labelled using the original image name, category and region position to conform a unique identifier. With this implementation, the image order is preserved and training can be performed without any trouble.

3.5. Training

Our system has been implemented in Keras [28] using Tensorflow as backend [29]. As mentioned in previous sections, our model input is the result of the concatenation of the reference and the foreground image along the depth axis. The model output is conformed by a vector with a length of 4096 elements. To obtain this desired structure, we have to process our original ground truth images. These images are provided by the CD2014 Dataset as a grayscale image. Black pixels represent unmodified regions. White pixels depict altered zones and grey indicate the border between an altered and an unaltered zone. The image processing has been performed using the ImageDataGenerator class from Keras and Numpy [30]. ImageDataGenerator allows Keras to select batches of data for training the model, instead of loading the complete dataset on memory and select the batches from memory. As a result, we obtain an iterative component called generator. These generator structures are widely implemented in Keras, and training methods are no exception. With this methodology, extensive datasets are easy to handle. These generators can be easily customized. In our case, we have performed two significant customizations. First, we have concatenated both input images into a unique structure with size of $64 \times 64 \times 6$ as our model input. Then, output images are transformed into vectors using Numpy API. Final dataset is detailed on Table 3. In summary, the training dataset is conformed by 130,476 patches, 43,492 for background, foreground and ground-truth respectively, with a proportion of 80% for training and 20% for evaluation purposes. 6750 additional patches from the IOM category have been included for metrics evaluation. Training loss is illustrated in Figure 5. The X-axis represents the steps (in thousands) of training. The Y-axis depict the loss value at a given step.

The graph has been obtained using Tensorboard from TensorFlow. Tensorboard has been employed to analyse the training process. The model has been trained on an NVIDIA 1080TI GPU for 12 epochs because of the implementation of early stopping [31].

Table 3. Description of the complete dataset used to train our model. Columns represent the category of the images, the number of complete images used, the size of each image and the resultant number of 64 × 64 patches used to train the algorithm.

Category	Images	Size	Total Patches
Blizzard (BW)	195	1216 × 704	40,755
Skating (BW)	177	1216 × 704	36,993
Snowfall (BW)	192	1216 × 704	40,128
Canoe (DBG)	210	640 × 384	12,600
Sofa (IOM) [1]	150	320 × 192	6750

[1] Only used for metrics evaluation.

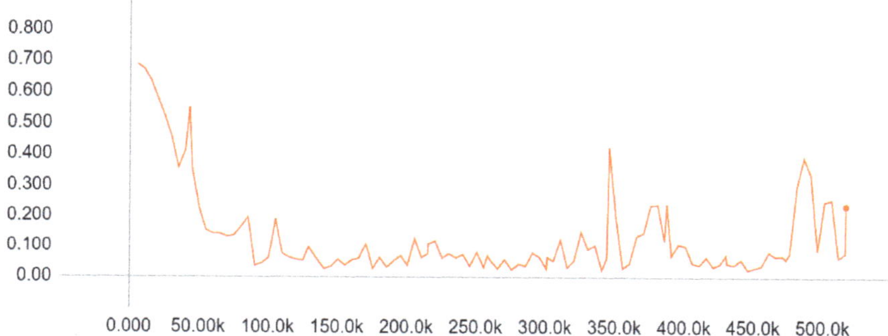

Figure 5. Training loss curve represented loss value per thousand steps.

3.6. Post Processing

After the deep learning model is applied to both images, the resultant grayscale pictures are processed. As stated before, the system outputs a set of 64 × 64 grayscale images. The length of the mentioned set is proportional to the image dimensions. It is a result of the sliding window algorithm applied to the inputs, as detailed in Section 3.2. If overlapping has been selected in the sliding window algorithm, the post-processing method divides the obtained values at the composition process by the overlapping factor. With this approach, the overlapping effect allows us to have a trade-off between precision and computational cost, to obtain a more flexible implementation. Therefore, the image is obtained by inverting the sliding window algorithm. That is to say, the 64 × 64 patches are placed on the equivalent position of the input patch at the original image in a new blank image. After all the patches have been included in the new image, we obtain a grayscale image with the dimensions of the inputs of the system. This image will be introduced into the filtering component of this post-processing module. Figure 6 represents an example output previous to the filtering component.

Pixel's intensity threshold is applied to filter possible noise effects such as blocking or insignificant changes detection in the image. Subsequently, a morphological dilation is performed on the resultant image. The objective of the dilation is to complete the possible gaps produced by noise in changed regions. Dilation has been selected as some information is typically removed by the filter to secure the complete elimination of disturbances. To compensate this, the dilation process expands the borders of the most relevant regions to complete the missed information after the previous threshold process. The dilation operator follows the formula:

$$(f \oplus b)(x) = \sup_{y \epsilon E}[f(y) + b(x-y)] \quad (8)$$

denoting by f(x) an image, and the structuring function by b(x). E is the Euclidean space into the set of real numbers.

The effect of this filtering on the images is depicted in Figure 7.

Figure 6. Image resultant of the combination of 64 × 64 grayscale patches obtained by CNN predictions.

Figure 7. Final output of our system after the filtering part of our post-processing module is applied to Figure 6.

4. Experimental Results

Model results are described in Section 4.1 along with the metrics used to analyse the model's performance. In Section 4.2, we have compared our solution with several state-of-the-art implementations for background subtraction and change detection, trained using the CD2014 dataset.

4.1. Evaluation Metrics

To perform a comparison between our system and several state-of-the-art methods, we have selected multiple metrics to characterize the performance of the models. All of them are based on four elemental concepts. We define true positives (TP) as the correctly classified changed pixel. True negatives (TN) represent the unmodified pixels which have been correctly predicted. "Let false positives" (FP) defines the incorrect classified changed pixels. Lastly, false negatives (FN) represent the incorrectly labelled background pixels. We have selected the metrics from [1], as all the methods based on this dataset compute them. These metrics, defined by the previous concepts, are:

- Recall (Re): $\frac{TP}{TP+TN}$
- Specificity (Sp): $\frac{TP}{TP+TN}$
- False Positive Rate (FPR): $\frac{FP}{FP+TN}$
- False Negative Rate (FNR): $\frac{FN}{TP+FN}$
- Percentage of Wrong Classifications (PWC): $100 * \frac{FP+FN}{TP+TN+FN+FP}$
- Precision (Pr): $\frac{TP}{TP+FP}$
- F-measure (FM): $2 * \frac{Re*Pr}{Re+Pr}$

In Figure 8 we have represented examples of the model's performance in different image categories. Similarly, Figure 9 depicts examples of our model's performance in UAV imagery. Ultimately, Table 4 shows the results of the stated metrics represented for each scenario of our dataset. It should be noted that recall and precision measures are included in F-measure. From the metrics obtained, we can observe that the system has obtained excellent scores in complex scenarios such as snow blizzard.

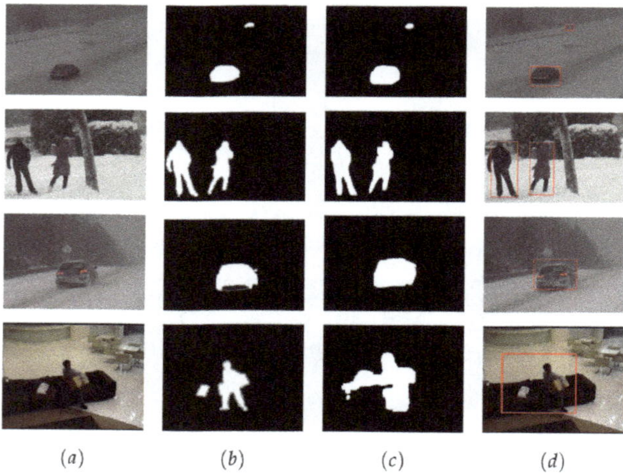

Figure 8. Results of our system applied to CD2014 images. Column (**a**) represents the image where change detection is applied, column (**b**) represents the ground truth images given by the CD2014 dataset, (**c**) the prediction images from our model and (**d**) the original images with the bounding boxes obtained by the prediction process.

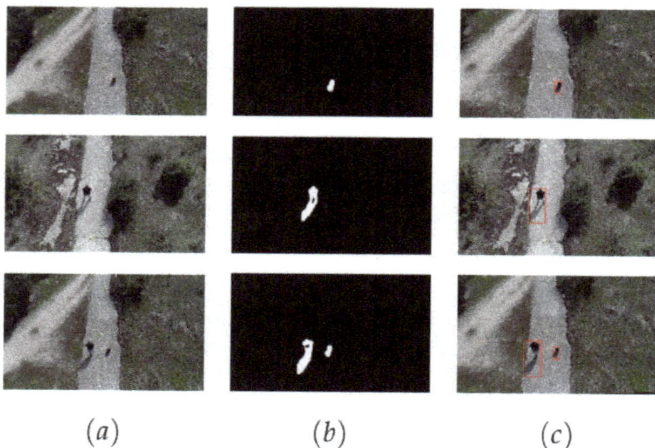

Figure 9. Results of our system applied to images acquired from an unmanned aerial vehicle (UAV), not included in the dataset. In this case, column (**a**) represents the foreground image, column (**b**) the binary image predicted and column (**c**) the original image bounding boxes as in previous table.

Table 4. Metrics scores of our proposed solution on each CD2014 selected category.

Category	Sp	FPR	FNR	PWC	FM
Bad Weather (BW)	0.999914	0.000086	0.022035	0.017216	0.977963
Dynamic Background (DBG)	0.999810	0.000190	0.048662	0.038013	0.951339
Intermittent Object Motion (IOM)	0.999790	0.000210	0.005232	0.004808	0.994768

4.2. Comparison with Other Change Detection Systems

As mentioned in the introductory part of this section, we have selected various implementations to compare our system. We have based our election in the use of the CD2014 dataset to have the most objective results as possible. Moreover, we have elected algorithms founded on various technologies such as traditional image processing techniques (SuBSENSE [8]), other traditional mathematical implementations (GMM [6]) and convolutional neural networks (ConvNet-IUTIS, ConvNet-GT [4]). Our purpose with this is to remark the performance of CNNs over traditional methods. As we compare a CNN based implementation, we can obtain valuable conclusions for our system performance. The mentioned traditional solutions have been selected because of their impact in background subtraction, as they are frequently compared in most papers dealing with this subject. In Table 5, the results of this comparison are represented, using the value of the F-measure as a reference for their performance. Only scenarios that have been tested with the model are compared. From the previous table, we can observe how the performance of our model denotes higher precision than other state-of-the-art implementations for static cameras. Therefore, our change detection component offers an accurate solution for the purpose. The combination of this development with the image alignment methods provides a state-of-the-art solution for remaking modified regions in images acquired by a moving camera.

Table 5. F-measure scores of our implementation and the other seven state-of-the-art implementations on the CD2014 dataset from categories "Bad Weather", "Dynamic Background" and "Intermittent Object Motion".

Implementation	$FM_{overall}$	FM_{BW}	FM_{DBG}	FM_{IOM}
Proposed	**0.9747**	**0.9779**	**0.9513**	**0.9947**
ConvNet-GT [4]	0.9054	0.9264	0.8845	Unknown
ConvNet-IUTIS [4]	0.8386	0.8849	0.7923	Unknown
CNN [3]	0.7718	0.8301	0.8761	0.6098
GMM [6]	0.6306	0.7380	0.6330	0.5207
SuBSENSE [8]	0,7788	0.8619	0.8177	0.6569
PBAS [32]	0.6749	0.7673	0.6829	0.5745
PAWCS [5]	0.8285	0.8152	0.8938	0.7764

5. Discussion

The primary purpose of this paper is to describe the development of a new system for change detection with UAV images using a combination of image alignment and CNNs. The most significant difference between our proposed solution and other state-of-the-art methods is the inclusion of the mentioned image alignment system to adjust images from moving cameras. Figure 10 depicts the improvement of the precision of our system caused by the use of the image alignment component. An additional difference in our system is represented by the improvement of precision it provides on the studied scenarios, as reflected in Table 5. Based on these reasons, we can affirm that our CNN architecture and training have been implemented effectively.

As mentioned before, the system employs a reference sequence to detect the changes. This implies restraints to the method as the reference must denote the ideal status of the recorded scenario. That is to say, all the elements included in this sequence are considered as background. Previous to this approach, a reconstruction method similar to [2] was implemented. Because of the unsatisfactory precision obtained and the computational cost involved, the approach was discarded.

An additional point to discuss would be the multiple variations that could appear between two UAV flights. The effect of flying at different heights is one of these possible variations that could occur on a real scenario. In this case, our system compensates this using the image alignment component to select the most relevant elements which are included on both reference and foreground images. At that point, it automatically selects the central region of both images to be analysed.

As can be observed in Table 5, implementations from [4] have an unknown F-measure score at the "Intermittent Object Motion" category. These methods do not consider this particular type of scenario. However, the results they obtain on the other two categories position them as a relevant state-of-the-art implementation to compare our system.

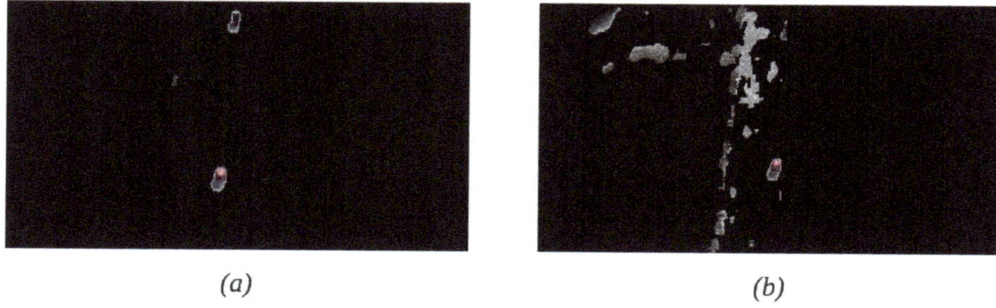

(a) (b)

Figure 10. Effect of the image alignment component on the system's output. (**a**) The resultant image with the image alignment applied, while (**b**) represent the results obtained without the use of the component. As can be observed, the aligned image detects changed elements (in this case, the two red boxes) more precisely.

6. Conclusions

In this paper we have presented a change detection system for static and moving cameras using image alignment based on ORB algorithm and convolutional neural networks. Because of the use of UAV imagery acquired by moving cameras, the problem of dynamic backgrounds has been addressed. As we have detailed along the paper, our mayor improvement from other state-of-the-art implementations consists on the use of an image alignment process. The objective of this element is to compensate the possible variations during UAVs flights described in previous sections. In addition to that, the inclusion of the sliding window algorithm reduces the computational cost of the CNN model by reducing the dimensions of the input images. Moreover, this method adds versatility to the system. The sliding window algorithm can be adjusted to provide overlapping sections to improve accuracy with an increment in computational cost. As far as we know, a moving camera scenario has not been taken into account in any of the state-of-the-art methods for change detection compared along the paper. Only dynamic backgrounds on static cameras have been studied on mentioned implementations. Our system is capable of adapting to these conditions using image alignment techniques and the idea of a reference video or image. Datasets with dynamic backgrounds have been selected to train the network to achieve meaningful outcomes for real-world applications. Results from experiments indicate a precise detection in scenarios with adverse weather such as a snowfall or a blizzard. The comparison with other state-of-the-art methods reflects that our system is the most accurate on the studied scenarios.

Future Work

The precision of the reference's acquisition is crucial for the system's performance. As a solution to this, we are working on GPS data processing for improving the alignment system as in [22]. In addition to that, we consider the option to include deep learning in the alignment process as another of our futures lines of work. Our objective with that is to compare the performance of deep learning against our current image alignment system based on ORB.

Author Contributions: Data curation, V.G.R.; Formal analysis, J.A.R.F. and J.M.M.G.; Funding acquisition, N.S.A. and J.M.L.M.; Investigation, V.G.R.; Methodology, V.G.R., J.A.R.F., J.M.M.G. and N.S.A.; Project administration, N.S.A. and J.M.L.M.; Resources, N.S.A. and J.M.L.M.; Software, V.G.R.; Supervision, J.A.R.F., J.M.M.G. and F.Á.; Validation, V.G.R.; Visualization, V.G.R.; Writing–original draft, V.G.R.; Writing—review & editing, J.A.R.F., J.M.M.G., N.S.A., J.M.L.M. and F.Á.

Funding: This research was funded by the 5G Public Private Partnership (5G-PPP) framework and the EC funded project H2020 ICT-762013 NRG-5 (http://www.nrg5.eu) grant number 762013.

Conflicts of Interest: The authors declare no conflict of interest.

References

1. Wang, Y.; Jodoin, P.; Porikli, F.; Konrad, J.; Benezeth, Y.; Ishwar, P. CDnet 2014: An Expanded Change Detection Benchmark Dataset. In Proceedings of the 2014 IEEE Conference on Computer Vision and Pattern Recognition Workshops, Columbus, OH, USA, 23–28 June 2014; pp. 393–400. [CrossRef]
2. Minematsu, T.; Shimada, A.; Uchiyama, H.; Charvillat, V.; Taniguchi, R.I. Reconstruction-Base Change Detection with Image Completion for a Free-Moving Camera. *Sensors* **2018**, *18*, 1232. [CrossRef] [PubMed]
3. Babaee, M.; Dinh, D.T.; Rigoll, G. A Deep Convolutional Neural Network for Video Sequence Background Subtraction. *Pattern Recogn.* **2018**, *76*, 635–649. [CrossRef]
4. Braham, M.; Van Droogenbroeck, M. Deep background subtraction with scene-specific convolutional neural networks. In Proceedings of the 2016 International Conference on Systems, Signals and Image Processing (IWSSIP), Bratislava, Slovakia, 23–25 May 2016; pp. 1–4.
5. St-Charles, P.; Bilodeau, G.; Bergevin, R. Universal Background Subtraction Using Word Consensus Models. *IEEE Trans. Image Process.* **2016**, *25*, 4768–4781. [CrossRef]
6. Stauffer, C.; Grimson, W.E.L. Adaptive background mixture models for real-time tracking. In Proceedings of the 1999 IEEE Computer Society Conference on Computer Vision and Pattern Recognition (Cat. No PR00149), Fort Collins, CO, USA, 23–25 June 1999; Volume 2, pp. 246–252. [CrossRef]
7. Elgammal, A.M.; Harwood, D.; Davis, L.S. Non-parametric Model for Background Subtraction. In Proceedings of the 6th European Conference on Computer Vision-Part II, Dublin, Ireland, 26 June–1 July 2000; Springer: London, UK, 2000; pp. 751–767.
8. St-Charles, P.; Bilodeau, G.; Bergevin, R. SuBSENSE: A Universal Change Detection Method With Local Adaptive Sensitivity. *IEEE Trans. Image Process.* **2015**, *24*, 359–373. [CrossRef] [PubMed]
9. Barnich, O.; Van Droogenbroeck, M. ViBe: A Universal Background Subtraction Algorithm for Video Sequences. *IEEE Trans. Image Process.* **2011**, *20*, 1709–1724. [CrossRef] [PubMed]
10. LeCun, Y.; Haffner, P.; Bottou, L.; Bengio, Y. Object Recognition with Gradient-Based Learning. In *Shape, Contour and Grouping in Computer Vision*; Springer: London, UK, 1999; pp. 319–345.
11. Szegedy, C.; Liu, W.; Jia, Y.; Sermanet, P.; Reed, S.E.; Anguelov, D.; Erhan, D.; Vanhoucke, V.; Rabinovich, A. Going Deeper with Convolutions. In Proceedings of the 2015 IEEE Conference on Computer Vision and Pattern Recognition (CVPR), Boston, MA, USA, 7–12 June 2014.
12. Zeng, D.; Zhu, M. Background Subtraction Using Multiscale Fully Convolutional Network. *IEEE Access* **2018**, *6*, 16010–16021. [CrossRef]
13. Liu, S.; Deng, W. Very deep convolutional neural network based image classification using small training sample size. In Proceedings of the 2015 3rd IAPR Asian Conference on Pattern Recognition (ACPR), Kuala Lumpur, Malaysia, 3–6 November 2015; pp. 730–734. [CrossRef]
14. Coombes, M.; McAree, O.; Chen, W.; Render, P. Development of an autopilot system for rapid prototyping of high level control algorithms. In Proceedings of the 2012 UKACC International Conference on Control, Cardiff, UK, 3–5 September 2012; pp. 292–297. [CrossRef]
15. Rublee, E.; Rabaud, V.; Konolige, K.; Bradski, G. ORB: An efficient alternative to SIFT or SURF. In Proceedings of the 2011 International Conference on Computer Vision, Barcelona, Spain, 6–13 November 2011; pp. 2564–2571. [CrossRef]

16. Calonder, M.; Lepetit, V.; Strecha, C.; Fua, P. Brief: Binary robust independent elementary features. In Proceedings of the European Conference on Computer Vision, Crete, Greece, 5–11 September 2010; Springer: Berlin, Germany, 2010; pp. 778–792.
17. Rosten, E.; Drummond, T. Machine learning for high-speed corner detection. In Proceedings of the European Conference on Computer Vision, Graz, Austria, 7–13 May 2006; Springer: Berlin, Germany, 2006; pp. 430–443.
18. Harris, C.; Stephens, M. A Combined Corner and Edge Detector. In Proceedings of the 4th Alvey Vision Conference, Manchester, UK, 31 August–2 September 1988; pp. 147–151.
19. Rosin, P.L. Measuring Corner Properties. *Comput. Vis. Image Underst.* **1999**, *73*, 291–307. [CrossRef]
20. Lowe, D.G. Distinctive Image Features from Scale-Invariant Keypoints. *Int. J. Comput. Vision* **2004**, *60*, 91–110. [CrossRef]
21. Bay, H.; Ess, A.; Tuytelaars, T.; Van Gool, L. Speeded-up robust features (SURF). *Comput. Vis. Image Underst.* **2008**, *110*, 346–359. [CrossRef]
22. Avola, D.; Cinque, L.; Foresti, G.L.; Martinel, N.; Pannone, D.; Piciarelli, C. A UAV Video Dataset for Mosaicking and Change Detection From Low-Altitude Flights. *IEEE Trans. Syst. Man, Cybern. Syst.* **2018**, 1–11. [CrossRef]
23. Laganiere, R. *OpenCV Computer Vision Application Programming Cookbook Second Edition*; Packt Publishing Ltd.: Birmingham, UK, 2014.
24. Hahnloser, R.; Sarpeshkar, R.; Mahowald, M.A.; Douglas, R.; Sebastian Seung, H. Digital selection and analogue amplification coexist in a cortex-inspired silicon circuit. *Nature* **2000**, *405*, 947–51. [CrossRef] [PubMed]
25. Srivastava, N.; Hinton, G.; Krizhevsky, A.; Sutskever, I.; Salakhutdinov, R. Dropout: A Simple Way to Prevent Neural Networks from Overfitting. *J. Mach. Learn. Res.* **2014**, *15*, 1929–1958.
26. Caruana, R.; Lawrence, S.; Giles, L. Overfitting in Neural Nets: Backpropagation, Conjugate Gradient, and Early Stopping. In Proceedings of the 13th International Conference on Neural Information Processing Systems, Hong Kong, China, 3–6 October 2000; MIT Press: Cambridge, MA, USA, 2000; pp. 381–387.
27. Ioffe, S.; Szegedy, C. Batch Normalization: Accelerating Deep Network Training by Reducing Internal Covariate Shift. In Proceedings of the 32Nd International Conference on International Conference on Machine Learning, Lille, France, 6–11 July 2015; Volume 37, pp. 448–456.
28. Keras: Deep Learning for Humans. Available online: https://github.com/fchollet/keras (accessed on 27 November 2018).
29. Abadi, M.; Barham, P.; Chen, J.; Chen, Z.; Davis, A.; Dean, J.; Devin, M.; Ghemawat, S.; Irving, G.; Isard, M.; et al. TensorFlow: A System for Large-scale Machine Learning. In Proceedings of the 12th USENIX Conference on Operating Systems Design and Implementation, Savannah, GA, USA, 2–4 November 2016; USENIX Association: Berkeley, CA, USA, 2016; pp. 265–283.
30. van der Walt, S.; Colbert, S.C.; Varoquaux, G. The NumPy Array: A Structure for Efficient Numerical Computation. *Comput. Sci. Eng.* **2011**, *13*, 22–30. [CrossRef]
31. Prechelt, L. Early stopping-but when? In *Neural Networks: Tricks of the Trade*; Springer: Berlin, Germany, 1998; pp. 55–69.
32. Hofmann, M.; Tiefenbacher, P.; Rigoll, G. Background segmentation with feedback: The pixel-based adaptive segmenter. In Proceedings of the 2012 IEEE Computer Society Conference on Computer Vision and Pattern Recognition Workshops, Providence, RI, USA, 16–21 June 2012; pp. 38–43.

Article

Multi-Variant Accuracy Evaluation of UAV Imaging Surveys: A Case Study on Investment Area

Grzegorz Gabara * and Piotr Sawicki *

Institute of Geodesy, University of Warmia and Mazury in Olsztyn, 10-719 Olsztyn, Poland
* Correspondence: grzegorz.gabara@uwm.edu.pl (G.G.); piotr.sawicki@uwm.edu.pl (P.S.); Tel.: +48-509-481-507 (G.G.)

Received: 28 October 2019; Accepted: 26 November 2019; Published: 28 November 2019

Abstract: The main focus of the presented study is a multi-variant accuracy assessment of a photogrammetric 2D and 3D data collection, whose accuracy meets the appropriate technical requirements, based on the block of 858 digital images (4.6 cm ground sample distance) acquired by Trimble® UX5 unmanned aircraft system equipped with Sony NEX-5T compact system camera. All 1418 well-defined ground control and check points were a posteriori measured applying Global Navigation Satellite Systems (GNSS) using the real-time network method. High accuracy of photogrammetric products was obtained by the computations performed according to the proposed methodology, which assumes multi-variant images processing and extended error analysis. The detection of blurred images was preprocessed applying Laplacian operator and Fourier transform implemented in Python using the Open Source Computer Vision library. The data collection was performed in Pix4Dmapper suite supported by additional software: in the bundle block adjustment (results verified using RealityCapure and PhotoScan applications), on the digital surface model (CloudCompare), and georeferenced orthomosaic in GeoTIFF format (AutoCAD Civil 3D). The study proved the high accuracy and significant statistical reliability of unmanned aerial vehicle (UAV) imaging 2D and 3D surveys. The accuracy fulfills Polish and US technical requirements of planimetric and vertical accuracy (root mean square error less than or equal to 0.10 m and 0.05 m).

Keywords: UAV imagery; bundle block adjustment; digital surface model; orthomosaic; data collection; accuracy; technical guidelines

1. Introduction

Due to significant advances in the research field of digital photogrammetry, computer vision, and unmanned aerial vehicles/unmanned aerial systems/remotely piloted aircraft systems (UAVs/UASs/RPASs) [1,2] construction, there has been an important change in geoinformation acquisition and the workflow of photogrammetric data collection [3]. This has been achieved by essential developments in the UASs components [4], in sensors production [5], and in the image-based surface reconstruction, e.g., structure-from-motion (SfM) [6,7], multi-view stereo (MVS) pipeline [8] and dense image-matching techniques [9].

The commercial off-the-shelf (COTS) software packages, focusing on UAV imagery applications in geomatics, are based on a large number of variously oriented digital images and work in the composed and often an autonomous processing chain [10–12]. They allow an automatic computation of spatial orientation of photos and the parameters of camera interior orientation with self-calibration [13] using the bundle-block adjustment (BBA) method, generation of dense point clouds [11], a 3D mesh model and, finally, the digital surface model (DSM) [10,14] and orthomosaic [15].

The spectrum of problems and research tasks in the new area of photogrammetry, the so-called "UAV Photogrammetry", is very wide [5,9,16–21]. The research and development mainly concerns, among other areas: small UAS [22], low-cost hardware solution [23], onboard sensors integration [24], sensors calibration [13,25,26], images orientation and direct georeferencing [27], data fusion obtained from optical sensors in various spectral ranges [24] and LiDAR (Light Detection and Ranging) [28], automation of flight planning [29], automation of digital images processing [30], testing and comparison of software dedicated to UASs [10,11]. An important research area is also testing the accuracy potential [3,31–34] of new measurement technologies based on UAV imagery, searching for new areas of its practical implementation [35–37], civil law regulations regarding the use of RPASs [18,38].

However, in our opinion, the operational use of UAV imagery acquired by means of single-lens reflex or compact system camera instead of medium or large format aerial digital cameras for data collection in order to undertake large-scale basic, thematic [39–42] and inventory mapping is one of the fundamental issues for the application of UAVs in geomatics and in particular in geodesy. For example, in Germany, the USA and also in Poland, there are no detailed technical guidelines regarding the use of UAV imagery in geodetic surveys. The only criterion for taking photogrammetric data into a geodetic database is the accuracy requirements [43–46] for the final products.

The use of RPASs in national mapping was described previously in state-of-art by the EuroSDR organization [38], in orthophoto quality assessment performed by Mesas-Carrascosa et al. [15] and Martínez-Carricondo et al. [47], in photogrammetric mapping evaluation by Agüera-Vega et al. [48] and in an investigation of the optimal number of ground control points (GCPs) for DSM by Tonkin and Midgley [49]. This current situation consequently limits the use of RPAS in geodesy. Researchers reported before that the regulation is the most limiting factor that hinders the thorough application of RPASs [18,38,50].

According to the authors' knowledge, the complex accuracy assessment of the computing pipeline of particular photogrammetric products (bundle-block adjustment with camera self-calibration, dense point cloud, digital surface model and orthomosaic) which we present in this paper, do not exist at present. Current studies concerning the accuracy assessment of UAV photogrammetric products are realized in incomparable technical conditions of projects and with the use of various statistical parameters of results assessment [39,40] relating to the particular stages of the processing [36,41,42,51]. The most commonly used metrical information of accuracy were root mean square errors RMSE (XYZ) on ground control and check points, which characterized bundle block adjustment results. The additional parameters of sensor interior orientation estimated in the self-calibration procedure were described by standard deviations. In addition, the quality and accuracy of DSMs and orthomosaics were analyzed using RMS (Z) and RMS (XY) deviations, respectively.

The issues mentioned above were the inspiration for defining the following main aims of this study:

- Definition of the computation pipeline methodology for measurement accuracy assessment under the real conditions of a posteriori ground control and check points measurements
- A multi-variant accuracy assessment of UAV imaging surveys performed in Pix4Dmapper suite [32,52]—a case of small investment area.
- Investigation of the impact of photos overlap and georeferences from UAV Global Navigation Satellite Systems (GNSS) receiver on the BBA results.
- Examination of the usefulness of UAV imagery acquired with the use of a medium spatial resolution [1] system camera for the development of full-featured photogrammetric products whose accuracy meets the appropriate technical requirements.
- Checking the possibilities of replacing geodetic field surveys by the UAV photogrammetric measurements.

The initial approach to the UAV mapping was signaled and presented by us in the publication [53], in which we found the possibility of using UAV imagery for inventory mapping in the case of the industrial estate. In the presented paper, we extend our considerations to the investment area and we discuss a new testing approach and other research results. The tests were conducted using a new input data set that contains a large number of low-altitude images and ground control points, additional ground check points and height ground check points. Also, we used another study methodology, which led us to pose new final results, and extended accuracy assessment (tabular summary of findings, histograms, screens) of respective stages of computation and image processing (multi-variant BBA, DSM and georeferenced orthomosaic generation).

The discussion of the results and the accuracy assessment takes into account the Polish [43], technical requirements (accuracy defined with errors $m_{XY} \leq 0.10$ m and $m_Z \leq 0.05$ m) for the land surveys of planar-elevation measurements of well-defined terrain details and other practical applications.

2. Materials and Methods

2.1. Study Area and Data Acquisition

The study site was an industrial area in the center of Poland in the Stryków municipality (51°52′38″ N, 19°35′21″ E), around 16 km northeast of the city of Łódź. This study site is representative for newly created industrial zones in the agricultural area because it contains different topographic objects, i.e., industrial buildings and their surroundings, access roads, a roundabout, and agricultural lands. In other study cases, the different localization and higher number of terrain objects may appear.

The flight mission design and photo capturing were made by the Geoplan Polish company in Zgierz. A block of archival low-altitude uncalibrated photos acquired on 15 April 2016, at 8:35–9:08 local time as photographic documentation of the industrial investment [53], was used in the case study. The images were acquired by Trimble® UX5 UAS equipped with Sony NEX-5T 16 megapixels digital compact system (mirrorless) camera with a CMOS (complementary metal-oxide semiconductor) sensor (size 23.5 × 15.6 mm, pixel count 4912 × 3264, pixel $p_{xy} = 4.78$ µm) and focal length f = 15.5 mm. Initial camera preferences were defined by shutter priority (1/2000 sec). The lens focus was set to infinity. The light source and white balance were set to Auto. The prevailing weather conditions on the day of flight mission were: mostly sunny sky with some clouds and the west wind speed was equal approximately to 5 m/sec and in gusts to 9 m/sec. Due to this fact, there were apparent drift and rotation (ω, φ, κ) of some images.

The image block was characterized by the following photogrammetric parameters: flight mission area was P = 1.0253 km^2, 858 photos in 41 photo strips oriented in direction N-S, crosswise to the long axis of the object, forward and side overlap $p_{\%} = q_{\%} = 85\%$, projected flight height h = 150 m, ground sample distance (GSD) = 4.6 cm and base-to-height ratio $\nu = 0.15$. The GNSS (single frequency receiver C/A, L1) of Trimble® UX5 UAS flight data were saved in LOG file. The flight path is shown in supplementary files (Figure S2 in Appendix 1–6).

The test field includes 5 artificially signalized GCPs, placed before the flight mission. For study purposes in 2019 the 295 natural, accurately defined ground points (water gate valves, manholes, curbstones) were identified on images, mainly used as check points (ChPs). The distribution of the GCPs and ChPs was a consequence of the field details placement, i.e., industrial buildings and elements of the utilities network. The GCPs and ChPs were distributed more densely in the area which was critical for matching (Figure 1a,b) and more sparsely in the remaining interest area (Figure 1c). Due to the thin sheet metal of the hall's roofing it was impossible to place GCPs there. For tests we used the previous researchers' experiences described in [27,49,54,55].

Furthermore, for the verification of the photogrammetric-generated DSM, the 1118 height ground ChPs in the form of dense cross-sections localized on scarps and terrain slopes were surveyed. The direct measurement of ground and height points was realized by applying the GNSS using the real-time network (RTN) method executed by virtual reference stations from the Polish Active Geodetic Network ASG EUPOS (virtual reference station solution). At every point, 10 fixed epochs were averaged in the RTN mode.

We decided to use the RTN solution because according to the ASG EUPOS (http://www.asgeupos.pl/index.php?wpg_type=serv&sub=gen) the advantage of RTN over solutions based on a single reference station RTK (real-time kinematic), is the possibility of better modeling of systematic errors related to, e.g., the work of satellite clocks and delays related to the signal propagation in atmosphere.

Additionally, RTN solutions enable to achieve the coordinates determination repeatability regardless of the distance between the receiver and physical station. In the RTK mode the error in coordinate determination increases along with the distance to the physical station from which the corrections originate. Also, ASG EUPOS recommends using the RTN in precise positioning wherever possible. Another argument for using the RTN was the distance of apploximately 16 km to the nearest physical station of ASG EUPOS (LODZ). According to the Geospatial Positioning Accuracy Standards, the independent source of higher accuracy should be the highest accuracy feasible and practicable to evaluate the accuracy of the dataset. The RTN solution fulfills these guidelines.

The coordinates of the points were determined using a Trimble SPS882 survey-grade GNSS receiver in the PL-2000 zone 6 coordinate system (EPSG: 2177), and in the PL-KRON86-NH elevation system, with the estimated accuracy $m_{XY} = 0.03$ m and $m_Z = 0.05$ m. The measured GNSS points fell within the range:

- Position Dilution of Precision (PDOP) 1.0 ÷ 4.8 (<5), mean PDOP = 1.6
- Horizontal Dilution of Precision (HDOP) 0.6 ÷ 3.3 (<4), mean HDOP = 0.9
- Vertical Dilution of Precision (VDOP) 0.8 ÷ 3.4 (<4), mean VDOP = 1.3

which proves a good configuration of satellite geometry with am effect on positioning accuracy of GNSS points: mean σN = 0.012 m, mean σE = 0.008 m, mean σH = 0.020 m and root mean square value RMS = 0.025 m.

Because of the surface properties and a variety of factors that may impact the accuracy and quality of digital photogrammetric processing, the test field was divided into the following sections (Figure 1):

- The major part which includes the area between industrial buildings (**a**);
- The terrain around the industrial buildings (**b**);
- The roundabout, access road, gas station (**c**).

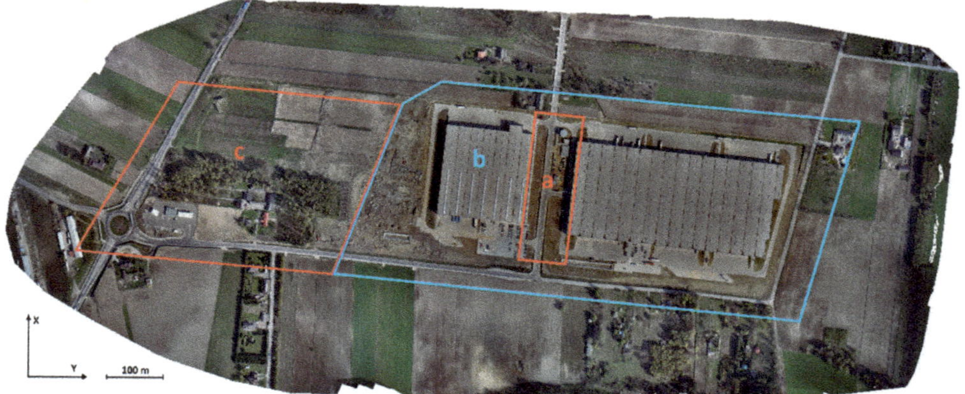

Figure 1. The test object Stryków with the defined study sections (**a**), (**b**), (**c**) on the orthomosaic generated in Pix4Dmapper.

2.2. Used Methods and Software

Due to the use of an archival image block in the study with the potential blur of images caused by wind gusts during Trimble® UX5 UAS flight, their radiometric quality verification was carried out. The graphical contents and a large number of images indicate that the detection of blur on images should be realized using automatic methods [56]. In this case, for preprocessing, the implementation of the Open Source Computer Vision Library (OpenCV) in Python was used. In the beginning, the variance of the absolute values using the Laplacian operator (linear high-pass filter) by Pech-Pacheco et al. [57] (Equation (1)) was calculated:

$$LAP\_VAR(I) = \sum_{m}^{M} \sum_{n}^{N} \left[|L(m,n)| - \overline{L} \right]^2 \tag{1}$$

where $\overline{L}$ is the mean of absolute values; $L(m, n)$ is the convolution of the image $Img(m, n)$ with the mask L; M, N is the function size.

Since some of the images were tagged as potentially blurred, the decision was made to use the second approach for blur detection, which involves the frequency domain. The greyscale intensity images were transformed using Fourier transform $F(u, v)$ accordingly to Equation (2):

$$F(u,v) = \frac{1}{MN} \sum_{x=0}^{M-1} \sum_{y=0}^{N-1} f(x,y) e^{-j2\pi\left(\frac{ux}{M} + \frac{vy}{N}\right)} \tag{2}$$

where M, N is the number of columns and rows, $f(x, y)$ is the pixel value [27]. The zero-frequency component has to be shifted to the center of the spectrum, and then the normalization from 0 to 255 has to be done (Equation (3)).

$$FF(u,v) = 255 \frac{\log\left(1 + |F(u,v)|\right)}{\max\left(\log\left(1 + |F(u,v)|\right)\right)} \tag{3}$$

where $FF(u, v)$ is Fourier transform component [27]. To determine the level of blur in the images, the skewness of the normalized 2D Fourier transform was computed according to Ribeiro-Gomes et al. [58] (Equation (4)):

$$skew = \frac{\frac{1}{M \cdot N} \sum_{u=0}^{M-1} \sum_{v=0}^{N-1} \left(FF(u,v) - \overline{FF(u,v)} \right)^3}{\left(\sqrt{\frac{1}{M \cdot N} \sum_{u=0}^{M-1} \sum_{v=0}^{N-1} \left(FF(u,v) - \overline{FF(u,v)} \right)^2} \right)^3} \tag{4}$$

where $FF(u, v)$ is each Fourier transform component shifted to the center of the spectrum and normalized and $\overline{FF(u,v)}$ is the mean value.

The photogrammetric measurements and advanced digital images processing were performed in the following software (in the order in which they were used):

- Matching (own application), Poland [59]
- IrfanView v. 4.41 (application by Irfan Skiljan), Vienna Austria (https://www.irfanview.com/).
- Pix4Dmapper Pro v. 4.3.31 (full license) of Pix4D S.A., Prilly Switzerland [32,52].
- RealityCapture v. 1.0.3.6310 (commercial license) of Capturing Reality s.r.o, Bratislava Slovakia [60,61].
- PhotoScan v. 1.5.1 build 7618 (commercial license) of Agisoft LLC, St. Petersburg Russia [10,62].
- CloudCompare v. 2.10.2 (GNU General Public License) of Telecom ParisTech and the research and development (R&D) division of EDF, Paris France [40,63].
- AutoCAD Civil 3D 2019 v.13.0.613.0 (academic license) of Autodesk, Inc., San Rafael USA (https://www.autodesk.pl).

Figure 2 demonstrates the proposed workflow of this study for the accuracy evaluation of data collection based on low-altitude images. This scheme is an application of the original methodology for complex photogrammetric data collection using Pix4Dmapper as the main software for processing.

The signalized ground control points and natural ground points were measured on images using the center-weighted method (centroid operator) by means of Matching, the original software [59] with mean subpixel accuracy $s_{x'y'} = 0.15$ pixel. On the images, where points could not be measured automatically, the edge filter detection using IrfanView application was applied, and in some cases, the measurement was performed manually. Each ground control point (GCP) was identified on average in 38 images.

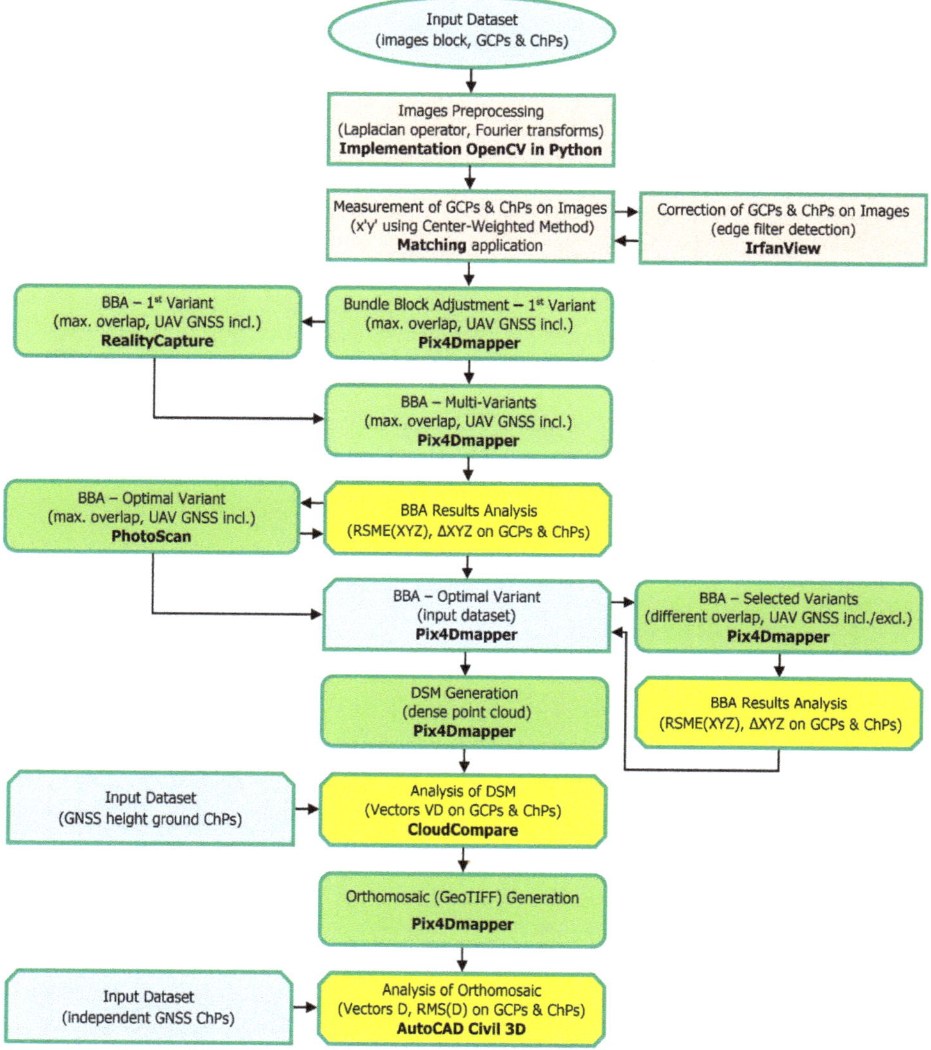

Figure 2. The workflow of measurement and processing proposed for unmanned aerial vehicle (UAV) surveys.

At the first stage of computation in Pix4Dmapper, the combined BBA with camera self-calibration, including 218 ground points (5 signalized GCPs and 213 ChPs in three interest areas, Figure 1) were carried out with an automatic calculation of interior (standard designation: c, x'_0, y'_0, K_1, K_2, K_3, P_1, P_2) and exterior (X_0, Y_0, Z_0, ω, φ, κ) orientation parameters [64]. In the adjustment, the approximated initial camera positions in the WGS84 coordinate system (*.LOG file) were used, and the output coordinate system was PL-2000 zone 6 (EPSG: 2177).

Due to the limited possibilities of changing settings in Pix4Dmapper, it is not discussed in this paper. The individual settings could be checked in reports from computations, which are in supplementary information files (Appendix 1–6).

The Pix4Dmapper calculates the global projection error for points measured on images without signalizing blunders. To avoid manual inspection of 218 points (each one separately) on mean 38 images in Pix4Dmapper, the RealityCapture software with its fast signalization of blunders was used. This step allows the identification of 4 faulty points numbering on images and their correcting.

To find the optimal solution of the BBA, 44 computing variants were executed in Pix4Dmapper with various numbers and configuration of GCPs and ChPs. In research, it was assumed that for each variant a new project was set up. This approach led to computing each input dataset independently. The 6 selected configuration variants of GCPs (no. 100 ÷ 103—artificially signalized, no. 201 ÷ 211—natural signalized points), optimal in terms of the number and adjustment accuracy, are presented below in Figure 3.

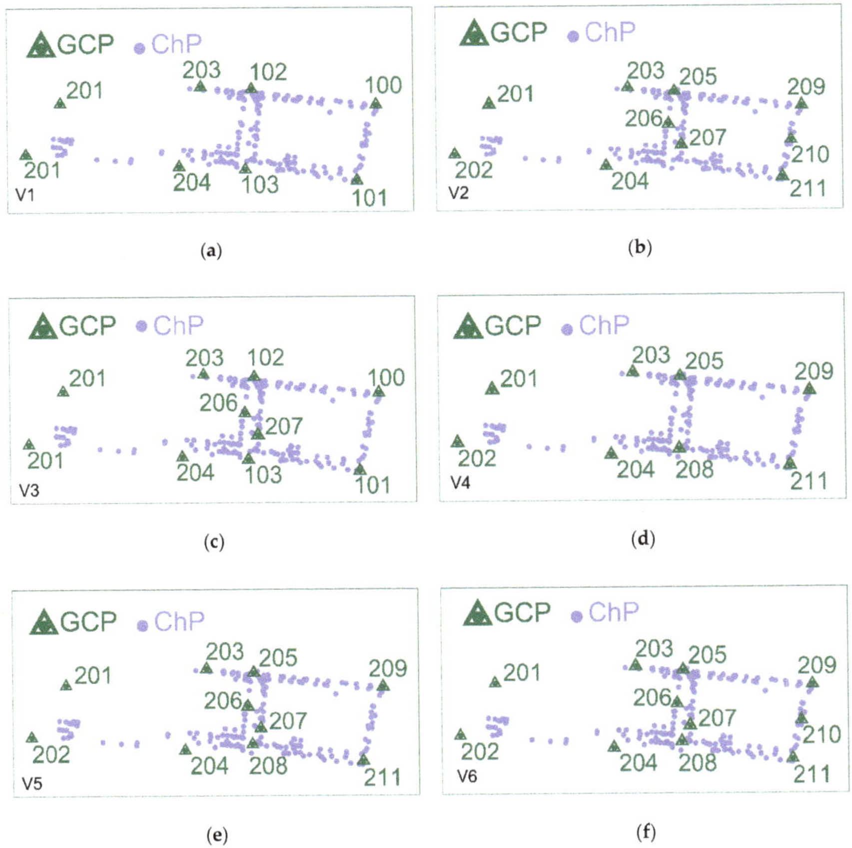

Figure 3. Configuration of ground control points (GCPs) and check points (ChPs) on the test object Stryków in the 6 BBA variants: (**a**) Variant 1 which contains 8 GCPs (4 signalized) and 210 ChPs; (**b**) Variant 2 which includes 10 GCPs and 208 ChPs; (**c**) Variant 3 consists of 10 GCPs (4 signalized) and 208 ChPs; (**d**) Variant 4 defined as 8 GCPs and 210 ChPs; (**e**) Variant 5 which is the extension of (**d**) by 2 GCPs between industrial buildings; (**f**) Variant 6 (it is characterized by 11 GCPs and 207 ChPs) which is the extension of (**e**) by 1 GCP at the right side of industrial building and close to it.

To prepare the accuracy assessment methodology, i.e., statistical test parameters, number and distribution of ground points, the guidelines described in [65] were used. As a statistical parameter the RMSE values are described, which are interpreted as a "square root of the average of the set of squared differences between dataset coordinate values and coordinate values from an independent source of higher accuracy for identical points". In this paper the 95% confidence level was not applied for accuracy assessment, because used values of RMSE are lower.

In the photogrammetry, various statistical tests are applied for accuracy evaluation. The statistical tests like Student's t-distribution and Fisher-Snedecor's F-distribution are used to verify the significance of camera interior orientation parameters computed by the bundle block adjustment with self-calibration or simultaneous (on-the-job) calibration [66]. The Cochran test is used to check the error significance of the 3D coordinates estimation (homogeneity of many standard deviations of independent samples). The blunder detection in the reference dataset, e.g., GCPs coordinates is possible using statistical tests, the Student *t*-test (used to assess the presence of bias) and chi-square test (used to determine whether random errors are adequate) [15,67]. In our study, for the measurement of GCPs and ChPs, we have used the Polish Active Geodetic Network ASG EUPOS, which is regularly tested and the correctness of its precision was proved before. According to commonly known adjustment computation principles [68] using statistical tests like Student *t*-test or chi-square was no justified in our case study.

To define reference values of limiting errors for planar-elevation measurements of well-defined terrain details, the ASPRS Positional Accuracy Standards for Digital Geospatial Data [44] and Polish [43] technical requirements were taken into account. The first document recommends the horizontal accuracy class RMSEx and RMSEy for orthoimagery purposes, and the vertical accuracy class $RMSE_Z$ for digital elevation data. The second regulation describes the point position errors in the cartesian and elevation coordinate systems. Polish technical requirements were adopted in the paper as more rigorous. To analyze the accuracy of photogrammetric products in this study, the horizontal accuracy RMSE(X) and RMSE(Y) and vertical accuracy RMSE(Z) respectively to 0.10 m and 0.05 m were used.

The results of the BBA in Pix4Dmapper were analyzed in terms of GCPs and ChPs number, RMSE(XYZ) on GCPs and RMSE(XYZ) on ChPs. Based on the accuracy assessment of the BBA for further works, variant 6 (Figure 3f) was chosen as optimal. The use of one GCP (210) in section b (Figure 1) and two GCPs (206, 207) in section a (Figure 1) was caused by significant matching errors on roof surfaces and in their immediate surroundings, and as a consequence, the significant errors of photogrammetrically calculated ChPs coordinates. Since it is possible to obtain different results in different "black-box" applications [10,11,69], for an independent verification of the results, variant 6 was computed in the PhotoScan application. It is a well-known suite and uses other computation algorithms for photogrammetric product generation.

The next aspect was the evaluation of the impact of less forward and side overlap, and reduction of image number on the significant decrease of the BBA accuracy. Based on the BBA optimal variant, the computations were carried out in two options: with and without approximated UAV GNSS coordinates of images' projection center (position accuracy of several meters). The forward overlap $p_{\%}$ = 65% was obtained by removing every second photo in each strip. The side overlap $q_{\%}$ = 65% and $q_{\%}$ = 45% were received by removing from the initial image block every second, and every second and third strips, respectively.

Advanced images processing in Pix4Dmapper included the generation of dense point clouds, the DSM and finally a georeferenced orthomosaic in GeoTIFF format.

The 1118 height points and dense point cloud generated in Pix4Dmapper were imported to the CloudCompare application [63] for comparison of Z component using the least squares planar fitting of 3D points [40]. For fitting the mathematical model on the nearest point and its 4 neighbors the function (Equation (5)) was applied, which assumes that the sum of squared errors between the Z_i and the plane values $Ax_i + By_i + C$ is minimized.

$$E(A,B,C) = \sum\nolimits_{i=1}^{m} [(Ax_i + By_i + C) - Z_i]^2 \quad (5)$$

As a result, the vertical distance from the compared point cloud to the computed DSM is determined. The assessment of vertical distances distribution was analyzed.

For the orthorectification in Pix4Dmapper, the same GSD size of the orthomosaic and acquired images (GSD_{ortho} = GSD = 4.6 cm) was assumed. The received georeferenced orthomosaic and XY coordinates of 11 GCPs and 289 ChPs (incl. 207 ChPs from the BBA) were imported to AutoCAD Civil 3D. The 2D data were compared, and the values and directions of deviation vectors were determined.

3. Results

The data processing in all of the used software was carried out on a workstation with the processor Intel® Core™ i9-7940X, 128 GB RAM DDR4-3200 MHz memory, MSI 1080 Ti graphic card and Samsung 960 Pro SSD hard drive.

3.1. Blur Detection on the Images

At the first step, Laplacian operator (Equation (1)) shows 196 images that were tagged as potentially blurred. The Fourier Transform (Equations (2)–(4)) indicated that 124 images could be blurred. The manual inspection was performed, to check if the blur occurred on the images. It was found that the images pointed out by algorithms included the roofs of industrial buildings, which were characterized by the same color and amorphous texture. There were no images that contain motion and optical blur, so all of the images were allowed for further computation.

3.2. Bundle Block Adjustment and Camera Self-Calibration

For our tests, the in-flight self-calibration, which describes the real acquisition condition and gives more accurate camera calibration [34], was used. The results of the 6 BBA solutions with reports of the camera standard parameter calibration are included in supplementary information files (Appendix 1–6). The main difference obtained in the 6 BBA variants were the values of reprojection error, which are understood as the distance between the marked and the reprojected point on one image. Due to very similar results (max. differences in variants 1 ÷ 6 are Δc = 1.3 pix, $\Delta x'_0$ = 0.3 pix, $\Delta y'_0$ = 0.2 pix, ΔK_1 = 0.000, ΔK_2 = 0.001, ΔK_3 = 0.000, ΔP_1 = 0.000, ΔP_2 = 0.000) of interior orientation parameters calculated in Pix4Dmapper suite, which does not allow the interference in computation process, and because camera self-calibration is well described in the literature [13,25,26], the parameters are not discussed here.

Table 1 presents only the results of the final 6 BBA solutions (218 GCPs and ChPs in various combinations) obtained in the Pix4Dmapper suite. The maximum differences between root mean square errors on GCPs obtained in 6 variants of the BBA in horizontal and vertical planes are respectively to RMSE(X) = 0.002 m, RMSE(Y) = 0.007 m, RMSE(Z) = 0.013 m. In the case of ChPs, the RMSE are following RMSE(X) = 0.003 m, RMSE(Y) = 0.001 m and RMSE(Z) = 0.013 m.

Table 1. Result of 6 variants of bundle block adjustment in Pix4Dmapper.

Parameters	V1	V2	V3	V4	V5	V6
No. of GCPs	8	10	10	8	10	11
No. of ChPs	210	208	208	210	208	207
RMSE(X) on GCPs [m]	0.011	0.011	0.011	0.009	0.010	0.010
RMSE(Y) on GCPs [m]	0.016	0.022	0.018	0.023	0.023	0.022
RMSE(Z) on GCPs [m]	0.022	0.035	0.030	0.025	0.032	0.033
RMSE(X) on ChPs [m]	0.025	0.024	0.025	0.022	0.022	0.023
RMSE(Y) on ChPs [m]	0.025	0.026	0.025	0.026	0.026	0.026
RMSE(Z) on ChPs [m]	0.048	0.037	0.040	0.049	0.037	0.036

The distribution of deviations ΔX, ΔY, ΔZ (differences between coordinates measured using GNSS and obtained in the BBA) on ChPs for 6 variants are presented in Figure 4.

Table 2 includes the analysis of ΔZ deviations in each computation BBA variants in more detail. In the analysis, the 5 intervals of the ΔZ deviations were assumed. The ΔZ thresholds are the result of the current Polish survey technical regulations, where the $N_{\Delta Z}$ is the number of absolute deviations values [ΔZ] for defined intervals.

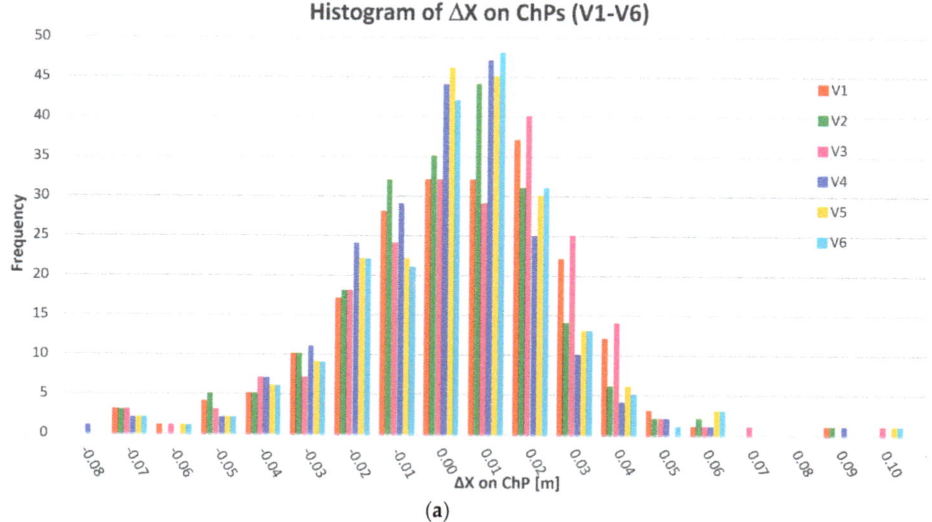

(**a**)

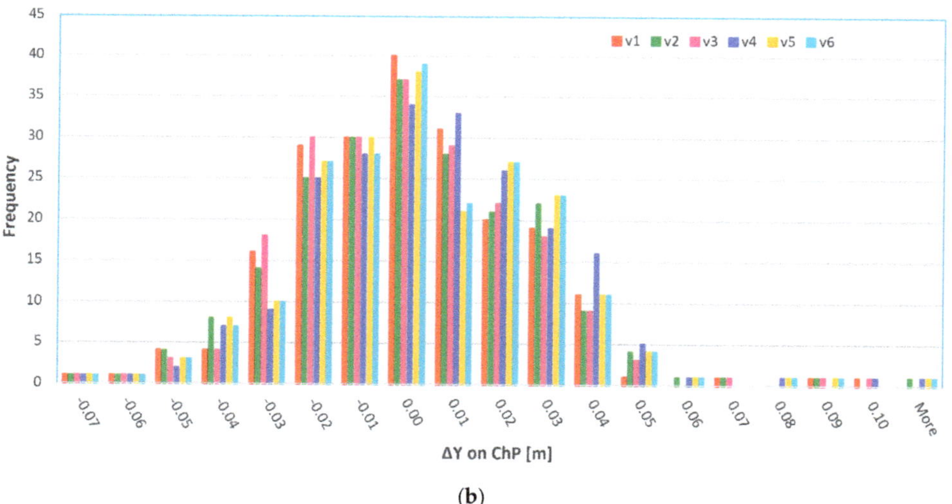

(**b**)

Figure 4. *Cont.*

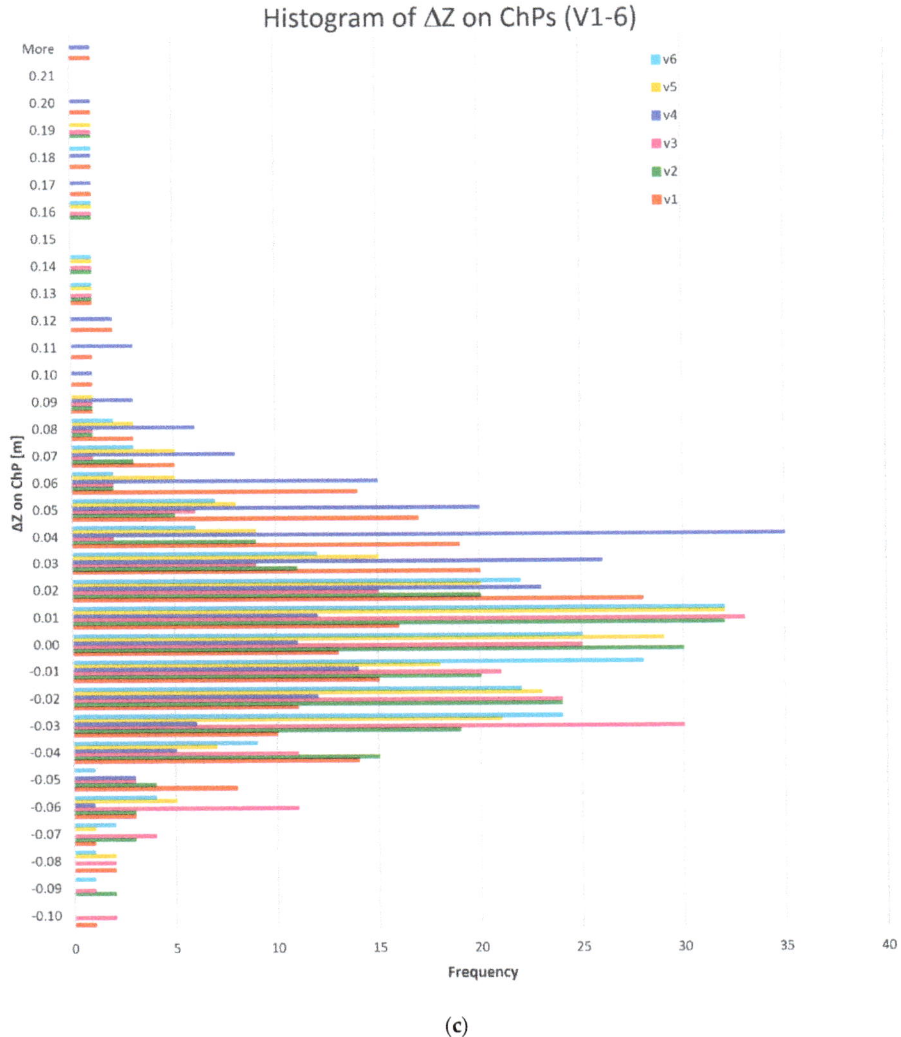

(c)

Figure 4. Histograms of deviations ΔX, ΔY, ΔZ on ChPs in computed variants (V1 ÷ V6) of the bundle-block adjustment (BBA): (**a**) ΔX deviation on ChPs; (**b**) ΔY deviation on ChPs; (**c**) ΔZ deviation on ChPs.

Table 2. Analysis of $N_{\Delta Z}$ in the ΔZ deviation intervals obtained from the BBA.

Parameters	$N_{\Delta Z}$ V1	$N_{\Delta Z}$ V2	$N_{\Delta Z}$ V3	$N_{\Delta Z}$ V4	$N_{\Delta Z}$ V5	$N_{\Delta Z}$ V6
No. of ChPs	210	208	208	210	208	207
[ΔZ] ≤ 0.05 m	163	185	176	164	182	187
0.05 < [ΔZ] ≤ 0.08 m	33	16	22	33	19	14
0.08 < [ΔZ] ≤ 0.10 m	5	3	4	4	3	2
[ΔZ] > 0.10 m	9	4	6	9	4	4
[%] of [ΔZ] > 0.05 m	22.4%	11.1%	15.4%	21.9%	12.5%	9.7%

The localization of the points (GCPs, ChPs), which belong to defined ΔZ intervals, is shown in Figure 5.

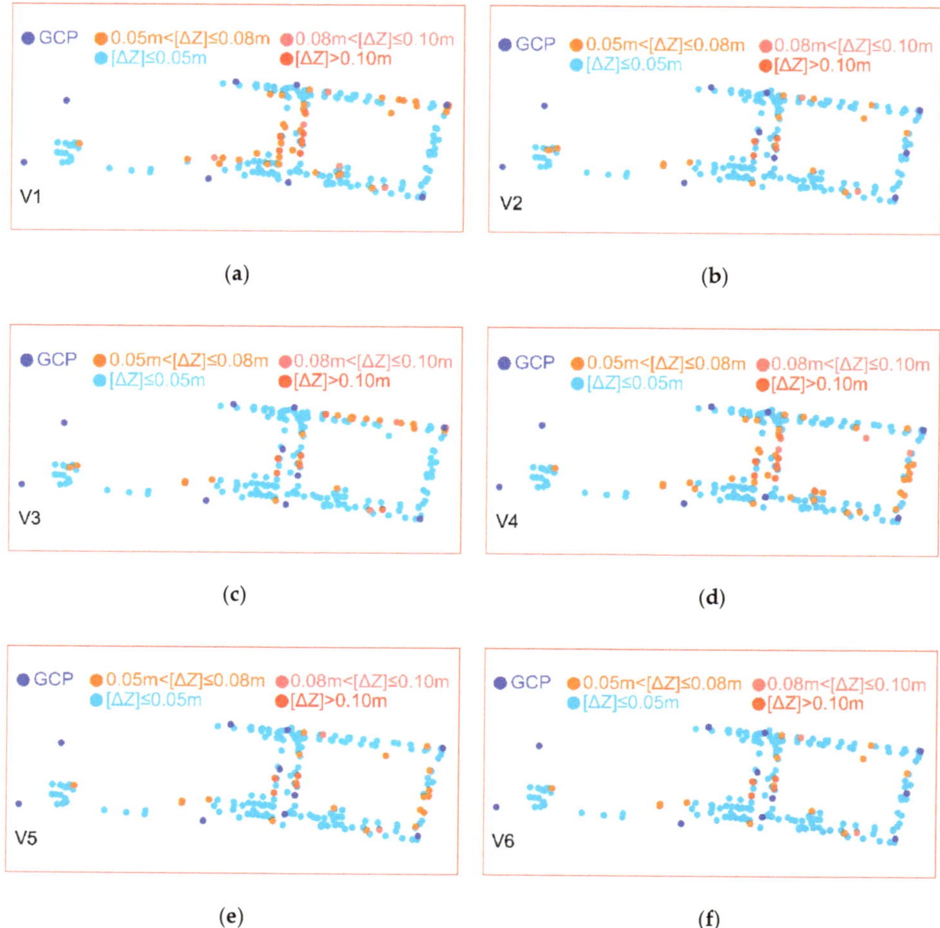

Figure 5. Localization of GCPs and ChPs in the BBA with representation of [ΔZ] in deviation intervals: (**a**) Variant 1; (**b**) Variant 2; (**c**) Variant 3; (**d**) Variant 4; (**e**) Variant 5; (**f**) Variant 6.

Upon the analysis of particular histograms (Figure 4) and the values of [ΔZ] deviation (Table 2) can be concluded that the most advantageous distribution of absolute values deviation, especially for [ΔZ] values occurred in variant 6. Furthermore, this variant poses the least percentage (9.7%) of [ΔZ] values higher than 0.05 m. The use of one GCP (210) in section b (Figure 1) and two GCPs (206, 207) in section a (Figure 1) enhanced the accuracy of Z coordinate determination. The differences of [ΔZ] deviation between variants 4 and 6 in individual sections a, b, c are, respectively, to 0.024 m, 0.008 m, 0.002 m.

To check the correctness and reliability of the optimal variant results (V6), the BBA was performed in an additional research and computation tool PhotoScan. This approach allows us to do an independent verification of digital processing results and to receive some additional information about applications' reliability and functionalities. Appendix **??** contains the report of computations.

Table 3 contains the comparison of variant 6 of the BBA results obtained in Pix4Dmapper and PhotoScan. The RMSE differences received on GCPs are respectively to RMSE(X_{GCP}) = 0.009 m, RMSE(Y_{GCP}) = 0.005 m and on ChPs amount to RMSE(X_{ChP}) = 0.002 m, RMSE(Y_{ChP}) = 0.002 m, RMSE(Z_{ChP}) = 0.004 m.

Table 4 presents the comparison of variant 6 of the BBA results with an extended analysis of the ΔZ deviations on ChPs. The main difference between the obtained outcomes is the number $N_{\Delta Z}$ of [ΔZ] deviations in the range of 0.05 ÷ 0.08 m to the detriment of the PhotoScan. The ChPs localization determined in the PhotoScan with the information about the [ΔZ] value is displayed in Figure 6. Figure 7 shows the histogram of ΔZ deviations on ChPs computed by the PhotoScan software.

Table 3. Results of the variant 6 of the BBA in Pix4Dmapper and PhotoScan.

Parameters	V6—Pix4Dmapper	V6—PhotoScan
No. of GCPs	11	11
No. of ChPs	207	207
RMSE(X) on GCPs [m]	0.010	0.019
RMSE(Y) on GCPs [m]	0.022	0.027
RMSE(Z) on GCPs [m]	0.033	0.033
RMSE(X) on ChPs [m]	0.023	0.025
RMSE(Y) on ChPs [m]	0.026	0.024
RMSE(Z) on ChPs [m]	0.036	0.040

Table 4. Analysis of $N_{\Delta Z}$ (variant 6) calculated in the BBA in Pix4Dmapper and PhotoScan.

Parameters	$N_{\Delta Z}$ V6—Pix4Dmapper	$N_{\Delta Z}$ V6—PhotoScan
No. of ChPs	207	207
[ΔZ] ≤ 0.05 m	187	169
0.050 < [ΔZ] ≤ 0.08 m	14	28
0.080 < [ΔZ] ≤ 0.10 m	2	4
[ΔZ] > 0.10 m	4	4
[%] of [ΔZ] > 0.05 m	9.7%	17.4%

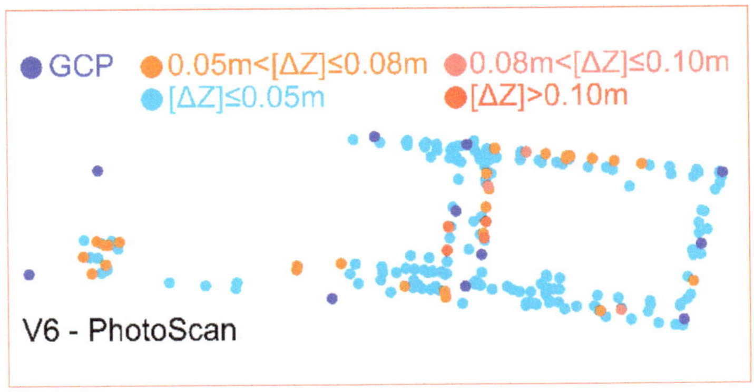

Figure 6. Localization of GCPs and ChPs in the variant 6 of the BBA performed in PhotoScan with the representation of number $N_{\Delta Z}$ in [ΔZ] deviation intervals.

The difference between number $N_{\Delta Z}$ of [ΔZ] deviations above threshold 0.05 m is 16 (ca. 8%). It is a consequence of a deviations number (6—Pix4Dmapper and 17—PhotoScan), which belongs in the 0.05 ÷ 0.06 m range. Therefore, the solutions in both applications are comparable.

In the authors' opinion, the differences in values and in the number $N_{\Delta Z}$ of [ΔZ] deviations which are higher than 0.05 m are probably caused by another stochastic (weight) model used in computation algorithms in both types of software, e.g., for tie points measured automatically.

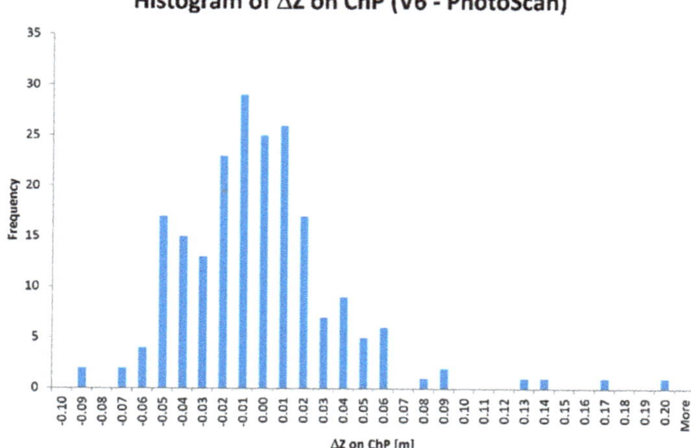

Figure 7. Histogram of ΔZ deviations on ChPs calculated in the BBA in PhotoScan application.

3.3. *Influence of Image Overlap and Global Navigation Satellite Systems (GNSS) Exterior Orientation*

The assessment of the influence of image overlap value and the use of UAV GNSS external orientation in the bundle adjustment was carried out in multi-variant (7 new variants) computation using the optimal BBA variant 6 as reference. The results of the reference and additional solutions with reports of the camera standard parameter calibration are included in supplementary information files (Appendix 6, 8–14). Table 5 contains the calculation results for four variants ($V6_{p\%\text{-}q\%}$) of image overlap, respectively to $p_{\%} = q_{\%} = 85\%$ ($V6_{85\text{-}85}$), $p_{\%} = 85\%$ and $q_{\%} = 45\%$ ($V6_{85\text{-}45}$), $p_{\%} = q_{\%} = 65\%$ ($V6_{65\text{-}65}$), $p_{\%} = 65\%$ and $q_{\%} = 45\%$ ($V6_{65\text{-}45}$) with additional observations of exterior orientation parameters included in the adjustment. Table 6 presents the same adjustment variants, but with observations of projection centers coordinates excluded in the adjustment.

Table 5. Results of the BBA with different image overlap and exterior orientation parameters of unmanned aerial vehicle position from Global Navigation Satellite Systems receiver (UAV GNSS) included in the adjustment.

Parameters	$V6_{85\text{-}85}$ incl. UAV GNSS	$V6_{85\text{-}45}$ incl. UAV GNSS	$V6_{65\text{-}65}$ incl. UAV GNSS	$V6_{65\text{-}45}$ incl. UAV GNSS
No. of images	858	293	205	138
No. of GCPs	11	11	11	11
No. of ChPs	207	207	207	207
RMSE(X) on GCPs [m]	0.010	0.016	0.010	0.012
RMSE(Y) on GCPs [m]	0.022	0.022	0.019	0.014
RMSE(Z) on GCPs [m]	0.033	0.028	0.020	0.024
RMSE(X) on ChPs [m]	0.023	0.029	0.024	0.037
RMSE(Y) on ChPs [m]	0.026	0.031	0.029	0.034
RMSE(Z) on ChPs [m]	0.036	0.048	0.046	0.055
[ΔZ] ≤ 0.05 m	187	155	159	134
0.050 < [ΔZ] ≤ 0.08 m	14	33	31	49
0.080 < [ΔZ] ≤ 0.10 m	2	10	9	10
[ΔZ] > 0.10 m	4	9	8	14
[%] of [ΔZ] > 0.05 m	9.7	25.1	23.2	35.3

In the case ($V6_{65\text{-}45}$) of minimal forward $p_{\%} = 65\%$ and side $q_{\%} = 45\%$ overlap (ca. 6.2-fold reduction of image number) the RMSE(Z_{ChP}) is a little bit over the critical value 0.05 m and the percentage of points with the absolute deviations values [ΔZ] > 0.05 m rise 3.5-fold. In two other variants ($V6_{85\text{-}65}$ and $V6_{65\text{-}65}$), the absolute deviations values [ΔZ] are lower than 0.05 m. The percentage of points with the absolute deviations values [ΔZ] > 0.05 m rise 2.5-fold. The exclusion of approximated coordinates of projection centers in the BBA process is not causing significant changes in obtained accuracy.

Table 6. Results of the BBA with different image overlap and exterior orientation parameters excluded in the adjustment.

Parameters	$V6_{85\text{-}85}$ excl. UAV GNSS	$V6_{85\text{-}45}$ excl. UAV GNSS	$V6_{65\text{-}65}$ excl. UAV GNSS	$V6_{65\text{-}45}$ excl. UAV GNSS
No. of images	858	293	205	138
No. of GCPs	11	11	11	11
No. of ChPs	207	207	207	207
RMSE(X) on GCPs [m]	0.010	0.015	0.010	0.012
RMSE(Y) on GCPs [m]	0.022	0.022	0.019	0.016
RMSE(Z) on GCPs [m]	0.029	0.027	0.019	0.023
RMSE(X) on ChPs [m]	0.023	0.029	0.024	0.038
RMSE(Y) on ChPs [m]	0.026	0.031	0.029	0.035
RMSE(Z) on ChPs [m]	0.036	0.047	0.047	0.056
[ΔZ] ≤ 0.05 m	182	156	154	133
0.050 < [ΔZ] ≤ 0.08 m	18	32	32	42
0.080 < [ΔZ] ≤ 0.10 m	2	10	13	18
[ΔZ] > 0.10 m	5	9	8	14
[%] of [ΔZ] > 0.05 m	12.1	24.6	25.6	35.7

In regard to variant $V6_{85\text{-}85}$ (incl. UAV GNSS) the max. differences of interior orientation parameters calculated for 3 variants of the BBA with UAV GNSS data and reduced overlap are the following: $\Delta c = -7.3$ pix, $\Delta x'_0 = 1.3$ pix, $\Delta y'_0 = 3.0$ pix, $\Delta K_1 = 0.000$, $\Delta K_2 = 0.000$, $\Delta K_3 = 0.001$, $\Delta P_1 = 0.001$, $\Delta P_2 = 0.000$. Max. differences of parameters between variant $V6_{85\text{-}85}$ (excl. UAV GNSS) and 3 variants of the BBA without UAV GNSS data and reduced overlap are respectively: $\Delta c = 9.3$ pix, $\Delta x'_0 = 1.4$ pix, $\Delta y'_0 = 3.0$ pix, $\Delta K_1 = 0.000$, $\Delta K_2 = 0.000$, $\Delta K_3 = 0.000$, $\Delta P_1 = 0.001$, $\Delta P_2 = 0.000$.

The value differences of parameters calculated between relevant variants with and without UAV GNSS data are very similar except for variant $V6_{65\text{-}45}$ ($p_{\%} = 65\%$ and $q_{\%} = 45\%$ overlap), respectively: $\Delta c = 3.3$ pix, $\Delta x'_0 = 0.1$ pix, $\Delta y'_0 = 2.4$ pix, $\Delta K_1 = 0.000$, $\Delta K_2 = 0.000$, $\Delta K_3 = 0.001$, $\Delta P_1 = 0.001$, $\Delta P_2 = 0.000$. The standard deviation of determined parameters are in range $s_c \in <0.5; 0.8>$ pix, $sx'_o = sy'_o \in <0.07; 0.14>$ pix.

3.4. Comparison of Photogrammetric-Generated Digital Surface Model (DSM) and GNSS Point Cloud

Due to the requirements of height accuracy (RMSE(Z) ≤ 0.05 m), the generated DSM in Pix4Dmapper (variant 6) was compared to the point cloud obtained using GNSS. The determination of DSM undulation is realized by Equation (5) and is presented in Figure 8.

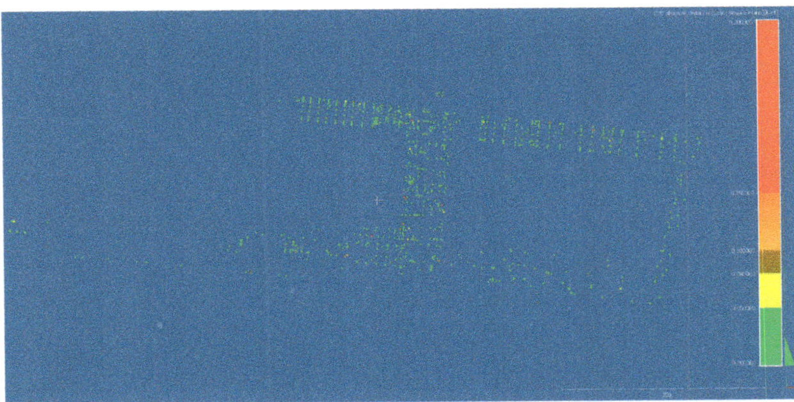

Figure 8. Localization and histogram of the vertical distances (VD) between computed digital surface model (DSM) in Pix4Dmapper and GNSS point cloud.

For most of the 1418 points (300 ChPs and 1118 height ground ChPs), the values of the vertical distance VD (height differences) belong to 0 ÷ 0.05 m range. Only for some random points, the estimated vertical distances are higher than 0.05 m. The vertical distances outliers comprise 4.3% (61) of the total number of points (1418). This fact confirms that DSM is correctly generated even in the area with the discontinuity of the terrain surface (scarps and terrain slope).

3.5. Comparison of the Orthomosaic and GNSS Data

The photogrammetric planar data collection was performed using orthomosaic. The outcome GeoTIFF ortomosaic (Figure 1) of Pix4Dmapper is 37058 × 15284 pixels (855 MB). To verify if the georeferenced orthomosaic comply with the mapping requirement (RMSE(XY) ≤ 0.10 m) of the point position in the cartesian coordinate system, the 2D reference coordinates of 11 GCPs, 207 ChPs and 82 supplementary ChPs measured by the GNNS method were applied. The 82 additional ChPs were the densification of the ChPs set and allowed an independent accuracy verification of the orthomosaic.

The photogrammetric 2D data collection (object vectorization) on the georeferenced orthomosaic can be treated as detailed land surveys. On the basis of the manual measurement on the GeoTIFF raster in AutoCAD Civil 3D, the deviation vectors $D_{GNSS\text{-}BBA}$ and $D_{GNSS\text{-}OM}$ were defined (GNSS—coordinates from GNSS measurement, BBA—coordinates computed in the bundle block adjustment, OM—coordinates from orthomosaic measurement). The a priori accuracy of manual pixel coordinates measurement was set at the level of 0.33 pixel size, which corresponds to 0.015 m in the terrain. Table 7 describes the accuracy evaluation of georeferenced orthomosaic. The length distributions of deviation vectors D in the plane are presented in Figure 9.

The measurement on the orthomosaic founds that 61 (20.3%) vectors exceed the threshold value of 0.10 m, where the calculated measurement accuracy for 300 ChPs was RMS($D_{GNSS\text{-}OM}$) = 0.082 m. In comparison with the reference data (2D coordinates) obtained from the BBA, the number of length vectors outliers increases up to 20-fold. The analysis of the histograms confirms that trend.

To check if the constant shift on the generated orthomosaic exists, the directions of deviation vectors were studied. The vectors' directions were assigned to 4 quarters of the coordinate system. Table 8 contains the results of the verification.

Table 7. Length analysis of deviation vectors D.

Parameters	V6—Pix4Dmapper
No. of vectors $D_{GNSS\text{-}BBA}$ > 0.10 m (218 ChPs)	3
No. of vectors $D_{GNSS\text{-}OM}$ > 0.10 m (218 ChPs)	50
No. of vectors $D_{GNSS\text{-}OM}$ > 0.10 m (82 ChPs)	11
No. of vectors $D_{GNSS\text{-}OM}$ > 0.10 m (300 ChPs)	61
[%] of vectors $D_{GNSS\text{-}BBA}$ > 0.10 m (218 ChPs)	1.4
[%] of vectors $D_{GNSS\text{-}OM}$ > 0.10 m (218 ChPs)	22.9
[%] of vectors $D_{GNSS\text{-}OM}$ > 0.10 m (82 ChPs)	13.4
[%] of vectors $D_{GNSS\text{-}OM}$ > 0.10 m (300 ChPs)	20.3
RMS($D_{GNSS\text{-}BBA}$) (218 ChPs) [m]	0.034
RMS($D_{GNSS\text{-}OM}$) (218 ChPs) [m]	0.084
RMS($D_{GNSS\text{-}OM}$) (82 ChPs) [m]	0.076
RMS($D_{GNSS\text{-}OM}$) (300 ChPs) [m]	0.082

Figure 10 presents the localization of ChPs with the information about vectors' length and directions, where each direction is marked with the following colors: N–E: green, N–W: red, S–W: blue, S–E: purple.

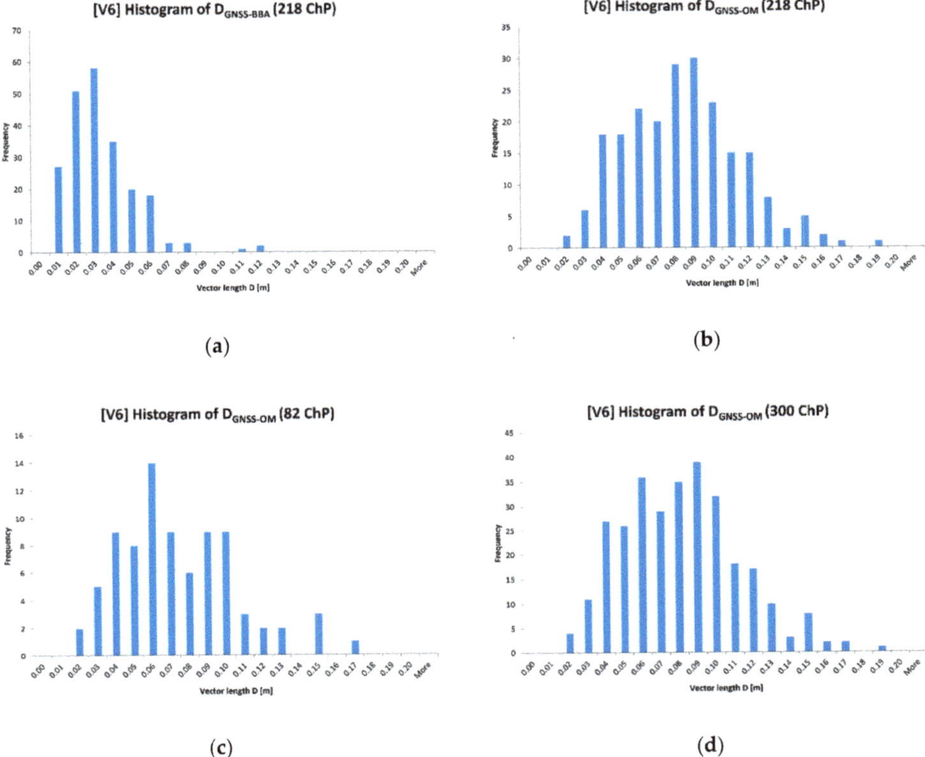

Figure 9. Length distribution of deviation vectors D: (**a**) $D_{GNSS\text{-}BBA}$ (218 ChPs); (**b**) $D_{GNSS\text{-}OM}$ (218 ChPs); (**c**) $D_{GNSS\text{-}OM}$ (82 ChPs); (**d**) $D_{GNSS\text{-}OM}$ (300 ChPs).

Table 8. Analysis of direction of deviation vectors D.

Vector Direction	↗	↖	↙	↘
No. of vectors $D_{GNSS\text{-}BBA}$ (218 ChPs)	69	53	54	42
No. of vectors $D_{GNSS\text{-}OM}$ (218 ChPs)	6	195	15	2
No. of vectors $D_{GNSS\text{-}OM}$ (82 ChPs)	3	67	10	2
No. of vectors $D_{GNSS\text{-}OM}$ (300 ChPs)	9	262	25	4

The analysis of the analytical BBA data presented a similar direction distribution of the vectors $D_{GNSS\text{-}BBA}$. On this distribution, the systematic errors are not found. However, directions distribution of vectors $D_{GNSS\text{-}OM}$ measured on georeferenced orthomosaic has a systematic shift in the N–W (↖) direction. The analysis of vectors ($D_{GNSS\text{-}BBA}$ and $D_{GNSS\text{-}OM}$) localization in both cases shows that the similarity of distributions does not exist. The values of deviation vectors on the orthomosaic were affected by the orthorectification process. In the authors' opinion, the direction of shadow projection and different lighting conditions on image sequences in flight strips may influence on the length and the direction of error vectors.

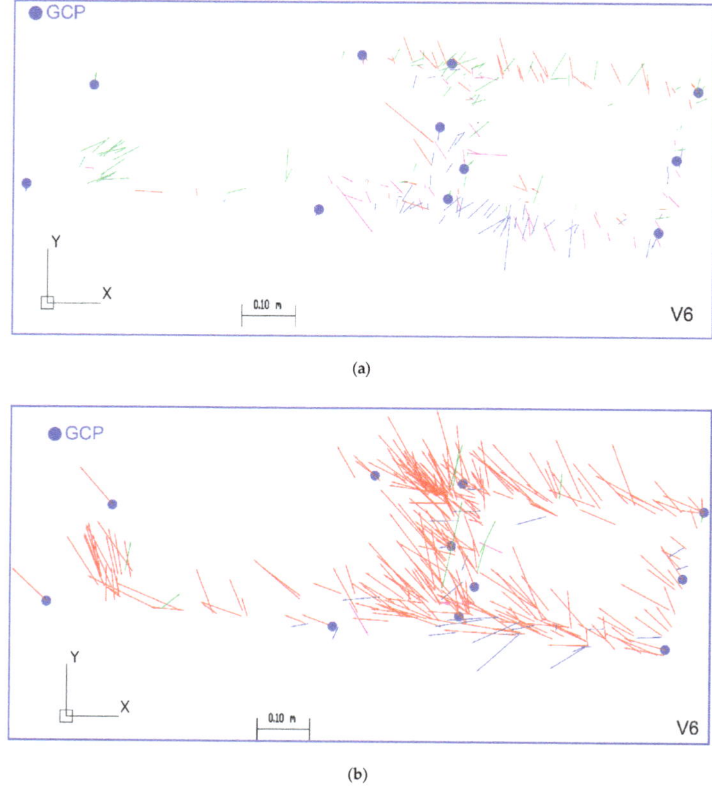

(**a**)

(**b**)

Figure 10. Localization of vectors D which include length and direction: (**a**) $D_{GNSS\text{-}BBA}$ (218 ChPs); (**b**) $D_{GNSS\text{-}OM}$ (300 ChPs).

3.6. Inspection of the Existing Digital Maps

The orthomosaic generated from UAV images is a relatively low-cost and quickly obtained photogrammetric product. Its actual photographic content allows the easy and fast visual detection of outdated information (Figure 11) and incompleteness (Figure 12) of existing vector digital maps, which are the result of geodetic detailed field surveys.

Figure 11. Visualization of the outdated information on the vector base map using orthomosaic in AutoCAD.

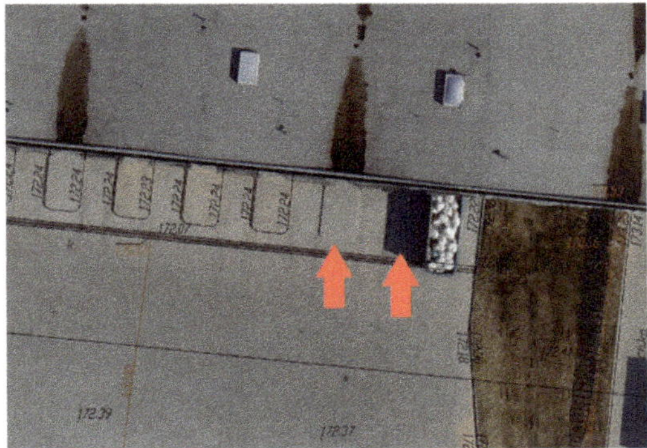

Figure 12. Visualization of the deficiencies on the vector base map using orthomosaic in AutoCAD.

4. Discussion

For accuracy analysis, the current Polish technical standards for the land surveys (planar-elevation measurements of well-defined terrain details) in order to undertake large-scale mapping, included in the Regulation of the Minister of the Interior and Administration in Poland, No. 263 of 9 November 2011 [43], were used as the reference. They precisely describe the accuracy of field geodetic surveys of the ground details, which are uniquely identifiable, retaining the long-term invariability of shape and position. The measurements should achieve the point position in the cartesian coordinate system with error $m_{XY} \leq 0.10$ m and point height position in the elevation system with error $m_Z \leq 0.05$ m. The geodetic control network points used as a reference for dedicated field surveys are treated as errorless.

The planimetric mapping (land survey) is performed on the assumption that the accuracy (error m_{XY}) of a point location of terrain details relative to the nearest horizontal geodetic control network is not less than:

- 0.10 m: Ist survey group, i.e., survey marks of the control network, boundary points, building objects, and construction equipment, including elements of the utilities network, directly available for measurement;
- 0.30 m: IInd survey group, i.e., buildings and earth devices in the form of embankments, excavations, dikes, dams, ditches, canals and artificial water reservoirs; invisible parts of buildings and construction equipment, including objects of covered infrastructural networks; land use objects, in particular: parks, green areas, lawns, playgrounds and rest, squares, single trees as well sport fields;
- 0.50 m: IIIrd survey group, i.e., land use contours and soil outcrops for the needs of the land soil classification; watercourses and water reservoirs with natural boundaries; section boundary in the forests and national parks.

The geodetic height measurement of terrain elements is performed on the assumption that the accuracy of point height (error m_Z) in relation to the nearest height (vertical) geodetic control network is not less than:

- 0.02 m: for underground pipes and sewage devices;
- 0.05 m: for building objects and construction equipment as well as height spots marked in the field;
- 0.10 m: for terrain structures and flexible lines or electromagnetically measured underground objects of the infrastructural network; and non-marked height spots of characteristic points of the ground elevation.

The specification of terrain detail types and their planar-elevation survey accuracy contained in the regulation [43] mentioned above in principle applies to the direct geodetic survey and does not strictly concern the photogrammetric measurement. Due to the lack of the technical guidelines for practical usage of UAVs images in geodesy, the listed max. errors for surveys of well-defined over ground points were adopted in tests for the accuracy assessment of 2D and 3D photogrammetric measurement.

In this paper, we have compared conventional survey quality considerations with a photogrammetric approach to evaluate the accuracy of UAV imagery data collection. Our case study was performed according to Federal Geographic Data Committee mapping accuracy standards [65]. For testing purposes, features with known horizontal position (high degree of accuracy) and position with respect to the geodetic datum were used. Also, the independent source of higher accuracy was acquired separately from data used in the aerotriangulation solution. According to [65], check points were distributed more densely in the area of interest and more sparsely in that of little or no interest.

In the study, the set of images used for processing was radiometric proper; they did not contain motion and optical blur. The process of the complex photogrammetric data collection was performed in Pix4Dmapper software in three stages. The first computation stage was the bundle block adjustment. Empirical tests have shown the need to apply 11 GCPs (variant 6 of the BBA) in the configuration (Figure 3f) related to the buildings localization. The use of 3 additional GCPs was necessary to minimize the reprojection errors on the stage of the BBA and to eliminate the influence of amorphous surfaces on the image matching. The interior orientation parameters computed in self-calibration depend significantly on the configuration of the image block (image number, block and strip geometry, forward and side overlap, drift, shift), GCPs (point signalize type, shape, size, color, number, distribution, resolution, accuracy) and UAV GNSS accuracy. The accuracy of camera calibration parameters determined using BBA affects the accuracy and reliability of estimated 3D ground point coordinates. The simultaneous self-calibration method most accurately corresponds to the real conditions of images acquisition and computation, and provides the optimal functional model of the solution by bundle method.

The 3D ground coordinates of all the pass points (natural signalized terrain details) estimated using the BBA can be the background for a 3D vector digital map production. The external accuracy measure, i.e., the RSME on 207 ground ChPs (XYZ) were respectively to RMSE(X) = 0.023 m, RMSE(Y) = 0.026 m and RMSE(Z) = 0.036 m. In respect of theoretical inhomogeneous horizontal and vertical accuracy of the estimated coordinates in the bundle adjustment, the ΔZ deviations were analyzed. It turned out that ca 90% of ΔZ deviations on 207 ChPs were to $\Delta Z \leq 0.05$ m and only ca. 2% to $\Delta Z > 0.10$ m. Barry and Coakley [39] used in their study 10 GCPs and 45 ChPs on the small field area (0.3250 ha). They acquired 95% reliably within 0.041 mm horizontally and 0.068 m vertically with the 1.17 cm size of GSD. Benassi et al. [70] in their tests used eBee RPAS with direct georeferencing on the test field (20 ha) with 12 GCPs, 14 ChPs. They declare standard deviations of planar camera positions between 0.01 ÷ 0.02 m and 0.03 ÷ 0.05 m in vertical direction, and acquired average horizontal RMSE on ChPs of 0.022 m and 0.055 m in elevation. In our opinion, the small sample of ChPs, their localization far from high objects and no independent points render them inadequate to do the accuracy assessment of RTK RPAS measurements. Besides with fewer GCPs used as a control, error magnitudes on control and check points diverged, which was confirmed by James et al. [41].

Another study aspect relative to the initial dataset was the assessment of the impact of less overlap and, as a consequence, a reduction of the images number in the bundle adjustment. We found that the optimal variant in terms of accuracy and computing time using standard grade workstation is that with reduced photos overlap at the level of $p_{\%} = q_{\%} = 65\%$ (ca. 4-fold decrease of image number). However, the optimal configuration of the GCPs is a condition for reducing the photos number, which is confirmed by the comparison of BBA results for variants V1 and V4 (Table 1), and $V6_{65\text{-}65}$ (Table 5). Tests shown that for block of nadir images with decreased overlap to $p_{\%} = q_{\%} = 65\%$, it is possible to achieve the horizontal and vertical accuracy of ground objects at the level of RMSE(X_{ChP},Y_{ChP}) = 0.5 GSD and RMSE(Z_{ChP}) = 1 GSD (0.046 m), but in this case the number $N_{\Delta Z}$ of [ΔZ] deviations higher than 0.05 m increases ca. 2.5-fold. As expected due to the low position accuracy (horizontal 5 m and vertical 10 m) of the GNSS single-frequency receiver (C/A, L1) located on the Trimble® UX5 UAS platform, the inclusion of approximated coordinates of projection centers in the bundle adjustment in all tested variants did not affect on the computation results. For the BBA results evaluation, the number and distribution of ChPs, which are treated as tie points in computation should be taken into consideration. Mostafa in his research [71] state that changing the forward and side overlap from $p_{\%} = q_{\%} = 80\%$ to $p_{\%} = 80\%$ and $q_{\%} = 40\%$ resulted in the same accuracy within the measurement noise. The investigations were conducted with GSD = 0.008 m and GSD = 0.016 m. Furthermore, it was indicated [71] that ground object positioning accuracy is about 2 ÷ 3 GSD value, and height accuracy is about 4 ÷ 5 GSD value.

The DSM which was generated in the second processing stage has been validated by 1418 GNSS height check points in CloudCompare application. The approach was realized successfully before [53]. The ca. 96% of the height differences are related to $\Delta Z \leq 0.05$ m and only ca. 4% are outliers $\Delta Z > 0.05$ m. Oniga et al. [55] compared results with terrestrial laser scanning (TLS) data and obtained a standard deviation equal to 0.098 m. Wierzbicki and Nienaltowski [42] created the triangulated irregular network (TIN) model based on points measured using the GNSS RTK technique, and they achieved a standard deviation of 0.090 m for height differences. To avoid the interpolation of the irregular shape of the ground, our study contains height differences only for points measured in the field.

Finally, the manual measurement of the well-defined points on the GeoTIFF orthomosaic (GSD_{ortho} = 4.6 cm) was carried out in AutoCAD Civil 3D. The photogrammetric 2D data collection (pointwise object vectorization) on the georeferenced orthomosaic can be treated as detailed land surveys. To take into account the limiting error ($m_L = 2\ m_0$) of the GeoTIFF manual measurement, it could be considered that ca. 90% of deviation vectors D on the set of 300 ChPs fulfill the planar accuracy requirements $RMSE(XY) \leq 0.10$ m. According to the authors' knowledge, this procedure has not been used before. For describing the accuracy of orthomosaics, researchers used the BBA results [36,72]. Hung et al. [12] and James et al. [41] analyzed the directions and length of deviations vectors for GCPs and ChPs from the BBA. In our study, we also checked them on the orthomosaic, which showed that the constant shift occurs despite the random distribution of vectors from the BBA.

It is worth noting that in a complex process of orthomosaic generation, error propagation occurs. The final geometric accuracy and radiometric quality of orthomosaic is influenced by the many factors caused by, e.g., target marking (GCPs) and coordinate measurement errors, incorrect stochastic (weighted) model in the BBA, image-matching errors, estimated parameters errors of interior and exterior orientation, insufficient correction of image distortions, densified point cloud noise, final density and quality of DSM, applied orthorectification and digital interpolation method during the resampling, radiometric correction of respective orthoimages, quality of mosaicking.

Analyzing the real, not ideal conditions of the input dataset, it can be assumed that the following factors could have a significant impact on the accuracy of the final results: a relatively large size of GSD, a small number and incorrect artificial signalization of GCPs, inappropriate orientation of the strips relative to the shape of the photo block, and difficult conditions of dense matching as a consequence of the amorphous texture of the industrial object surface.

The positive experimental results of the presented study encourage the authors to continue research on the accuracy evaluation of photogrammetric products based on UAV imagery. Therefore, further work will be carried out on a topographically diverse test field, a representative for the general case of study, and the negative factors will be eliminated, which decreases the accuracy of the UAVs survey.

5. Conclusions

The use of unmanned aerial vehicles (UAV) for image-based surveys is becoming increasingly widespread across the geosciences and in industries such as engineering and surveying. However, the increasing availability of such data has not been paralleled by an equivalent increase in research about the data quality. The aim of the study is to maximize the accuracy and reliability of UAV survey results, whilst optimizing the input dataset and digital-processing requirements.

The presented study evaluates the accuracy of photogrammetric 2D and 3D data collection based on processed medium resolution images acquired from Trimble®UX5 UAS production flight mission, in the case of a small investment area with a focus on the influence of ground control points configuration. The study site is representative for other areas with similar characteristic and location of terrain objects.

The initial dataset was characterized by a block of 858 digital images with 16 megapixel resolution, 4.6 cm ground sample distance, 5 artificially signalized ground control points. For the verification of the results, the 295 ground check points and 1118 height check points were measured applying the GNSS using the real-time network method with the mean sigma (NEH) respectively to 0.012 m, 0.008 m, 0.020 m.

The photogrammetric data collection for well-defined points was carried out using the Pix4Dmapper suite in three processing stages: in the bundle block adjustment, on the generated digital surface model and the georeferenced orthomosaic. Furthermore, the bundle block adjustment was performed using RealityCapure and PhotoScan packages. In addition, the extended accuracy analyses were achieved on the digital surface model using CloudCompare application and on the orthomosaic supported by AutoCAD Civil 3D.

The main original research contributions of this paper are the following:

- Creating a new methodology for complex accuracy assessment of computing pipeline of particular photogrammetric products;
- Proving high accuracy and significant statistical reliability of photogrammetric 2D and 3D data collection based on UAV medium spatial resolution imagery with lower than 5 cm ground sample distance;
- Stating the absolute need for a multi-variant bundle block adjustment and error assessment to search for optimal ground control points configuration for further advanced digital processing chain, which allows minimizing image-matching errors caused by the amorphous texture;
- Proving the high horizontal and vertical accuracy (root mean square error less than or equal to 0.10 m and 0.05 m) for well-defined terrain points of photogrammetric products based on UAV imagery, which fulfills Polish and US technical requirements;
- Finding the optimal photos overlap in terms of accuracy and computing time;
- Proving the lack of influence of low-accuracy image georeferences on the bundle block adjustment results;
- Confirming the appropriateness of UAV imagery and dedicated photogrammetric software packages for 2D and 3D data collection in vector, raster and hybrid form in order to undertake large-scale basic, thematic and inventory mapping;
- Proving the possibilities of replacing direct geodetic field measurements by the UAVs photogrammetric measurements with the accuracy for large-scale mapping;
- Verifying the usefulness of UAV imagery and their products for regular updates existing digital base maps and geodetic database.

The accuracy of the achieved products shows the full practical utility of a digital compact system camera for geodata collection. High accuracy could be obtained only under the conditions of multi-variant image processing and extended error analysis, according to the proposed workflow. In the case of a smaller ground sample distance, the accuracy of data collection will be significantly higher.

When the Pix4Dmapper package is applied for the bundle-block adjustment, additional software should be used in the post processing to find the image coordinates measurement blunders and to verify primary adjustment results. For the accuracy assessment of the georeferenced orthomosaic, we propose to analyze the distribution and length of deviation vectors.

The more frequent application of UAV imagery in geodesy and geomatics is, unfortunately, limited by the lack of dedicated national technical regulations and guidelines.

Further study will be carried out on the specially designed test field characterized by a high number of artificially signalized and evenly distributed ground control and check points which have been measured with high accuracy using static GNSS method and precise leveling. In addition, the research area should have a block of nadir and oblique images with a small ground sample distance (1 ÷ 2 cm level), acquired with a medium or large-format digital camera with high-resolution and a digital surface model based on dense point clouds obtained from airborne laser scanning, which will be used as a reference. A new test field will allow for testing the quality and accuracy of standard products (bundle-block adjustment, digital surface model, georeferenced orthomosaic) and other products derived from UAV imagery that were not the subject of the described study, i.e., true orthophoto maps, 3D object reconstruction, and 3D city models in the two-level of detail (LOD3 and LOD4).

Supplementary Materials: The following are available online at http://www.mdpi.com/1424-8220/19/23/5229/s1.

Author Contributions: Conceptualization, G.G. and P.S.; methodology, G.G. and P.S.; software, G.G. and P.S.; validation, G.G. and P.S; formal analysis, G.G. and P.S. investigation G.G. and P.S.; resources, G.G. and P.S.; data curation, G.G. and P.S.; writing—original draft preparation, G.G. and P.S.; writing—review and editing, G.G. and P.S.; visualization, G.G. and P.S.

Funding: This research received no external funding, and the APC was funded by the University of Warmia and Mazury in Olsztyn.

Acknowledgments: We gratefully thank the Geoplan company who provided data used in this study.

Conflicts of Interest: The authors declare no conflict of interest.

References

1. Granshaw, S.I. Photogrammetric Terminology: Third Edition. *Photogramm. Rec.* **2016**, *31*, 210–252. [CrossRef]
2. Granshaw, S.I. RPV, UAV, UAS, RPAS ... or just drone? *Photogramm. Rec.* **2018**, *33*, 160–170. [CrossRef]
3. Colomina, I.; Molina, P. Unmanned aerial systems for photogrammetry and remote sensing: A review. *ISPRS J. Photogramm. Remote Sens.* **2014**, *92*, 79–97. [CrossRef]
4. Austin, R. *Unmanned Aircraft Systems: UAVS Design, Development and Deployment*; Wiley: Chippenham, UK, 2010; ISBN 9780470058190.
5. Lin, Z.-J. UAV for Mapping—Low Altitude Photogrammetric Survey. *Int. Arch. Photogramm. Remote Sens. Spat. Inf. Sci.* **2008**, *37*, 1183–1186.
6. Hirschmüller, H. Accurate and Efficient Stereo Processing by Semi-Global Matching and Mutual Information. Comput. Vis. Pattern Recognition, 2005. CVPR 2005. *IEEE Comput. Soc. Conf.* **2005**, *2*, 807–814. [CrossRef]
7. Hirschmüller, H. Stereo processing by semiglobal matching and mutual information. *IEEE Trans. Pattern Anal. Mach. Intell.* **2008**, *30*, 328–341. [CrossRef] [PubMed]
8. Vu, H.H.; Labatut, P.; Pons, J.P.; Keriven, R. High accuracy and visibility-consistent dense multiview stereo. *IEEE Trans. Pattern Anal. Mach. Intell.* **2012**, *34*, 889–901. [CrossRef] [PubMed]
9. Remondino, F.; Spera, M.G.; Nocerino, E.; Menna, F.; Nex, F. State of the art in high density image matching. *Photogramm. Rec.* **2014**, *29*, 144–166. [CrossRef]
10. Jaud, M.; Passot, S.; Le Bivic, R.; Delacourt, C.; Grandjean, P.; Le Dantec, N. Assessing the accuracy of high resolution digital surface models computed by PhotoScan® and MicMac® in sub-optimal survey conditions. *Remote Sens.* **2016**, *8*, 465. [CrossRef]
11. Remondino, F.; Nocerino, E.; Toschi, I.; Menna, F. A critical review of automated photogrammetric processing of large datasets. In Proceedings of the International Archives of the Photogrammetry, Remote Sensing and Spatial Information Sciences, Ottawa, ON, Canada, 28 August–1 September 2017. [CrossRef]
12. Hung, I.-K.; Unger, D.; Kulhavy, D.; Zhang, Y. Positional Precision Analysis of Orthomosaics Derived from Drone Captured Aerial Imagery. *Drones* **2019**, *3*, 46. [CrossRef]
13. Oniga, V.E.; Pfeifer, N.; Loghin, A.M. 3D calibration test-field for digital cameras mounted on unmanned aerial systems (UAS). *Remote Sens.* **2018**, *10*, 2017. [CrossRef]
14. Uysal, M.; Toprak, A.S.; Polat, N. DEM generation with UAV Photogrammetry and accuracy analysis in Sahitler hill. *Meas. J. Int. Meas. Confed.* **2015**, *73*, 539–543. [CrossRef]

15. Mesas-Carrascosa, F.J.; Rumbao, I.C.; Berrocal, J.A.B.; Porras, A.G.F. Positional quality assessment of orthophotos obtained from sensors onboard multi-rotor UAV platforms. *Sensors* **2014**, *14*, 22394–22407. [CrossRef] [PubMed]
16. Nex, F.; Remondino, F. UAV for 3D mapping applications: A review. *Appl. Geomat.* **2014**, *6*, 1–15. [CrossRef]
17. Pajares, G. Overview and current status of remote sensing applications based on unmanned aerial vehicles (UAVs). *Photogramm. Eng. Remote Sens.* **2015**, *81*, 281–330. [CrossRef]
18. Crommelinck, S.; Bennett, R.; Gerke, M.; Nex, F.; Yang, M.Y.; Vosselman, G. Review of automatic feature extraction from high-resolution optical sensor data for UAV-based cadastral mapping. *Remote Sens.* **2016**, *8*, 689. [CrossRef]
19. Gevaert, C.; Sliuzas, R.; Persello, C.; Vosselman, G. Opportunities for UAV mapping to support unplanned settlement upgrading. *Rwanda J.* **2017**, *1*. [CrossRef]
20. D'Oleire-Oltmanns, S.; Marzolff, I.; Peter, K.D.; Ries, J.B. Unmanned aerial vehicle (UAV) for monitoring soil erosion in Morocco. *Remote Sens.* **2012**, *4*, 3390–3416. [CrossRef]
21. Yao, H.; Qin, R.; Chen, X. Unmanned Aerial Vehicle for Remote Sensing Applications—A Review. *Remote Sens.* **2019**, *11*, 1443. [CrossRef]
22. Lippitt, C.D.; Zhang, S. The impact of small unmanned airborne platforms on passive optical remote sensing: A conceptual perspective. *Int. J. Remote Sens.* **2018**, *39*, 4852–4868. [CrossRef]
23. Shahbazi, M.; Sohn, G.; Théau, J.; Menard, P. Development and evaluation of a UAV-photogrammetry system for precise 3D environmental modeling. *Sensors* **2015**, *15*, 27493–27524. [CrossRef] [PubMed]
24. Lehmann, J.R.K.; Prinz, T.; Ziller, S.R.; Thiele, J.; Heringer, G.; Meira-Neto, J.A.A.; Buttschardt, T.K. Open-source processing and analysis of aerial imagery acquired with a low-cost Unmanned Aerial System to support invasive plant management. *Front. Environ. Sci.* **2017**, *5*, 44. [CrossRef]
25. Harwin, S.; Lucieer, A.; Osborn, J. The impact of the calibration method on the accuracy of point clouds derived using unmanned aerial vehicle multi-view stereopsis. *Remote Sens.* **2015**, *7*, 11933–11953. [CrossRef]
26. Remondino, F.; Fraser, C. Digital camera calibration methods: Considerations and comparisons. *Int. Arch. Photogramm. Remote Sens. Spat. Inf. Sci.* **2006**, *36*, 266–272.
27. Forlani, G.; Diotri, F.; Cella, U.M.; Roncella, R. Indirect UAV Strip Georeferencing by On-Board GNSS Data under Poor Satellite Coverage. *Remote Sens.* **2019**, *15*, 1765. [CrossRef]
28. Vo, A.V.; Truong-Hong, L.; Laefer, D.F.; Tiede, D.; Doleire-Oltmanns, S.; Baraldi, A.; Shimoni, M.; Moser, G.; Tuia, D. Processing of Extremely High Resolution LiDAR and RGB Data: Outcome of the 2015 IEEE GRSS Data Fusion Contest-Part B: 3-D Contest. *IEEE J. Sel. Top. Appl. Earth Obs. Remote Sens.* **2016**, *9*, 5560–5575. [CrossRef]
29. Koch, T.; Körner, M.; Fraundorfer, F. Automatic and Semantically-Aware 3D UAV Flight Planning for Image-Based 3D Reconstruction. *Remote Sens.* **2019**, *11*, 1550. [CrossRef]
30. Agarwal, S.; Snavely, N.; Simon, I.; Seitz, S.M.; Szeliski, R. Building Rome in a day. In Proceedings of the 2009 IEEE 12th International Conference on Computer Vision, Kyoto, Japan, 29 September–2 October 2009; pp. 72–79. [CrossRef]
31. Haala, N.; Cramer, M.; Weimer, F.; Trittler, M. Performance Test On Uav-Based Photogrammetric Data Collection. *ISPRS Int. Arch. Photogramm. Remote Sens. Spat. Inf. Sci.* **2012**, *38*. [CrossRef]
32. Remondino, F.; Barazzetti, L.; Nex, F.; Scaioni, M.; Sarazzi, D. UAV Photogrammetry for Mapping and 3D Modeling—Current Status and Future Perspectives. *ISPRS Int. Arch. Photogramm. Remote Sens. Spat. Inf. Sci.* **2012**, *38*, C22. [CrossRef]
33. Harwin, S.; Lucieer, A. Assessing the accuracy of georeferenced point clouds produced via multi-view stereopsis from Unmanned Aerial Vehicle (UAV) imagery. *Remote Sens.* **2012**, *4*, 1573–1599. [CrossRef]
34. Zhou, Y.; Rupnik, E.; Faure, P.H.; Pierrot-Deseilligny, M. GNSS-assisted integrated sensor orientation with sensor pre-calibration for accurate corridor mapping. *Sensors* **2018**, *18*, 2783. [CrossRef] [PubMed]
35. Gindraux, S.; Boesch, R.; Farinotti, D. Accuracy assessment of digital surface models from Unmanned Aerial Vehicles' imagery on glaciers. *Remote Sens.* **2017**, *9*, 186. [CrossRef]
36. Mesas-Carrascosa, F.J.; García, M.D.N.; De Larriva, J.E.M.; García-Ferrer, A. An analysis of the influence of flight parameters in the generation of unmanned aerial vehicle (UAV) orthomosaicks to survey archaeological areas. *Sensors* **2016**, *16*, 1838. [CrossRef] [PubMed]

37. Zhang, S.; Lippitt, C.D.; Bogus, S.M.; Loerch, A.C.; Sturm, J.O. The accuracy of aerial triangulation products automatically generated from hyper-spatial resolution digital aerial photography. *Remote Sens. Lett.* **2016**, *7*, 160–169. [CrossRef]
38. Cramer, M.; Bovet, S.; Gültlinger, M.; Honkavaara, E.; McGill, A.; Rijsdijk, M.; Tabor, M.; Tournadre, V. On the Use of RPAS in National Mapping—The EuroSDR Point of View. *ISPRS Int. Arch. Photogramm. Remote Sens. Spat. Inf. Sci.* **2013**, *XL-1/W2*, 93–99. [CrossRef]
39. Barry, P.; Coakley, R. Field Accuracy Test of RPAS Photogrammetry. *ISPRS Int. Arch. Photogramm. Remote Sens. Spat. Inf. Sci.* **2013**, *40*. [CrossRef]
40. James, M.R.; Robson, S.; Smith, M.W. 3-D uncertainty-based topographic change detection with structure-from-motion photogrammetry: Precision maps for ground control and directly georeferenced surveys. *Earth Surf. Process. Landf.* **2017**, *42*, 1769–1788. [CrossRef]
41. James, M.R.; Robson, S.; d'Oleire-Oltmanns, S.; Niethammer, U. Optimising UAV topographic surveys processed with structure-from-motion: Ground control quality, quantity and bundle adjustment. *Geomorphology* **2017**, *281*, 51–66. [CrossRef]
42. Wierzbicki, D.; Nienaltowski, M. Accuracy Analysis of a 3D Model of Excavation, Created from Images Acquired with an Action Camera from Low Altitudes. *ISPRS Int. J. Geo Inf.* **2019**, *8*, 83. [CrossRef]
43. Regulation of the Minister of the Interior and Administration in Poland, No. 263 of 9 November 2011 Dz.U.2011.263.1572 2011. Available online: http://prawo.sejm.gov.pl/isap.nsf/download.xsp/WDU20112631572/O/D20111572.pdf (accessed on 20 October 2019).
44. ASPRS. Positional Accuracy Standards for Digital Geospatial Data. *Photogramm. Eng. Remote Sens.* **2015**, *3*, 1–26. [CrossRef]
45. DIN 18740-3 Photogrammetrische Produkte—Teil 3: Anforderungen an das Orthobild 2015. Available online: https://www.din.de/de/mitwirken/normenausschuesse/nabau/normen/wdc-beuth:din21:235160913 (accessed on 20 October 2019).
46. DIN 18740-6 Photogrammetrische Produkte—Teil 6: Anforderungen an digitale Höhenmodelle 2014. Available online: https://www.din.de/de/mitwirken/normenausschuesse/nabau/normen/wdc-beuth:din21:222023697 (accessed on 20 October 2019).
47. Martínez-Carricondo, P.; Agüera-Vega, F.; Carvajal-Ramírez, F.; Mesas-Carrascosa, F.J.; García-Ferrer, A.; Pérez-Porras, F.J. Assessment of UAV-photogrammetric mapping accuracy based on variation of ground control points. *Int. J. Appl. Earth Obs. Geoinf.* **2018**, *72*, 1–10. [CrossRef]
48. Agüera-Vega, F.; Carvajal-Ramírez, F.; Martínez-Carricondo, P. Assessment of photogrammetric mapping accuracy based on variation ground control points number using unmanned aerial vehicle. *Meas. J. Int. Meas. Confed.* **2017**, *98*, 221–227. [CrossRef]
49. Tonkin, T.N.; Midgley, N.G. Ground-control networks for image based surface reconstruction: An investigation of optimum survey designs using UAV derived imagery and structure-from-motion photogrammetry. *Remote Sens.* **2016**, *8*, 786. [CrossRef]
50. Stöcker, C.; Bennett, R.; Nex, F.; Gerke, M.; Zevenbergen, J. Review of the current state of UAV regulations. *Remote Sens.* **2017**, *9*, 459. [CrossRef]
51. Murtiyoso, A.; Grussenmeyer, P.; Börlin, N.; Vandermeerschen, J.; Freville, T. Open Source and Independent Methods for Bundle Adjustment Assessment in Close-Range UAV Photogrammetry. *Drones* **2018**, *2*, 3. [CrossRef]
52. Cramer, M. The UAV@LGL BW Project—A NMCA Case Study. *Photogr. Week* **2013**, *2011*, 165–179.
53. Gabara, G.; Sawicki, P. Application of UAV Imagery for Inventory Mapping—A Case of Industrial Estate. In Proceedings of the 2018 Baltic Geodetic Congress (BGC Geomatics), Olsztyn, Poland, 21–23 June 2018. [CrossRef]
54. Sanz-Ablanedo, E.; Chandler, J.H.; Rodríguez-Pérez, J.R.; Ordóñez, C. Accuracy of Unmanned Aerial Vehicle (UAV) and SfM photogrammetry survey as a function of the number and location of ground control points used. *Remote Sens.* **2018**, *10*, 1606. [CrossRef]
55. Oniga, E.; Breaban, A.; Statescu, F. Determining the optimum number of ground control points for obtaining high precision results based on UAS images. *Proceedings* **2018**, *2*, 352. [CrossRef]
56. Sieberth, T.; Wackrow, R.; Chandler, J.H. Automatic detection of blurred images in UAV image sets. *ISPRS J. Photogramm. Remote Sens.* **2016**, *122*, 1–16. [CrossRef]

57. Pech-Pacheco, J.L.; Cristobal, G.; Chamorro-Martinez, J.; Fernandez-Valdivia, J. Diatom autofocusing in brightfield microscopy: A comparative study. In Proceedings of the 15th International Conference on Pattern Recognition, Barcelona, Spain, 3–7 September 2000. [CrossRef]
58. Ribeiro-Gomes, K.; Hernandez-Lopez, D.; Ballesteros, R.; Moreno, M.A. Approximate georeferencing and automatic blurred image detection to reduce the costs of UAV use in environmental and agricultural applications. *Biosyst. Eng.* **2016**, *151*, 308–327. [CrossRef]
59. Sawicki, P.; Ostrowski, B. Research on chosen matching methods for the measurement of points on digital close range images. *Ann. Geomat.* **2005**, 135–144.
60. Jancosek, M.; Shekhovtsov, A.; Pajdla, T. Scalable multi-view stereo. In Proceedings of the 2009 IEEE 12th International Conference on Computer Vision Workshops, ICCV Workshops 2009, Kyoto, Japan, 27 September–4 October 2009. [CrossRef]
61. Heller, J.; Havlena, M.; Jancosek, M.; Torii, A.; Pajdla, T. 3D reconstruction from photographs by CMP SfM web service. In Proceedings of the 14th IAPR International Conference on Machine Vision Applications, MVA 2015, Tokyo, Japan, 18–22 May 2015; pp. 30–34. [CrossRef]
62. Gabara, G.; Sawicki, P. Accuracy Study of Close Range 3D Object Reconstruction Based on Point Clouds. In Proceedings of the 2017 Baltic Geodetic Congress (BGC Geomatics), Gdansk, Poland, 22–25 June 2017; pp. 25–29. [CrossRef]
63. Girardeau-Montaut, D. CloudCompare: 3D Point Cloud and Mesh Processing Software. Available online: http://www.cloudcompare.org (accessed on 3 March 2018).
64. Luhmann, T.; Robson, S.; Kyle, S.; Boehm, J. *Close Range Photogrammetry and 3D Imaging*; De Gruyter: Berlin, Germany, 2013; ISBN 978-3110302691.
65. Authority, T.V. Federal Geographical Data Committee Geospatial Positioning Accuracy Standards Part 3: National Standard for Spatial Data Accuracy. National Aeronautics and Space Administration: Virginia, NV, USA, 1998.
66. Sawicki, P. Simultaneous Calibration of a Digital Camera Kodak DC4800 in Close Range Photogrammetric Point Measurement. *Arch. Photogramm. Cartogr. Remote Sens.* **2003**, *13*, 457–466. (In Polish)
67. American Society of Civil Engineers. Committee on Cartographic Surveying. In *Map Uses, Scales and Accuracies for Engineering and Associated Purposes*; ASCE: New York, NY, USA, 1983.
68. Wiśniewski, Z. *Adjustment Computation in Geodesy (with Examples)*; Springer: Berlin/Heidelberg, Germany, 2016; ISBN 978-83-8100-057-4. (In Polish)
69. Nikolov, I.; Madsen, C. Benchmarking close-range structure from motion 3D reconstruction software under varying capturing conditions. In *Lecture Notes in Computer Science*; Springer: Berlin/Heidelberg, Germany, 2016. [CrossRef]
70. Benassi, F.; Dall'Asta, E.; Diotri, F.; Forlani, G.; di Cella, U.M.; Roncella, R.; Santise, M. Testing accuracy and repeatability of UAV blocks oriented with GNSS-supported aerial triangulation. *Remote Sens.* **2017**, *9*, 172. [CrossRef]
71. Mostafa, M.M.R. Accuracy Assessment of Professional Grade Unmanned Systems for High Precision Airborne Mapping. *ISPRS Int. Arch. Photogramm. Remote Sens. Spat. Inf. Sci.* **2017**, *42*, 257–261. [CrossRef]
72. Manfreda, S.; Dvorak, P.; Mullerova, J.; Herban, S.; Vuono, P.; Arranz Justel, J.; Perks, M. Assessing the Accuracy of Digital Surface Models Derived from Optical Imagery Acquired with Unmanned Aerial Systems. *Drones* **2019**, *3*, 15. [CrossRef]

MDPI
St. Alban-Anlage 66
4052 Basel
Switzerland
Tel. +41 61 683 77 34
Fax +41 61 302 89 18
www.mdpi.com

Sensors Editorial Office
E-mail: sensors@mdpi.com
www.mdpi.com/journal/sensors

www.ingramcontent.com/pod-product-compliance
Lightning Source LLC
LaVergne TN
LVHW071610170726
843515LV00008B/2366
* 9 7 8 3 0 3 6 5 1 5 9 0 8 *